W. Maysenhölder

Lecture Notes in Physics

Edited by H. Araki, Kyoto, J. Ehlers, München, K. Hepp, Zürich
R. Kippenhahn, München, H. A. Weidenmüller, Heidelberg
and J. Zittartz, Köln
Managing Editor: W. Beiglböck

249

Trends in Applications of Pure Mathematics to Mechanics

Proceedings of the Sixth Symposium on
Trends in Applications of Pure Mathematics to Mechanics
Held at the Physikzentrum of the German Physical Society
Bad Honnef, October 21–25, 1985

W. Maysenhölder

6/86

Edited by E. Kröner and K. Kirchgässner

Springer-Verlag
Berlin Heidelberg New York Tokyo

Editors

Ekkehart Kröner
Institut für Theoretische und Angewandte Physik, Universität Stuttgart
Pfaffenwaldring 57, D-7000 Stuttgart 80

Klaus Kirchgässner
Mathematisches Institut A, Universität Stuttgart
Pfaffenwaldring 57, D-7000 Stuttgart 80

ISBN 3-540-16467-7 Springer-Verlag Berlin Heidelberg New York Tokyo
ISBN 0-387-16467-7 Springer-Verlag New York Heidelberg Berlin Tokyo

Printed in Germany

Printing and binding: Beltz Offsetdruck, Hemsbach/Bergstr.
2153/3140-543210

PREFACE

The "Sixth Symposium on Trends in Applications of Pure Mathematics to Mechanics" was held October 21-25, 1985 at the Physikzentrum of the German Physical Society in Bad Honnef, under the auspices of the International Society for the Interaction of Mechanics and Mathematics, as a continuation of the series of meetings held successively in Lecce (1975), Kozubnik (1977), Edinburgh (1979), Bratislava (1981), and Paris (1983).

The purpose of the Society is to promote and enhance exchanges between mathematics and mechanics. This symposium, as the previous ones, was a vivid illustration of this purpose. Forty-two speakers from fifteen different countries delivered lectures that perfectly exemplified the interplay between the two sciences. They covered the most recent advances in the mathematical analysis of the equations of mechanics as well as their mechanical and physical aspects, in particular in the field of statistical mechanics, non-linear waves, solitons, gauge theories in mechanics, hydrodynamic stability and non-Newtonian behaviour of fluids, with a pervading emphasis on non-linearity.

If this symposium was a success, then this was mainly thanks to the endeavour of all the lecturers whose inspiring contributions were responsible for the high quality of this meeting. The support of the following institutions is deeply appreciated: Deutsche Forschungsgemeinschaft, Max-Planck-Gesellschaft zur Förderung der Wissenschaften, Universität Stuttgart and, last but not least, the Deutsche Physikalische Gesellschaft whose Physikzentrum provided us with the relaxing atmosphere which is so important for stimulating intensive discussions. Our special thanks go to the members of the Scientific Committee, Professors E.G.D. Cohen, V. Fiszdon, D.D. Joseph, I. Müller, and A. Seeger.

Finally we are grateful to our secretaries, Ms. K. Möss and Ms. A. Hackbarth for their great commitment.

Stuttgart, April 1986

E. Kröner, K. Kirchgässner
Universität Stuttgart

The International Society for the Interaction of Mechanics and Mathematics mourns the death of the Honorary Members

F.K.G. Odquist, Stockholm,

O. Onicescu, Bucuresti,

and the Ordinary Members

E. Becker, Darmstadt, M. Kac, New York, W.D. Kupradze, Tbilisi, C. Miranda, Napoli.

Ernst Becker
1929-1985

Marc Kac
1914-1984

Wiktor Kupradze
1903-1985

Carlo Miranda
1912-1982

Folke Odquist
1899-1984

Octav Oniscescu
1892-1983

TABLE OF CONTENTS

Session I:

STATISTICAL MECHANICS

EIGENMODES OF CLASSICAL FLUIDS IN THERMAL EQUILIBRIUM

E. G. D. Cohen
The Rockefeller University
1230 York Avenue
New York, NY 10021

Abstract

The eigenmodes of a classical fluid in thermal equilibrium are discussed. For long wavelengths and times, they can be computed from linear hydrodynamic equations. They are then the hydrodynamic modes, in particular, the heat mode, which describes the diffusion of heat in the fluid and two sound modes. For short wavelengths and times they can be derived from linear kinetic operators. For low densities, the linear Boltzmann operator can be employed and the three most important eigenmodes are direct extensions of the kinetic analogues of the heat and sound modes. For high densities, a generalization of the Boltzmann operator is used. The most important eigenmode is the extended heat mode, while next in importance come two eigenmodes that are extensions of the sound modes. These three extended hydrodynamic modes can be used to obtain the light and neutron spectra of fluids and vice versa.

I. Introduction

In this paper, I am concerned with the transition from classical mechanics to classical statistical mechanics for macroscopic systems, consisting of very many particles. The basic question then is: how does one make a connection between, on the one hand, the dynamics of the many particles in the system, and on the other hand, the physically observed properties of the system in the laboratory? I restrict myself here to a discussion of this question for classical fluids, i.e., gases and liquids, and what follows has been written with these systems in mind.

The most familiar macroscopic description involves the identification of observed macroscopic properties of a system with statistically averaged microscopic properties. Thus the observed local number density at a particular point in the fluid is identified with the average number of particles at this point or the local velocity is connected with the average momentum density at this point and similarly for other local thermodynamic properties. A finer description involving not only positions but also velocities uses distribution functions, like the single particle distribution function, which gives the average number of particles at a particular point with a certain velocity or the pair distribution function, that gives the average number of pairs of particles at two points with certain velocities. In systems in thermal equilibrium, these average quantities are independent of position and time and therefore their local and global values are the same. In non-equilibrium systems, they change with position and time and obey hydrodynamic or kinetic equations.

In the last thirty years, another description, in terms of correlation functions of the fluctuations of these local macroscopic properties around their average values, has come to the foreground [1-4]. In particular, correlation functions in a fluid, which is on the average in thermal equilibrium, have played a dominant role. In this paper, I will capitalize on the fact that the same hydrodynamic or kinetic equations that govern the time evolution of the average densities or distribution functions, also govern that of the correlation functions of fluctuations. I will make use of two simplifications for the time evolution of the correlation functions: 1) linearized forms of the evolution equations are used, and 2) no boundary value problems are considered.

The correlation functions I will consider are equilibrium time correlation functions:

$$M_{AB}(r,t;n,T) = \mathrm{ThLim}\ \langle \delta A(\vec{r}_1,\Gamma) e^{iL_N(\Gamma)t} \delta B(\vec{r}_2,\Gamma)\rangle_{eq} \qquad (1.1)$$

Here $\Gamma \equiv \vec{R}_1 \ldots \vec{R}_N, \vec{p}_1 \ldots \vec{p}_N$ is the phase of the N particles of the system in the phase space of the entire system and $\langle\ \rangle_{eq}$ is an average over an equilibrium (e.g., canonical) ensemble, with probability density $\sim\exp-[H_N(\Gamma)/k_BT]$, where $H_N(\Gamma)$ is the Hamilton function of the system, k_B Boltzmann's constant and T the absolute temperature. $H_N(\Gamma) = \sum_{i=1}^{N} \frac{p_i^2}{2m} + \sum_{i<j}^{N} \phi(R_{ij})$ is the sum of the kinetic energy and the potential energy of the system, where the latter is assumed to be additive, i.e., a sum of pair potentials $\phi(R_{ij})$. In addition, I have assumed that the interparticle potential is spherically symmetric and depends only on the distance $R_{ij}=|\vec{R}_i-\vec{R}_j|$ between the two interacting particles i and j. $iL_N(\Gamma) \equiv [\ \ ,H_N]$ defines the Liouville operator $L_N(\Gamma)$, where [,] are the Poisson brackets. $\delta A(\vec{r}_1;\Gamma)=A(\vec{r}_1;\Gamma)-\langle A(\vec{r}_1;\Gamma)\rangle_{eq}$ is the fluctuation of the quantity $A(\vec{r}_1,\Gamma)$, i.e., the difference between the actual value $A(\vec{r}_1,\Gamma)$ of the quantity $A(\Gamma)$ at the position $\vec{r}_1$ and its average value $\langle A(\vec{r}_1;\Gamma)\rangle_{eq}=\langle A(\Gamma)\rangle_{eq}$ and similarly for $\delta B(\vec{r}_2;\Gamma)$. Because of the spatial isotropy of the thermal equilibrium state, M_{AB} will depend on $r=|\vec{r}_1-\vec{r}_2|$ only. $e^{iL_N(\Gamma)t}\delta B(\Gamma)=\delta B(\Gamma_t)$, where Γ_t is the phase of the system at time t, if Γ is the phase at time t=0. Note that the fixed points $\vec{r}_1$ and $\vec{r}_2$ in space have to be distinguished from the coordinates $\vec{R}_1 . . . \vec{R}_N$ of the particles of the system and that L_N only acts on the latter. To compute Γ_t, one obviously has to solve the equations of motion, i.e., Newton's equations, for N particles. For the physically relevant quantity M_{AB} the bulk (or usually called thermodynamic) limit, ThLim, is needed, where the number of particles N and the volume V both go to infinity, such that an infinite system is considered, at a given density n=N/V and a given temperature T.

A special case of (1.1) are the equal time correlation functions $M_{AB}(r;n,T)$ with t=0, in terms of which the thermodynamic and all other static equilibrium properties can be expressed. An example is the pair correlation G(r), defined by [1]:

$$nG(r) = \mathrm{ThLim}\ \langle \delta n(\vec{r}_1;\Gamma)\delta n(\vec{r}_2;\Gamma)\rangle_{eq} \qquad (1.2)$$

where the density fluctuation $\delta n(\vec{r};\Gamma)$ is defined by:

$$\delta n(\vec{r};\Gamma) = \sum_{i=1}^{N} \delta(\vec{r}-\vec{R}_i)-n \tag{1.3}$$

i.e., as the difference between the actual and the average number of particles at $\vec{r}$, since

$$<\sum_{i=1}^{N} \delta(\vec{r}-\vec{R}_i)>_{eq} = \frac{N}{V} = n. \tag{1.4}$$

Then:

$$G(r) = \delta(r) + n[g(r)-1] = \delta(r) + nh(r). \tag{1.5}$$

Here g(r) is the radial distribution function, which gives the average number of particles at a distance r from a given particle. The dependence of G(r) on n and T has not been indicated explicitly. All equal time correlation functions can be computed using equilibrium statistical mechanics. In doing so, the dynamics has been completely eliminated.

This is not so for the time dependent correlation functions ($t\neq0$), such as those related to the transport properties [4] or the differential cross-sections for the inelastic scattering of light [5] and neutrons [2,3] by the fluid. An example of the latter is the so-called van Hove function, or (unequal time) density-density correlation function:

$$\begin{aligned} nG(r,t) &= \text{ThLim} <\delta n(\vec{r}_1,\Gamma)\delta n(\vec{r}_2,\Gamma_t)>_{eq} \\ &= \text{ThLim} <\delta n(\vec{r}_1,\Gamma)e^{iL_N t}\delta n(\vec{r}_2,\Gamma)>_{eq} \end{aligned} \tag{1.6}$$

For $t=0$, G(r,t) reduces to the pair correlation function G(r) defined in (1.2). In this paper I will restrict myself to a discussion of G(r,t) for fluids but the basic problem of how to deal with the two N-body problems--the dynamical one in $e^{iL_N t}$ and the statistical one in $< >_{eq}$--in the bulk limit is similar for other correlation functions.

I remark that instead of G(r,t) one often considers its spatial and temporal Fourier transforms: the intermediate scattering function F(k,t):

$$F(k,t) = \int d\vec{r}\; e^{i\vec{k}\cdot\vec{r}} G(r,t) \tag{1.7}$$

and the dynamic structure factor $S(k,\omega)$:

$$S(k,\omega) = \int_{-\infty}^{\infty} dt\; e^{i\omega t} F(k,t) \tag{1.8}$$

F(k,t) is usually obtained in computer simulations, while $S(k,\omega)$ can be measured by light or neutron scattering, since it is proportional to the intensity of scattered light or neutrons, when a momentum transfer ~k and an energy transfer ~ω has taken place from the fluid to the incident light or neutron.

One can distinguish between two different approaches in the computation of G(r,t), or for that matter, $M_{AB}(r,t;n,T)$. One approach is a numerical approximation method [6,7], that applies the operator $e^{iL_N t}$ and then the average $< >_{eq}$ directly to a finite system, using large electronic computers. By employing periodic boundary conditions--thus avoiding some obvious aspects of rigid walls--and by doing the calculation for a varying number of particles, one extrapolates to $N\to\infty$ and assumes then that a result representative of an infinite system has been obtained. This computer based method avoids in a way the basic N-body problem of statistical mechanics, by using the computer as a deus--or perhaps more appropriately demon--ex machina, since it really behaves like a Maxwell demon, observing the motion of all the particles in the system. Therefore, the results, even if correct, are on the level of experimental results and do not, in general, provide a solution to the basic problem of statistical mechanics: the connection between the microscopic interparticle forces on the one hand and the macroscopically observed properties, on the other hand.

The second way to obtain time correlation functions is that of statistical mechanics or kinetic theory. It replaces, in some manner, the two very difficult N-body problems by a simpler statistical few body problem. Before I discuss this in more detail, I should point out that even if one could solve exactly the dynamical N-body problem, one is really interested--in view of the bulk limit--in the N, V-dependence of the solution. In terms of eigenmodes: even if one knew all the eigenmodes of the Liouville operator--the eigenvalues of which all lie on the real axis--one is really interested in an asymptotic property of these eigenmodes for the case the system becomes infinitely large.

The methods of statistical mechanics or kinetic theory, alluded to above, all capitalize on the fact that one has to consider dynamical (N-body) problems for a distribution of initial states. I will discuss a very restricted class of approximation methods: the hydrodynamic approximation and two kinetic approximations, one, for a dilute gas, based on the Boltzmann equation and one, for a dense fluid, based on an equation given by Enskog for a dense fluid of hard spheres. In all these cases, the dynamical and statistical N-body problem in the bulk limit is replaced by a simpler problem: that of finding the eigenmodes of a linear kinetic operator, which determine then the connection between the microscopic and macroscopic properties of the fluid. I note that one is concerned here with the determination of (approximate) eigenmodes of a strongly interacting, highly anharmonic many particle system.

In this formulation, one makes contact with a number of approximation procedures that have been discussed in the literature before. In fact, the eigenmodes of classical fluids have been considered in the following two cases:

(1) The macroscopic quantities vary slowly in space and time, so that their variation can be characterized by long wave lengths λ (or small wave numbers $k=2\pi/\lambda$) and long times t (or small frequencies ω). The time evolution of these quantities can then be determined for all fluid densities from the eigenmodes of linear hydrodynamic equations.

(2) The quantities of physical interest vary in space (time) over characteristic lengths (times) approaching the mean free path ℓ_0 (time t_0) between two successive binary collisions in a dilute gas. The time evolution is then determined by the eigenmodes of the linear Boltzmann equation.

In the following, I first discuss, in Section II, the more familiar hydrodynamic case and then, in Section III, for dilute gases, the linear Boltzmann case. An extension will be made in Section IV to the case of dense fluids, using a generalization of the Boltzmann equation to dense hard sphere fluids. Some results for dense fluids will be discussed in Section V.

II. Hydrodynamic modes in hydrodynamics

In case (1), mentioned above, the time evolution of the fluid is determined by the local conservation laws of number, momentum and energy, i.e., by the hydrodynamic equations linearized around thermal equilibrium. This means that instead of the phase space of the entire fluid, one considers only a five-dimensional space, using the fluctuations of the five hydrodynamic quantities [1-3, 5, 8, 9]. Many formal derivations of this transition have been given [10]. Rigorous proofs are discussed in a recent review [11].

If $\overline{\delta n}(\vec{r},t)$, $\overline{\delta T}(\vec{r},t)$ and $\overline{\delta \vec{u}}(\vec{r},t)$ are the average fluctuations in the local density (cf. eq. (1.3)), temperature and velocity, respectively, at the position $\vec{r}$ at time t in the fluid, for a given value of $\delta n(\vec{r}',0)$ at t=0, then one considers

$$\overline{\delta \underset{\sim}{a}}(\vec{k},t) \equiv \{\overline{\delta n}(\vec{k},t),\ \overline{\delta T}(\vec{k},t),\ \overline{\delta \vec{u}}(\vec{k},t)\}, \tag{2.1}$$

where one has used the Fourier representation:

$$\overline{\delta \underset{\sim}{a}}(\vec{r},t) = \int \overline{\delta \underset{\sim}{a}}(\vec{k},t)e^{i\vec{k}\cdot\vec{r}}d\vec{k} \tag{2.2}$$

The time-evolution of the $\overline{\delta \underset{\sim}{a}}(\vec{k},t)$ is then given by a five by five matrix that can be derived from the linearized hydrodynamic equations. This matrix can be put in a simpler form by separating the velocity $\overline{\delta \vec{u}}(\vec{k},t)$ into a longitudinal part $\overline{\delta u}(\vec{k},t)=\hat{k}\cdot\overline{\delta \vec{u}}(\vec{k},t)$ and two transversal parts $\overline{\delta \vec{u}_{\perp}^{(i)}}(\vec{k},t)=\hat{k}_{\perp}^{(i)}\cdot\overline{\delta \vec{u}}(\vec{k},t)$ (i=1,2), where $\hat{k}$, $\hat{k}_{\perp}^{(1)}$, $\hat{k}_{\perp}^{(2)}$ form an orthonormal set of unit vectors. The hydrodynamic modes are the eigenmodes of this matrix. Two of these eigenmodes--the two transversal velocity eigenmodes--decouple from the others and can be found immediately: they are the two viscous modes. In the following, I will not consider these modes and restrict myself to a discussion of the eigenmodes of the remaining matrix, referring to $\overline{\delta n}(\vec{k},t)$, $\overline{\delta T}(\vec{k},t)$, $\overline{\delta u}(\vec{k},t)$. The relevant equations read then, after a transition to appropriate dimensionless variables [12]:

$$\frac{\partial \delta a_i(\vec{k},t)}{\partial t} = \sum_j \left(\underset{\approx}{H}(\vec{k})\right)_{ij}\delta a_j(\vec{k},t) \qquad (i,j=1,2,3) \tag{2.3}$$

with

$$\delta a_1(\vec{k},t)=(mc^2/\gamma n^2 k_B T)\overline{\delta n}(\vec{k},t);\ \delta a_2(\vec{k},t)=(m\alpha^2c^2/\gamma(\gamma-1)k_BT)^{1/2}\overline{\delta T}(\vec{k},t);$$

$$\delta a_3(\vec{k},t)=(m/k_BT)^{1/2}\overline{\delta u}(\vec{k},t). \tag{2.4}$$

Here α is the thermal expansion coefficient, c is the velocity of sound and the symmetric matrix $\underset{\approx}{H}(\vec{k})$ is given by:

$$\underset{\approx}{H}(\vec{k}) = \begin{pmatrix} 0 & 0 & -ikc/\gamma^{1/2} \\ 0 & -\gamma D_T k^2 & -ikc[(\gamma-1)/\gamma]^{1/2} \\ -ikc/\gamma^{1/2} & -ikc[(\gamma-1)/\gamma]^{1/2} & -D_\ell k^2 \end{pmatrix} \tag{2.5}$$

where $k=|\vec{k}|$, $\gamma=c_p/c_v$ with c_p, c_v the specific heats per unit mass at constant pressure and density, respectively, $D_\ell = 4\eta/3\rho + \zeta/\rho$, with η the shear viscosity and $\rho=nm$ the mass density, where m the mass of a fluid particle and ζ the bulk viscosity. I note that in $\underset{\approx}{H}(\vec{k})$ the diagonal elements are due to the dissipative part of the hydrodynamic equations. They are all negative (since the transport coefficients and γ are non-negative), implying a decay of any perturbation from equilibrium, in accordance with the second law of thermodynamics. The off-diagonal terms are of O(k), purely imaginary and contain only thermodynamic but no transport properties of the fluid. They arise from the Euler or ideal fluid equations alone and represent, physically, elastic or restoring forces, like, for instance, $c=[(\partial p/\partial\rho)_s]^{1/2}$ in the (31)-element, where s is the entropy density. The hydrodynamic equations are only valid for small k, such that $k\ell \ll 1$, where the characteristic length ℓ is the mean free path ℓ_0 in the case of a dilute gas or the effective molecular size σ_{eff} in the case of a dense gas or a liquid. Since $\ell_0 \sim 10^{-5}$cm and $\sigma_{eff} \sim 10^{-8}$cm, the hydrodynamic equations can be used to describe the scattering of visible light (with $\lambda > 10^{-5}$cm) by not too dilute gases and by liquids. They cannot be used, however, for X-ray or neutron scattering of liquids, since then $k\sigma \approx O(1)$.

For small k, the eigenmodes of $\underset{\approx}{H}(k)$ can easily be determined by perturbation theory. They are [12]:

Table I

	Eigenvalue to $O(k^2)$	Eigenvectors to $O(k^0)$
1 heat mode	$z_h = -D_T k^2$	$\Psi_h = \{(\frac{\gamma-1}{\gamma})^{1/2}, -\gamma^{-1/2}, 0\}$
2 sound modes	$z_\pm = \pm ick - \Gamma_s k^2$	$\Psi_\pm = 2^{-1/2}\{\gamma^{-1/2}, (\frac{\gamma-1}{\gamma})^{1/2}, \pm 1\}$

Here the sound damping $\Gamma_s = [D_\ell + (\gamma-1)D_T]/2$.

The dynamic structure factor $S(k,\omega)$ can be computed on the basis of the hydrodynamic equations as long as $k\ell \ll 1$ and $\omega\tau \ll 1$, with $\tau = O(\ell/c)$. Then F(k,t) is given by a sum of exponentials, while $S(k,\omega)$ is given by a sum of three Lorentzians:

$$S(k,\omega) = S(0) \cdot 2\mathbf{Re} \sum_{j=h,\pm} \frac{A_j}{i\omega - z_j(k)} . \tag{2.6}$$

Here $A_j = [\Psi_j^{(1)}]^2$, where $\Psi_j^{(1)}$ is the first component of the eigenvector Ψ_j and S(0) is the k=0 value of the static structure factor $S(k) = \int_{-\infty}^{\infty} (d\omega/2\pi) S(k,\omega)$, where S(k) is related to the Fourier transform of the pair correlation function G(r) of the fluid:

$$S(k) = 1 + n \int d\vec{r}\, e^{i\vec{k}\cdot\vec{r}} h(r). \tag{2.7}$$

In the hydrodynamic regime, one can use Table I for the $z_j(k)$ and A_j and that $S(k) = S(0) = nk_BT\chi_T$ where $\chi_T = (\partial n/\partial p)_T/n$ is the isothermal compressibility. One then obtains the well-known Landau-Placzek formula for $S(k,\omega)$ [2,5]:

$$S(k,\omega)=nk_BT\chi_T\left[\frac{\gamma-1}{\gamma}\frac{2D_Tk^2}{\omega^2+(D_Tk^2)^2}+\frac{1}{\gamma}\frac{\Gamma_sk^2}{(\omega-ck)^2+(\Gamma_sk^2)^2}+\frac{1}{\gamma}\frac{\Gamma_sk^2}{(\omega+ck)^2+(\Gamma_sk^2)^2}\right] \qquad (2.8)$$

Therefore, for small k and ω the heat and sound eigenmodes determine theoretically the $S(k,\omega)$ and vice-versa these eigenmodes can be deduced from experiment by fitting three Lorentzians to the observed $S(k,\omega)$.

III. Hydrodynamic Modes in Dilute Gases

In case the variations in space and time are on the scale of the mean free path and time, respectively, a more detailed description than provided by linearized hydrodynamics is necessary. For a dilute gas one can base this on a linearized form of the Boltzmann equation [2,13]. In that case one considers fluctuations of the single particle distribution function $f(\vec{r},\vec{v},t)$ in $(\vec{r},\vec{v})$-space. Small fluctuations $\overline{\delta f}$ of f satisfy the Boltzmann equation linearized around thermal equilibrium, which can be written in the following general form:

$$\frac{\partial\overline{\delta f}}{\partial t} = -\vec{v}\cdot\frac{\partial\overline{\delta f}}{\partial\vec{r}} + n\bar{\Lambda}_B(\vec{v})\overline{\delta f} \qquad (3.1)$$

Here $\overline{\delta f}=\overline{\delta f}(\vec{r},\vec{v},t)$ is the average fluctuation of the number of particles at the position $\vec{r}$ with velocity $\vec{v}$ at time t, given the fluctuation of $\overline{\delta f}$ at t=0. The rate of change of $\overline{\delta f}$ is determined by two processes: free streaming of the particles $(-\vec{v}\cdot\partial\overline{\delta f}/\partial\vec{r})$ and binary collisions between pairs of particles $\left(\bar{\Lambda}_B(\vec{v})\overline{\delta f}\right)$. Here $\bar{\Lambda}_B$, a linear operator acting on the velocity $\vec{v}$ in $\overline{\delta f}$, involves (a) a statistical ansatz (molecular chaos) and (b) the dynamical two-body problem, the solution of which depends on the interparticle potential. Writing $\overline{\delta f}(\vec{r},\vec{v},t) = f_M(v)(\vec{r},\vec{v},t)$, where $f_M(v)=n(\beta m/2\pi)^{3/2}\exp(-\beta mv^2/2)$ is the equilibrium single particle distribution function $(v=|\vec{v}|)$ and using a Fourier representation:

$$h(\vec{r},\vec{v},t) = \int h(\vec{k},\vec{v},t)e^{i\vec{k}\cdot\vec{r}}d\vec{k}, \qquad (3.2)$$

eq. (3.1) reads:

$$\frac{\partial h}{\partial t} = -i\vec{k}\cdot\vec{v}h + n\Lambda_B(\vec{v})h = L_B(\vec{v})h, \qquad (3.3)$$

where $\Lambda_B(\vec{v})$ is the linear Boltzmann collision operator. The $\vec{k}$-dependence of $L_B(\vec{v})$ has not been indicated explicitly. Many derivations of the eq. (3.3) exist (see for instance [13]). Rigorous proofs for the eq. (3.3) are discussed in [14]. Since Λ_B is proportional to the total cross-section for binary collisions, σ^2_{eff}, one sees that the relative importance of the free streaming term $(-i\vec{k}\cdot\vec{v})$ and the collision term (Λ_B) is determined by the parameter $k\ell_0$, the dimensionless combination of the two basic lengths that occur in L_B: $\lambda=2\pi/k$ and $\ell_0\sim1/n\sigma^2_{eff}$. A characteristic time is the mean free time $t_0=\ell_0/\langle v\rangle$, where $\langle v\rangle=2\sqrt{2}/(\beta m\pi)^{1/2}$ is an average velocity.

(1) Chapman-Enskog method

For small k, i.e., $k\ell_0\ll1$, one can use that the effect of $-i\vec{k}\cdot\vec{v}$ on h is small compared to that of $\Lambda_B(\vec{v})$ and find the eigenmodes of L_B from those of Λ_B, using $-i\vec{k}\cdot\vec{v}$ as a perturbation. In other words, for large wavelengths, where $\lambda\gg\ell_0$ and many particles are contained in a wavelength, the collisions (Λ_B) mainly determine the time evolution of h, i.e., $\overline{\delta f}$.

Little is known about the eigenmodes of Λ_B for general interparticle potentials. However, without going into details, one knows that for a class of interparticle potentials, that includes those discussed here that [14]: (1) there is a five-fold degenerate zero-eigenvalue and (2) all other eigenvalues are negative and (3) are separated from the zero eigenvalue by a finite distance. For spherically symmetric interparticle potentials $\phi(R)$, only for so-called Maxwell molecules, where $\phi(R)\sim R^{-4}$, are the eigenvalues and eigenfunctions exactly known. In that case, the eigenvalues are all discrete and the eigenfunctions can be expressed in terms of associated Laguerre or Sonine polynomia [15]. For hard spheres $\left(\phi(R) = \infty \text{ for } R<\sigma,\ \phi(R) = 0 \text{ for } R\geq\sigma\right.$, where σ is the hard sphere diameter$\left.\right)$, there are a number of discrete eigenvalues near zero and from a finite distance onwards there is a continuum of eigenvalues. Approximate eigenvalues and eigenfunctions have been computed in this case [16].

Using now $-i\vec{k}\cdot\vec{v}$ as a perturbation on the five eigenmodes of Λ_B with zero eigenvalue, gives to $O(k\ell_0)^2$, parabolae for the eigenvalues of $L_B(\vec{k})$ as a function of k. Of the five perturbed zero eigenmodes, three can be identified as the kinetic analogues of the heat and sound modes and two of the viscous modes of the linear hydrodynamic equations (2.3). In fact, for small k, the eigenvalues of the heat and sound modes reduce to those given in Table I, except that the thermodynamic and transport properties have their low density values. The corresponding five eigenfunctions are to $O(k^0)$ identical to those of $\Lambda_B(\vec{v})$ and equal to the five functions of $\vec{v}$ that are conserved in a binary collision. These five conserved quantities are proportional to the mass (or number), the three components of the velocity $\vec{v}$ and the (kinetic) energy v^2 of a particle, respectively. The following linear combinations of these conserved quantities form an orthonormal set: $\psi_1(\vec{v}) = 1$; $\psi_2(\vec{v}) = (\beta m)^{1/2}v_z$; $\psi_3(\vec{v})=(\beta mv^2-3)/\sqrt{6}$; $\psi_4(\vec{v}) = (\beta m)^{1/2}v_x$; $\psi_5(\vec{v}) = (\beta m)^{1/2}v_y$. Here $\beta = 1/k_BT$ and $\langle\psi_i\psi_j\rangle = \delta_{ij}$, where the brackets denote an average with the Maxwell velocity distribution function $f_M(v)/n$ and the z-axis of the coordinate system is chosen parallel to $\vec{k}$. Going to higher order in the perturbation calculation, gives extensions of these (kinetic) hydrodynamic modes to higher values of k. The eigenmodes of L_B associated with the non-zero eigenvalues of Λ_B are called kinetic modes. We will call this method of finding the eigenmodes of the dilute gas for increasingly large k, the Chapman-Enskog method, since the same eigenvalues can be derived from the linearized hydrodynamic equations--the Euler, Navier-Stokes, Burnett, etc., equations--for n, T and $\vec{u}$, obtained by solving the (non-linear) Boltzmann equation using this method [2,17].

(2) Moment Method

Another way of finding eigenmodes of $L_B(\vec{v})$, that is not restricted to small k, can be based on the so-called moment method due to H. Grad [18]. This method leads to a finite matrix $(L_B)_{ij} = \langle\psi_i L_B \psi_j\rangle$, by using a finite number of a complete (orthonormal) set $\{\psi_i\}$ of functions of $\vec{v}$. The original moment method of Grad used 13-Hermite polynomia, the first five of which are identical to the ψ_i (i=1,...,5) given above. Later, Chang and Uhlenbeck [15] used the more efficient Sonine polynomia, exploiting the cylindrical symmetry in $\vec{v}$-space of $L_B(\vec{v})$. In fig. 1, the first few eigenvalues of $(L_B)_{ij}$ are sketched as a function of $k\ell_0$ for a gas of hard spheres, using 55 polynomia [19].

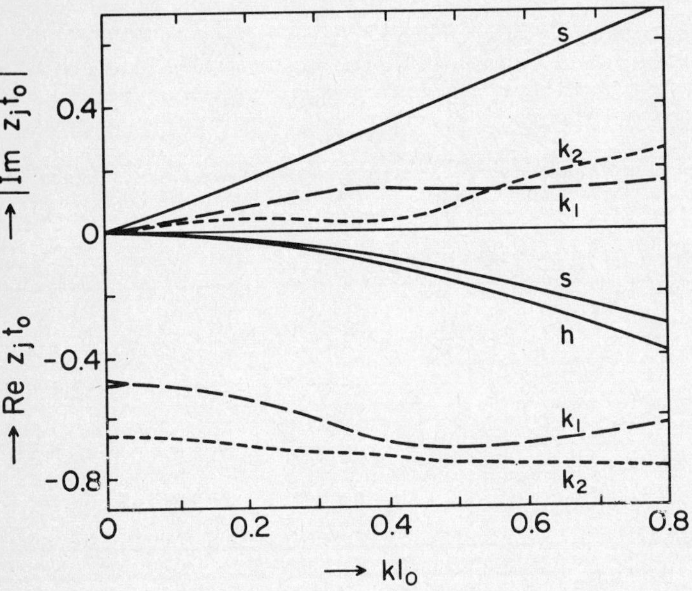

Figure 1: Reduced real and magnitude of imaginary parts of eigenvalues of extended heat (h), sound (s) and nearest kinetic modes (k_1, k_2) for dilute gas of hard spheres (V_0/V=0.01) as a function of reduced wavenumber $k\ell_0$, using 55 polynomia.

(3) B-G-K method

Finally, there is the Bhatnagar-Gross-Krook method (B-G-K method) [2] of treating the infinite matrix $(L_B)_{ij}$ by using a finite M×M block of matrix elements $(\Lambda_B)_{ij}$ (i,j=1,...,M) for Λ_B and replacing the remaining matrix elements of Λ_B all by zero except along the diagonal, where all elements are chosen equal to the (M+1, M+1) diagonal element. The free streaming term $-i\vec{k}\cdot\vec{v}$ is not approximated. Also in this case can the eigenmodes be determined, and the results do not differ too much from the moment method as long as $k\ell_0<1$.

(4) Eigenmodes

The most striking feature of the eigenvalues sketched in fig. 1 is the separation of the extended hydrodynamic mode eigenvalues from the extended kinetic mode eigenvalues. Fig. 1 is representative for densities V_0/V as low as 0.0001 to densities as high as 0.1. Here $V_0=N\sigma^3/\sqrt{2}$, is the volume of N hard spheres at close packing. For Maxwell molecules a very similar behavior obtains [19]. This behavior of the eigenvalues implies that for dilute gases an extended hydrodynamic description, i.e., a description using the three extended hydrodynamic modes alone is possible for k values where, say, $k\ell_0 \leq 0.5$, i.e., for k-values of the order of $10^5 cm^{-1}$. The eigenmodes for small values of k are summarized in Table II.

Table II

$L_B(\vec{k})$	Eigenvalue to $O(k^2)$	Eigenfunction to $O(k^0)$
1 heat mode	$z_h=-D_{T,0}k^2$	$\Psi_h=(2/5)^{1/2}(\beta mv^2-5/2)$
2 sound modes	$z_\pm=\pm ic_0k-\Gamma_{s,0}\,k^2$	$\Psi_\pm=(3/10)^{1/2}[\beta mv^2/3\pm(5\beta m/3)^{1/2}v_z]$

Here c_0, $D_{T,0}$ and $\Gamma_{s,0}$ are the low density values of c, D_T and Γ_s, respectively.

(5) Results

The eigenmodes of dilute gases beyond the hydrodynamic regime, i.e., for $0<<k\ell_0<1$, have been studied experimentally in two ways: by ultrasound and by light scattering. In the first case, the (extended) sound modes can be studied by observing the propagation and absorption of ultra sound (cf. ref. [9]). In the second case, $S(k,\omega)$ can be computed and compared with light scattering experiments. $S(k,\omega)$ becomes then a sum of Lorentzians, just as in hydrodynamics, except that the sum includes, in principle, <u>all</u> eigenmodes of $L_B(\vec{v})$, i.e., not only the (extended) hydrodynamic modes, but also the kinetic modes:

$$S(k,\omega) = 2\,\mathbf{Re}\sum_{j=1}^{\infty}\frac{A_j(k)}{i\omega-z_j(k)}\,. \tag{3.4}$$

Here the eigenvalues $z_j(k)$ and the amplitudes $A_j(k) = \langle\Psi_j\rangle^2$ (where Ψ_j are the eigenfunctions) are obtained from the matrix $(L_B)_{ij}$ and depend on the approximation considered. Up until $k\ell_0 \approx 0.7$, $S(k,\omega)$ can be very well approximated by the three (extended) hydrodynamic modes alone. Eq. (3.4) reduces in the hydrodynamic regime, $k\ell_0<<1$, $\omega t_0<<1$ to the Landau-Placzek formula (2.8), since $S(0)=1$ for a dilute gas and the kinetic modes do not contribute. As we shall see later, a slight modification of formula (3.4) can be used for dense fluids, where $k\sim 10^8 cm^{-1}$.

IV. Dense Fluids

(1) Introduction

For a description of dense fluid properties, such as $S(k,\omega)$, on the basis of a linear kinetic operator, there is the fundamental difficulty that no kinetic equation of comparable stature as the Boltzmann equation is available for non-dilute gases, let alone for liquids, because of the very complicated nature of the multiple particle collisions that take place in such fluids.

Nevertheless, we will sketch an attempt to extend the kinetic treatment of a dilute gas to dense fluids, albeit in a very approximate manner. Using this extension and essentially the same three approximation methods that have been discussed above to obtain the eigenmodes of the linearized Boltzmann equation, approximate eigenmodes and their relation to $S(k,\omega)$ for dense fluids can be obtained.

(2) Kinetic Operator for Dense Fluids

As in the case of a dilute gas, one considers fluctuations $\overline{\delta f}(\vec{r},\vec{v},t)$ of the single particle distribution function in $(\vec{r},\vec{v})$-space. The kinetic operator used is a generalization of the linearized Boltzmann operator, obtained with two important approximations: (1) the interparticle potential is simplified to that between hard spheres; (2) the linear kinetic operator $L_E(\vec{v})$ that describes the time evolution of fluctuations from equilibrium is a linearization of a non-linear operator occurring in a generalization of a transport theory of dense hard sphere fluids, that was proposed by Enskog in the 1920's [2,17]. For a derivation of $L_E(\vec{v})$ and closely related operators, I refer to the literature [20-25]. No rigorous results are known for the operator $L_E(\vec{v})$. The linear operator $L_E(\vec{v})$ can be written in the following general form [26]:

$$L_E(\vec{v}) = -i\vec{k}\cdot\vec{v} + ng(\sigma)\Lambda_{\vec{k}}(\vec{v}) + nA_{\vec{k}}(\vec{v}), \qquad (4.1)$$

where the $\vec{k}$-dependence of $L_E(\vec{v})$ has not been indicated explicitly. Here the binary collision operator $\Lambda_{\vec{k}}$, which replaces Λ_B of the Boltzmann equation, contains--like Λ_B--only binary collision dynamics, but the statistical ansatz of the Boltzmann equation has been modified in two respects: (1) the collision frequency for binary collisions has been increased by a factor $g(\sigma)$, the equilibrium radial distribution for two hard spheres at contact; (2) the difference in position of the two hard spheres at collision has been take into account, leading in the collision operator $\Lambda_{\vec{k}}$ to a factor $e^{-i\vec{k}\cdot\vec{\sigma}}$, which is a Fourier transform of a factor $\delta(\vec{r}-\vec{\sigma})$ in the $\vec{r}$-space representation of $\Lambda_{\vec{k}}$. The vector $\vec{\sigma}=\sigma\hat{\sigma}$, where the unit vector $\hat{\sigma}$ defines the geometry of the binary collision. The "mean field" operator $A_{\vec{k}}$ depends on the static structure factor $S(k)$ and takes into account the average influence of the other particles on the free motion of a hard sphere. For continuous potentials, $A_{\vec{k}}$ is easy to understand and for an electron gas it yields the mean field term in the linearized Vlasov equation. For hard spheres $A_{\vec{k}}$ involves excluded volume corrections, due to the non-overlapping condition of hard spheres. Both $\Lambda_{\vec{k}}$ and $A_{\vec{k}}$ depend on $\vec{k}$ through the parameter $k\sigma$ only.

The most important physical difference between the operators $L_E(\vec{v})$ and $L_B(\vec{v})$ is that the former incorporates--through the factor $e^{-i\vec{k}\cdot\vec{\sigma}}$ in $\Lambda_{\vec{k}}$--the instantaneous collisional transfer of momentum and energy, while the latter does not. Both contain the translational transfer of momentum and energy through free flight. With increasing density, the collisional transfer of momentum and energy quickly becomes more important than translational transfer, so that the latter can be completely neglected in dense fluids.

For low densities, i.e, $n\sigma^3 \ll 1$, and small k, i.e., $k\sigma \ll 1$, $\Lambda_{\vec{k}} \approx \Lambda_B$, since then $g(\sigma)\approx 1$ and $A_{\vec{k}} \approx 0$.

For all densities and large k, i.e., $k\sigma >> 1$, $A_{\vec{k}} \approx 0$ and $L_E(\vec{v})$ approaches the Lorentz-Boltzmann operator $L_E^s(\vec{v})$. This operator describes the motion of a tagged hard sphere in a fluid of identical hard spheres. The operator $L_E^s(\vec{v})$ is like that used for the kinetic description of a heavy particle in a fluid in thermal equilibrium: the so-called Rayleigh problem [27]. Therefore, a better name for $L_E^s(\vec{k})$ would be the Rayleigh-Boltzmann operator. I note that $L_E(\vec{v})$ contains three basic lengths: λ, ℓ_E and σ and that there are therefore two characteristic parameters: $k\ell_E$ and $k\sigma$. Here $\ell_E = \ell_0/g(\sigma)$ is the mean free path of a dense hard sphere fluid, where $\ell_0 = 1/\pi n \sigma^2 \sqrt{2}$ is the mean free path of a dilute hard sphere gas. A corresponding characteristic time is $t_E = \ell_E/\langle v \rangle$.

(3) Eigenmodes

No systematic study of the mathematical properties of the operators $\Lambda_{\vec{k}}(\vec{v})$, $A_{\vec{k}}(\vec{v})$ or $L_E(\vec{v})$ has been made. Nevertheless, $L_E(\vec{v})$ does possess a number of desirable mathematical properties [26].

As mentioned above, the same three methods to obtain eigenmodes, of $L_B(\vec{v})$, can also be applied to $L_E(\vec{v})$ (or $L_E^s(\vec{v})$). For small k, i.e., in the hydrodynamic regime, when $k\ell_E << 1$ and $k\sigma << 1$, the eigenmodes can be determined by using the streaming term $-i\vec{k}\cdot\vec{v}$ as a perturbation on $ng(\sigma)\Lambda_B = (ng(\sigma)\Lambda_{\vec{k}} + nA_{\vec{k}})_{k=0}$. The zero eigenvalue of $ng(\sigma)\Lambda_B$ then gives the hydrodynamic modes, shown in Table III (cf. Table II) [26]:

Table III

$L_E(\vec{v})$	Eigenvalue to $O(k^2)$	Eigenfunction to $O(k^0)$
1 heat mode	$z_h = -D_{T,E}k^2$	$\Psi_h = \tilde{c}\psi_1 - (\beta m)^{-1/2}\psi_3$
2 sound modes	$z_\pm = \pm ick - \Gamma_{s,E}k^2$	$\Psi_\pm = [(\beta m)^{1/2}S(0)]^{-1}\psi_1 \pm c\psi_2 + \tilde{c}\psi_3$
$L_E^s(\vec{k})$	Eigenvalues to $O(k^2)$	Eigenfunction to $O(k^0)$
1 self-diffusion mode	$z_{D,E} = -D_E k^2$	$\Psi_D = 1$

In addition, there are two viscous modes. We have also given the single hydrodynamic mode for $L_E^s(\vec{v})$: the self-diffusion mode, where D refers to self-diffusion. The subscript E indicates that the eigenvalues are determined by the transport coefficients $D_{T,E}$, D_E, etc., as obtained from the Enskog dense hard sphere fluid transport theory, mentioned before. $\tilde{c} = c[(\gamma-1)/\gamma]^{1/2}$, and the eigenfunctions of $L_E(\vec{v})$ given in Table III are only the right eigenfunctions, since $L_E(\vec{v})$, (unlike $L_B(\vec{v})$), is not a symmetric operator. For low densities these eigenmodes reduce to those given in Table II. One can also say that the eigenvalues in Table III are identical to those obtained from the linearized hydrodynamic equations that follow from the Enskog transport theory. The moment method [19] and the BGK method [25, 28-30] have both been applied to $L_E(\vec{v})$. Using the BGK method with ten functions ψ_i, yields eigenmodes of $(L_E(\vec{v}))_{ij}$, the first few eigenvalues of which are sketched in fig. 2 as a function of $k\sigma$ for two densities, $V_0/V = 0.45$ and 0.625. These densities correspond roughly to the critical and triple point densities, respectively, of liquid Argon. All eigenvalues and eigenfunctions have to be determined numerically. The convergence of the heat mode with increasing number of polynomia is good as long as $k\ell_0 \leq 1$ but for the sound and kinetic modes much less clear. It could, therefore, well be, that higher approximations, using more polynomia, will give a more complicated picture, especially for the sound and kinetic modes. This should be borne in mind in all comparisons that follow, since the theoretical results quoted are based on the BGK-method with ten polynomia.

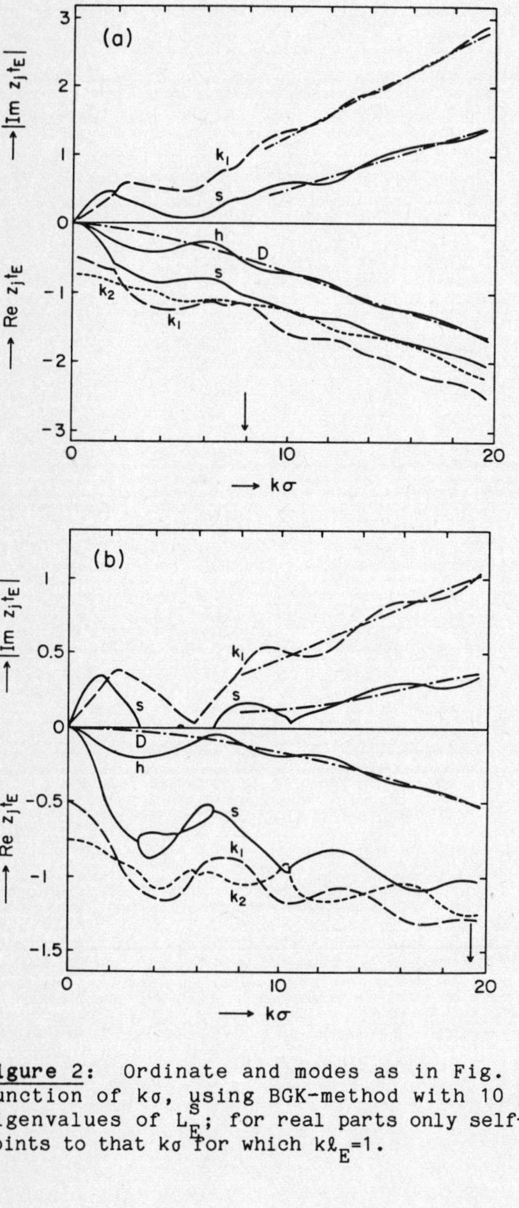

Figure 2: Ordinate and modes as in Fig. 1, for V_0/V=0.45 (a) and 0.625 (b) as a function of kσ, using BGK-method with 10 polynomia. Dash-dot lines indicate eigenvalues of L_E^S; for real parts only self-diffusion mode (D) is shown. The arrow points to that kσ for which $k\ell_E$=1.

V. Results

(1) Comparing figs. 1, 2a and 2b, one notices that with increasing density an increasing separation of the extended heat mode eigenvalue from the other eigenvalues occurs. In addition, for $k\ell_E \leq 0.5$, the three extended hydrodynamic modes are separated from the kinetic modes; for small k they reduce to the values given in Table III.

(2) We note that for large k, i.e., $k\sigma >> 1$, all the eigenvalues oscillate around those of $L_E^S(\vec{v})$ (indicated by dash-dot lines in the figures). These oscillations are due to oscillations in $A_{\vec{k}}$ (because of the presence of S(k)) as well as in $\Lambda_{\vec{k}}$ (because of the presence of $e^{-i\vec{k}\cdot\vec{\sigma}}$). Since both these terms are absent in $L_E^S(\vec{v})$, no oscillations are found in the eigenvalues of $L_E^S(\vec{k})$.

(3) In the approximate hard sphere theory based on $L_E(\vec{v})$, the intermediate scattering function F(k,t) is given by a sum of exponentials, while S(k,ω) is given by a sum of Lorentzians:

$$S_E(k,\omega) = S(k) \cdot 2\mathrm{Re} \sum_{j=1}^{\infty} \frac{A_j(k)}{i\omega - z_j(k)} \tag{5.1}$$

Here the eigenvalues $z_j(k)$ and the amplitudes $A_j(k) = \langle\Phi_j^*(k)\rangle\langle\Psi_j(k)\rangle$ ($\Phi_j(k)$ is a left eigenfunction of $L_E(k)$) are obtained from the matrix $(L_E)_{ij}$. In general, the $A_j(k)$ and $z_j(k)$ are complex functions of k. In all cases investigated, the extended heat mode eigenvalue $z_h(k)$ is a real function of k, while the two sound mode eigenvalues $z_\pm(k)$ are complex conjugate functions of k: $z_\pm(k) = \pm i\omega_s(k) + z_s(k)$. Here $\omega_s(k)$ represents the sound dispersion, characterizing the propagation of the extended sound mode, while $z_s(k)$ represents the damping of the extended sound modes. The sum runs, in principle, over all eigenmodes of $L_E(\vec{v})$, but in view of the above, only three eigenmodes suffice for $k\ell_E \leq 0.5$. Thus S(k,ω) for a fluid, as given by eqs. (3.4) or (5.1), can be considered as a direct generalization of the Landau-Placzek formula (2.8), as long as $k\ell_E \leq 0.5$.

(4) For sufficiently high densities $V_0/V \geq 0.5$ and $k\sigma \geq 0.1$, the heat mode dominates S(k,ω) for all k, since the heat mode eigenvalue is then by far the lowest. In addition, the magnitude of the heat mode eigenvalue, $|z_h(k)|$, exhibits a pronounced minimum at $k\sigma \approx 2\pi$ or $\lambda \approx \sigma$. In fact, $|z_h(k)|$ almost vanishes at $k\sigma \approx 2\pi$, i.e., $\lambda \approx \sigma$. The behavior of $z_h(k)$ for high densities can physically be understood in terms of the local conservation laws [28,31]. As was mentioned earlier, three basic lengths appear in the kinetic problem: ℓ_E, σ, λ, where λ characterizes the length scale of the small disturbance in the fluid from equilibrium.

For high densities $\ell_E << \sigma$ and as long as $\lambda > \sigma$ and at least some particles are included in λ, a sufficient number of collisions will still take place, that the local number, momentum and energy densities are approximately conserved inside λ. In that case, the extended hydrodynamic modes will dominate. However, when $\lambda \leq \sigma$ only one particle is included in λ and only the local number density, but not the local momentum or energy densities will be conserved. Then the extended heat mode degenerates into a

self-diffusion-like mode, the very approximate conservation of momentum and energy still manifesting itself (via $e^{-i\vec{k}\cdot\vec{\sigma}}$) in oscillations around this mode. The pronounced minimum of $|z_h(k)|$ at $k\sigma\approx 2\pi$ can be understood as follows. The heat mode started for $k\sigma\ll 1$, i.e., large λ, as $|z_h(k)|=D_{T,E}k^2$, while for $k\sigma>1$, i.e., small λ, $|z_h(k)|\sim D_E k^2$. Now at high densities $D_E \ll D_{T,E}$, because D_E only contains translational and no collisional transfer contributions, since in self-diffusion only the mass of a particle is involved, which cannot be transferred in a collision. Therefore, $|z_h(k)|$ has to make a transition from a behavior $\approx D_{T,E}k^2$ at $k\sigma\ll 1$ to a very different behavior $\sim D_E k^2$ at $k\sigma\gg 1$. This transition is accompanied by an overshoot at $k\sigma\approx 2\pi$, where S(k) has a pronouned maximum, so that $|z_h(k)|$ almost vanishes at $k\sigma\approx 2\pi$.
To understand this transition behavior of $z_h(k)$ around $k\sigma\approx 2\pi$ in more detail, we can use a perturbation theory with $-i\vec{k}\cdot\vec{v}+n A_{\vec{k}}$ as a perturbation on the collision term $ng(\sigma)\Lambda_{\vec{k}}$. The expansion parameter is then $k\ell_E$, and the procedure is valid as long as $0<k\ell_E<1$. We note, however, that $k\sigma$ can be of $O(1)$ (i.e., $\approx 2\pi$), since $\ell_E/\sigma<1$ for $V_0/V>0.15$. This way the following expression is obtained for $|z_h(k)|$ around $k\sigma\approx 2\pi$:

$$|z_h(k)| = \frac{D_E k^2}{S(k)}\, d(k\sigma) + O(k\ell_E)^4 \qquad (5.2)$$

where $d(k\sigma)$ is a dimensionless oscillating function of order 1. Eq. (5.2) shows that $|z_h(k)|$ will oscillate around $|z_D(k)|\sim D_E k^2$ and that the strong maximum in S(k) will cause a pronounced minimum in $|z_h(k)|$.
(5) With increasing density, the magnitude of the <u>sound damping</u> $z_s(k)$ increases, while the <u>sound propagation</u> dispersion curve $\omega_s(k)$ bends over more and more. In fact, from an unremarkable smooth behavior at low densities (cf fig. 1), a "Landau-like" sound dispersion curve--similar somewhat in form to that of superfluid helium--can be seen around $V_0/V\approx 0.45$ (cf. fig. 2a), while for $V_0/V\geq 0.52$ sound propagation gaps, appear around $k\sigma\approx 2\pi$, in which no propagation of sound occurs at all and two overdamped sound modes are present (cf. fig. 2b, where two sound propagation gaps are present).
The disappearance of sound propagation at high densities around $k\sigma\approx 2\pi$ can also be understood on the basis of the local conservation laws, but it has nothing to do with the maximum of S(k) at $k\sigma\approx 2\pi$. In fact, setting S(k) equal to 1 or to S(0), i.e., a constant, gives sound dispersion curves, with propagation gaps, that are hardly distinguishable from those using the full S(k). The physical origin of this effect is related to a competition of elastic (restoring) forces--contained in the imaginary part of $L_E(\vec{k})$--and the dissipative forces--contained in the real part of $L_E(\vec{k})$. In this sense, kinetic theory is similar to hydrodynamics, in that also there [cf. $\underline{\underline{H}}(\vec{k})$ of eq. (2.5)], the off-diagonal elastic terms are all imaginary, while the diagonal dissipative terms are all real. The influence of these two different types of terms on the propagation of sound can already be studied by considering the 2×2 matrix $(L_E(\vec{k}))_{ij}$ (i,j=2,3) [29, 31]. This can be motivated by remarking that in a very simplified manner sound propagation involves an exchange of (longitudinal) velocity

(related to ψ_2) and energy (related to ψ_3). One has then to diagonalize a matrix of the form:

$$\begin{pmatrix} -f_1(k) & ig(k) \\ ig(k) & -f_2(k) \end{pmatrix}$$

with $f_1(k)=\frac{2}{3}[1-j_0(k\sigma)+2j_2(k\sigma)]$, $f_2(k)=\frac{2}{3}[1-j_0(k\sigma)]$ and $g(k)=(\frac{\pi}{12})^{1/2}[k\ell_E+\sqrt{2}j_1(k\sigma)]$, where the j_ℓ are spherical Bessel functions. This leads to eigenvalues:

$$z_\pm(k)t_E = -\frac{1}{2}[(f_1+f_2)] \pm \{(f_1-f_2)^2 - (2g)^2\}^{1/2}] \qquad (5.3)$$

The crucial observation now is that although the dissipation, i.e., f_1 and f_2, can be very large, the propagating character of the solution, determined by its imaginary part, is determined by the difference $|f_1-f_2|$ compared to $|2g|$. Since this difference can become comparable to $|2g|$, this interference of dissipative (i.e., f_1, f_2) and elastic (i.e., g) forces can cause a propagation gap in the sound dispersion curve, viz. when the square root in (5.3) is real. We remark that for $k\sigma<<1$, i.e., in the hydrodynamic regime, propagation always occurs. However, when $k\sigma\geqq1$ and $V_0/V>0.59$, $\Delta=(f_1-f_2)^2-(2g)^2$ can be positive (cf. Fig. 3), so that then no sound propagation occurs and two overdamped sound modes are present.

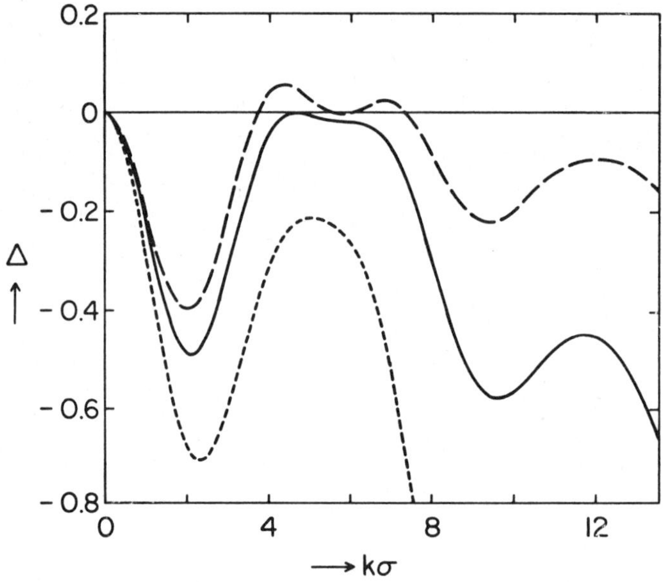

Figure 3: Δ as a function of $k\sigma$ at $V_0/V=0.45$ (----), 0.59 (___) and 0.70 (- - -).

(6) A comparison of the three extended hydrodynamic eigenvalues of a hard sphere fluid determined [32] from computer simulations by Alley and Alder [33] for a hard sphere fluid at $V_0/V=0.625$ and those computed on the basis of the BGK-method, is given in fig. 4. The intermediate scattering function F(k,t) [34] and the half width at half height ω_H [32]--defined by the equation $S(k,\omega_H)=\frac{1}{2}S(k,0)$-- are compared with

the theory, using the three extended hydrodynamic modes only, in figs. 5 and 6, respectively.

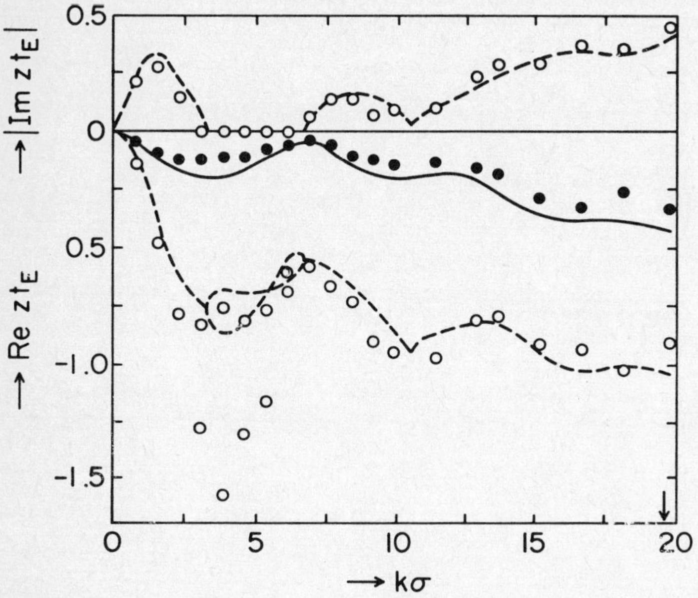

Figure 4: Ordinate as in Fig. 1 for extended heat (●) and sound (o) modes, deduced from computer simulations for hard spheres at $V_0/V=0.625$ and theory (heat ____, sound -----), as a function of kσ.

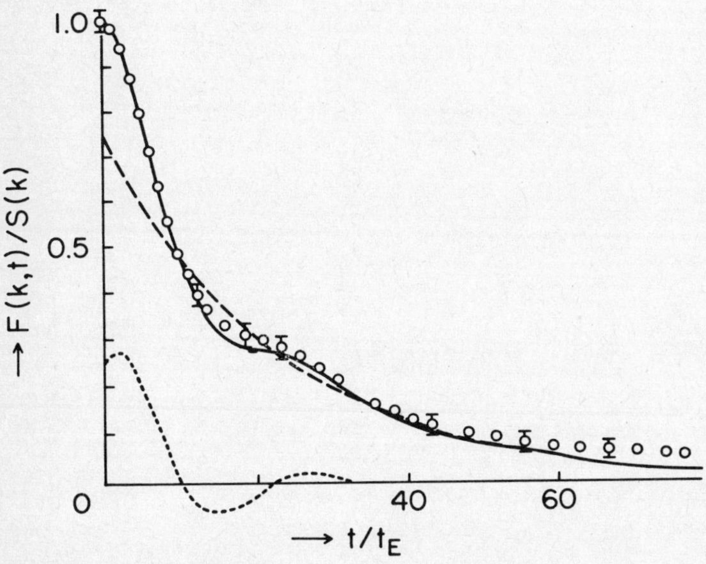

Figure 5: Reduced intermediate scattering function F(k,t)/S(k) (o) for hard sphere fluid at $V_0/V=0.625$ and kσ=0.76 (with error bars), compared with contribution of three extended hydrodynamic modes (____) (heat— — —, sound -----), as a function of kσ.

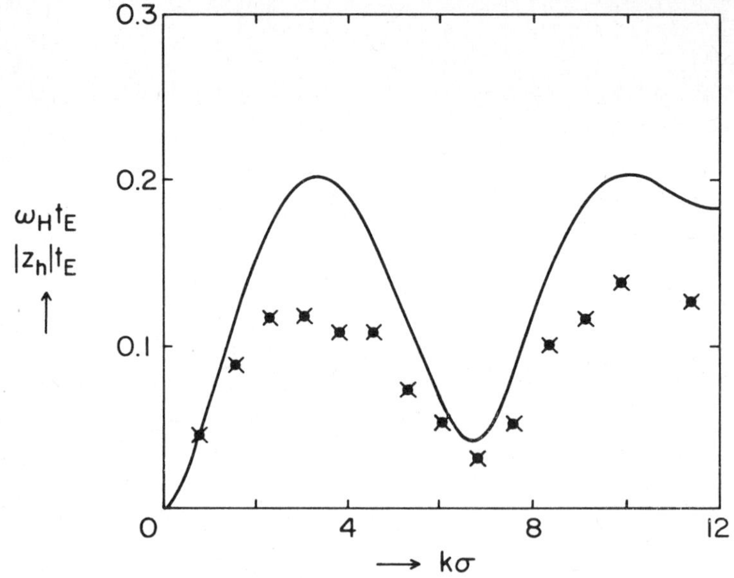

Figure 6: Reduced half width at half height $\omega_H t_E$ and $|z_h| t_E$ for hard sphere fluid at $V_0/V=0.625$ from computer simulations (•,x) and theory (___).

Thus the minimum in $|z_h(k)|$ at $k\sigma \approx 2\pi$ is clearly reflected in $S(k,\omega)$, in that a considerable narrowing of $S(k,\omega)$ around $k\sigma \approx 2\pi$ occurs--the so-called de Gennes narrowing.

(7) Once eigenmodes are known, not only the density-density correlation function, but, in principle, a whole class of other time correlation functions can be computed. In fact, from Alley and Alder's work follows, at least for dense hard sphere fluids, that three extended hydrodynamic modes not only describe the density-density correlation function, i.e., $S(k,\omega)$, but also all other correlation functions of the fluctuations of the hydrodynamic quantities [32].

(8) These results for hard sphere fluids have lead to attempts to interpret the neutron spectra of real fluids also in terms of eigenmodes. In that view, $S(k,\omega)/S(k)$ is given in terms of the eigenvalues $z_j(k)$ and the eigenfunctions, which determine the $A_j(k)$, by eq. (5.1). One should keep in mind though that the importance of the contributions of the eigenmodes of the fluid to $S(k,\omega)$ may be quite different for hard sphere and real fluids. This may obtain, in particular, for the extended sound and kinetic modes of the fluid. I mention two kinds of attempts:

(a) One replaces the real fluid by an equivalent hard sphere fluid by introducing--at a given temperature--an effective hard sphere diameter σ_{eff}, representative of the fluid considered. Such a σ_{eff} could, for example, be determined by letting the first maximum of the S(k) of the real and the effective hard sphere fluid coincide at the same value of $k\sigma$ [28,30,35].

(b) Conversely, one can use sufficiently accurate experimental $S(k,\omega)$ data either for real liquids or for computer generated fluids, such as hard sphere fluids or Lennard-Jones fluids--i.e., fluids consisting of particles interacting with a 12-6 Lennard-Jones potential-- to <u>derive</u> their extended hydrodynamic eigenmodes. This can be done by a fit of the data with three Lorentzians, i.e., representing $S(k,\omega)/S(k)$ by the right hand side of eq. (5.1) with three unknown (complex) $z_j(k)$ and $A_j(k)$. Without going into the details of this procedure, consistent values of $z_j(k)$ and $A_j(k)$, where j represents each of the extended hydrodynamic modes, have been determined for liquid Ar [36-39], Ne [40,41] as well as Lennard-Jones fluids [42]. The heat mode eigenvalues of liquid Argon and a corresponding Lennard-Jones fluid, together with their half widths ω_H, have been plotted in fig. 7.

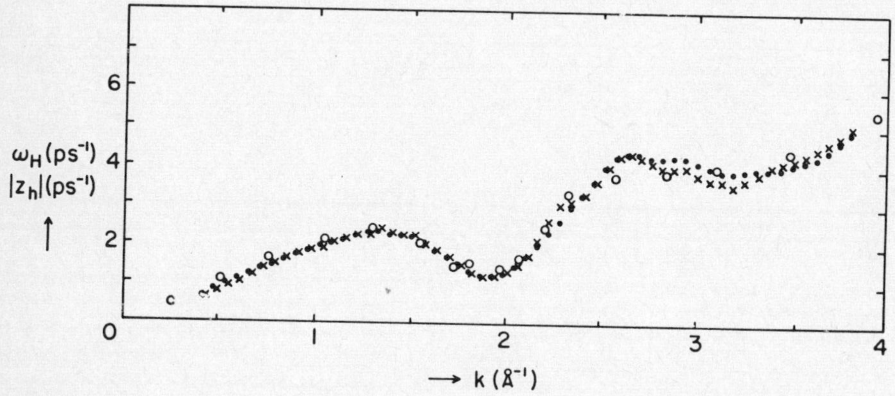

Figure 7: ω_H (•) and $|z_h|$ (x) for liquid Argon at 120°K and 115 bar and $\omega_H \approx |z_h|$ for corresponding Lennard-Jones fluid (o).

One notes the close similarity of the results for the Lennard-Jones liquid and liquid Ar and the striking common (de Gennes) minimum at $k\sigma \approx 2\pi$, if a value σ_{eff}=3.43Å is used for liquid Argon. In Fig. 8, the magnitude of the imaginary part of the sound mode eigenvalue, ω_s, for liquid Argon and a corresponding Lennard-Jones fluid have been plotted as a function of k. The presence of a common propagation gap (where $\omega_s(k)$=0) around values of $k \approx 2Å^{-1}$ is apparent. Present neutron scattering experiments do not show any identifiable contributions of kinetic, i.e., non (extended) hydrodynamic modes.

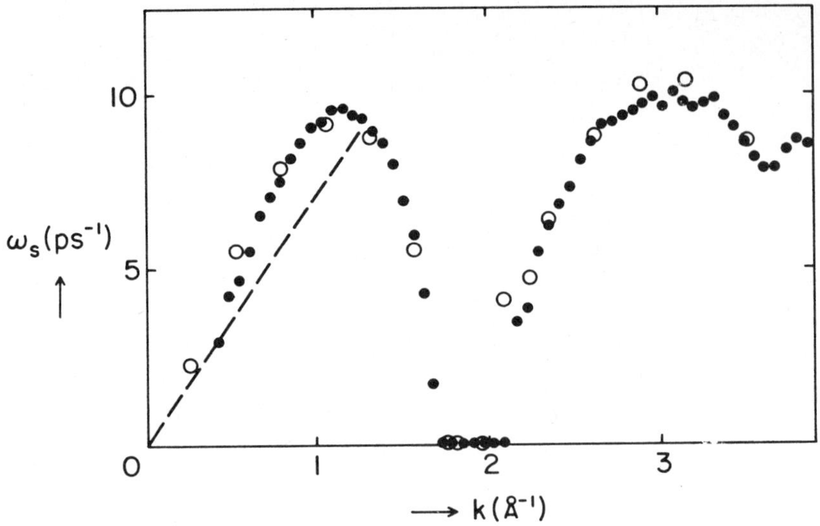

Figure 8: ω_s for liquid Argon and a Lennard-Jones fluid, as in Fig. 7. The dashed line represents the hydrodynamic result $\omega_s = ck$.

(9) The real and imaginary parts of the eigenvalues of hard sphere, Lennard-Jones and real fluids all appear to approach straight lines for large k. This linear behavior is consistent with that of an ideal gas, which is expected to occur for very large k. It is unclear, however, how the transition from collective (i.e., collision dominated) behavior to individual particle (i.e., free streaming) behavior takes place.

Acknowledgement

The author is very much indebted to Dr. B. Kamgar-Parsi for his help in preparing this manuscript and to Mrs. Diane Lott for typing it. Critical comments by T. R. Kirkpatrick are also gratefully acknowledged. This work was supported by the Department of Energy under contract No. DE-AC02-81ER10807.006.

References

[1] Boon, J. P. and Yip, S., Molecular Hydrodynamics, (McGraw Hill, New York, 1980).

[2] Résibois, P. and de Leener, M., Classical Kinetic Theory of Fluids, (Wiley-Interscience, New York, 1977).

[3] Hansen, J. P. and McDonald, I. R., The Theory of Simple Liquids, (Academic Press, London, 1976).

[4] Zwanzig, R., Ann. Rev. Phys. Chem. 16 (1965) 67.

[5] Berne, B. J. and Pecora, R., Dynamic Light Scattering, (Wiley-Interscience, New York, 1976).

[6] Wood, W. W., in: Fundamental Problems in Statistical Mechanics, III, Cohen, E. G. D., ed., (North-Holland, Amsterdam, 1975).

[7] Wood, W. W. and Erpenbeck, J. J., Ann. Rev. Phys. Chem. 27 (1976) 319.

[8] Mountain, R. D., Rev. Mod. Phys. 38 (1966) 205.

[9] Foch, J. D. and Ford G. W., in: Studies in Statistical Mechanics, V, de Boer, J. and Uhlenbeck, G. E., eds., (North-Holland, Amsterdam, 1970), Part B, Chap. II.

[10] See, for instance, Ernst, M. H. and Dorfman, J. R., J. Stat. Phys. 12 (1975) 311.

[11] DeMasi, A., Ianiro, N., Pellegrinotti, A. and Presutti, in: Studies in Statistical Mechanics, XI, Lebowitz, J. L. and Montroll, E. W., eds. (North-Holland, Amsterdam, 1984).

[12] de Schepper, I. M., van Rijs, J. C. and Cohen, E. G. D., Physica A, 1986.

[13] Ref. [9], Part B, Ch. III.

[14] Greenberg, W., Polewczak, J. and Zweifel, P. F., in: The Boltzmann Equation in Studies in Statistical Mechanics, X, Montroll, E. W. and Lebowitz, J. L., eds., (North-Holland, Amsterdam, 1983).

[15] Wang Chang, C. S. and Uhlenbeck, G. E., in: Studies in Statistical Mechanics, V, de Boer, J. and Uhlenbeck, G. E., eds., (North-Holland, Amsterdam, 1970), Part A, Ch. IV.

[16] Alterman, Z., Frankowski, K. and Pekeris, C. S., Am. Astrophys. J. Suppl. 69, VII, (1962) 291.

[17] Chapman, S. and Cowling, T. G., The Mathematical Theory of Nonuniform Gases, (Cambridge University Press, Cambridge, 1970).

[18] Grad, H., Commun. Pure and Appl. Math. 2 (1949) 331.

[19] Kamgar-Parsi, B. and Cohen, E. G. D., to be published.

[20] Lebowitz, J. L., Percus, J. K. and Sykes, J., Phys. Rev. 188 (1969) 487.

[21] Mazenko, G. F., Phys. Rev. A7 (1973) 209, 222; 9 (1974) 360.

[22] Konijnendijk, H. M. U. and van Leeuwen, J. M. J., Physica 64 (1973) 232.

[23] Beijeren, H. and Ernst, M. H., Physica 68 (1973) 43, J. Stat. Phys. 21 (1979) 125.

[24] Dorfman, J. R. and Cohen, E. G. D., Phys. Rev. A12 (1975) 292.

[25] Mazenko, G. F. and Yip, S., in: Mondern Theoretical Chemistry, Vol. 6, (Plenum Press, New York, 1977).

[26] de Schepper, I. M. and Cohen, E. G. D., J. Stat. Phys. 27 (1982) 223.

[27] Ref. [15], Ch. V.

[28] de Schepper, I. M., Cohen, E. G. D. and Zuilhof, M. J., Phys. Lett. 101A (1984) 399.

[29] Zuilhof, M. J., Cohen, E. G. D. and de Schepper, I. M., Phys. Lett. 103A (1984) 120.

[30] Cohen, E. G. D., de Schepper, I. M. and Zuilhof, M. J., Physica 127B (1984) 282.

[31] Kirkpatrick, T. R., Phys. Rev. A32 (1985).

[32] Cohen, E. G. D., de Schepper, I. M. and Kamgar-Parsi, B., Phys. Letts., (1986).

[33] Alley, W. E. and Alder, B. J., Phys. Rev. A27 (1983) 3158.

[34] Alley, W. E., Alder, B. J. and Yip, S., Phys. Rev. A27 (1983) 3174.

[35] de Schepper, I. M. and Cohen, E. G. D., Phys. Rev. A22 (1980) 287.

[36] de Schepper, I. M., Verkerk, P., van Well, A. A. and de Graaf, L. A., Phys. Rev. Lett. 50 (1983) 974.

[37] van Well, A. A., de Graaf, L. A., Verkerk, P., Suck, J. B. and Copley, J. R., Phys. Rev. A31 (1985) 3391.

[38] Verkerk, P., Ph. D. Thesis, University of Technology, Delft (1985).

[39] van Well, A. A. and de Graaf, L. A., Phys. Rev. A32 (1985).

[40] van Well, A. A. and de Graaf, L. A., Phys. Rev. A32 (1985).

[41] van Well, A. A., Ph. D. Thesis, University of Technology, Delft (1985).

[42] de Schepper, I. M., van Rijs, J. C., van Well, A. A., Verkerk, P., de Graaf, L. A. and Bruin, C., Phys. Rev. 29A (1984) 1602.

STUDY OF THE INVERSE PROBLEM IN RANDOM MEDIA USING COHERENCE THEORY

Mark J. Beran
School of Engineering
Tel Aviv University

Abstract

We begin by reviewing the basic definitions of the coherence functions used in studies of propagation in random media. The governing moment equations, and a summary of relevant solutions, is then given for the second and fourth-order coherence functions. The inverse problem of determining the two-point correlation function of the index of refraction fluctuations is discussed in the context of these coherence functions. We shall treat plane wave, point source and finite source boundary conditions. In addition the effect of media anisotropy will be considered. Finally we shall discuss a case in which an inverse algorithm has been implemented.

INTRODUCTION

We consider here propagation of radiation through a medium with random spatial index of refraction fluctuations. For convenience, the radiation is taken to be periodic with frequency, f. We restrict our attention to small fluctuations so that we may consider the radiation to be confined to small angles about a mean propagation direction. The theory to be described is used in studying the propagation of laser light in a turbulent atmosphere and acoustic radiation in the ocean.

Using the small angle approximation it is possible to show that the governing equation is

$$2ik\,\partial u/\partial z + \nabla^2_T u + k^2 \mu(x) u = 0 \qquad (1)$$

Here u is the field variable, k is a mean radiation wave number and $\mu(x)$ represents the spatial variation in the index of refraction. We chose z as the mean propagation direction and x,y as the transverse coordinates. The symbol, ∇^2_T is the transverse Laplacian.

Definition of the Coherence functions

To treat the problem from a stochastic point of view we define the following coherence functions

$$\{\Gamma(x_{T1},x_{T2},z)\} = \{u(x_{T1},z)u^*(x_{T2},z)\} \qquad (2)$$

$$\{L^1(x_{T1},x_{T2},x_{T3},x_{T4},z)\} = \{u(x_{T1},z)u^*(x_{T2},z)u^*(x_{T3},z)u(x_{T4},z)\} \tag{3}$$

Here u^* denotes the complex conjugate of u and the transverse vector x_T, has the components x and y. The brackets, { }, denote an ensemble average.

The intensity, $\{I(x_T,z)\}$, is found from $\{\Gamma(x_{T1},x_{T2},z)\}$ by setting $x_{T1} = x_{T2}$. The intensity fluctuations, $\{I^2(x_T,z)\}$, and the coherence of intensities, $\{R_I(x_{T1},x_{T2},z)\}$, are found from $\{L^1\}$ from the following relations

$$\{I^2(x_T,z)\} = \{L^1(x_T,x_T,x_T,x_T,z)\} \tag{4}$$

$$\{R_I(x_{T1},x_{T2},z)\} = \{L^1(x_{T1},x_{T2},x_{T2},x_{T1},z)\} \tag{5}$$

We write $\mu(x)$ in the form

$$\mu(x) = \bar{\mu}(x) + \mu'(x) \tag{6}$$

where $\bar{\mu}(x)$ represents the mean background profile and $\mu'(x)$ represents the random index of refraction fluctuations. We shall assume here that the statistics of the index of refraction field are homogeneous. The two-point correlation function, $\sigma(x_1,x_2)$, is defined as

$$\sigma(x_1,x_2) = \{\mu'(x_1)\,\mu'(x_2)\} \tag{7}$$

and depends upon $r = x_1 - x_2$. In addition we define the function

$$\bar{\sigma}(x_{T1} - x_{T2}) = (1/4)\int_{-\infty}^{\infty}\sigma(x_{T1} - x_{T2},r_z)\,dr_z \tag{8}$$

Moment Equations

The equations governing $\{\Gamma\}$ and $\{L^1\}$ (See Refs. (1) - (5)) are

$$\begin{aligned}\partial\{\Gamma\}/\partial z = {}& (i/2k)[\nabla^2_{T1} - \nabla^2_{T2}]\{\Gamma\} - \\ & k^2[\bar{\sigma}(0) - \bar{\sigma}(x_{T1} - x_{T2})]\{\Gamma\} + \\ & ik[\bar{\mu}(x_{T1}) - \bar{\mu}(x_{T2})]\{\Gamma\}\end{aligned} \tag{9}$$

$$\begin{aligned}\partial\{L^1\}/\ z = {}& (i/2k)[\nabla^2_{T1} - \nabla^2_{T2} - \nabla^2_{T3} + \nabla^2_{T4}]\{L^1\} + \\ & k^2[\bar{\sigma}(x_{T1} - x_{T2}) + \bar{\sigma}(x_{T1} - x_{T3}) + \\ & \bar{\sigma}(x_{T2} - x_{T4}) + \bar{\sigma}(x_{T3} - x_{T4}) - \\ & \bar{\sigma}(x_{T1} - x_{T4}) - \bar{\sigma}(x_{T2} - x_{T3}) - \\ & 2\bar{\sigma}(0)]\{L^1\} + \\ & ik[\bar{\mu}(x_{T1}) - \bar{\mu}(x_{T2}) - \bar{\mu}(x_{T3}) + \\ & \bar{\mu}(x_{T4})]\{L^1\}\end{aligned} \tag{10}$$

For both the second and fourth-order coherence functions it is convenient to use a different set of independent coordinates. For the second-order case we transform to sum and difference coordinates defined as

$$p = (x_{T1} + x_{T2})/2$$

$$\mathbf{s} = (x_{T1} - x_{T2}) \tag{11}$$

In terms of $\mathbf{s}$ and p, Eq. (9) becomes

$$\partial\{\Gamma\}/\partial z = (i/k)\,\partial/\partial \mathbf{s}\cdot\partial/\partial p\{\Gamma\} - k^2[\bar{\sigma}(0) - \bar{\sigma}(\mathbf{s})]\{\Gamma\}$$
$$+ ik[\bar{\mu}(p + \mathbf{s}/2) - \bar{\mu}(p - \mathbf{s}/2)]\{\Gamma\} \tag{12}$$

For the fourth-order case we transform to the coordinates

$$\mathbf{s} = (1/2)(x_{T1} - x_{T2} + x_{T3} - x_{T4})$$

$$p = (1/2)(x_{T1} + x_{T2} - x_{T3} - x_{T4})$$

$$q = (x_{T1} - x_{T2} - x_{T3} + x_{T4})$$

$$R = (1/4)(x_{T1} + x_{T2} + x_{T3} + x_{T4}) \tag{13}$$

where

$$r_{12} = \mathbf{s} + q/2 \qquad r_{13} = p + q/2$$

$$r_{14} = \mathbf{s} + p \qquad r_{23} = -\mathbf{s} + p$$

$$r_{24} = p - q/2 \qquad r_{34} = \mathbf{s} - q/2 \tag{14}$$

Using Eqs. (13) and (14), Eq. (10) may be written as

$$\begin{aligned}\partial\{L^1\}/\partial z = (i/k)[(\partial/\partial \mathbf{s})\cdot(\partial/\partial p) +\\ (\partial/\partial R)\cdot(\partial/\partial q)]\{L^1\} +\\ k^2[\bar{\sigma}(r_{12}) + \bar{\sigma}(r_{13}) +\\ \bar{\sigma}(r_{24}) + \bar{\sigma}(r_{34}) -\\ \bar{\sigma}(r_{14}) - \bar{\sigma}(r_{23}) -\\ 2\bar{\sigma}(0)]\{L^1] +\\ 1k[\bar{\mu}(x_{T1}) - \bar{\mu}(x_{T2}) - \bar{\mu}(x_{T3}) +\\ \bar{\mu}(x_{T4})]\{L^1\}\end{aligned} \tag{15}$$

Except for a few remarks about the effect of a linear profile we do not consider here the inverse problem when there is a mean background profile, $\bar{\mu}$, present. We have included these terms so that the reader may see see how they appear in Eqs. (12) and (15). Referring to Eq. (12) we see that including the $\bar{\mu}$ terms turns this equation into a difference-differential equation and thus considerably complicates any analytic developments. Most work that has been done using the background profile makes use of what has been called the quadratic approximation. Here the $\bar{\mu}$ difference term is expanded in a Taylor series in $\mathbf{s}$ and the following approximation is used

$$\bar{\mu}(p + \mathbf{s}/2) - \bar{\mu}(p - \mathbf{s}/2) = d\bar{\mu}/dp\cdot\mathbf{s} \qquad (16)$$

If Eq. (16) is used in Eq. (12) and we take the Fourier transform of this equation with respect to $\mathbf{s}$ we find a radiation transport equation in terms of the function

$$\{\hat{\Gamma}(p,k_T,z)\} = \iint_{-\infty}^{\infty} \exp(ik_T\cdot\mathbf{s})\{\Gamma(p,\mathbf{s},z)\}d\mathbf{s} \qquad (17)$$

A similar simplification may used in the fourth-order case but terms of second-order must be retained in the $\bar{\mu}$ expansion.

Solution to Eqs. (12) and (15)

Plane Wave Boundary Condition

When there is no mean profile (i.e. $\bar{\mu} = 0$), Eq. (12) may be readily solved for an initial plane wave. In the plane wave case the term $(\partial/\partial p)\{\Gamma\} = 0$ and the solution is

$$\{\Gamma(\mathbf{s},z)\} = I\exp(-k^2[\bar{\sigma}(0) - \bar{\sigma}(\mathbf{s})]z) \qquad (18)$$

where I is the initial wave intensity. [If there is a linear profile, the right-hand side of Eq. (18) is multiplied by a phase term of the form exp(ikAsz) where the constant A is determined by the slope of the profile.]

The intensity is independent of z and this is readily confirmed by setting $\mathbf{s} = 0$. As $|\mathbf{s}|$ approaches infinity we find

$$\{\Gamma(\infty,z)\} = I\exp[-k^2\bar{\sigma}(0)z] \qquad (19)$$

We may interpret $\{\Gamma(\infty,z)\}$ as the intensity of the coherent field that remains after scattering. The quantity $l_+ = 1/[k^2\bar{\sigma}(0)]$ is the characteristic distance over which significant scattering occurs.

The solution of Eq. (15) does not have a simple analytic solution similar to that given in Eq. (18). Recent work has shown, however, that the solution may be reduced to the numerical evaluation of a double integral (See Refs. (6) - (9)). To our knowledge, noone has yet considered using these solutions and developing algorithms for use in the inverse problem. Rather, the perturbation solution of $\{R_I(x_{T1},x_{T2},z)\}$, obtained for $k^2\bar{\sigma}(0)z \ll 1$, has been shown to be suitable (10).

The perturbation solution for $\{R_I(x_{T1},x_{T2},z)\}$ is

$$\{R_I(x_{T1},x_{T2},z)\} = I^2[1 + (k^2z/2) \iint_{-\infty}^{\infty} \bar{\bar{\sigma}}(k_{12})\cdot$$
$$\exp(ik_{12}\cdot r_{12})\cdot$$
$$(1 - (k/k_{12}^2z)\sin[k_{12}^2z/k])dk_{12}] \tag{20}$$

where

$$\bar{\bar{\sigma}}(k_{12}) = (1/2\pi)^2 \iint_{-\infty}^{\infty} \bar{\sigma}(r_{12})\exp(ik_{12}\cdot r_{12})dr_{12}$$

A linear mean profile does not change the nature of this solution.

Point Source Boundary Condition

When the initial plane wave is replaced by a point source (all other parameters remaining the same) Eq. (18) is replaced by

$$\{\Gamma(\mathbf{s},z)\} = (I_s/z^2)\exp(-k^2z[\bar{\sigma}(0) - \int_0^1 \bar{\sigma}(a\mathbf{s})da]) \tag{21}$$

Here I_s is a normalization constant.

Gaussian Beam Boundary Condition

For $\bar{\mu} = 0$, Eq. (12) may be solved analytically for a gaussian beam boundary condition (11). For this case we have

$$\{\Gamma(x_{T1},x_{T2},0)\} = I\exp[-(|x_{T1}|^2 + |x_{T2}|^2)/2b^2]$$
$$= I\exp[-(|\mathbf{p}|^2 + |\mathbf{s}|^2/4)/b^2] \tag{22}$$

We shall not require here the full solution for $\{\Gamma\}$, but rather only the intensity. The intensity, $\{I(\mathbf{p},z)\}$, is found to be

$$\{I(\mathbf{p},z)\} = (Ib^2/4\pi) \iint_{-\infty}^{\infty} d\mathbf{w}\exp(-i\mathbf{w}\cdot\mathbf{p})\cdot$$
$$\exp[-w^2b^2(1 + z^2/k^2b^4) + \bar{Q}(\mathbf{w},z)] \tag{23}$$

where

$$\bar{Q}(\mathbf{w},z) = (k^3/w) \int_0^{wz/k} [\bar{\sigma}(w_xt/w,w_yt/w) - \bar{\sigma}(0,0)]dt \tag{24}$$

The only effect that a linear mean profile will have is to shift the maximum value of the intensity.

When the fluctuation field is statistically isotropic then $\bar{\sigma}$ is a function only of the separation distance and Eqs. (23) and (24) become

$$\{I(p,z)\} = (Ib^2/2) \int_0^\infty dw J_0(wp) w \exp[-w^2 b^2 (1 + z^2/k^2 b^4)/4 + \bar{Q}(w,z)] \tag{25}$$

where

$$\bar{Q}(w,z) = (k^3/w) \int_0^{wz/k} [\bar{\sigma}(t) - \bar{\sigma}(0)] dt \tag{26}$$

INVERSE PROCEDURE

We shall now restrict our attention to statistically, homogeneous isotropic statistics but the methods are applicable to the anisotropic case. In fact, when the medium is highly anisotropic, as it often is in the ocean, the resultant expressions are sometimes simpler than in the isotropic case.

Initial Plane Wave - $\{\Gamma\}$

From an analytical point of view, perhaps the simplest inversion procedure is obtained from Eq. (18). Taking the logarithm of Eq. (18) we find

$$\bar{\sigma}(s) = \bar{\sigma}(0) + (1/k^2 z) \log[\{\Gamma(s,z)\}/I] \tag{27}$$

In the limit of s approaching infinity $\bar{\sigma}(s) = 0$. Thus

$$\bar{\sigma}(s) = (1/k^2 z) \log[\{\Gamma(s,z)\}/\{\Gamma(\infty,z)\}] \tag{28}$$

The quantity (s) is very valuable in itself, since it represents an effective transverse correlation function and is the function required in making any statistical calculations. If the function $\sigma(s,r_z)$ is required, it may be determined from the defining relation (see Eq. (8))

$$\bar{\sigma}(s) = (1/4) \int_{-\infty}^{\infty} \sigma(\sqrt{s^2 + r_z^2})\, dr_z \tag{29}$$

by converting it to an Abel's integral equation. We note, however, that there is a Bessel transform relation between $\bar{\sigma}(s)$ and the spatial spectrum

$$\Phi(v) = (1/2\pi)^3 \iiint_{-\infty}^{\infty} \exp(-iv\cdot r)\, \sigma(r)dr \tag{30}$$

This relationship is, (for isotropic statistics),

$$\Phi(v) = (1/\pi^2) \int_0^{\infty} \bar{\sigma}(s) J_o(vs) s ds \tag{31}$$

<u>Initial Plane Wave - $\{R_I\}$</u>

We may also obtain information about $\bar{\sigma}(s)$ and $\Phi(v)$ by measuring the coherence of intensities, $\{R_I(s,z)\}$. In principal measuring either $\{\Gamma\}$ or $\{R_I\}$ will yield the same information. As a practical matter, however, measurement of $\{R_I\}$ is much simpler but inversion of $\{R_I\}$ has so far only been studied in the perturbation region. We have in this region

$$\{R_I(s,z)\} = I^2[1 + 2(\pi k)^2 z \int_0^{\infty} \Phi(v) J_o(vs) v \, (1 - (k/v^2 z)\sin(v^2 z/k)) dv] \tag{32}$$

From Eq. (32) we see that $\{R_I\}$ is determined by a rather complicated integral of the spectrum $\Phi(v)$. However, this integral reduces to a Bessel transform in two limits: (1) in the geometric optics region where $z \ll ks_c^2$, s_c being the smallest characteristic length associated with the coherence function and (2) when $z \gg ks_c^2$. In the geometric optics region we find

$$\{R_I(s,z)\} = I^2[1 + (\pi^2/3) z^3 \int_0^{\infty} \Phi(v) v^5 J_o(sv) dv] \tag{33}$$

When $z \gg ks_c^2$ we have

$$\{R_I(s,z)\} = I^2[1 + (\pi/4) k^2 z \int_0^{\infty} \Phi(v) v J_o(sv) dv]$$

$$= I^2[1 + k^2 z\, \bar{\sigma}(s)/4\pi] \tag{34}$$

The inversion of Eqs. (33) and (34) yield respectively

$$\Phi(v) = (3/\pi^2 z^3 v^4) \int_0^{\infty} [\{R_I(s,z)\} - I^2] J_o(sv) s ds \tag{35}$$

$$\bar{\sigma}(s) = (4\pi/k^2 z)[R_I(s,z) - I^2] \tag{36}$$

Both Eqs. (35) and (36) have limitations. As a result of the v^4 factor in Eq. (35) we see that in the geometric optics region $\{R_I(s,z)\}$ is strongly determined by the high spatial frequency components of the fluctuation spectrum. Thus in a real measurement it will be very difficult to determine the full spectrum. On the other hand, in the region $z \gg ks_c^2$ where $\bar{\sigma}(s)$ is determined simply from an algebraic formula, we require that the perturbation solution be valid at very large distances. This later condition requires very weak fluctuations with attendant difficulties in measurement.

Initial Point Source - $\{\Gamma\}$

If we take the logarithm of both sides of Eq. (21) we find

$$\int_0^1 \bar{\sigma}(as)da = \bar{\sigma}(0) + (1/k^2z)\log[\{\Gamma(s,z)\}/(I_z/z^2)] \tag{37}$$

Next noting that $d/ds[s\int_0^1 \bar{\sigma}(as)da] = \bar{\sigma}(s)$ we find the following expression for $\bar{\sigma}(s)$

$$\bar{\sigma}(s) = \bar{\sigma}(0) + d/ds[(s/k^2z)\log(\{\Gamma(s,z)\}/(I_z/z^2))] \tag{38}$$

Gaussian Beam Boundary Condition - $\{I\}$

From Eq. (25) we see that $\{I\}$ is represented as a Bessel transform of an exponential quantity that contains the function $\bar{\sigma}(t)$. Inverting the transform and taking the logarithm of the resulting expression we find

$$\bar{Q}(w,z) = w^2b^2(1 + z^2/k^2b^4)/4 + \log[2/(Ib^2)\int_0^\infty\{I(p,z)\}pJ_o(wp)dp] \tag{39}$$

From Eq. (26) we then find

$$\bar{\sigma}(wz/k) = (1/k^2z)[\partial w\bar{Q}(wz/k)/\partial w - \partial w\bar{Q}(wz/k)/\partial w|_{w\to\infty}] \tag{40}$$

The inversion procedure in this case is difficult since it involves a differentiation as a final step. However, experimentally we have only to measure the mean intensity rather than $\{\Gamma\}$ or $\{R_I\}$.

IMPLEMENTATION

As an example of how the above algorithms may be used in a real problem we refer to the work of Coles and Frehlich (12) in which they determine $\bar{\sigma}(s)$ using $\{\Gamma\}$ and a point source boundary condition. We provide here only a summary of their work. The reader is referred to the original article for a detailed discussion.

A 4mW He-Ne laser transmitter with a divergence of .01 rad was used to simulate a point source. The range of the experiment was 1163 m carried out 2 m over a very flat dry lake bed. At the receiving site a 20 cm telescope with an angular field of view of .001 rad and a resolution of less than 10^{-5} rad was used to measure the angular spectrum of the received radiation. The exposure times were 10 seconds.

From the angular spectrum the coherence function $\{\Gamma\}$ was found using a Fourier transform. From the point of view of this article $\{\Gamma\}$ may be considered to be the measured data. Coles and Frehlich discuss the experimental errors that occur in determining $\{\Gamma\}$. A significant error may occur as a result of truncation at large angles; an effect which manifests itself in an incorrect behavior of $\{\Gamma\}$ when the separation distance is small.

The authors first determined a quantity proportional to

$\int_0^1 \bar{\sigma}(as)ds - \bar{\sigma}(0)$ using an expression equivalent to

Eq. (37). They then performed the differentiation indicated in Eq. (38) to find $\bar{\sigma}(s) - \bar{\sigma}(0)$. Because of experimental limiations they were only able to determine $\bar{\sigma}(s) - \bar{\sigma}(0)$ in the Kolmogorov region where $\bar{\sigma}(s) - \bar{\sigma}(0)$ is proportional to $s^{5/3}$. At small separation distances the results were limited by angular truncation errors while at large separation distances they were limited by errors introduced by finite integration times.

There was nothing intrinsic about the inability of the authors to obtain $\bar{\sigma}(s)$ accurately over the entire range of s. To do so would have required a much more elaborate and expensive experiment. The problems they encountered, however, are probably representative of those that will appear in most inversion attempts. In order to find $\bar{\sigma}(s)$ over the entire range of separation distances will require very great effort. On the other hand, we feel that such effort will often be justified. For example, if in the above experiment $\bar{\sigma}(s)$ had been determined accurately for small separation distances, information about the inner scale would have been found. At present we do not have sufficient information about the inner scale but we know that it plays an important role in intensity fluctuations.

SUMMARY

Expressions have been derived which allow us to determine $\bar{\sigma}(s)$ and $\Phi(v)$ in terms of the measured quantities $\{I\}$, $\{\Gamma\}$ or $\{R_I\}$ for a homogeneous, isotropic random medium. In Eq. (28) the measured quantity is $\{\Gamma\}$ and the boundary condition is represented by a plane wave. In Eqs. (35) and (36) the measured quantity is $\{R_I\}$ and the boundary condition is again represented by a plane wave. In Eq. (38) the measured quantity is $\{\Gamma\}$ and the boundary condition is a point source. In Eq. (40) the measured quantity is {I} and the boundary condition is a finite guassian beam.

Each combination has advantages and disadvantages and the method used depends upon the experimental environment. It is, of course, desirable to use more than one method so that the results may be compared for consistency.

The inversion algorithms may be extended to a homogeneous anisotropic medium. In general, a mean linear index of refraction profile has no significant effect on the inversion procedure.

In the last section we have described an atmospheric experiment in which $\bar{\sigma}(s)$ has been determined from $\{\Gamma\}$ using a point source as a boundary condition. As was stated in the discussion, only the central portion of $\bar{\sigma}(s)$ was determined with acceptable accuracy. In this experiment determination of $\bar{\sigma}(s)$ for very large and very small values of s would have required much more elaborate equipment.

References

1. Shishov, V. I.: Theory of wave propagation in random media. IVUZ-Radiophysics (Russian). 11, 866 (1968)

2. Beran, M. J. and T. L. Ho: Propagation of the fourth-order coherence function in a random medium (a nonperturbative formulation). J. Opt. Soc. Amer., 59, 1134 (1969)

3. V. I. Tatarski: The Effects of The Turbulent Atmosphere on Wave Propagation, translated from the Russian and available from the National Technical Information Service (Springfield, Virginia 1971)

4. B. J. Uscinski: The Elements of Wave Propagation in Random Media (McGraw Hill, New York 1977)

5. A. Ishimaru: Wave Propagation and Scattering in Random Media, Vols. 1 and 2 (Academic Press, New York 1978)

6. B.J. Uscinski: Intensity fluctuations in a multiple scattering medium. Solution of the fourth moment equation. Proc. Roy. Soc. Lond., A380, 137, (1982)

7. C. Macaskill: An improved solution to the fourth moment equation for intensity fluctuations. Proc. Roy. Soc. Lond., A386, 461, (1983)

8. S. Frankenthal, A. M. Whitman and M. J. Beran: Two-scale solutions for intensity fluctuations in strong scattering. J. Opt. Soc. Amer., 1, 585, (1984)

9. A. M. Whitman and M. J. Beran: Two scale solution for atmospheric scintillation: J. Opt. Soc. Amer., To appear Dec. 1985

10. M. J. Beran: The inverse problem in a random medium. SPIE, Proc. Inverse Optics, 413, 61, (1983)

11. A. M. Whitman and M. J. Beran: Beam spread of laser light propagating in a random medium. J. Opt. Soc. Amer., 60, 1595, (1970)

12. W. A. Coles and R. G. Frehlich: Simultaneous measurements of angular scattering and intensity scintillation in the atmosphere. J. Opt. Soc. Amer., 72, 1042, (1982)

HALF-SPACE PROBLEMS IN THE KINETIC THEORY OF GASES

C.Cercignani
Dipartimento di Matematica
Politecnico di Milano

Milano, Italy

Abstract

A survey of the state-of-the art on half-space problems in kinetic theory is given, with particular concern for existence, uniqueness and closed form representations of the solutions.

1. INTRODUCTION

The importance of half space problems in kinetic theory stems from their role of boundary layer problems for more complicated situations. In fact, the basic equation of kinetic theory is the Boltzmann equation established by Boltzmann in 1872 (1-3)

$$\frac{\partial f}{\partial t} + \underline{\xi} \cdot \frac{\partial f}{\partial \underline{x}} + \underline{X} \cdot \frac{\partial f}{\partial \underline{\xi}} = Q(f,f) \tag{1.1}$$

where $f = f(\underline{x},\underline{\xi},t)$, the unknown, gives the probability density of finding a molecule at a position $\underline{x}$ with velocity $\underline{\xi}$ at time t. $\underline{X}$ is the body force per unit mass acting on the molecule, while $Q(f,f)$ is a quadratic expression, describing the effect of the binary collisions (i.e. short range interactions) between molecules. In the case of monatomic gases $Q(f,f)$, usually called the collision term, is given by a rather complicated fivefold integral (1-3). A basic concept connected with the Boltzmann equation is that of mean free path, i.e. the average distance travelled by a molecule between a collision and the next one. Although its precise definition depends upon the definition of average that is adopted, its qualitative meaning and its order of magnitude are independent of these details. In particular if $\bar{\xi}$ is a typical molecular speed, the order of magnitude of the collision term is $\bar{\xi}f/\ell$, where ℓ is the mean free path.

In the most usual situations the mean free path is very small: compared with a typical macroscopic length in normal conditions (atmospheric pressure and room temperature) is of the order of 10^{-5} cm. This explains why for a long time, by some suitable *ansatz* in the last century, by systematic formal expansions in powers of ℓ in the early part of this century, people looked at solutions that may claim validity only when the distribution function has very small changes on the scale of a mean free path (4). They were pleased to see that, according to the results of these expansions, the Boltzmann equation agreed with the Euler equations for a compressible fluid at the lowest level of expansion, with the compressible Navier-Stokes equations at the next step, while deviations from these equations, already known from continuum mechanics, would appear only with terms having the order of the square of the mean free path. In addition the Boltzmann equation furnished very simple methods for evaluating the viscosity and heat conduction coefficients once the law of interaction between molecules is assigned. For a monatomic gas this reduces to assigning the interaction potential as a function of the distance between the molecules. As a limiting case of this interaction law, one may assume that the molecules are hard elastic spheres that collide like billiard balls.

It took much more time before people became interested in boundary value problems for the Boltzmann equation: then it is clear that any solution obtained by perturbation methods must take into account the singular position of ℓ in Eq.(1), i.e. the fact that sufficiently general solutions must contain terms having a significant change on the scale of the mean free path. I.e. even if we do not enter into the typical problems of rarefied gas dynamics, involving e.g. a sphere moving through a gas having a diameter comparable with the mean free path, we encounter boundary layers, different from the viscous boundary layers introduced by Prandtl and having the thickness of a few mean free paths. These layers are called kinetic layers or Knudsen layers. Then even if the boundary is not flat, but has the radius of curvature much larger than the mean free path, the behavior of the distribution near the boundary can be expected to be describable, at least at the lowest level, by the solution of the Boltzmann equation in a half space.

The need for such a solution seems to have been felt by astrophysicists working on the propagation of radiation in stellar or planetary atmospheres. In that case we are basically in the presence of two gases: the atmosphere molecules and the photons. The former are assumed to have reached some sort of statistical equilibrium, at least locally, and accordingly have an assigned distribution function, while the latter essentially interact with the molecules of the former through the processes of emission, absorption and scattering. These processes replace the intermolecular collisions. The half space problem originating from this physical situation is called the Milne problem (5).

The same situation was met later in neutron transport theory. Here again one studies a gas, the neutrons in a nuclear reactor, that interacts with particles having a fixed distribution, the atoms of the moderator, through processes of emission (fission) absorption and scattering, and it is possible to transfer the solutions valid for the neutron case (6). It is to be remarked that, in these situations, when the particles collide with particles of another species, the medium, the problems appear from the beginning in a linear form.

The first appearance of half space problems in the kinetic theory of a gas of molecules, the one Boltzmann had in mind when he derived his equation, seems to date back to 1949 and to be due to H.Kramers (7). Hence people working in gas dynamics talk of Kramers problem when they deal with half space problems in kinetic theory. Recently there has been an attempt (12) to differentiate between a Milne and a Kramers problem in terms of boundary conditions, but this attempt does not seem to be well founded neither historically nor from the point of view of current practice.

A systematic analysis of perturbation procedures for the Boltzmann equation was effected by H. Grad (8) who remarked that when the mean free path is sufficiently small one may expect that simple perturbation procedures, connected with the names of Hilbert and Chapman-Enskog, loose validity in initial layers (i.e. for times of the order of $\ell/\bar{\xi}$), in the kinetic boundary layers and in the shock layers. The latter are regions having the thickness of a mean free path occurring inside the gas and located where the Euler equations would predict the discontinuity surfaces commonly known under the name of shock waves.

The analysis of the initial layer is rather simple, and was performed by H.Grad

(8) in full generality. The study of kinetic boundary layers and shock layers was performed in particular cases and/or by approximate methods, starting in the late fifties. In the next sections we shall present some results on the Knudsen layers, i.e. on half space problems.

2. GENERAL PROPERTIES AND UNIQUENESS

While the shock layers involve the full nonlinear Boltzmann equation, most of the kinetic layers can be studied by means of the linearized Boltzmann equation, with the notable exception of the problems of strong evaporation and condensation, to be considered in a later section.

The linearized Boltzmann equation can be formally obtained from Eq.(1.1) by letting

$$f = f_\circ(1+h) \tag{2.1}$$

where $f_\circ$ is a Maxwellian distribution

$$f_\circ = \frac{\rho_\circ}{(2\pi RT_\circ)^{3/2}} \exp(-(\underline{\xi} - \underline{u}_\circ)^2/2RT_\circ) \tag{2.2}$$

and h a "small perturbation". Here $\rho_\circ$, $\underline{u}_\circ$, $T_\circ$ are constants having the meaning of density, bulk velocity and temperature of the equilibrium state described by Eq. (2.2); R is the gas constant, i.e. the ratio between Boltzmann's universal constant k and the molecular mass. Inserting Eq.(2.1) into Eq.(1.1) and neglecting powers of h higher than first gives:

$$\frac{\partial h}{\partial t} + \underline{\xi}\cdot\frac{\partial h}{\partial \underline{x}} + \underline{X}\cdot\frac{\partial h}{\partial \underline{\xi}} = L\,h \tag{2.3}$$

where L is the so called linearized collision operator, defined by

$$L\,h = 2Q(f_\circ, f_\circ h)/f_\circ \tag{2.4}$$

where $Q(f,g)$ is the bilinear symmetric operator uniquely related to the nonlinear collision operator $Q(f,f)$.

In connection with the linearized Boltzmann equation it is useful to introduce the Hilbert space of functions of $\underline{\xi}$ which are square integrable with respect to the weight $f_\circ$. In this space the scalar product is, of course, given by

$$(g,h) = \int f_\circ \bar{g}\, h\, d\underline{\xi} \tag{2.5}$$

and the operator L turns out to be symmetric and also self-adjoint in all the cases that have been analyzed in detail (2-3).

The precise shape and properties of L depend on the chosen law of interaction between the molecules. If they are assumed to be hard spheres then

$$L h = K h - \nu(\underline{\xi})h \tag{2.6}$$

where K is a compact operator and ν a function satisfying

$$\nu_\circ(1 + |\underline{\xi} - \underline{u}_\circ|) < \nu(\underline{\xi}) < \nu_1(1 + |\underline{\xi} - \underline{u}_\circ|) \tag{2.7}$$

$\nu_\circ$ and ν_1 being positive constants. If the molecules interact with a power law central force extending to infinity, then the operator L turns out to have the structure of a pseudo differential operator, which makes the corresponding theory rather complicated. For this reason, Grad (9) introduced a cutoff in the angle of scattering in the interaction, in such a way that grazing collisions (which produce a very little change in the velocities of the molecules) are discarded. Under this assumption the form (2.6), with K compact, is recovered but $\nu(\underline{\xi})$ does no longer satisfy Eq. (2.7) which is replaced by

$$\nu_\circ(1 + |\underline{\xi} - \underline{u}_\circ|)^\alpha < \nu(\underline{\xi}) < \nu_1(1 + |\underline{\xi} - \underline{u}_\circ|)^\alpha \tag{2.8}$$

with $\alpha<1$. The present author (10,2,3) pointed out that it would more physical to cut the interactions occurring when the molecules are rather far from each other; in this case Eq.(2.6) and (2.7) hold, but K is a complicated integral operator, that has not been shown to be compact.

The operator L is non positive

$$(h, L h) \leq 0 \tag{2.9}$$

and has five eigenfunctions corresponding to the eigenvalues $\lambda = 0$

$$\psi_\circ = 1 \, , \quad \psi_i = \xi_i \quad (i=1,2,3), \quad \psi_4 = \xi^2 \tag{2.10}$$

These eigenfunctions are called collision invariants and are related to the conservation of mass, momentum and energy in a collision.

The remaining part of the spectrum is not easy to analyze in general, but one can show (10,2,3) that for interaction force not softer than the inverse fifth power (corresponding to the so called Maxwellian molecules) the origin is an isolated point of the spectrum. The simplest half space problem for the linearized Boltzmann equation corresponds to solving

$$\xi_1 \frac{\partial h}{\partial x} = L h \qquad (0 < x < \infty) \tag{2.11}$$

where x replaces x_1 for simplicity and no body forces are assumed to act on the molecules, with the boundary condition, prescribing the distribution of particles entering the half space

$$h = g \qquad x = 0 \ , \ \xi_1 > 0 \tag{2.12}$$

Further in this and next section we shall assume $\underline{u}_\circ = 0$ in the Maxwellian (2.2). If $g = 0$ no solution bounded in x is expected to exist, but the search for solutions growing linearly with x can be reduced to solving the problem of finding a bounded solution with $g \neq 0$.

As a matter of fact there is a three parameter family of exact solutions of Eq. (2.11) having the form

$$\bar{h} = [a\xi_2 + b\xi_3 + c(\xi^2 - 5RT_\circ)]\ x + \\ + L^{-1}\ [a\xi_1\xi_2 + b\xi_2\xi_3 + c\xi_1(\xi^2 - 5RT_\circ)] \tag{2.13}$$

where a, b, c are arbitrary constants and, of course, L^{-1} means the inverse in the subspace orthogonal to the collision invariants. Then if h satisfies Eq.(2.11) with $h = 0$ at $x = 0$ $(\xi > 0)$, $h' = h - \bar{h}$ will satisfy Eq.(2.11) with $h' = -\bar{h}$ at $x = 0$, $(\xi_1 > 0)$ and we are reduced to the inhomogeneous boundary condition (2.12). Accordingly we shall restrict ourselves to inhomogeneous boundary conditions and bounded (in x) solutions. Further the decomposition (2.6) will be assumed.

We shall look for a solution h of Eq.(2.11) with boundary condition (2.12)) such that ξ_1 h, $[\nu(\underline{\xi})]^{\frac{1}{2}}\ h \in L^\infty(R_x^+,\ L^2(R_{\underline{\xi}}^3))$ and $\xi_1(\partial f/\partial x) \in L^2_{loc}(R_x^+,\ L^2(R_{\underline{\xi}}^3))$ where loc means $x \in [a,b]$, $a \neq 0$, $b \neq \infty$.

Then we have the following property

$$\frac{d}{dx}\,(\xi_1\psi_\alpha, h) = 0 \qquad \alpha = 0,1,2,3,4 \tag{2.14}$$

In particular for $\alpha = 0$ we obtain that the particle net flow

$$j = (\xi_1, h) \tag{2.15}$$

is constant in space. j is one component of h in the five dimensional space spanned by the collision invariants. It is useful to substract this component to get

$$\tilde{h} = h - (\xi_1, h)n\xi_1 = h - nj\xi_1 \tag{2.16}$$

where n is a normalization constant given by

$$n = \|\xi_1\|^{-2} \tag{2.17}$$

Note that $\tilde{h}$ satisfies the same equation and boundary condition as h, with g replaced by $\tilde{g} = g - nj\xi_1$. g is assumed to satisfy

$$\int_{\xi_1>0} \xi_1 g^2 f_\circ \, d\underline{\xi} < B \tag{2.18}$$

and an inequality of the same kind will be satisfied by $\tilde{g}$.

We now decompose $\tilde{h}$ in its "fluid dynamic part" $q \in F$ and its "kinetic part" w belonging to the orthogonal complement W:

$$\tilde{h} = q + w \tag{2.19}$$

We remark that q has only four terms because the contribution along ξ_1 has been subtracted and hence

$$q = b_\circ + b_2\xi_2 + b_3\xi_3 + b_4\xi^2 \tag{2.20}$$

This implies that

$$(\xi_1 \; q, q) = 0 \tag{2.21}$$

because

$$(\xi_1\psi_\alpha, q) = 0 \qquad (\alpha = 0,2,3,4) \tag{2.22}$$

Another consequence of (2.22) is that Eq.(2.14) becomes

$$\frac{d}{dx}(\xi_1\psi_\alpha, w) = 0 \qquad (\alpha = 0,2,3,4) \tag{2.23}$$

A study of the general solution of Eq. (2.10) can be easily performed (11,3) by studying the eigenvalue problem

$$L\,g - \lambda\xi_1 g = 0 \tag{2.24}$$

and leads to the result that the general solution is the sum of $\bar{h}$, given by (2.13) and where a, b, c are arbitrary constants plus an arbitrary linear combination of the collision invariants ψ_α, plus a contribution from the spectrum of Eq.(2.24) with $\lambda \neq 0$. The contribution from half of this spectrum ($\lambda > 0$) grows exponentially, that from the second half ($\lambda < 0$) decreases exponentially. If we look for bounded solutions then $a = b = c = 0$ and the first half of the spectrum contributes nothing. Further the nonfluiddynamic part w has contribution only from the modes decaying to zero at infi nity. Therefore the four scalars$(\xi_1\psi_\alpha,w)$ must go to zero at infinity and being constant because of Eq.(2.23) they must be identically zero. One can also prove this property by a direct argument. In fact the result is true by construction for $\alpha = 0$. Accordingly only three quantities remain to be considered

$$I_\alpha = (\xi_1\hat{\psi}_\alpha,w) \qquad (\alpha = 2,3,4) \tag{2.25}$$

where for convenience we use $\hat{\psi}_i = \xi_i$ and $\hat{\psi}_4 = \xi^2 - 5RT_0$ rather than ψ_α. Assume $I_\alpha \neq 0$; then scalar multiplication of the equation

$$\xi_1 \frac{\partial\tilde{h}}{\partial x} = L\tilde{h} = Lw \tag{2.26}$$

by $L^{-1}(\hat{\psi}_\alpha\xi_i)$ $(\alpha = 2,3,4)$ gives

$$\frac{d}{dx}(\xi_1 L^{-1}(\hat{\psi}_\alpha\xi_1),\tilde{h}) = (L^{-1}(\hat{\psi}_\alpha\xi_1),Lw) = (\hat{\psi}_\alpha\xi_1,w) = I_\alpha \tag{2.27}$$

Then $I_\alpha \neq 0$ would imply a linear growth of the scalar product in the left hand side, against the assumptions on $\tilde{h}$. Thus $I_\alpha = 0$ and

$$(\xi_1\hat{\psi}_\alpha,w) = 0 \qquad (\alpha = 2,3,4) \tag{2.28}$$

Eq.(2.28) implies

$$(\xi_1 q,w) = 0 \tag{2.29}$$

and together with Eq.(2.21)

$$(\xi_1\tilde{h},\tilde{h}) = (\xi_1 w,w) \tag{2.30}$$

From what we said about the general solution of Eq.(2.10), $(\xi_1 w,w)$ must tend to zero when $x \to \infty$. For a direct proof we scalarly multiply Eq.(2.25) by $\tilde{h}$ and using Eq. (2.29), obtain

$$\frac{d}{dx}(\xi_1 w,w) = (w,Lw) \leq 0 \tag{2.31}$$

This shows that $(\xi_1 w,w)$ is decreasing. It must also be bounded in x from below. Hence it tends to a limit ℓ when $x \to \infty$. We want to show that ℓ is zero. As a matter of fact if we apply L^{-1} to both sides of Eq.(2.25) and multiplied by $\tilde{h}\xi_1$ we obtain

$$(\xi_1\hat{h},L^{-1}\xi_1 \frac{\partial\hat{h}}{\partial x}) = (\xi_1\hat{h},w) = (\xi_1 w,w) \tag{2.32}$$

or

$$\frac{d}{dx}(\xi_1\tilde{h},L^{-1}\xi,\tilde{h}) = (\xi_1 w,w) \tag{2.33}$$

If $(\xi_1 w,w)$ tends to a nonzero limit when $x\to\infty$, the scalar product in the left hand side must be unbounded against the assumptions. Hence

$$(\xi_1 w,w) \to 0 \qquad \text{when} \quad x\to\infty \tag{2.34}$$

If the positive part of the spectrum of the problem (2.24) is bounded away from zero, the decay to zero in Eq.(2.34) will be exponential. This occurs in the case of rigid spheres, where the point of the continuous spectrum closer to the origin is $\lambda_0 = \lim_{\xi_1\to\infty} \nu(\underline{\xi})/\xi_1 > 0$ (3). For a direct proof see (12).

Eqs. (2.31), (2.34) and (2.18) lead to

$$0 \le (\xi_1 w,w)_{x=0} = (\xi_1\tilde{h},\tilde{h})_{x=0} = (\xi_1 h,h)_{x=0} \;-n\,(\xi_1,h)^2 <$$

$$\leqslant \int_{\xi_1>0} \xi_1\, g^2(\xi)\; {}^2f_0\, d\underline{\xi} + nj^2 \leqslant B + nj^2 \tag{2.35}$$

Hence the scalar product $(\xi_1 w,w)$ exist for all x.

Eq.(2.31) together with the fact that the spectrum of L is bounded away from zero gives

$$0 \leqslant -\int_0^\infty (w,Lw)\,dx = (\xi_1 w,w)_{x=0} \leqslant B + nj^2 \tag{2.36}$$

Hence the integral $\int_0^\infty (w,Lw)\,dx$ exists.

We are now ready to prove the following

Uniqueness theorem. For a given $j \in R$ and a given g satisfying (2.18), there is only one solution of (2.10) + (2.12) + (2.15) with $\xi_1 h$ and νh in $L^\infty(R_x^+, L^2(R_1^3))$ and $\xi_1\,\partial h/\partial x \in L^2_{loc}(R_x^+,L^2(R^3))$. In fact given two solutions h_1 and h_2, their difference h will satisfy Eq.(2.10)+(2.12) + (2.15) with $g = j = 0$.

Scalarly multiply Eq.(2.10) by h and integrate with respect to x. Since $j = 0$, $h = \tilde{h}$ and use of Eqs.(2.30) and (2.34) leads to

$$-(\xi_1 w,w)_{x=0} = \int_0^\infty (w,Lw)\,dx \leqslant 0 \tag{2.37}$$

but, since $g = 0$

$$(\xi_1 h,h)_{x=0} = \int_{\xi_1<0} \xi_1 h^2 f_0\, d\underline{\xi} \tag{2.38}$$

and hence

$$\int_{\xi_1<0} \xi_1 h^2 f_\circ \, d\underline{\xi} \leqslant 0 \tag{2.39}$$

and

$$\int_0^\infty (w, L w)\, dx = 0 \tag{2.40}$$

The latter equation implies $w = 0$. Hence $h = q$ and Eq.(2.10) now reduces to

$$\xi_1 \frac{dq}{dx} = 0$$

which implies $q = \text{constant}$ and because of Eq.(2.39), $q = 0$. Thus $h = 0$ and the theorem is proved. We remark that a uniqueness argument of this form appears to have sketched for the first time by the Author (13,2,3) for problems in unbounded domains in any dimension.

3. EXISTENCE

The only proofs of existence that deal directly with the half space problem for a sufficiently general collision operator are due to Beals (14), Greenberg and Van der Meer (15) and make use of operator techniques for an abstract version of the linearized equation. The restrictions imposed on the operators should be verified in each case.

For a more elementary and direct approach, we first examine the solution of the problem in a slab $(0,\ell)$ and then look for the limit $\ell \to \infty$. As remarked by Bardos, Caflish and Nicolaenko (12), it is much easier to look at this limit if a specular reflection condition

$$h(\ell,\xi_1,\xi_2,\xi_3) = h(\ell, -\xi_1,\xi_2,\xi_3) \tag{3.1}$$

is assumed for $x = \ell$. Note, however, that this is exactly the same as looking for a solution in $(0,2\ell)$ with the boundary condition $h = g(-\xi_1,\xi_2,\xi_3)$ at $x = 2\ell$; in fact the solution of this problem will has the symmetry property $h(x,\xi_1,\xi_2,\xi_3) = h(2\ell - x,-\xi_1,\xi_2,\xi_3)$ and, in particular, will satisfy Eq.(3.1)

The first existence theorem in a slab of arbitrary size for the general linearized Boltzmann equation appears to be due to the present author (16). It is located in an $L^2([0,\ell]R^3)$ with weight $\{ \nu(\xi)^2 + c^2\xi_1^2/\ell^2\}^{\frac{1}{2}} f_\circ$ (where c is a constant independent of ℓ) and makes use of the contraction mapping theorem. On the basis of this proof Y.P.Pao extended the proof to other function spaces and even to the nonlinear case (with data close to equilibrium) (17).

Willis, Zweifel and Van der Mee (18) have recently pointed out that for a certain number of cases existence in a slab is equivalent to a problem to which the Fredholm alternative theorem applies. Hence uniqueness implies existence.

Bardos, Caflish and Nicolaenko (12) for the case of rigid sphere molecules pro ve existence with sufficiently strong estimates to be able to go to the limit of a half space.

They essentially use, independently of the above authors (18), the alternative theorem for a problem where the small values of ξ_1 are cutoff (this permits a derivative in $L^2(R_x^+, R^3)$). Then a weak compactness argument based on uniform estimates is used to remove the cutoff and the restriction to bounded ℓ.

We remark that it is easy to obtain uniform estimates by means of the arguments in the previous section. In fact, first of all, Eq.(2.36) together with the fact that the continuous spectrum of L is bounded away of zero, implies that the norm of w in $L^2(R_x^+ \times R^3)$ is uniformly bounded. The same result for $q - \bar{q}$ ($\bar{q}$ constant in x) follows from the remark that Eq.(2.29), because of Eq.(2.28), implies

$$(\xi_1 L^{-1}(\hat{\psi}_\alpha \xi_1), \tilde{h}) = k_\alpha \qquad (\alpha = 2,3,4) \tag{3.2}$$

where k_α $(\alpha = 2,3,4)$ are constants, and hence

$$\sum_{\beta=2}^{4} (\xi_1 L^{-1}(\hat{\psi}_\alpha \xi_1), \psi_\beta) b_\beta - k_\alpha = -(\xi_1 L^{-1}(\hat{\psi}_\alpha \xi_1), w) \tag{3.3}$$

This implies that there are constants $\bar{b}_\beta$ such that $b_\beta - \bar{b}_\beta \in L^2(R_x^+)$ $(\beta = 2,3,4)$. As for $b_\circ$, Eq.(2.14) with $\alpha = 1$ gives

$$b_\circ \|\psi_1\|^2 + b_4(\xi_1^2, \psi_4) + (\xi_1^2, w) = j \tag{3.4}$$

and again there is a constant $\bar{b}_\circ$ such that $b_\circ - \bar{b}_\circ \in L^2(R_x^+ \times R^3)$.

Hence the solution h_ℓ corresponding to a finite ℓ for the boundary conditions (2.12) and (3.1) has a norm in $L^2(R_x^+ \times R^3)$ bounded by a constant. Thus if we let $\ell \to \infty$, there is a subsequence $\{h_\ell\}$ such that

$$\begin{aligned} \bar{q}_\ell &\to q_\infty && \text{in} \quad L^2(R_\xi^2) \\ h_\ell - \bar{q} &\to h - q_\infty && \text{weakly in} \quad L^2(R_x^+ \times R_{\underline{\xi}}^3) \end{aligned} \tag{3.5}$$

In order to show that h solves our problem, it is expedient to use the integral form of our problem, obtained by integration of

$$\xi_1 \frac{\partial h}{\partial x} + \nu(\xi) h = K h \tag{3.6}$$

The equation for h_ℓ is

$$h_\ell(x, \underline{\xi}) = g(\underline{\xi}) e^{-x\nu/\xi_1} + \frac{1}{\xi_1} \int_0^x e^{-(x-y)\nu/\xi_1} K h_\ell(y, \underline{\xi}) dy \qquad (\xi_1 > 0)$$

$$h = e^{-(\ell - x)\nu/|\xi_1|}\{g(\underline{\xi}_R)e^{-\ell\nu/|\xi_1|} + \frac{1}{\xi_1}\int_0^{\ell} e^{-(\ell - y)\nu/|\xi_1|} K h_\ell(y,\underline{\xi}_R)\} +$$

$$+ \frac{1}{|\xi_1|}\int_x^{\ell} e^{-(y - x)\nu/|\xi_1|} K h_\ell(y,\underline{\xi})dy \qquad (\xi_1<0) \qquad (3.7)$$

The same equation holds for $h_\ell - \overline{q}_\ell$ (because $K\overline{q}_\ell = \nu\overline{q}_\ell$) and hence, after subtracting $\overline{q}_\ell$, we can take the weak limit in $L^2(R^+_x \times R^3)$. After that we can add again q_∞ to obtain

$$h(x,\underline{\xi}) = g(\underline{\xi})e^{-x\nu/\xi_1} + \frac{1}{\xi_1}\int_0^x e^{-(x - y)\nu/\xi_1} Kh(y,\underline{\xi})dy \qquad (\xi_1>0)$$

$$h(x,\underline{\xi}) = \frac{1}{|\xi_1|}\int_x^\infty e^{-(y - x)\nu/|\xi_1|} Kh(y,\underline{\xi})dy \qquad (\xi_1<0) \qquad (3.8)$$

We remark that the singularity in $\xi_1 = 0$ is a minor one because of the exponential multiplying $1/\xi_1$. From Eq.(3.8) due to the smoothing effect of the integral, actually follows that $|\xi_1| h$ is in $L^\infty(R^+_x, L^2(R^3))$ differentiable with derivative in $L^2_{loc}(R^+_x, L^2(R^3))$ and h satisfies the integrodifferential equation (3.6) and, obviously, the boundary condition at $x = 0$. We remark that, by construction, the solution will have $j = 0$, because such is the case for h^ℓ, as a consequence of Eq.(3.1). In order to obtain a solution with $j \neq 0$ it is sufficient to replace g by $\tilde{g} = g - nj\xi_1$ and compute the corresponding solution $\tilde{h}$; then $h = \tilde{h} + nj\xi_1$ will provide a solution with $j \neq 0$, for any j. We have thus obtained the following

Existence theorem. For any $j \in R$ and g satisfying Eq.(2.18), there is a solution g with $|\xi_1|^{\frac{1}{2}} g \in L^\infty(R^+_x, L^2(R^3))$ $\nu^{\frac{1}{2}} h \in L^\infty(R^+_x, L^2(R^3))$, $h_x \in L^2_{loc}(R^+_x, L^2(R^3))$ of Eq.(2.11) with boundary conditions (2.12) and (2.15).

This theorem for the particular case of rigid sphere molecules has been already given by Bardos, Caflish and Nicolaenko (12).

4. GENERALIZATIONS

In the previous treatment we assumed the bulk velocity $\underline{u}_\circ$ in the Maxwellian (2.2) to be zero. Now it is easily seem that no change arises if $\underline{u}_\circ$ has components along the axes orthogonal to x, but a significant change may arise if $\underline{u}_\circ$ has a component, say $u_\circ$, along the x axis. It is then convenient to shift the velocities, by changing ξ_1 into $\xi_1 + u_\circ$.

This implies that the collision term L will be the same has before but Eq.(2.11) will be replaced by

$$(\xi_1 + u_\circ)\frac{\partial h}{\partial x} = L h \qquad (4.1)$$

with boundary conditions

$$h(0,\underline{\xi}) = g(\underline{\xi}) \qquad \text{for} \qquad \xi_1 > -u_o \tag{4.2}$$

In a sense the new equation can be thought of as incorporating the free parameter j into the equation ($u_o = j/\rho_o$); in fact it would be a little queer from a physical point of view to look for solutions of Eq.(4.1) which do not satisfy

$$\lim_{x\to\infty} (\xi_1, h) = 0 \tag{4.3}$$

If Eq.(4.3) is assumed, one free parameter is eliminated from the solution; we can then expect that when u_o is sufficiently close to zero there is, for any u_o, just one solution of Eq.(4.1) + (4.2). Looking at certain numerical results and motivated by both an asymptotic analysis of Eq.(4.1) and physical arguments, the present author was led, however, to conjecture (19,20), that this result would break down for $u_o = \sqrt{5RT_o/3}$) (The number 5/3 is related to the number of dimensions of velocity space, $n = 3$; for a general n it is replaced by $(n+2)/n$). People familiar with gas dynamics will immediately recognize this condition to mean that M_o, the Mach number of the unperturbed Maxwellian, equals unity. The proof of this conjecture was offered by M.Arthur and C.Cercignani (20) and by Siewert and Thormas (21,22) by solving in closed form a particularly simple model where ν is constant and K is of finite rank. A more general proof was given by Greenberg and Van der Mee (23) by means of operator techniques; although the only concrete examples they gave are those considered by the previous authors (20-22), there is little doubt that their results have a wider generality.

In fact they find that once u_o is fixed and Eq.(4.3) holds, then if $\tilde{\psi}_\alpha$ are the collision invariants chosen in such a way that

$$(\tilde{\psi}_\alpha(\xi_1 + u_o), \tilde{\psi}_\beta) = 0 \qquad (\alpha \neq \beta) \tag{4.4}$$

then the numbers

$$N_\alpha = (\tilde{\psi}_\alpha(\xi_1 + u_o), \tilde{\psi}_\alpha) \tag{4.5}$$

determine the possibility of solving the problem. In fact the number of negative values among the N_α gives the number of the additional conditions such as (4.3) which can be imposed. A simple calculation indicates that one can take

$$\tilde{\psi}_o = 1, \quad \tilde{\psi}_1 = \xi^2 - 3u_o\xi_1, \quad \tilde{\psi}_2 = \xi_2, \quad \tilde{\psi}_3 = \xi_3, \quad \tilde{\psi}_4 = \xi^2 - 3RT_o \tag{4.6}$$

and

$$N_o = u_o, \quad N_1 = 9u_oRT_o(u_o^2 - \tfrac{5}{3}RT_o), \quad N_2 = N_3 = u_oRT, \quad N_4 = 6u_o(RT_o)^2 \tag{4.7}$$

Obviously if $u_\circ > 0$, then there is one negative value for $u_\circ < \sqrt{5RT_\circ/3}$ and none if $u_\circ > \sqrt{5RT_\circ/3}$ and, in agreement with the aforementioned conjecture, condition (4.3) cannot be imposed.

The case $u_\circ < 0$ has not been discussed in the literature, but it is clear that, since the number of negative values is four for $|u_\circ| < \sqrt{5RT_\circ/3}$ and five for $|u_\circ| > \sqrt{5RT_\circ/3}$, one can impose four conditions in the first case and five in the second one.

These results have a bearing on the problem of evaporation from or condensation on a flat plate bounding a half space. They indicate that evaporation is governed by only one parameter ($u_\circ$) and can exist only for a subsonic flow of the vapor in the Knudsen layer, while condensation is governed by four parameters in the subsonic case, by five in the supersonic case. If we leave out two parameters having to do with the transverse components of velocity, there are still two parameters in the subsonic case, three in the supersonic case. Although this matter is not completely clarified, it seems that one of the parameters must specify the vapor pressure-temperature relation for the vapor coming from infinity; this is not required in the case of evaporation because the vapor comes from the plate and the vapor pressure is specified there when assigning $g(\underline{\xi})$.

The additional parameter in the supersonic case seems to indicate that when a vapor flows supersonically toward a condensing surface it must first slow down to subsonic speeds through a shock layer.

Of course all these results have been obtained through a linearized analysis and should by confirmed by a treatment of the corresponding nonlinear problems. However there seems to be little doubt that the nonlinear analysis should confirm the qualitative picture provided by the linearized treatment, because the number of auxiliary conditions seems to be dictated by the asymptotic behavior far from the plate, where linearization is valid because the solution tends toward the unperturbed Maxwellian, when $x \to \infty$ (19).

This discussion leads us to a remark on the problem of the shock wave structure: the simplest shock layer of the kind mentioned in the first section occurs in an unbounded space when one looks at a solution depending on one space variable x and tending to two Maxwellians one supersonic and the other subsonic for $x \to \pm\infty$. In this problem the Boltzmann equation shows up its nonlinear nature, without the complication of the boundary conditions, which describe the interaction between molecules and solid surfaces. Yet, one could look at this problem as a pair of half space problems and try to apply our results on the number of free parameters. An elementary result is that the gas must have a bulk flow from the supersonic side to the subsonic side and not the other way around, because there is no supersonic solution with $u_\circ > 0$ in the half space $x > 0$; this is, of course, well in agreement with what is known about steady normal shock waves.

Another kind of generalization is found in time dependent half space problems. These arise from initial value problems or from the study of steady oscillations of frequency ω. Very little has been done in general, although the linearized problem is very similar to the steady one in the sense that they reduce to solving Eq.(3.6) with $\nu + i\omega$ in place of ν in the case of steady oscillations and the same equation with $\nu + s$ (s complex) in place ν if the initial value problem is treated with the Laplace transform or resolvent techniques.

This kind of problems has been treated only with special models of the linearized Boltzmann equation. The simplest one is the so called Bhatnagar - Gross - Krook (BGK) model where Lh is replaced by $\nu(\underline{P}h - h)$ where ν is a constant and P the orthogonal projector onto the five dimensional space of collision invariants. A modifica tions of this model allowing for a velocity dependent collision frequency ν is

$$L = \nu(\underline{\xi})\left[\sum_{\alpha=0}^{4} \hat{\psi}_\alpha(\hat{\psi}_\alpha \nu, h) - h\right] \qquad (4.8)$$

where the collision invariants $\hat{\psi}_\alpha$ are normalized in such a way that

$$(\hat{\psi}_\alpha \nu, \hat{\psi}_\beta) = \delta_{\alpha\beta} \qquad (4.9)$$

For these models it is possible to construct the generalized eigenfunctions of the eigenvalue problem (2.24) in both the steady (24-26,2,3) and unsteady (27-28,2, 3) cases. This in turn leads to an explicit representation of the general solution of the equation. In order to solve a specific half space problem it is however necessary to determine the arbitrary functions and constants appearing in the general solutions.

It appears that there are classes of problems simpler than the others. One can in fact split the problems into three symmetry classes:

a) the boundary data g are of the form $\phi_1(\xi_1,\xi_2^2,\xi_3^2)$
b) the boundary data g are of the form $\xi_2\phi_2(\xi_1,\xi_2^2,\xi_3^2)$
c) the boundary data g are of the form $\xi_3\phi_3(\xi_1,\xi_2^2,\xi_3^2)$

It is clear that general data can be split into these three types.

Classes b) and c) are the same except for an exchange between ξ_2 and ξ_3 and will be accordingly referred to as class b only. This is the easier class; physically it corresponds to perturbations due to motions transversal to the plane bounding the half space. It turns out that the half space problems of this class can be solved in closed form by solving a Riemann-Hilbert problem in the steady case (24-26,2, 3); in the unsteady case the same result applies to the BGK model (27) but for the more general model (4.8) with nonconstant ν the problem can be solved (28) by techniques borrowed from generalized analytic functions (29). This is related to the appearance of a continuous spectrum occupying an area of the complex plane in the spectral problem (2.24).

If one passes to class a), which describes perturbations of density and temperature, we find difficulty even for the simplest model and steady problems. In fact the solution of the problem reduces now to a matrix Riemann-Hilbert problem: now, although we know that these problems have a solution (30) there are no general algorithms for producing a closed form solution. For the steady problems arising in connection with the BGK model such an algorithm was introduced by the present author in 1977 (26,31). Essentially the matrix Riemann-Hilbert problem can be diagonalized, but, when doing so, new (artificial) singularities of the branch cut type are introduced in the complex plane. If one tries to eliminate these in a naive fashion, he introduces at least a new singularity of the essential type. It is necessary to introduce modifications that although do not change the fact that we have a solution, compensate the aforementioned singularity. The determination of the integration extremes in an integral appearing in this modification reduces to the so called Jacobi inversion problem (32,33) which can be solved analytically. In the simplest case this problem can be solved by means of elliptic functions (26) but in general it requires more powerful algorithms based on the Riemann's theta function (32).

C.E. Siewert and coworkers improved upon this author's presentation and exten ded it to other problems (34-36). C.Cercignani and C.E.Siewert (37) introduced a standard method for constructing the canonical $\underline{\underline{X}}$ matrix in the sense of Muskhelishvili (30), a problem which had been left open in the previous papers.

Aoki and Cercignani (38,39) extended the method to unsteady problems and indicated for the first time in an explicit fashion how to use Riemann's theta function to give a closed form solution under all respects. They also solved the problem of propagation of sound in a half space bounded by a vibrating plate in closed form.The same techniques has been applied to the scattering of polarized light in an atmosphere (40).

5. CONCLUDING REMARKS

It is clear that the results discussed in this paper can be extended into various directions. We omit to mention the obvious extension in the linear field, to mention the need to have some results in the nonlinear field; so far only the case of the weakly nonlinear problem for a gas having zero temperature at infinity has been treated (41).

This work should be extended to more general problems. Finally, the truly nonlinear problem should be dealt with. It is true that no general results are known for the initial value problem with space dependence, but the steady half space problems is formally similar to an initial value problem for space independent solutions; the main difference is that the data are assigned for the half range $\xi_1 > 0$ at $x = 0$. Accordingly it is not hopeless to try to show the existence and uniqueness. Approximate and numerical solutions indicate that no anomalous behavior of the solutions is to be expected.

REFERENCES

1. L. Boltzmann, Sitzungsberichte der Akademie der Wissenschaften, Wien 66, 275 (1872).
2. C. Cercignani, "Mathematical Methods in Kinetic Theory", Plenum Press, New York and McMillan, London (1969).
3. C. Cercignani, "Theory and Application of the Boltzmann Equation", Scottish Academic Press, Edinburgh, and Elsevier, New York (1975).
4. S. Chapman and T. Cowling, "The Mathematical Theory of Nonuniform Gases", Cambridge University Press, Cambridge (1952).
5. S. Chandrasekhar, "Radiative Transfer", Oxford University Press, Oxford (1950).
6. B. Davison, "Neutron Transport Theory", Oxford University Press, Oxford (1957).
7. H.A. Kramers, Nuovo Cimento Suppl., 6, 297 (1949).
8. H. Grad, Phys. Fluids, 6, 147 (1963).
9. H. Grad, in "Rarefied Gas Dynamics", J.A. Laurman, Ed., Vol.I, 26, Academic Press, New York (1963).
10. C. Cercignani, Phys. Fluids, 10, 2097 (1967).
11. C. Cercignani, in "Rarefied Gas Dynamics", M. Becker and M. Fiebig, eds., Vol.I A.9, DFVLR - Press, Porz - Wahn (1974).
12. C. Bardos, R.E. Caflish and B. Nicolaenko, to appear (1985).
13. C. Cercignani, Phys. Fluids, 11, 303 (1968).
14. R. Beals, J. Funct. Anal. 34, 1 (1979).

15. W. Greenberg and C.V.M. Van der Mee, Transp. Theor. Stat. Phys. 11, 155 (1982).
16. C. Cercignani, J. Math. Phys., 8, 1653 (1967).
17. Y.P. Pao, J. Math. Phys., 8, 1893 (1967).
18. B.L. Willis, P.F. Zweifel and C.V.M. Van der Mee, to appear in Transp. Theor. Stat. Phys. (1985).
19. C. Cercignani, in "Mathematical Problems in the Kinetic Theory of Gases", D.C. Pack and H. Neunzert, 129, P. Lang, Frankfurt (1980).
20. M.D. Arthur and C. Cercignani, Z. Angew. Math. Phys., 31, 634 (1980).
21. C.E. Siewert and J.R. Thomas, Z. Angew. Math. Phys., 32, 421 (1981).
22. C.E. Siewert and J.R. Thomas, Z. Angew. Math. Phys., 33, 202 (1982).
23. W. Greenberg and C.V.M. Van der Mee, Z. Angew. Math. Phys., 35, 156 (1984).
24. C. Cercignani, Ann. Phys. (NY), 20, 219 (1962).
25. C. Cercignani, Ann. Phys. (NY), 40, 469 (1966).
26. C. Cercignani, Transp. Theor. Stat. Phys., 6, 29 (1977).
27. C. Cercignani and F. Sernagiotto, Ann. Phys. (NY), 30, 154 (1964).
28. C. Cercignani, Ann. Phys. (NY), 40, 454 (1966).
29. I.N. Vekua, "Generalized Analytic Functions", Pergamon Press, Oxford (1962).
30. N.I. Muskhelishvili, "Singular Integral equations", Noordhoff, Gromingen (1953).
31. C. Cercignani, Nuclear Sci. Eng., 64, 882 (1977).
32. B. Riemann, "Collected Works", H. Weber, Ed., p.88, Dover Publications, New York (1953).
33. G. Springer, "Introduction to Riemann Surfaces", Addison - Wesley, Reading (1957).
34. C.E. Siewert and C.T. Kelley, Z. Angew. Math. Phys., 31, 344 (1980).
35. C.E. Siewert, C.T. Kelley and R.D.M. Garcia, J. Math. Anal. Appl., 84, 509 (1981).
36. C.E. Siewert and J.R. Thomas, Jr., Z. Angew. Math. Phys., 33, 473 (1982).
37. C. Cercignani and C.E. Siewert, Z. Angew. Math. Phys., 33, 297 (1982).
38. K. Aoki and C. Cercignani, Z. Angew. Math. Phys., 35, 127 (1984).
39. K. Aoki and C. Cercignani, Z. Angew. Math. Phys., 35, 345 (1984).
40. K. Aoki and C. Cercignani, Z. Angew. Math. Phys., 36, 61 (1985).
41. R.E. Caflish, to appear in Comm. Pure Appl. Math. (1985).

VIRIAL COEFFICIENTS FROM EXTENDED THERMODYNAMICS

I-Shih Liu*

Instituto de Matemática
Universidade Federal do Rio de Janeiro

Abstract

The virial equation of state is a convenient and useful expression for the calculation of thermodynamic properties of gases. Although statistical mechanical considerations permit the theoretical determination of virial coefficients in terms of certain hypothetical intermolecular potential energy, the actual calculations are so tedious that many empirical expressions have been proposed for practical purpose.

This work presents a different theoretical deviation of virial coeffcients in analytical forms. It is based on the extended thermodynamics recently proposed by Liu & Müller [1].

1. Extended Thermodynamics of Real Gases**

Extended thermodynamics is a phenomenological theory whose balance equations are strongly motivated by the moment equations of the kinetic theory of gases. In the absence of external forces, they can be written in the following form:

$$\frac{\partial \rho}{\partial t} + \frac{\partial}{\partial x_i}(\rho v_i) = 0, \tag{1}$$

$$\frac{\partial}{\partial t}(\rho v_i) + \frac{\partial}{\partial x_j}(\rho v_i v_j + M_{ij}) = 0, \tag{2}$$

$$\frac{\partial}{\partial t}(\rho v_i v_j + m_{ij}) + \frac{\partial}{\partial x_k}\left[(\rho v_i v_j + m_{ij})v_k + 2v_{(i}M_{j)k} + M_{ijk}\right] = \ell_{ij}, \tag{3}$$

$$\frac{\partial}{\partial t}(\rho v^2 v_k + 3v_{(j}m_{j)k} + m_{jjk}) + \frac{\partial}{\partial x_\ell}\Big[(\rho v^2 v_k + 3v_{(j}m_{j)k} + m_{jjk})v_\ell +$$

$$+ 3v_{(j}v_j M_{k)\ell} + 3v_{(j}M_{jk)\ell} + M_{jjk\ell}\Big] = \ell_{jjk} + 2v_j \ell_{jk}, \tag{4}$$

---

* Presently a Humboldt fellow at Hermann-Föttinger-Institut, TU Berlin

** See [2] for notations and details of this theory

where ρ is the mass density and v_i is the velocity. $M_{i_1\cdots i_n i_{n+1}}$ and $\ell_{i_1\cdots i_n}$ are the flux and the production density respectively for the Nth order central moment $m_{i_1\cdots i_n}$.

For monatomic ideal gases, we have $M_{i_1\cdots i_n} = m_{i_1\cdots i_n}$ which reduces the system (1) through (4) to the well-known Grad's thirteen moment equation [3].For dense gases, this relation no longer holds. Consequently, even though $\underline{m}$ is a completely symmetric Nth order rensor, it is not necessarily so for $\underline{M}$.

In the present theory, we shall regard $\{\rho, v_i, m_{ij}, m_{jji}\}$ as 13 basic field quantities which completely characterize a state of the gas. The fluxes and the productions are regarded as constitutive quantities which are functions of $\{\rho, v_i, m_{ij}, m_{jji}\}$ in a materially dependent manner. Moreover, they are required to be restricted by several universal physical principles such as the principle of material frame indifference and the entropy principle. This theory has been formulated in [2]. In the following, we shall only list the resulting linear constitutive equations relevant to the present work:

$$M_{ij} = p\,\delta_{ij} + \nu\, m_{<ij>}, \tag{5}$$

$$M_{ijk} = \frac{1}{3}\alpha_0\, \delta_{ij}\, m_{\ell\ell k} + \alpha\, m_{\ell\ell<i}\,\delta_{j>k}, \tag{6}$$

$$M_{jj<k\ell>} = \left(\frac{10}{3}\varepsilon + \frac{\zeta}{p}\right)\nu\, m_{<ij>}, \tag{7}$$

where p the pressure, ε the internal energy, as well as other material parameters are functions of ρ and the temperature T. Moreover they must satisfy

$$\frac{\partial \varepsilon}{\partial \rho} = \frac{p}{\rho^2} - \frac{T}{\rho^2}\frac{\partial p}{\partial T}, \tag{8}$$

$$\frac{\partial \varepsilon}{\partial T}\frac{\partial p}{\partial \rho} = \frac{\partial \varepsilon}{\partial \rho}\left(\frac{\partial p}{\partial T} - \frac{\alpha_0}{\alpha}\frac{p}{T}\right) + \frac{3}{5}\frac{\alpha_0}{\alpha\nu}\frac{p}{\rho T}\frac{\partial p}{\partial \rho} + \frac{3}{10}\frac{\zeta}{T}\frac{\partial}{\partial \rho}\left(\frac{\alpha_0}{\alpha}\right), \tag{9}$$

$$\frac{\partial \zeta}{\partial \rho} = \frac{2}{\nu}\frac{p}{\rho}\frac{\partial p}{\partial \rho} - \frac{10}{3}\frac{p}{\rho^2}\left(p - T\frac{\partial p}{\partial T}\right). \tag{10}$$

In particular, for ideal gases with the equation of state given by $p = \rho R T$, where R is the gas constant, and assuming α_0, α, ν being constant, we can integrate (8) and (9) to obtain

$$\varepsilon = \frac{3}{5}\frac{\alpha_0}{\alpha\nu} RT + c\,. \tag{11}$$

This relation enables us to interpret the constant $\frac{3}{5}\frac{\alpha_0}{\alpha\nu}R$ as the specific heat at constant volume c_v^0 . Such data are readily available in the literature.

2. Virial Expansion

For a moderately dense gas, the equation of state is usually expressed as a power series in density, known as the virial expansion

$$p = \rho RT\left(1 + B(T)\rho + C(T)\rho^2 + \cdots\right). \tag{12}$$

B(T) and C(T) are called the second and the third virial coefficients respectively.

We shall assume that the following material parameters can also be expanded into power series in ρ ,

$$\frac{2}{5}\frac{\alpha_0}{\alpha} = \chi_0 + \chi_1\rho + \chi_2\rho^2 + \cdots\,, \tag{13}$$

$$\frac{3}{5}\frac{\alpha_0}{\alpha\nu} = r_0 + r_1\rho + r_2\rho^2 + \cdots\,. \tag{14}$$

According to (11), we must have $r_0 = c_v^0/R$. Moreover, like monatomic ideal gases, we shall assume that M_{ijk} will become completely symmetric in the ideal condition for other gases as well. From (13) and (6), this assumption implies that $\chi_0 = 1$.

By the use of (12) through (14), we can integrate (10) for

$$\zeta = (RT)^2\left\{\frac{4}{3}r_0\rho + \frac{2}{3}\left(r_1 - r_0\chi_1 + 3r_0B + \frac{5}{2}TB'\right)\rho^2 + \cdots\right\}, \tag{15}$$

where we have set the integration constant to be zero (see [1]).

Substituting (12) through (15) into (8) and (9), we can obtain $\frac{\partial\varepsilon}{\partial\rho}$ and $\frac{\partial\varepsilon}{\partial T}$ in their power series expansions. The integrability condition requires that the mixed second derivative of ε calculated from these two expansions must be consistent. This condition leads to the following two equations for the zeroth and the first order terms in :

$$T^2B'' + \frac{7}{2}TB' + r_0B = -(r_0\chi_1 + r_1)\,, \tag{16}$$

$$T^2C'' + 5TC' + 2r_0C = (r_0\chi_1 - 2r_1)B - \frac{15}{2}\chi_1TB' + 3TBB' \tag{17}$$

$$+ 2T^2B'^2 + r_0\chi_1^2 - r_1\chi_1 - 4r_0\chi_2 - 2r_2\,.$$

These are two linear differntial equations for the determination of the virial coefficients B(T) and C(T). Differential equations for the higher virial coefficients can also be obtained in the same manner by retaining higher order terms in the series expansions.

3. Determination of Virial Coefficients

For gases with nearly constant specific heat, one can easily solve the equations (16) and (17).

Monatomic gases for instance, $c_v^o = \frac{3}{2}R$, i.e. $r_0 = \frac{3}{2}$. We obtain

$$B(T) = b_1 T^{-3/2} + b_2 T^{-1} - (\chi_1 + \tfrac{2}{3} r_1) , \tag{18}$$

$$\begin{aligned} C(T) = {} & c_1 T^{-3} + c_2 T^{-1} - 21 b_1 \chi_1 T^{-3/2} + 6 b_2 \chi_1 T^{-1} \ln T \\ & + 2 b_1 b_2 T^{-5/2} + b_2^2 T^{-2} - 2 b_1 (\chi_1 + \tfrac{2}{3} r_1) T^{-3/2} + \tfrac{4}{9} r_1^2 - (2\chi_2 + \tfrac{2}{3} r_2) . \end{aligned} \tag{19}$$

For many other gases, the specific heat varies very little in a wide range of moderate temperature, so that we can regard it as a constant. For such cases, we obtain

$$B(T) = T^{-5/4} (b_1 \sin(\theta_1 \ln T) + b_2 \cos(\theta_1 \ln T)) - (\frac{r_1}{r_0} + \chi_1) , \tag{20}$$

$$C(T) = T^{-2} (c_1 \sin(\theta_2 \ln T) + c_2 \cos(\theta_2 \ln T)) + h(T) , \tag{21}$$

where h(T) is the non-homogeneous solution of (17), which can be obtained explicitly by the use of B(T) from (20), (see [4]), and

$$\theta_1 = (r_0 - \tfrac{25}{16})^{1/2} , \qquad \theta_2 = (2 r_0 - 4)^{1/2} . \tag{22}$$

In the above solutions, we have also taken r_1, r_2, χ_1, χ_2 as constants, and b_1, b_2, c_1, c_2 are integration constants. They are material constants subject to experimental determination.

Virial coefficients of gases with non-constant specific heat and comparison with experimental data are treated elsewhere [4].

References

[1] Liu, I-Shih, Müller, I. Extended Thermodynamics of Classical and Degenerate Ideal Gases. Arch. Rational Mech. Anal. 83, 285-332 (1983)

[2] Liu, I-Shih Extended Thermodynamics of Fluids and Virial Equations of State. Arch. Rational Mech. Anal. 88, 1-23 (1985)

[3] Grad, H. On the Kinetic Theory of Rarefied Gases. Comm. Pure and Appl. Math. 2, 331-407 (1949)

[4] Liu, I-Shih Determination of Virial Coefficients from the Extended Theory of Thermodynamics (submitted for publication)

Keywords: equation of state, virial coefficients, extended thermodynamics, balance equations, constitutive equations.

ON THE TRANSIENT BEHAVIOUR OF STRUCTURED SOLIDS

D.R. Axelrad
Micromechanics Research Laboratory
McGill University
Montreal, Canada

I. Introduction:

In previous publications a random theory of deformation for structured solids has been developed that permits the inclusion of interaction effects between elements of the microstructure. In this theory the evolution of the occurring deformations has been expressed on the basis of Markov theory and the changes from a given state of the material to a neighbouring one by the corresponding probability transition functions. For a stable transition mechanism to exist, these probabilities were shown to be time independent and to exhibit semi-group properties. For the transient behaviour in which structural changes of states occur, the transition probabilities cannot be regarded as time independent and hence a wider class of Markov processes must be considered. This class of random processes involves distribution of states at various instants of time during the evolution of deformations and is known as jump Markov processes (see for instance Dynkin [1], Gihman and Skorohod [2],. Since changes of the states of the material microstructure are caused by certain internal mechanisms that are induced by some unobserved random variables, the corresponding evolution of states may be regarded as a partially observed jump process. The latter processes are extensively used in control theory and are employed in the present stochastic analysis of the transient behaviour of structured solids. Throughout this paper the axiomatic definitions of the relevant random variables given in [3] will be maintained and used in the present analysis.

II. State-space analysis:

Generally, the evolution of a random phenomenon such as structural changes that occur during the transient response of a structured solid can be characterized by a family of random variables $\{x_t\}, t \geq 0$ usually taking values in $\mathbb{R}^n$. Such a family can be recognized as a stochastic process $x_t \in \mathfrak{X}$, where $\mathfrak{X}$ is a probabilistic function space. For the state-space analysis of the random phenomena two concepts are important, i.e., the state of the structure and that of observable quantities. A specific state $z \in Z = \{z_0, z_1, \ldots\}$, where Z is the state-space identified with $\mathfrak{X}$ (see [3]), can be characterized by a single deformation u for instance, but more often by a set $U_i = \{u, u_1, u_2 \ldots\}; U_i \in \mathcal{U} \subset \mathfrak{X}$. Thus given $u \in U_i$ and the corresponding state $z \in Z$, the distribution $P(u, z, E)$ can be interpreted as the probability of the observable u having values in an event set E (Borel), when the structure is in the specific state $z \in Z$. In the stochastic deformation theory identifying Z with the probabilistic function space

$\mathfrak{X}$, a state vector ${}^{\alpha}\underset{\sim}{z}$ represents an outcome or elementary event in $\mathfrak{X}$ as a result of a statistical experiment α . Thus the event E is defined by a set of state vectors within an experimental range of measurements, i.e.,

$$E = \{{}^{\alpha}z^i : z^i < {}^{\alpha}z^i < z^i + \Delta z^i\}, (i=1 \ldots r); E \in \mathfrak{F}_z \tag{1}$$

where $\mathfrak{F}_z$ is the σ-algebra of the events. The fundamental condition that relates the random process in the state-space $z(t)$ in Z or equivalently $\{z(t_i) = x_i ; t_i = 0,1,2,\ldots\}$ to the measure $\mathcal{P}_z\{z(t) \in E \mid z(s) = x\}$ is the well-known Markov principle.

In the present analysis these events and hence the process z_t are considered to be controlled by parameters that are responsible for the changes of states of the structure. For instance in a polycrystalline solid one may consider the ratio of the normal component of the microstress to the microshear stress at the interface of two neighbouring crystals involved in the formation of slip bands and in the ensuing structural changes, as a control parameter. Similarly in fibrous structures a possible controlling factor of the transiency may be seen in the partial or total bond-breakage of overlapping fibres in the network.

Using the notion of transition probability functions of Markov theory and the measure $\mathcal{P}_z$ from above, the function of $P(s,x,t,E)$ is a transition function with probability one, if

$$P(s,x,t,E) = \mathcal{P}\{z(t) \in E \mid z(s) = x\} \tag{2}$$

satisfying $P(\cdot/\cdot) = 1$ for $x \in E$ and 0 for $x \notin E$, $P(\cdot/\cdot) \leq 1$. For a fixed s,t and $x \in E$, $P(\cdot/\cdot)$ is also a measure on Z, $E \in \mathfrak{F}_z$ and a Z -measurable function of $x \in Z$. Hence the functional relation

$$P(s,x,t,E) = \int_Z P(s,x,\tau,d\eta)\, P(\tau,\eta,t,E) \; ; \; s \leq \tau \leq t \tag{3}$$

will be valid. It represents the well-known Chapman-Kolmogorov relation.

In the phenomenological description by specifying a particular state z during the transiency, one can use a <u>state function</u> of the form:

$$g\{z(t+1)\} = f\{t, z(t), \theta(t)\} \tag{4}$$

in which $f\{\cdot/\cdot\}$ contains the discrete variables $t = 0,1,2\ldots \in \mathfrak{T}$, $z \in Z$, $\theta \in \Theta$ and where Θ is the control parameter space as a subspace of $\mathfrak{X}$. For an arbitrary time instant $\tau > t$, $\tau \in \mathfrak{T}$ and a given initial state:

$$z(\tau) = G_{t,\tau}\left\{z(t),\ \theta_t,\ \theta_{t+1}\ldots\ldots\theta_{\tau-1}\right\} \tag{5}$$

where $G_{t,\tau}$ is assumed to be $\mathcal{F}_t \times \mathcal{F}_\tau$ measurable and where $[\Theta, \mathcal{F}_t, \mathcal{P}_t]$ designates a control parameter probability space, $\mathcal{F}_\tau$ the corresponding σ-algebra and $\mathcal{P}_\tau$ the associated measure for which the random elements are defined.

For the process $z(t)$ as stated in (5) one can write the transition probability function as the expected value of the state function $G_{t,\tau}$:

$$P(s,x,t,E) = E\left\{\varphi_E\left[G_{t,s}(x, \theta_t\ldots\ldots\theta_{s-1})\right]\right\} \tag{6}$$

in which φ_E is the characteristic function of the set E .

From a thermodynamics point of view, one can use for the description of critical phenomena such as <u>structural-phase changes</u> potentials or a family of potentials $\{\Phi(x,\theta)\}$, which depends on n-state variables x , that can be identified with <u>order parameters $\varphi \in \mathbb{R}^n$</u> and <u>k-control parameters $\theta \in \mathbb{R}^k$</u>. The state of the structure is then determined by the value of x that minimizes the potential $\Phi(x;\theta)$ locally. Thus it becomes necessary to establish some criteria for equilibrium conditions (stationarity of $\Phi(x;\theta)$) and the stability characteristics of the microstructure, i.e., $\nabla^2\Phi(x,\theta) \geq 0$. Phase transitions will occur when the chosen state variables $x \in \mathbb{R}^n$ characterizing the state of the structure jump from one critical branch in a corresponding phase diagram to another. Transitions may also occur from variation in the control parameter θ .

In the present analysis the phenomena concerning structural-phase changes are considered to be <u>subcritical</u> and hence the changes of states are designated by <u>metabatic-state changes</u>.

III. Approximation to the transient behaviour by jump processes:

(a) General remarks

There are two ways of applying <u>partially observed jump processes</u> for the approximation to structural state changes and hence the transient response of solids. One can either employ the concepts of stochastic control theory relating to discontinuous processes and Martingale theory or use the conditional distributions of states of Markov processes given the observations of a state function indicated previously by equ. (5).

Thus considering the jump process $\{x_t\ ; t \geq 0\}$ taking on values in the space $\mathcal{X}$, it can be defined by a countable sequence of random variables:

$$\{\tau_0, z_0, \tau_1, z_1, \ldots\ldots\ldots \tau_n, z_n\} \tag{7}$$

which is defined on a probability space $[Z, \mathcal{F}_t, \mathcal{P}_t]$, where $\{\tau_i\}$ are the jump times and $\{z_i\}$ the states. The sample path of the process is then given by:

$$x_t = z_i \ , \ t \in [\tau_i , \tau_{i+1}) \ , \ i = 0,1,2\ldots.. \tag{8}$$

It is evident that the condition $0 = \tau_0 < \tau_1 < \ldots\ldots \tau_n \to \infty$ a.s. holds. The σ-algebra $\mathcal{F}_t$ is the family of σ-algebras $\mathcal{F}\{x_s , s \leq t\}$ and hence one can use a family of counting processes associated with x_t for which the following probability can be given:

$$P(t,E) = \sum_{\substack{s \leq t \\ x_s \neq x_{s-}}} I_{(x_s \in E)} = \sum_{\tau_i \leq t} I_{(z_i \in E)} \tag{9}$$

where I is the indicator function of the set E .

Consider the map $\mu : R^1 \times X \times Z \to [0,1]$ to be such that $\mu(\cdot, E)$ is measurable for each $E \in Z$ and $\mu(t,x,\cdot)$ is a probability measure on $[X, Z]$ for each $(t,x) \in R^1 \times X$, one can construct for each $x \in X \subset \mathfrak{X}$ a base measure $\mathcal{P}_x$ on $[X, \mathcal{F}_\infty]$ so that $\mathcal{P}_x(x_0 = x) = 1$ (see Prohorov and Rozanov [4]). Thus a local description of the jump process x_t with measure $\mathcal{P}_x$ can be given by using the pair $[t, \mu(t, x_t, E)]$ and where a probability $\tilde{P}(t,E)$ is defined by:

$$\tilde{P}(t,E) = \int_0^t \mu(s, x_s, E)\, ds \tag{10}$$

Hence taking

$$q(t,E) = P(t,E) - \tilde{P}(t,E),$$

it becomes an $\mathcal{F}_t$ -Martingale for each $E \in X$ and the base measure $\mathcal{P}_x$ (see also M.H.A. Davis [5]).The jump process x_t is then a regular step Markov process ([2,5] and Blumenthal and Getoor [6]). A comprehensive treatment of such processes and Martingale dynamics is due to P. Brémaud [7].

In the second approach to incompletely observed jump processes, one can describe the process of metabatic state changes during a certain period of time or transiency by:

$$z = \{z(t) \ ; \ t \in [0,T] \in \mathcal{T} \} \tag{11}$$

where the process starts at the beginning of the transient response and stops at the beginning of steady-state conditions. It is thus defined on a closed time interval $[0,T] \in \mathcal{T}$ with values in a n-dimensional Euclidean space.

(b) Analysis in the stress-strain space:

In the formulation of the transient response the Euclidean space R^n is considered as a product space $\Sigma^{n_1} \times \mathcal{E}^{n_2}$ such that with probability one, z has piece-wise right continuous paths. Σ^{n_1} denotes a stress-space and $\mathcal{E}^{n_2}$ the corresponding strain-space for the structured solid.

The significance of these subspaces for establishing a material operator or constitutive relations for the material as well as the required decomposition of the state vector $\underset{\sim}{z} \in Z \subset \mathcal{X}$ have been discussed in detail in ref [3].

Since z has values in the produce space $R^{n_1+n_2}$ or $\Sigma^{n_1} \times \mathcal{E}^{n_2}$, it can be expressed in terms of two component processes i.e.:

$$z = (\sigma, \epsilon) \tag{12}$$

where $\sigma \in \Sigma$ is the vector of the first n_1 component of z and $\epsilon \in \mathcal{E}$, the vector of the last n_2 component of z. It is readily recognized that ϵ is the component of z that can be observed, whilst σ is unobservable. Noting that the σ-algebra generated by the past process z_t is:

$$\mathcal{F}_t = \mathcal{F}\{z(s) : 0 \leq s \leq t\} \tag{13}$$

and that of the past measurements ϵ_t:

$$\mathcal{G}_t = \mathcal{G}\{\epsilon(s) : 0 \leq s \leq t\} \tag{14}$$

one can define a functional $\Phi(t)$ on the past process as a measurable real valued stochastic process, which for each time instant t is $\mathcal{F}_t$-measurable with an expected value $E\{|\Phi(t)|\} < \infty$. This representation will not be persued here. It may be remarked however, that for computational purposes it is necessary to consider the conditional expectation of the functional, i.e., $E\{\Phi(t) | \mathcal{G}_t\}$ (see also [5]).

Since the process $z = (\sigma, \epsilon)$ has piece-wise continuous paths with probability one, there is a sequence of successive metabatic states:

$$Z_n := (z_0, \tau_0, \tau_1, z_1, \tau_2, z_2 \ldots\ldots \tau_n, z_n) \tag{15}$$

at times τ_i at which the process jumps from z_{i-1} to z_i and correspondingly a sequence of random elements:

$$F_n := (\sigma_0, \epsilon_0, \tau_1, \sigma_1, \epsilon_1, \ldots\ldots \tau_n, \sigma_n, \epsilon_n) \tag{16}$$

where the correspondence between (15) and (16) is one-to-one. Thus given the sequence in (15) or (16), z_t on a closed time-interval $[0,T] \in \mathcal{G}$ will be defined by:

$$z_0(t)=[\sigma(0),\varepsilon(0)]=(\sigma_0,\epsilon_0) \quad \text{for} \quad 0 \leq t \leq \tau_1 \; ; \; \tau_0 \equiv 0 \qquad \text{(a)}$$

$$z(t)=z_t=[\sigma(\tau_n),\epsilon(\tau_n)]=(\sigma_n,\epsilon_n) \quad \text{for} \quad \tau_n \leq t < \tau_{n+1} \, , \qquad \text{(b)} \tag{17}$$

$$\text{if} \quad \tau_n = t = T$$

The initial conditions (a) , i.e., the beginning of the transient response of the solid are thus given by an unobserved stress σ_0 , but an observable strain ϵ_0 at τ_0 of the jump process z_t . The transient will end when the jump times $\tau_n = T$. For the description of z_t two conditional probability distribution functions are required, i.e.:

$$P_\tau \left\{ \tau_{n+1} \leq t \,\middle|\, (\sigma_0,\epsilon_0,\tau_1,\sigma_1,\epsilon_1 \ldots\ldots \tau_n,\sigma_n,\epsilon_n) \right\} \tag{18}$$

which is the conditional probability of the next jump to occur given the history up to the nth jump and another distribution :

$$P_n \left\{ \sigma_{n+1}, \tau_{n+1} \in Z \,\middle|\, (\sigma_0,\epsilon_0,\tau_1,\sigma_1,\epsilon_1, \ldots\ldots \tau_n,\sigma_n,\epsilon_n,\tau_{n+1}) \right\} \tag{19}$$

which represents the position of the jump given the history up to the nth jump and that the $(n+1)^{st}$ jump has just occurred. In terms of the abbreviated notation of (15),(16) the distribution of jump times can also be written as:

$$P_\tau(\tau_{n+1} \leq t \mid F_n) = \int_{\tau_n}^{t} p(s \mid F_n)\, ds \tag{20}$$

where $p(s \mid F_n)$ designates the conditional probability density on $[\tau_n, T)$. Using also a shortened notation for the history of the observed variables, i.e., the strains ϵ and corresponding jump times so that the sequence $(\epsilon_0 \ldots\, \epsilon_n) \rightarrow (\epsilon_0' \ldots\, \epsilon_n')$ with repetitions omitted and $(\tau_1' \ldots\ldots \tau_k')$ is a subset of $\{\tau_1 \ldots\ldots \tau_n\}$ for which there is a jump at each τ_i' , then

$$G_k := (\epsilon_0', \tau_1', \epsilon_1' \ldots\ldots \tau_k', \epsilon_k') \tag{21}$$

To derive this correspondence, it is assumed that the number of jumps $n(t)$ of z in $[0,t]$ and that of the strains ϵ denoted by $k(t)$ in $[0,t]$ are such that, when $n(t)=n$, $k(t)=k$, $k \leq n$ so that ϵ on $[0,t]$ corresponds to $\{\epsilon_0' \ldots\, \epsilon_k'\}$. The conditional distribution in (19) can also be written in shorter form as:

$$P_n \{ Z \mid F_n, \tau_{n+1} \} \quad \text{for any Borel set } Z \in \mathbb{R}^n \text{ or } \Sigma^{n_1} \times \mathcal{E}^{n_2} \tag{22}$$

In the analysis of partially observed jump processes another quantity is of considerable importance, i.e., the conditional jump rate of z_t. It has a basic role in control theory and its application in the present study is equally significant. It can be defined in terms of the density function $p(s \mid F_n)$ on the interval $\tau_n \leq t < T$ for z_t as follows ([8,9]):

$$q(t \mid F_n) = \frac{p(t \mid F_n)}{\int_{\tau_n}^{t} p(s \mid F_n) + P_\tau(\tau_{n+1} = T \mid F_n)} \tag{23}$$

from which one can compute on the assumption that the distribution is of the exponential type :

$$p(t \mid F_n) = q(t \mid F_n) \exp\left[-\int_{\tau_n}^{t} q(s_n, F_n)\, ds\right] \tag{24}$$

As mentioned earlier the jump process is defined as a conditional Markov process in accordance with the sequence of jump times indicated by (21). It has been shown by R. Rishel [9] that there is a mapping A of the past history F_n and the observed history G_k such that:

$$A(F_n) = G_k \tag{25}$$

Hence the jump rates can be taken as:

$$q(t, F_n) = q(t, G_k, \sigma_n) \tag{26}$$

and the distributions as:

$$P_\tau\{Z \mid F_n, \tau_{n+1}\} = P_n\{Z \mid G_k, \sigma_n, \tau_{n+1}\} \quad \text{for } Z \in \mathbb{R}^n \tag{27}$$

This indicates that for conditional Markov jump processes, the jump rate and state jump distribution depend only on the measurement history G_k and the current value σ_n of the unobserved component of state.

IV. Conclusions:

Since the metabatic state changes are caused by some internal mechanism controlling the location of jumps and the jump times, it is apparent that a description of the transient behaviour could be based on partially observed controlled jump Markov processes. In this case the analysis will include a controlled jump rate, a controlled conditional state jump distribution and the specification of a control parameter or a function of such parameters. Thus a control may be regarded as a function θ_t or a family of functions $\{\theta_t, G_k\}$ of time and various measurement histories.

If the values of the observed and unobserved components of the current state z_n of F_n, e.g., of (σ_n, ϵ_n) are briefly denoted by (i,j), one can define the conditional distribution of time of the next jump and the conditional distribution of the next jump for the controlled process z_t as follows:

$$P_\tau \left\{ \tau_{n+1} > t \mid F_n \right\} = \exp\left[- \int_{\tau_n}^{t} q(s,i,j,\theta[t,\theta_k])\, ds \right] \tag{28}$$

and

$$P_\epsilon \left\{ \epsilon_{n+1}, \sigma_{n+1} = (k,\ell) \mid F_n, \tau_{n+1} \right\}$$

$$= P\left\{ (k,\ell) \mid (i,j), \tau_{n+1}, \theta[\tau_{n+1}, G_k] \right\} \tag{29}$$

showing that these distributions depend on the past measurements. Hence the evolution of the controlled process describing the transient response may also depend on G_k and the process non-Markovian. This would also be the case, if the restriction on the conditional distribution to be of the exponential type in (24) is removed, which leads to a semi-Markov process. However the latter can still be approximated by a Markov process (see also [10]).

Finally, it may be recognized that the conditional jump rate is important in the representation of the transient response behaviour in terms of Markov jump processes. If it increases rapidly the subcritical changes of state of the microstructure may become critical. In this case a stochastic approach to critical phenomena is more appropriate [11]), which is based on the strong dependence of the involved variables and the central limit theorem of probability theory. A numerical analysis of the above approximation to the transient behaviour for a particular class of solids, i.e., fibrous networks will be given in a forthcoming publication.

References

[1] E. DYNKIN : Markov Processes, Vol. I and II., Springer Verlag, Berlin (1965).

[2] I.I. GIHMAN and A.V. SKOROHOD : The theory of Stochastic Processes, Vol. I and II, Springer Verlag, Berlin (1975).

[3] D.R. AXELRAD : Foundations of the Probabilistic Mechanics of Discrete Media, Pergamon Press, Oxford (1984).

[4] Yu, V. PROHOROV and Yu. A. ROZANOV : Probability Theory, Springer Verlag, Berlin, (1969).

[5] M.H.A. DAVIS : The Representation of Martingales of Jump Processes, SIAM, Journ. Control and Optimization, Vol. 14, No. 4, July (1976).

[6] R.M. BLUMENTHAL and R.K. GETOOR : Markov Processes and Potential Theory, Academic Press, N.Y. (1968).

[7] P. BREMAUD : Point Processes and Queues,Martingale Dynamics, Springer Verlag, Berlin (1981).

[8] W.H. FLEMING and R.W. RISHEL : Deterministic and Stochastic Optimal Control, Springer Verlag, Berlin (1975).

[9] R. RISHEL : A minimum Principle for Controlled Jump Processes, Lect. Notes in Economics and Math. Systems, No. 107, Springer Verlag (1975).

[10] M. RUDEMO : State Estimation for Partially Observed Markov Chains, Journ. Math. Analysis and Applications, Vol. 44, pg. 581-611 (1973).

[11] D.R. AXELRAD : Seminars in the probabilistic Mechanics of Discrete Media, Publ. Université de Genève, UGVA-DEP. 1985/04-461, (1985).

ON SPECTRAL ANALYSIS OF NON-SELFADJOINT OPERATORS IN MECHANICS

J. Brilla

Institute of Applied Mathematics and
Computing Technique, Comenius University
842 15 Bratislava, Czechoslovakia

1. Introduction

Many time dependent problems of mechanics lead to analysis of differential equations

$$Au = \sum_{k=0}^{n} A_k \frac{\partial^k}{\partial t^k} u = f, \qquad \text{in } \Omega, \tag{1}$$

or to systems of such equations, where A_k are symmetric strongly elliptic operators of order 2m. We assume that the domain of definition Ω is bounded and the boundary $\partial\Omega$ is sufficiently smooth. We consider homogeneous boundary conditions and nonhomogeneous initial conditions.

We arrive at an equation (1) when dealing with quasistatic and dynamic problems of linear viscoelastic continuum and structures and when linearizing different nonlinear problems as buckling, dynamic stability etc. In the case of dynamic problems the operator A_n or operators A_n and A_{n-1} have to be replaced by the identity operator. Thus equations (1) include also parabolic and hyperbolic equations.

We assume that $f(t) \in L_2(0,\infty)$ and applying the Laplace transform we arrive at

$$A(p)\tilde{u} = \sum_{k=0}^{n} p^k A_k \tilde{u} = \tilde{f}^*, \tag{2}$$

where a tilde denotes the Laplace transform and $\tilde{f}^*$ includes initial conditions. The operator $A(p)$ is a complex symmetric non-self-adjoint elliptic operator.

After the Laplace transform we arrive to similar equations when considering

$$\int_0^t G(t-\tau) \frac{\partial w}{\partial \tau} d\tau + \frac{\partial^2 w}{\partial t^2} = f, \tag{3}$$

where G is a symmetric strongly elliptic operator of order 2m. For analysis of equations (2) we have introduced [1 - 2] spaces of analytic functions valued in Sobolev spaces which are isomorphic to weighted anisotropic Sobolev spaces which we have proposed for analysis of equations (1) .

Now we shall deal with spectral analysis of complex symmetric elliptic operators A (p) and show that it is possible to prove theorems on existence of eigenvalues and completeness of sets of eigenvectors similar as in the case of real symmetric elliptic operators.

2. Spectral analysis

Operators A (p) are complex symmetric operators. Thus it holds

$$A(p) = \overline{A(p)} \tag{4}$$

and

$$(Ax, \bar{x}) = (x, \overline{Ax}) \tag{5}$$

When $A_k A_e \neq A_e A_k$, i.e. when operators A_k are noncommutative, $A^* A \neq A A^*$ and A (p) is a nonnormal operator. Thus for its spectral analysis it is not possible to apply the spectral theory of symmetric elliptic operators. However we can generalize some of its results.
We consider the equation

$$A(p)\ e(p) = \sum_{k=0}^{n} p^k A_k\ e(p) = \lambda(p)\ e(p) . \tag{6}$$

Then λ (p) for which the solutions of (6)exist are eigenvalues and the corresponding solutions e (p) are eigenvectors of (6). In general, eigenvalues and eigenvectors are functions of p.

For nonnegative real values of p A (p) is a symmetric strongly elliptic operator. Thus it has a discrete spectrum and a complete pairwise orthogonal set of eigenvectors.

We can prove :
Theorem 1. A complex symmetric strongly elliptic operator has at least one non-zero eigenvalue and its eigenvalues and eigenvectors are solutions of the variational problem

$$\min \max \{ |(Ae, \bar{e})| - |\lambda| |(e, \bar{e})| \} \tag{7}$$

Proof : There exists a neigbourhood $\Omega_{p_1^+}$ of the positive real semiaxis p_1^+ where A (p) has the compact inverse A^{-1} (p) with positive or negative real and imaginary parts, the traces of which are not equal to zero. Therefore A^{-1} (p) is not a quasi - nilpotent operator and has at least one non-zero eigenvalue. Then also A (p) has at least one eigenvalue. Further the first Gateaux derivative of (7) leads to the condition

$$\frac{1}{|(Ae,\overline{e})|}\{(Ae,\overline{h})(\overline{Ae},e)+(Ae,\overline{e})(\overline{Ae},h)\}-$$
$$-|\lambda|\frac{1}{|(e,\overline{e})|}\{(e,\overline{h})(\overline{e},e)+(e,\overline{e})(\overline{e},h)\}=0\,, \tag{8}$$

which is satisfied by

$$A\,e=\lambda\,e\,. \tag{9}$$

Analysis of the second Gateaux derivative shows that (9) is a saddle point of (8) .

This variational formulation is convenient for an approximate solution of eigenvalues and eigenvectors.

Points p, at which it holds $(e(p),\overline{e(p)})=0$ will be called exceptional points of the operator A (p) . Using estimates derived by T. Kato [3] it is possible to prove that there exists a neighbourhood Ωp_1^+ of p_1^+ and a right-hand half-plane of p without exceptional points of A(p).

Then we have :

<u>Theorem 2.</u> Eigenvectors of a complex symmetric strongly elliptic operator A (p) and eigenvectors of its adjoint A (p) form, with exception of exceptional points, complete biorthogonal systems which can be biorthonormalized.

Proof : It holds $(Ae_k,\overline{e}_1)=\lambda_k\,(e_k,\overline{e}_1)$
and

$$(Ae_k,\overline{e}_1)=(Ae_1,\overline{e}_k)=\lambda_1\,(e_1,\overline{e}_k)=\lambda_1\,(e_k,\overline{e}_1)\quad.\ \text{Then}$$
$$(\lambda_k-\lambda_1)\,(e_k,\overline{e}_1). \tag{10}$$

Hence for $\lambda_k\neq\lambda_1$ it holds $(e_k,\overline{e}_1)=0$ and eigenvectors e_k , $\overline{e}_1$ form biorthogonal systems.

The construction of eigenvalues and eigenvectors and the proof of biorthogonality and completeness can be done similarly as in the case of symmetric operators applying the variational formulation (7).

Then for $f \; A^{-1} h$ we have

$$f = \sum_{k=1}^{\infty} (f, \bar{e}_k)\, e_k = \sum_{k=1}^{\infty} (f, e_k)\, \bar{e}_k \qquad (11)$$

what corresponds to covariant and contravariant expansions of vectors. Then it holds

$$||f||^2 = \sum_{k=1}^{\infty} (f, \bar{e}_k)(\bar{f}, \bar{e}_k) \qquad (12)$$

Finally it is possible to prove the theorem on analycity of eigenvectors and eigenvalues of complex symmetric positive definite elliptic operators. We have

Theorem 3. There exists such a neighbourhood Ωp_1^+ of the nonnegative real semiaxis p_1^+ where a complex symmetric positive definite operator $A(p)$ has a regular discrete spectrum with eigenvalues $\lambda_1(p)$, $\lambda_2(p), \ldots$ and eigenvectors $e_1(p)\; e_2(p), \ldots$ of $D_{A(p)}$ all regular in this neigbourhood, such that for every $p \in \Omega p_1^+$ the following holds :

1 $\quad A(p)\, e_n(p) = \lambda_n(p)\, e_n(p), \quad n = 1, 2, \ldots \qquad (13)$

2 $\quad e_1(p), e_2(p), \ldots$ is a complete system biorthonormal to $\overline{e_1(p)}, \overline{e_2(p)}, \ldots$ what are eigenvectors of $A(p)$. Thus it holds

$$(e_k(p), \overline{e_l(p)}) = \delta_{kl} \qquad (14)$$

3 $$\lim_{n \to \infty} \lambda_n(p) = \infty . \qquad (15)$$

The proof can be done by generalization of results on the perturbation theory of eigenvalues due to F. Rellich [4].
Firstly it is obvious that the operator

$$B(p - p_o) = \sum_{k=0}^{n} (p - p_o)^k B_k = \sum_{k=0}^{n} p^k A_k \qquad (16)$$

is regular in $p-p_o$ for each real nonnegative p_o. Thus it is possible to apply Rellich s results for every real nonnegative p_o. Secondly we have introduced biorthonormal system of eigenvectors what enables to apply Weierstrass preparation theorem in the proof due to Rellich also to complex values of the parameter p.

Further we have proved that complex symmetric operators are semisimple with exception at exceptional points and we can assume that the analytic continuation is possible to a right-hand halfplane of p.

References

1. Brilla, J. New functional spaces and linear nonstationary problems of mathematical physics. Proceedings of Equadiff 5, Bratislava 1981, Teubner, Leipzig 1982, 64 - 71.

2. Brilla, J. Novye funkcionaľnye prostranstva i linejnye nestacionarnye problemy matematičeskoj fyziki. Proceedings of the 7th Soviet - Czechoslovak Conference, Yerevan State University 1982, 49 - 58.

3. Kato, T. Perturbation Theory for Linear Operators. Springer, Berlin-Heidelberg-New York, 1966.

4. Rellich, F. Perturbation Theory of Eigenvalue Problems. Gordon and Breach, New York - London - Paris, 1969.

ON THE PHOTOELASTIC EFFECT IN A HEMITROPIC DISSIPATIVE DIELECTRIC

Ph.Boulanger
Département de Mathématique
Université Libre de Bruxelles - Campus Plaine C.P.218/1
Boulevard du Triomphe - 1050 Bruxelles - Belgium

1. Basic Equations

In this work, I consider a deformable, polarizable, non magnetizable dielectric. For such a dielectric, the field equations relating the motion $\underset{\sim}{x}(\underset{\sim}{X},t)$, the Cauchy stress tensor $\underset{\sim}{t}$, the mass density ρ, the polarization $\underset{\sim}{P}$, and the electromagnetic field $\underset{\sim}{E}$, $\underset{\sim}{B}$ are [1] :

$$\rho\ddot{\underset{\sim}{x}} = \operatorname{div}[\underset{\sim}{t} + \underset{\sim}{E} \otimes \underset{\sim}{E} + \underset{\sim}{B} \otimes \underset{\sim}{B} - \tfrac{1}{2}(E^2 + B^2)\underset{\sim}{1}],$$

$$\frac{1}{c}\,\partial_t \underset{\sim}{B} + \operatorname{rot}\underset{\sim}{E} = 0 \quad , \quad \operatorname{div}\underset{\sim}{B} = 0, \tag{1}$$

$$\operatorname{rot}(\underset{\sim}{B} + \frac{1}{c}\,\dot{\underset{\sim}{x}} \times \underset{\sim}{P}) - \frac{1}{c}\,\partial_t(\underset{\sim}{E} + \underset{\sim}{P}) = 0 \quad , \quad \operatorname{div}(\underset{\sim}{E} + \underset{\sim}{P}) = 0,$$

where the mass density ρ at time t is related to the mass density ρ_o in the reference configuration by $\rho = \rho_o|(\underset{\sim}{x}/\underset{\sim}{X})|$.

Constitutive relations must be added to these equations. I here assume that the stress tensor $\underset{\sim}{t}$ and the electromotive intensity $\underset{\sim}{E}' = \underset{\sim}{E} + \frac{1}{c}\,\dot{\underset{\sim}{x}} \times \underset{\sim}{B}$ are functions of the deformation gradient, the polarization and their first time derivatives. Owing to a requirement of invariance under superimposed rigid body motions, the constitutive relations read (components of vectors and tensors are written in a rectangular Cartesian coordinate system) :

$$t^{ij} = \rho x^i{}_{,A} x^j{}_{,B}\, \Sigma^{AB}[\underset{\sim}{C},\dot{\underset{\sim}{C}},\underset{\sim}{\Pi},\dot{\underset{\sim}{\Pi}}],$$

$$E'_i = \rho_o X^A{}_{,i}\, \mathcal{E}_A[\underset{\sim}{C},\dot{\underset{\sim}{C}},\underset{\sim}{\Pi},\dot{\underset{\sim}{\Pi}}], \tag{2}$$

where $\underset{\sim}{C}$ is the Cauchy strain tensor and $\underset{\sim}{\Pi}$ the material measure of polarization :

$$C_{AB} = \delta_{ij} x^i{}_{,A} x^j{}_{,B} \quad , \quad \Pi^A = |(\underset{\sim}{x}/\underset{\sim}{X})|\, X^A{}_{,i} P^i .$$

Moreover, the equilibrium values of the functions $\underset{\sim}{\Sigma}$, $\underset{\sim}{\mathcal{E}}$ are assumed to be related to the derivatives of an internal energy function ε by :

$$\overset{o}{\Sigma}{}^{AB} = \Sigma^{AB}[\underset{\sim}{C},0,\underset{\sim}{\Pi},0] = 2\,\frac{\partial\varepsilon}{\partial C_{AB}}$$

$$\overset{o}{\mathcal{E}}_A = \mathcal{E}_A[\underset{\sim}{C},0,\underset{\sim}{\Pi},0] = \frac{\partial\varepsilon}{\partial \Pi^A}\,, \tag{3}$$

where ε is a scalar valued function of $\underset{\sim}{C}$ and $\underset{\sim}{\Pi}$.

For a purely transparent elastic dielectric, $\underset{\sim}{\Sigma}$ and $\underset{\sim}{\mathcal{E}}$ are functions of $\underset{\sim}{C}$ and $\underset{\sim}{\Pi}$ only, and the relations (3) imply that the energy balance is identically satisfied with a vanishing heat flux [1].

I now assume that the dielectric is <u>hemitropic</u> in its reference state, which means that the tensor, vector, scalar-valued functions $\underset{\sim}{\Sigma}$, $\underset{\sim}{\mathcal{E}}$, ε of the two symmetric tensors $\underset{\sim}{C}$, $\dot{\underset{\sim}{C}}$ and the two vectors $\underset{\sim}{\Pi}$, $\dot{\underset{\sim}{\Pi}}$ are invariant under the <u>proper</u> orthogonal group. With the assumption that these functions are polynomial, a representation theorem of G.F.Smith [2] can be used in order to express them in terms of a certain number of basic invariants, tensors and vectors.

2. <u>The Equations for Small Displacements and Fields Superimposed on a state of Finite Deformation without Electromagnetic Field</u>

Let

$$\overset{\circ}{\underset{\sim}{x}}\ ,\ \overset{\circ}{\underset{\sim}{t}}\ ,\ \overset{\circ}{\underset{\sim}{E}} = \overset{\circ}{\underset{\sim}{B}} = \overset{\circ}{\underset{\sim}{P}} = 0, \tag{4}$$

denote an equilibrium solution of the field and constitutive equations (1) (2), characterized by a finite deformation and the absence of electromagnetic field. The linearized equations for small time-dependent displacements and fields superimposed on this equilibrium state can be written in the form

$$\begin{aligned} &\overset{\circ}{\rho}\ddot{\underset{\sim}{u}} = \mathrm{div}\,\underset{\sim}{\sigma}, \\ &\frac{1}{c}\dot{\underset{\sim}{b}} + \mathrm{rot}\,\underset{\sim}{e} = 0 \quad , \quad \mathrm{div}\,\underset{\sim}{b} = 0 \\ &\mathrm{rot}\,\underset{\sim}{b} - \frac{1}{c}(\dot{\underset{\sim}{e}} + \dot{\underset{\sim}{p}}) = 0 \quad , \quad \mathrm{div}(\underset{\sim}{e} + \underset{\sim}{p}) = 0, \end{aligned} \tag{5}$$

$$\begin{aligned} \sigma^{ij} &= \overset{\circ}{t}{}^{jk} u^{i}{}_{,k} + \overset{\circ}{C}{}^{ijk\ell} u_{k,\ell} + \overset{\circ}{D}{}^{ijk\ell} \dot{u}_{k,\ell} + \overset{\circ}{S}{}^{ij}{}_{k} p^{k} + \overset{\circ}{R}{}^{ij}{}_{k} \dot{p}^{k}, \\ e_{i} &= \overset{\circ}{T}_{ij} p^{j} + \overset{\circ}{Z}_{ij} \dot{p}^{j} + \overset{\circ}{S}{}_{i}^{k\ell} u_{k,\ell} + \overset{\circ}{Q}{}_{i}^{k\ell} \dot{u}_{k,\ell}, \end{aligned} \tag{6}$$

where $\underset{\sim}{u}$, $\underset{\sim}{\sigma}$, $\underset{\sim}{e}$, $\underset{\sim}{b}$, $\underset{\sim}{p}$ denote the additional displacement, stress, electromagnetic field, and polarization defined as in [1]. The superimposed dot now denotes the partial time derivative (at fixed $\underset{\sim}{x}$); the superimposed zeros refer to the equilibrium state (4). Using the expressions of $\underset{\sim}{\Sigma}$, $\underset{\sim}{\mathcal{E}}$, ε derived from the representation theorem of G.F.Smith [2], one can obtain expressions for the coefficients of the linearized constitutive equations (6) as functions of the Finger strain tensor of the equilibrium state

$$c^{ij} = \delta^{AB}\, \overset{\circ}{x}{}^{i}{}_{,A}\, \overset{\circ}{x}{}^{j}{}_{,B}.$$

Those expressions are

$$\overset{\circ}{t}_{ij} = \alpha_o \delta_{ij} + \alpha_1 c_{ij} + \alpha_2 c^2_{ij},$$

$$\overset{\circ}{C}_{ijk\ell} = - \alpha_o(\delta_{ik}\delta_{\ell j} + \delta_{\ell i}\delta_{jk}) + \alpha_2(c_{ik}c_{j\ell} + c_{jk}c_{i\ell})$$
$$+ (\alpha_3 + 2\alpha_o)\delta_{ij}\delta_{k\ell} + \alpha_4(c_{ij}\delta_{k\ell} + c_{k\ell}\delta_{ij}) + (\alpha_5 - 2\alpha_2)c_{ij}c_{k\ell}$$
$$+ \alpha_6(c^2_{ij}\delta_{k\ell} + c^2_{k\ell}\delta_{ij}) + \alpha_7(c_{ij}c^2_{k\ell} + c_{k\ell}c^2_{ij}) + \alpha_8 c^2_{ij}c^2_{k\ell},$$

$$\overset{\circ}{D}_{ijk\ell} = \beta_3\delta_{ij}\delta_{k\ell} + \beta_4(c_{ij}\delta_{k\ell} + c_{k\ell}\delta_{ij}) + \beta_4'(c_{ij}\delta_{k\ell} - c_{k\ell}\delta_{ij})$$
$$+ \beta_5 c_{ij}c_{k\ell} + \beta_6(c^2_{ij}\delta_{k\ell} + c^2_{k\ell}\delta_{ij}) + \beta_6'(c^2_{ij}\delta_{k\ell} - c^2_{k\ell}\delta_{ij})$$
$$+ \beta_7(c_{ij}c^2_{k\ell} + c_{k\ell}c^2_{ij}) + \beta_7'(c_{ij}c^2_{k\ell} - c_{k\ell}c^2_{ij}) + \beta_8 c^2_{ij}c^2_{k\ell}$$
$$+ \beta_9(\delta_{ik}\delta_{\ell j} + \delta_{\ell i}\delta_{jk}) + \beta_{10}(c_{ik}c_{j\ell} + c_{jk}c_{i\ell})$$
$$+ \beta_{11}(\delta_{ik}c_{j\ell} + \delta_{\ell j}c_{ik} + \delta_{jk}c_{i\ell} + \delta_{i\ell}c_{kj}), \qquad (7)$$

$$\overset{\circ}{T}_{ij} = \lambda_o\delta_{ij} + \lambda_1 c_{ij} + \lambda_2 c^2_{ij} \;, \quad \overset{\circ}{Z}_{ij} = \mu_o\delta_{ij} + \mu_1 c_{ij} + \mu_2 c^2_{ij} \;, \quad \overset{\circ}{S}_{kij} = 0,$$

$$\overset{\circ}{R}_{kij} = \rho_1(\varepsilon_{ik\ell}c_{\ell j} + \varepsilon_{jk\ell}c_{\ell i}) + \rho_2(c_{ik\ell}c^2_{\ell j} + \varepsilon_{jk\ell}c^2_{\ell i})$$
$$+ \rho_3\varepsilon_{k\ell m}(c_{\ell i}c^2_{mj} + c_{\ell j}c^2_{mi}),$$

$$\overset{\circ}{Q}_{kij} = \gamma_1(\varepsilon_{ik\ell}c_{\ell j} + \varepsilon_{jk\ell}c_{\ell i}) + \gamma_2(\varepsilon_{ik\ell}c^2_{\ell j} + \varepsilon_{jk\ell}c^2_{\ell i})$$
$$+ \gamma_3\varepsilon_{k\ell m}(c_{\ell i}c^2_{mj} + c_{\ell j}c^2_{mi}),$$

where the coefficients $\alpha, \beta, \lambda, \mu, \rho, \gamma$ are functions of the three invariants of the tensor $\tilde{c}$. It is remarkable that the usual coefficient $\overset{\circ}{S}_{kij}$ of piezo-electric interaction identically vanishes, while the analogous coefficients $\overset{\circ}{R}_{kij}$, $\overset{\circ}{Q}_{kij}$ with derivatives of $\tilde{p}$ and $\tilde{u}$ do not vanish. The presence of these interaction coefficients is due to the non holotropy of the dielectric in its reference state, and to the deformation of equilibrium state (4). They disappear when this equilibrium state is undeformed or hydrostatically deformed.

3. Propagation of Plane Sinusoidal Waves

I now assume that the deformation of the equilibrium state (4) is homogeneous. There are then solutions of the system (5) (6) of the type

$$\underset{\sim}{a}(\underset{\sim}{x},t) = \mathrm{Re}\{\bar{\underset{\sim}{a}} e^{i\omega(t-\frac{n}{c}\underset{\sim}{s}.\underset{\sim}{x})}\}, \quad (\underset{\sim}{a} = \underset{\sim}{u},\underset{\sim}{\sigma},\underset{\sim}{e},\underset{\sim}{b},\underset{\sim}{p}), \tag{8}$$

where $\bar{\underset{\sim}{a}}$ is the complex amplitude of the quantity $\underset{\sim}{a}$, $\underset{\sim}{s}$ the unit vector in the propagation direction, n the complex refractive index, and ω the angular frequency (real).

Introducing (8) in (5) (6), one obtains an algebraic system in the amplitudes. Taking $\overset{\circ}{S}_{kij} = 0$ into account, and eleminating $\bar{\underset{\sim}{\sigma}}$, $\bar{\underset{\sim}{b}}$, $\bar{\underset{\sim}{p}}$ from this system, one obtains

$$\begin{aligned} &(n^2\underset{\sim}{S}^2 + \underset{\sim}{1} + \underset{\sim}{\chi})\bar{\underset{\sim}{e}} - \omega^2\frac{n}{c}\,\underset{\sim}{\chi}\underset{\sim}{J}\bar{\underset{\sim}{u}} = 0, \\ &-\frac{c}{n}\,\underset{\sim}{I}\underset{\sim}{\chi}\bar{\underset{\sim}{e}} + (\underset{\sim}{A} + \omega^2\underset{\sim}{I}\underset{\sim}{\chi}\underset{\sim}{J} - \overset{\circ}{\rho}\frac{c^2}{n^2}\underset{\sim}{1})\bar{\underset{\sim}{u}} = 0, \end{aligned} \tag{9}$$

where $\underset{\sim}{S}$ is defined by $S_{ij} = \varepsilon_{ikj}s_k$, and where the tensors $\underset{\sim}{A}$, $\underset{\sim}{\chi}$, $\underset{\sim}{I}$, $\underset{\sim}{J}$ are defined and called as follows :

$A_{ik} = (\overset{\circ}{t}_{j\ell}\delta_{ik} + \overset{\circ}{C}_{ijk\ell} + i\omega\overset{\circ}{D}_{ijk\ell})s_j s_\ell$, acoustical tensor in the direction $\underset{\sim}{s}$,

$\chi^{-1}_{ij} = \overset{\circ}{T}_{ij} + i\omega\overset{\circ}{Z}_{ij}$, inverse of the complex susceptibility tensor,

$I_{ik} = \overset{\circ}{R}_{kij}s_j$, $J_{ik} = \overset{\circ}{Q}_{ikj}s_j$, interaction tensors in the direction $\underset{\sim}{s}$.

The condition for (9) to yield non trivial solutions is an equation for $1/n^2 = (U/c)^2$, U being the complex propagation speed of the sinusoidal plane waves. Each solution of this equation with its corresponding solutions of (9) defines a propagation mode.

In the absence of interaction terms (isotropic case : $\underset{\sim}{I}$, $\underset{\sim}{J} = 0$) equations (9) are uncoupled. Using (7), one then get known results about the photoelastic effect [3] [4] (electromagnetic waves), and small amplitude-waves in deformed viscoelastic solids [5] (mechanical waves).

The interaction tensors $\underset{\sim}{I}$, $\underset{\sim}{J}$ introduce an electro-mechanical coupling effect. In order to be consistent with the non-relativistic approximation inherent to this theory, and to be able to distinguish between fast (electromagnetic) and slow (mechanical) waves, I assume that

$$\frac{A_{ik}}{\overset{\circ}{\rho}c^2}, \quad \omega^2\frac{I_{ik}J_{j\ell}}{\overset{\circ}{\rho}c^2} \tag{10}$$

are small (relativistic) parameters, negligeable with respect to 1.

The electromagnetic waves, with $1/n^2$ of order zero in the small parameters (10) can then be obtained from

$$
\begin{aligned}
&(n^2 \underset{\sim}{S}^2 + \underset{\sim}{1} + \underset{\sim}{\chi})\bar{\underset{\sim}{e}} = 0, \\
&\bar{\underset{\sim}{u}} = - \frac{n}{\overset{\circ}{\rho} c} \underset{\sim}{I}\underset{\sim}{\chi}\bar{\underset{\sim}{e}}.
\end{aligned}
\qquad (11)
$$

The mechanical waves, with $1/n^2$ of order one in the small parameters (10) can be obtained from

$$
\begin{aligned}
&\left[\underset{\sim}{A} + \omega^2 \underset{\sim}{I}\underset{\sim}{\chi}\underset{\sim}{J} - \omega^2 \frac{\underset{\sim}{I}\underset{\sim}{\chi}(\underset{\sim}{1} + \underset{\sim}{S}^2)\underset{\sim}{\chi}\underset{\sim}{J}}{\underset{\sim}{s}\cdot(\underset{\sim}{1} + \underset{\sim}{\chi})\cdot\underset{\sim}{s}} - \overset{\circ}{\rho} \frac{c^2}{n^2} \underset{\sim}{1}\right]\bar{\underset{\sim}{u}} = 0 \\
&\bar{\underset{\sim}{e}} = \omega^2 \frac{n}{c} \frac{\underset{\sim}{s}\cdot\underset{\sim}{\chi}\underset{\sim}{J}\cdot\bar{\underset{\sim}{u}}}{\underset{\sim}{s}\cdot(\underset{\sim}{1} + \underset{\sim}{\chi})\cdot\underset{\sim}{s}} s + \frac{\omega^2}{c^2} \frac{c}{n} \underset{\sim}{S}^2 \underset{\sim}{\chi}\underset{\sim}{J}\bar{\underset{\sim}{u}}.
\end{aligned}
\qquad (12)
$$

Equations (11) show that the values of the complex refractive index n of the electromagnetic waves are not influenced by the electromechanical coupling. On the contrary, equations (12) show that the values of the propagation speed U of mechanical waves are influenced by this coupling, since terms are added to the usual acoustical tensor $\underset{\sim}{A}$. More specific results about this influence are presented in the next sections.

4. Propagation of Mechanical Waves in Principal Directions

The coordinate axis are now chosen along the principal directions of the homogeneous equilibrium deformation. Let c_1, c_2, c_3 denote the principal values of the Finger strain tensor $\underset{\sim}{c}$ (squared principal extension ratios) and χ_1, χ_2, χ_3 the principal values of the susceptibility tensor $\underset{\sim}{\chi}$ (principal complex susceptibilities).

Along a principal direction x_i three mechanical waves may propagate with complex speed U_{ij} $(j = 1,2,3)$ and amplitude $\bar{\underset{\sim}{u}}$ along the x_j-axis $(j = 1,2,3)$. Using (7), I obtain

$$
\begin{aligned}
&U_{11}^2 = \overset{*}{U}{}_{11}^2 \qquad \text{(longitudinal wave)} \\
&\bar{\underset{\sim}{u}} = (\bar{u}_1, 0, 0) \quad , \quad \bar{\underset{\sim}{e}} = 0,
\end{aligned}
\qquad (13)
$$

and

$$
\begin{aligned}
&U_{12}^2 = \overset{*}{U}{}_{12}^2 + \omega^2 \frac{\chi_3}{\overset{\circ}{\rho}} (c_1 - c_2)^2 [\rho_1 + \rho_2(c_1 + c_2) - \rho_3 c_1 c_2][\gamma_1 + \gamma_2(c_1 + c_2) - \gamma_3 c_1 c_2], \\
&\bar{\underset{\sim}{u}} = (0, u_2, 0), \qquad \text{(transverse wave)} \\
&\bar{\underset{\sim}{e}} = - \frac{\omega^2}{c^2} U_{12} \chi_3 (c_1 - c_2)[\gamma_1 + \gamma_2(c_1 + c_2) - \gamma_3 c_1 c_2](0, 0, u_2),
\end{aligned}
\qquad (14)
$$

where $\overset{*}{U}_{ij}$ denotes the speed of those waves in the absence of interaction tensors (isotropic case [5]). One notes that only the propagation speed of transverse waves is modified by the electro-mechanical coupling.

5. Propagation of Mechanical Waves in a Principal Plane

In the direction $\underset{\sim}{s} = (s_1, s_2, 0)$ of the principal x_1x_2-plane three mechanical waves may propagate : one is purely transverse with its amplitude $\underset{\sim}{\bar{u}}$ along the x_3-axis, and the two other ones are, in general, polarized elliptically with the amplitude $\underset{\sim}{\bar{u}}$ in the x_1x_2-plane.

In the absence of interaction tensors (isotropic case), the squared speed $\overset{*2}{U_T}$ of the purely transverse wave is given by

$$\overset{*2}{U_T} = s_1^2 \overset{*2}{U}_{13} + s_2^2 \overset{*2}{U}_{23}, \tag{15}$$

while in the presence of the interactions tensors (hemitropic case), the squared speed U_T^2 of this wave is given by

$$U_T^2 = s_1^2 U_{13}^2 + s_2^2 U_{23}^2 - \omega^2 \frac{RG}{1 + \chi_1 s_1^2 + \chi_2 s_2^2} s_1^2 s_2^2, \tag{16}$$

with

$$R = \chi_1(c_2 - c_3)[\rho_1 + \rho_2(c_2 + c_3) - \rho_3 c_2 c_3] + \chi_2(c_3 - c_1)[\rho_1 + \rho_2(c_3 + c_1) - \rho_3 c_1 c_3],$$

$$G = \chi_1(c_2 - c_3)[\gamma_1 + \gamma_2(c_2 + c_3) - \gamma_3 c_2 c_3] + \chi_2(c_3 - c_1)[\gamma_1 + \gamma_2(c_3 + c_1) - \gamma_3 c_1 c_3].$$

The difference between (15) and (16) could be used in order to determine experimentally whether some real materials exhibit the electro-mechanical coupling presented in this paper or not.

References

[1] Toupin, R.A., Int.J.Eng.Sci., 1, 101 (1963)

[2] Smith, G.F., Arch.Rational Mech.Anal., 17, 282 (1965)

[3] Boulanger, Ph., Mayné, G., Hermanne, A., Kestens, J., Van Geen, R. Cahiers du Groupe Français de Rhéologie, 2, n°5 (1971)

[4] Smith, G.F. and Rivlin, R.S., ZAMP, 21, 101 (1970)

[5] Hayes, M.A. and Rivlin, R.S., J.Acoust.Soc.Amer., 46, 610 (1969)

Session II:

NONLINEAR WAVES - SOLITONS

SPACE-TIME COMPLEXITY IN SOLID-STATE AND STATISTICAL PHYSICS MODELS

A. R. Bishop, R. Eykholt, and E. A. Overman II†
Theoretical Division and Center for Nonlinear Studies
Los Alamos National Laboratory
Los Alamos, NM 87545, USA

Abstract

Representative examples of space-time complexity motivated by solid-state and statistical physics models are discussed. The importance of competing interactions in defining classes of space-time attractors is emphasized for partial differential equations and related cellular automata or coupled-map lattices.

1. Introduction.

It has become widely appreciated that soliton-like coherent structures are important excitations and are intrinsic defect patterns in condensed-matter materials, especially in low dimensions. They can play crucial roles structurally, energetically, and in transport (e.g., [1]). By now, the (near) equilibrium statistical mechanics of models containing soliton-like excitations has received a great deal of analytic and simulation attention, particularly in one- and two-dimensional cases, building on the pioneering work of Seeger et al. [2], of Kosterlitz and Thouless [3], and of Krumhansl and Schrieffer [4]. The basic ingredients of mode-mode interactions and phase space sharing in these nonlinear systems has been well incorporated into theoretical descriptions [4,5,6] and directly interpreted in terms of Bethe Ansatz or literal soliton formulations for special "integrable" or "exactly solvable" cases [6]. Although challenging questions remain in calculations of statistical mechanical (as opposed to thermodynamic) properties such as dynamical correlation functions, the basic phenomena are fairly clear, with responses from localized collective (particle-like) structures in addition to extended ("phonon"-like) modes. The current situation is further reviewed elsewhere in these proceedings.

However, most previous studies have focused on linear-response and weak-perturbation regimes. For many problems now coming into focus, it is essential to gain experience with new phenomena which can dominate the physics in nonequilibrium and nonperturbative situations. We focus on some of these issues here. In view of the current interest in dynamical systems [7], one natural question is how strong coherence affects, or is affected by, strong (static or dynamic) perturbations which might promote (spatial or temporal) complexity, including "chaos". Certainly, the persistence of coherent spatial structures in a temporally chaotic environment may be very important to transport in solid state devices, biological materials, etc.

From a solid-state and statistical physics perspective, there are, perhaps, three main areas of involvement with trends in research on dynamical systems and chaos [8]: (i) Intrinsically inhomogeneous ground states are predicted (and experimentally observed) in equilibrium Hamiltonian systems with competing interactions. These are situations of purely structural (i.e. space) complexity, and are found now in many solid state contexts, including surfaces, ferroelectrics, magnets, and charge-density-wave materials [9,10]. Combined with disorder, this is also the essence of the complex metastable states and associated hysteresis in spin glasses and related materials (random-field magnets, charge-density waves with frustrating impurities, etc.). In systems with competing interactions, the basic ground state patterns may be spatially uniform ("commensurate") or incommensurate, with commensurate regions separated by dislocation-like discommensurations. These discommensurations may be regularly spaced, or, if pinning by a discrete lattice or impurities dominate inter-discommensuration interactions, the discommensurations may trap into irregular (metastable) arrays -- these are the analogs of chaotic trajectories in a discrete-time mapping [10]. (ii) Driven, damped arrays of coupled nonlinear oscillators arise almost inevitably in modeling solid-state and condensed-matter systems. These then fit into the growing field of perturbed partial differential equations (pde's) and many coupled ordinary differential equations (ode's) and reveal many patterns ("attractors") in space-time, as described in more detail below. Furthermore, the lattice discreteness characteristic of the solid state can lead to "chaos" in time and space because of the combination of influences from (i) and (ii) [11,12]. As we will describe below, competing interactions do, indeed, appear to be the unifying key to "space-time complexity" very generally, as we have suggested previously [12]. (iii) Finally, solid-state physics may provide interesting new models with which to study the poorly understood concept of "quantum chaos," i.e., quantum and semiclassical behavior in integrable and nonintegrable quantum Hamiltonians. Recent examples may be found in Refs. [13,14]. In particular they have suggested possible connections with the important solid-state concept of Anderson localization found in extrinsically disordered materials [15].

In this article, for purposes of illustration, we focus on the driven and damped sine-Gordon (SG) equation in one and two spatial dimensions ((ii) above). This is representative of many quasi-one- and two-dimensional (1D and 2D) systems that may be modeled in terms of coupled nonlinear oscillators [1], e.g., Josephson junction arrays and transmission lines, magnetic chains and layers, anisotropic charge-density-wave compounds, and epitaxial systems. A wide range of investigations have lead us to appreciate that there are typical ways that complexity in space and time can be manifested, and we will illustrate some of them here. A central feature in the behaviors reported below is a "competition" between space and time, which severly alters responses compared with single oscillators, and leads to interesting space-time intermittency, etc. A synergetic "mapping" to an equilibrium (higher-dimensional) system embeds this competition in the same type of Hamiltonian models that are known to yield defected and spatially complex ground states (as mentioned above) because of static competing interactions. A preliminary discussion of this class of mappings is given in §3, since we believe that these mappings represent the conceptual basis for classifying space-time patterns from most currently disparate approaches.

The equations and phenomena we wish to illustrate are central in disciplines far beyond solid-state physics, so that it is worthwhile setting the emerging field of "space-time complexity" in a wide context -- it clearly represents a cutting edge of dynamical systems research, and an interdisciplinary appreciation (both experimentally and theoretically) is of paramount importance.

In recent years considerable attention has been given to the properties of low-dimensional maps as models for complicated dynamics in higher-dimensional dynamical systems [7]. This attention has been merited by the proof of "universal" properties in classes of one-dimensional maps [7]. However, with few exceptions, the low dimensionality has been introduced explicitly by restricting consideration to models with a very small number of degrees of freedom. On the other hand, equally active research has focused on the subject of spatial pattern selection in nonequilibrium nonlinear systems with many degrees of freedom (e.g., convection cells and reaction-diffusion systems). In these cases, mode-locking is very strong, and a small number of modes can dominate the spatial structure and temporal evolution in a nonlinear pde, or a large system of coupled ode's.

The perspective we emphasize here is that the phenomena of pattern formation, low-dimensional chaos, and coexisting coherence and chaos can be intimately connected in perturbed, dissipative dynamical systems with many degrees of freedom. More specifically, chaotic dynamics may develop by chaotic motions of the collective coordinates identifying the dominant (determining) patterns in the quiescent regimes, or by chaotic motions of a few radiation modes accompanying the coherent structures. In this way, only a small loss of mode-locking is responsible for the temporal chaos coexisting wih spatial coherence -- only a small number of modes become unstable, and the many remaining ones retain strong coherence.

There are many physical examples of this scenario (below), which gives the problem of identifying and testing mode-reduction schemes a general mathematical and physical importance. Physically, it is essential in experiments to gain information on both time and space correlations; chaotic diffusion may be dominated by motion of the coherent pattern; possible consequences for spatial fractal structure of the pattern need to be investigated. It is important to emphasize that the few unstable modes signaling the onset of chaos may be long-wavelength "radiation" or short-wavelength, collective degrees of freedom labeling the nonlinear coherent structures (~ "solitons"). We will show examples of both situations. In either case, the remaining "slave" modes will also be driven to a chaotic evolution, but with quantitative differences. Likewise, in both cases, current, restricted dimension measures (below) can indicate chaos in a low dimension.

The range of physical problems in which "space-time complexity," or "self-organization," or "pattern selection," or "coexisting coherence and chaos" play evidently important, but poorly understood, roles is, indeed, huge. They include clumps and cavitons in turbulent plasmas; filamentation in lasers and laser plasmas; large-scale structures in turbulent fluids (e.g., modon "blocking" patterns influencing atmospheric flow; oceanographic, gulf-stream vortex rings; and even, perhaps, the red spot of Jupiter!); and instabilities of moving interfaces [16] (in phase separation [17], flame fronts [18], etc.). Typically, the influence of the long-lived, coherent structures on transport and predictability in space and time is a primary concern. More recently, there has been a growing experimental concern with precise measurements on laboratory-scale systems drawn from condensed matter, electronics, optics, etc. -- or on simpler systems which model these. Among these systems are included: convection cells [19] and similar fluid-dynamical systems; charge-density-wave materials [20]; spin systems (including spatial and parametric instabilities) [21]; Josephson junctions, arrays, and fluxon oscillators [22]; electron-hole plasmas in semiconductors [23]; incommensurate systems (e.g., ferroelectrics) [24]; bistable optical rings [25]; oscillating water tanks [26]; and acoustic oscillators [27]. We anticipate increasingly precise space-time measurements on an increasing number of such systems. In addition, we can expect the development of detailed connections between space-time dynamical-systems theory and dynamical (as well as purely spatial) behavior in frustrated-disordered materials (including spin glasses, random-field magnets, random incommensurate materials, and disordered charge-density-wave compounds, as well as the conceptually related fields of pattern recognition, neural networks, and adaptation). Low-dimensional, condensed-matter materials (quasi-one- and two-dimensional) have a special role in theoretical studies of space-time complexity in real experiments: they provide real, well-controlled, realizable systems involving many degrees of freedom in crucial ways, and, yet, they do not involve the formidable complexity of fully-developed, <u>three</u>-dimensional turbulence -- they may be viewed as a welcome stepping stone.

From a broad, theoretical perspective, the study of complexity in real materials and classes of pde's is only one approach to the central concern with patterns and complexity in space and time. Complementary attacks include investigations of cellular automata [28] and coupled-map lattices [29], as well as neural-network, pattern-recognition, and adaptation-learning models. It is also important to appreciate the value of state-of-the-art computational facilities to acquire large amounts of space-time data, to visualize that data, and to diagnose it using parallel and array processing, dedicated chips, video feedback, analog-digital machines, interactive color graphics, etc.

Perhaps the most important observation from <u>all</u> the studies of space-time complexity so far is that considerable regimes of <u>typical</u> behavior are becoming apparent. Not only are there clear classes of systems showing the same phenomena within all the approaches mentioned above, but the phenomena (qualities of patterns, chaos, intermittency, etc.) are increasingly common to all of the approaches. We anticipate that rigorous <u>mappings</u> between approaches will become available in the near future, and the possibility of "universal" basins of attraction is tantalizing. For the moment we can mention a few examples:
(i) Classes of mappings have been established between deterministic cellular automata and Ising Hamiltonians with Glauber dynamics, or to <u>equilibrium</u> Hamiltonians in one higher spatial dimension [30]. For purely diffusive dynamics, these mappings provide explicit Hamiltonian principles ("Lyapunov free energy") to discriminate between attractors, and they are equivalent to "supersymmetric" situations in a continuum limit with a definite potential function. The same mappings have been used in stochastic quantization schemes [31], and they have been used to embed <u>dynamical</u> critical behavior in anisotropic, higher-dimensional, <u>equilibrium</u> critical models [32]. An alternative application is to

classify the long-time attractor patterns of perturbed pde's (or discretized versions in space and time). This is potentially extremely important, and §3 is devoted to a more complete introduction. (ii) Similarly, we can anticipate mappings of (classes of) cellular automata and pde's. An exciting recent example is an apparently minimal set of automaton rules which reproduce incompressible Navier-Stokes fluid flow [33]. (iii) Since automata are discretized pde's of various sorts, and well-known dynamical systems maps (single hump, circle, etc.) can be related to classes of ode's (e.g., by suspension theory [34]), it is natural to anticipate relations betwen (classes of) coupled map lattices [29], cellular automata, and pde's; although, this is technically more difficult than for ode's. A result which we expect will gain a central importance is that mappings from nonequilibrium (steady-state) to equilibrium Hamiltonians <u>typically</u> induce <u>competing</u> <u>interactions</u>. Spatial complexity in the presence of such frustrating interactions are now well known [9,10], as alluded to earlier, and provide a natural language with which to unify approaches to space-time complexity, as well as phenomena such as space-time intermittency, which appears as "discommensurations" (see §2). Recently, nonequilibrium experimental systems have elegantly demonstrated this kind of mapping -- e.g., commensurate-incommensurate transitions and discommensuration-lattice melting observed in liquid-crystal convection cells, where roll density can compete with the periodicity of an external electric field [19].

2. Examples of Space-Time Phenomena in Driven Sine-Gordon Systems.

Comprehensive descriptions and numerous examples of our studies of driven pde's (derived from various physical systems) have been given previously [11,12,25,26,27]. As we stress elsewhere in this article, the trend towards identification of typical behaviors in space-time is striking (in classes of pde's <u>and</u> related dynamical systems). To illustrate a few important features we will focus here on examples taken from the driven, damped sine-Gordon equation,

$$\ddot{\psi} - \nabla^2\psi + \sin\psi = F(x,y;t) - \varepsilon\dot{\psi} \quad .$$

Here, ε is a damping constant, and F is a forcing term for the scalar field ψ defined in one (x) or two (x,y) spatial dimension. In §3, we discuss the importance of the dispersive term ($\ddot{\psi}$), space-dependence in F, and boundary conditions in generating non-trivial behavior in space <u>or</u> time. Here we focus on numerical examples with finite ε, periodic boundary conditions, and spatially homogeneous and time-periodic driving. In these cases, the basic phenomenon observed [12,25,26,27] as we increase the driving strength (with other parameters fixed) is that a modulational instability develops, unlocking doubly-degenerate, locked degrees of freedom and saturating in spatially localized, highly coherent and time-dependent structures. Complexity may then develop in either space or time (or both) as we further change control parameters.

2(i). One Dimension with Homogeneous AC Driving [12,25,26].

Here we consider $F(x;t) = \Gamma \sin\omega_d t$ and a one-dimensional line of length $L = 24$ with 120 lattice points. The initial data is a static "pulse" profile -- the actual shape is generally quite unimportant unless $\varepsilon \ll 1$ (i.e., the number of coexisting attractors and hysteresis effects are surprisingly small). There is, generally, the greatest sensitivity to the phase of the driving field [12,26].

For small Γ, the long-time attractor is spatially uniform and simply periodic in time, entrained to the driving frequency ω_d. Above a critical value $\Gamma(\varepsilon,\omega_d,L)$, modulational instability develops and saturates to a single (for $\omega_d < 1$), spatially-localized, "breathing" structure with breathing at the frequency ω_d: the time-dependence remains simply periodic, since the uniform background (or "long-wavelength" phonon) on which the breather rides also responds at the frequency ω_d. Increasing Γ further typically leads to one of two basic scenarios, which are essentially dictated by the amplitude (compared to 2π) of the synchronized breather structure; since this breather is close to the unperturbed SG breather [28], the amplitude is large for $\omega_d \ll 1$ and small for $\omega_d \lesssim 1$. We find the typical sequences:

A: $\omega_d \lesssim 1$, $\varepsilon \ll 1$ [26,29]

Spatially flat; time-periodic → Spatially period-1; time-periodic

→ Spatially period-1; time-quasi-periodic → Spatially period-1; time-chaotic (intermittent) → ...

B: $\omega_d \ll 1$ [12,25]

spatially flat; time-periodic → spatially period-1; time-periodic

→ Spatially period-$\frac{1}{2}$; time-periodic → Spatially period-1 (chaotic KK̄); time-chaotic (intermittent) → ...

Scenario A represents a route to chaos now commonly observed in nonlinear-Schrödinger-like systems perturbed in various ways [30,31,32]. This is not surprising, because the structures synchronized with ω_d close to the natural frequency (unity) are low-amplitude, so that the SG equation is well-approximated by the cubic Schrödinger equation [28]. In this case angular variations on the scale of 2π do not occur and, as described in §3, breathers do not break into kink-antikink ($K\bar{K}$) structures. Rather, increasing Γ modulationally destabilizes a second long-wavelength "phonon" -- specifically the longest wavelength that can be supported by the length L, with a (generally incommensurate) frequency determined by the (discrete) linearized dispersion relationship [29]. This phonon, together with the breather and zero-wavevector, uniform background, typically results in quasi-periodicity and defines a natural mode reduction -- they form a specific, low-dimensional dynamical system which will describe the quasi-periodic regime, the chaotic transition, and the chaotic dynamics. By tuning L or ω_d we can induce high-order "lock-ins" of the two frequencies (i.e., various subharmonic frequencies) [29]; this behavior is similar to that found in abstract low-dimensional maps (e.g., circle maps) derived from two-frequency problems [7].

Typical field configurations for case A are shown in Fig. 1, together with standard Poincaré and power-spectral diagnostics and the time dependence of spatial field averages. Notice that there is little evidence for large-scale spatial chaos in this case. From the spatial averages, we may conclude that the chaos is intermittent, with laminar bursts of the pre-chaotic quasi-periodicity. This conclusion is further supported by estimates of the divergence of nearby trajectories, i.e., the leading Lyapunov exponent, as illustrated in Fig. 2. Fourier analysis of modal content (i.e. Fourier transform of space as well as time) is generally ineffective for detailed studies of modal evolution where nonlinear modes are concerned. However, we have found [26,29] that a nonlinear spectral decomposition of the field is an effective way to assess the nonlinear mode content and to follow it in time. Specifically, we have projected the field onto the "soliton" modes of the unperturbed, periodic SG equation. This has proven successful even in the chaotic regime, and, together with Fourier time-spectral analysis (Fig. 1), is the basis for our assignment of modes (above) in both the pre-chaotic and chaotic regimes. Furthermore, it directly motivates the finite (small) number of determining modes which constitute the low-dimensional, mode-reduced dynamical system for this situation. An example of the nonlinear spectral analysis is shown in Fig. 3. Using this tool we are able to diagnose the dynamics separately for the localized, coherent

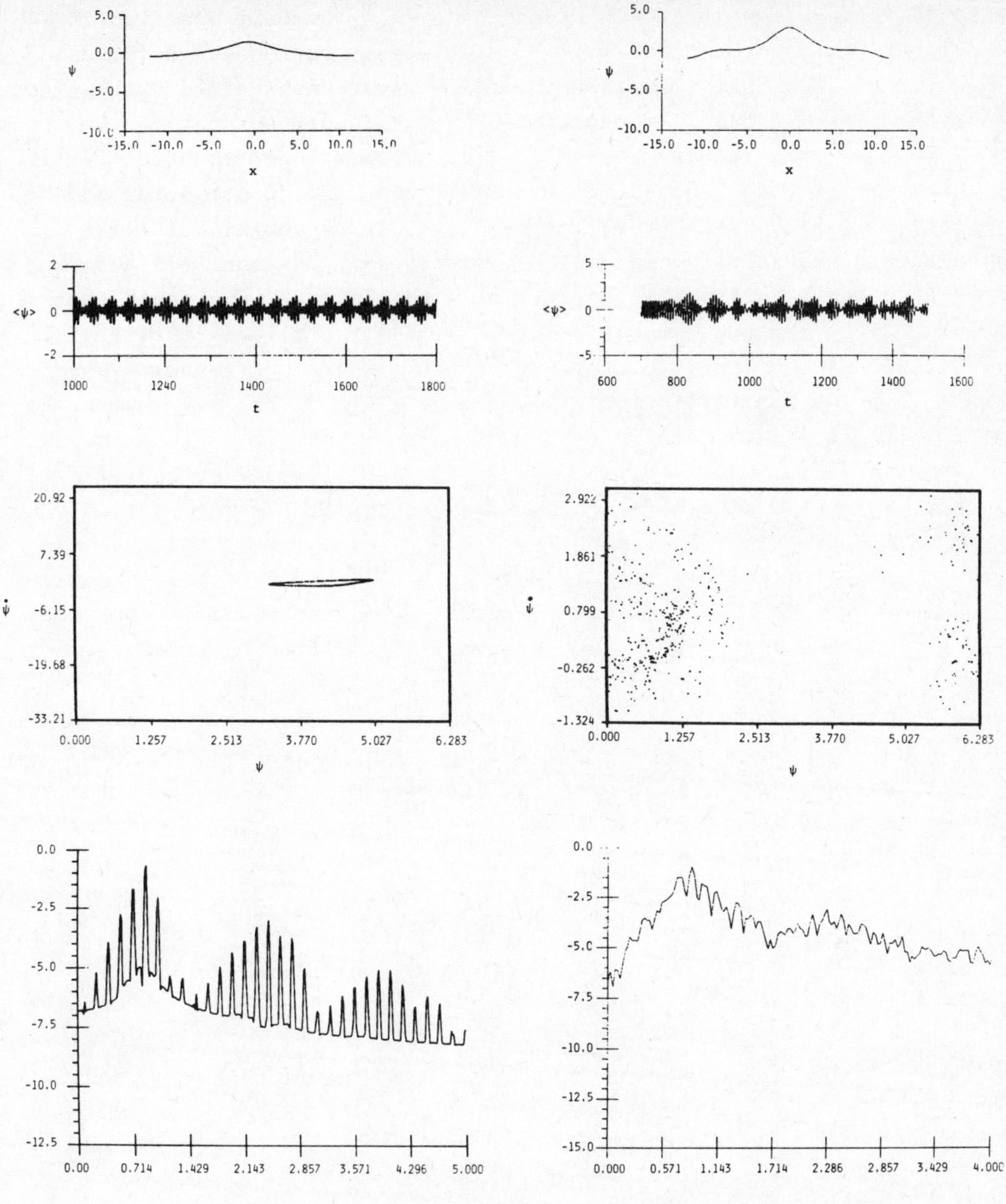

Fig. 1. Attractor diagnostics for the SG chain with $\omega_d = 0.87$, $\varepsilon = 0.04$, and $\Gamma = 0.105$ (left column, quasi-periodic), $\Gamma = 0.107$ (right column, intermittent chaos). From top to bottom, diagnostics are: instantaneous field configuration $\psi(x;t)$; time-dependence of spatially-averaged field; Poincaré section; and power spectrum $S(\omega)$ at center of chain on $\log_{10}$ scale.

(~ "soliton") and extended (~ "radiation" or "phonon") modes, and to assess which are most active in the observed chaos [29].

Scenario B is quite different (although also typical) and has been described in some detail elsewhere [12,25,26]. The most important difference from scenario A is that the synchronized spatial structures are of a large enough amplitude (on the scale of 2π and relative to the spatial average) that the full 2π nonlinearity of the SG field is experienced, and increasing Γ leads to breather break-up into $K\bar{K}$ separating (and coalescing) pairs before the simple modulational instability of case A. Breather break-up is also dynamically important in dc driving [33]. With the present ac driving, before the onset of chaos, it leads to higher spatial symmetry (but simple periodicity in time) with multiple coherent (breathing) structures, e.g., spatial period-$\frac{1}{2}$, as illustrated in Fig. 4(a). In these cases, quasi-periodicity does not occur. Rather, initial transients diverge above a critical $\Gamma(\varepsilon,\omega_d,L)$, leaving a space-time pattern which is chaotic in both space and time (see Fig. 4(b)), but with strong remnants of the pre-chaotic spatial coherence and clear space-time intermittency of a type which is now familiar in chaotic extended systems (pde's [12,25,30,32], cellular automata [28], coupled-map lattices [29], etc.) -- namely, the temporally chaotic bursts are period-1 in space; whereas, the laminar (temporally periodic) regimes are higher symmetry in space.

As is suggested by Fig. 4(b), diffusion in the chaotic region (e.g., the evolution of the spatially-averaged field value) is dominated by

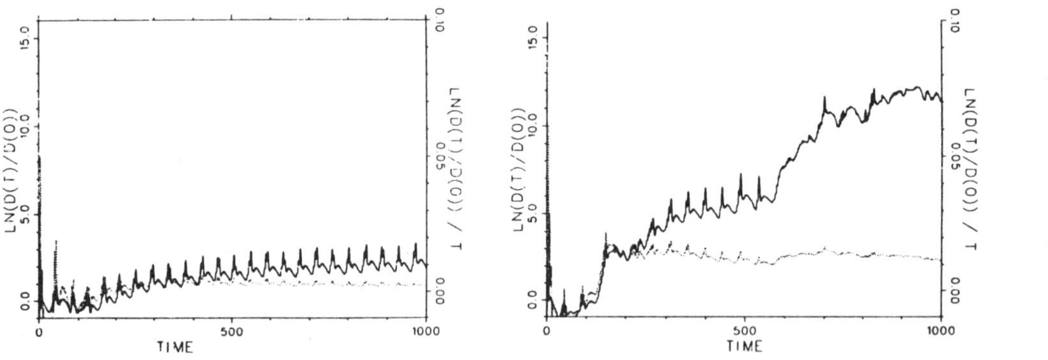

Fig. 2. Lyapunov exponents for the same quasi-periodic and chaotic cases as in Fig. 1. Here, the measure $D \equiv \int dx[(\partial\psi/\partial x)^2 + (\partial\psi/\partial t)^2]$. The solid lines (left scale) indicate the divergence of nearby initial data, and the dotted lines (right scale) give the associated Lyapunov exponent estimate. Notice the detailed structure in time; although, only the asymptotic limit has a rigorous interpretation (Lyapunov exponent).

the motion of the (dislocation-like) kinks and antikinks. However, the diffusion is taking place self-consistently in a nonequilibrium "sea" of extended, phonon-like modes -- both linear and nonlinear spectral analysis [26] suggests that many radiation-like modes accompany the $K\bar{K}$ break-ups and collisions. However, it is important to point out that estimates of the "correlation dimension" (a bound, due to Grassberger and Procaccia [34], on the fractal dimension of the space in which the (strange) attractor lives) are much the same in both cases A and B [12,29]. Typically, the dimension is in the range 2-4; an example is shown in Fig. 5. Although dimension estimates of this sort provide quite limited and crude information in space-time, we may conclude that a severe mode reduction could describe the chaotic regimes in both scenarios A and B. The problem (which is typical of most chaotic

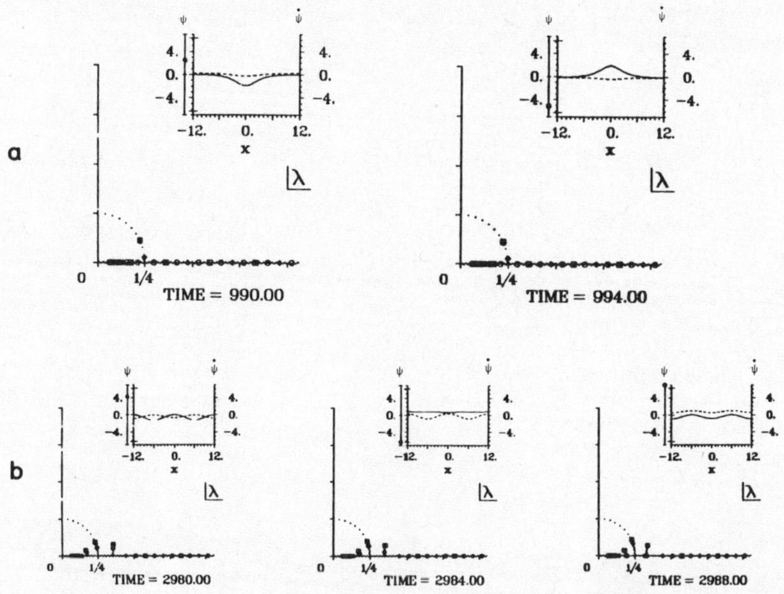

Fig. 3. Transition to chaos, as seen in a nonlinear spectral analysis [26], for the "cubic Schrödinger limit" of the ac-driven SG chain with $\omega_d = 0.9$, $\varepsilon = 0.04$: (a) $\Gamma = 0.05$ (simply periodic); (b) $\Gamma = 0.09$ (intermittent chaos). Instantaneous snapshots are shown of the field profiles ψ and $\dot{\psi}$ (insets), and of the discriminant for the "soliton-transform operator" in the complex eigenvalue plane λ. Details of this transform and the notation used in this figure are given in Ref. [26]. In case (a), this diagnostic confirms the single, low-amplitude, soliton (breather) decomposition of the asymptotic field with no "radiation" (~ "phonon") modes other than the uniform background (k = 0). Both modes evolve periodically. In the chaotic case (b), we observe that the same breather and k = 0 radiation are accompanied by a single (longest-wavelength) additional radiation mode. Further details are given in Ref. [26].

systems) is how to identify the modes in which an efficient mode reduction can be made. This is evidently a subtle matter. Limiting the nonlinear basis to those modes which describe the pre-chaotic locked, periodic states is not adequate to describe the chaotic regime. Examples within scenario A are close enough to the unperturbed system that the nonlinear SG spectral analysis (above) provides an adequate low-dimensional basis [29]; although, even here, both localized and extended modes are essential (Fig. 3). Examples falling into scenario B are far more typical in space-time-complex systems. Here the transition to chaos occurs sufficiently far from the unperturbed limit that spectral decomposition in unperturbed modes is not able to identify an optimal mode reduction basis in which a sufficiently low-dimensional dynamical system can be defined; although, the importance of a small number of localized structures is confirmed [26]. Since alternative, distorted, localized modes which might provide a better basis in the chaotic regime are extremely difficult to obtain, even in simple one-dimensional problems, we are left with a non-optimal basis of localized and (many) extended modes. This basis still recognizes the localized structures which may dominate important physics (e.g. transport), but a dynamical-systems approach is hardly practical. Rather a self-consistent statistical treatment of the approximate (nonlinear) modes is probably most appropriate -- for instance, a renormalization of the collective variables for the localized structures by the sea of extended modes (treated, e.g., in a random-phase approximation). This is, of course, just the kind of statistical physics problem we have learned to face in equilibrium

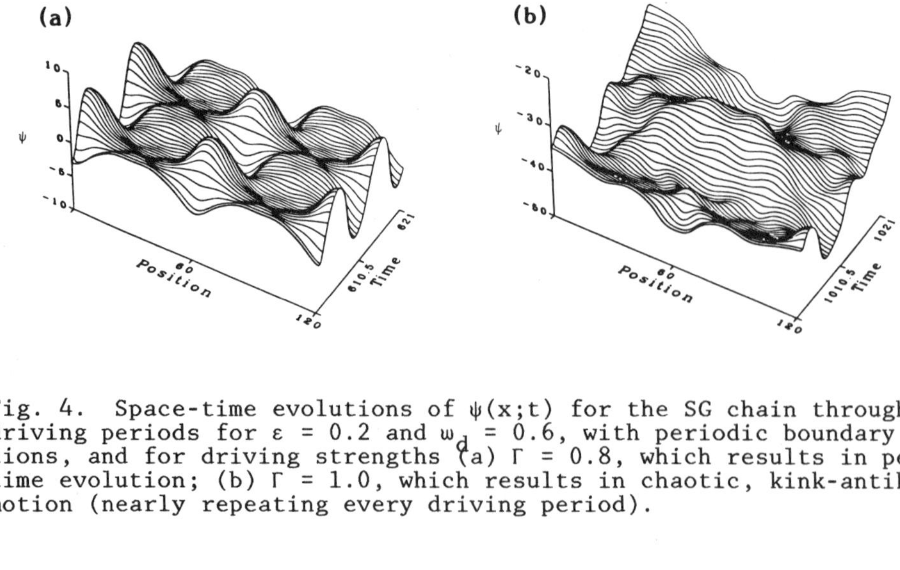

Fig. 4. Space-time evolutions of $\psi(x;t)$ for the SG chain through two driving periods for $\varepsilon = 0.2$ and $\omega_d = 0.6$, with periodic boundary conditions, and for driving strengths (a) $\Gamma = 0.8$, which results in periodic time evolution; (b) $\Gamma = 1.0$, which results in chaotic, kink-antikink motion (nearly repeating every driving period).

nonlinear models (§1). The difference here is that the statistics are nonequilibrium, and, quite probably, there is a strong degree of mode-locking. Various approaches to this fascinating area of nonlinear, nonequilibrium statistical mechanics are being considered and will be reported elsewhere.

2(ii). Two Dimensions with AC Driving [27].

The phenomena of pattern selection and low-dimensional chaos are not restricted to one dimension. Similar behaviors are observed in two dimensions, as well as some qualitatively new effects. To illustrate this, we consider [27] a 10 × 10 square with 161 × 161 particles, periodic boundary conditions, and initial data composed of a spherical pulse plus a random background. $\varepsilon = 0.2$ and $F(x,y;t) = \Gamma \sin \omega_d t$ with $\omega_d = 0.6$ (scenario B in §2(i)). The final attractors, as functions of

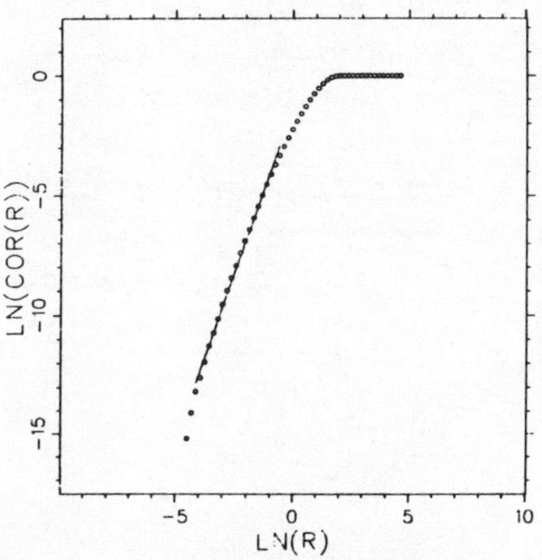

Fig. 5. Correlation dimension (Grassberger-Procaccia scheme [34]) for SG chain with $\omega_d = 0.9$, $\varepsilon = 0.04$, $\Gamma = 0.09$ (intermittent chaos). We have systematically varied the embedding dimension with both time delays and spatial lattice points. In the example shown, four time delays were used at one spatial point, and the fit indicated yields a dimension estimate of 2.8 ± 0.2. (Realistic error bars for the algorithm are somewhat greater.)

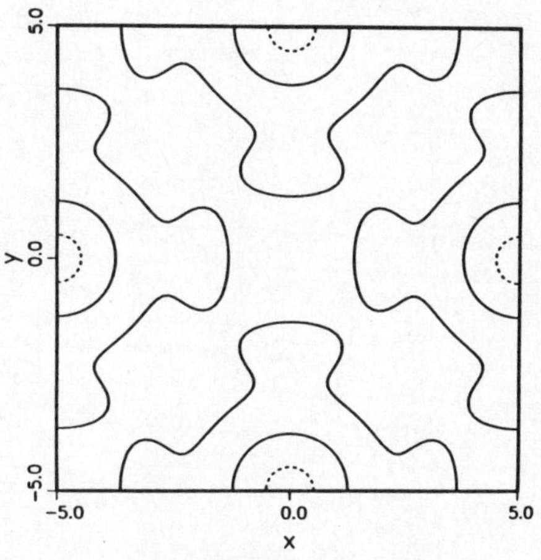

Fig. 6. Homogeneously driven, damped SG system in two space dimensions (a 10 × 10 square) with periodic boundary conditions. Parameters are $\varepsilon = 0.2$ and $\omega_d = 0.06$, and a random pulse profile was used as initial data [37]. Here $\Gamma = 0.9$. The attractor is simply periodic in time, but with a spontaneously higher spatial symmetry -- period - 1/2 on a $\pi/4$-rotated, $\sqrt{2} \times \sqrt{2}$ lattice. (Lines are ψ-intensity contours: solid and dashed lines signify values of ψ above and below the spatial average, respectively, in units of 0.04π.)

Γ, fall into a few simple classes, as in Cases 2(i), and are reported elsewhere [27]. We show illustrative examples here.

Figure 6 shows a typical, non-chaotic, higher-spatial-symmetry, precursor attractor found at relatively low values of Γ. For $1.0 \lesssim \Gamma \lesssim 1.2$, the strong mode-locking of this pattern is broken, and low-dimensional (intermittent) chaos ensues (correlation dimension ~ 2.5). Again, there is strong local coherence in the form of the remnant localized structures, but these flow chaotically, dominate diffusion, and are "dressed" by a sea of extended modes. The higher spatial dimension also admits new classes of phenomena [27]. Most interestingly, for $1.4 \lesssim \Gamma \lesssim 1.9$, after a chaotic initial transient, there is a transition to an attractor where the response is time-periodic, but with a spontaneously

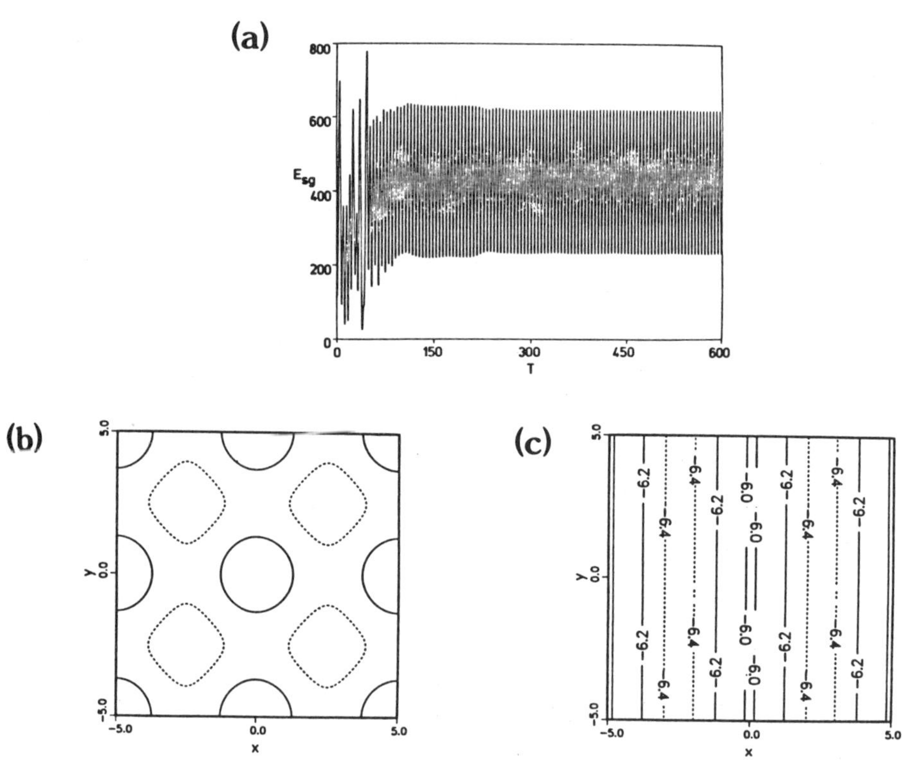

Fig. 7. Same as in Fig. 6, but with Γ = 1.6. After a "chaotic" initial transient, the time evolution is simply periodic and has a spontaneously higher spatial symmetry. However, there is also a pattern conversion from a period-1/2 pattern on a $\sqrt{2} \times \sqrt{2}$ lattice (Fig. 7(a)) to a purely one-dimensional, period-1/2 pattern (Fig. 7(b)). Figure 7(c) exhibits this pattern discrimination in terms of the total SG energy as a function of time -- note that the one-dimensional attractor has a slightly lower average dissipation rate.

higher spatial symmetry (as in Fig. 6). However, this attractor is only metastable, and, after a second transient period, there is another transition to the final, apparently stable, attractor. This is also time-periodic, but variations now occur in only one spatial direction (which depends on the specifics of the initial data), with a spatially period-$\frac{1}{2}$ pattern in that direction. An example of this sequence is shown in Fig. 7. These patterns are excellent examples of the synergetic mappings to competing-interaction, equilibrium Hamiltonian systems: the two- and one-dimensional time-periodic attractors correspond dramatically to interpenetrating and striped discommensuration arrays observed, e.g., in incommensurate, physisorbed, epitaxial surface layers [9].

3. An Example of Mappings.

Consider the ode

$$\ddot{\psi} + \varepsilon\dot{\psi} + U'(\psi) = \zeta \quad , \qquad [1]$$

where $\dot{\psi} = \frac{d\psi}{dt}$, $U'(\psi) = \frac{dU}{d\psi}$, and ζ is a noise term with a distribution

$$P[\zeta] = \left(\frac{\Delta}{2\pi\sigma}\right)^{1/2} \exp(-\frac{\Delta}{2\sigma}\sum_{n=1}^{N}\zeta_n^2) \sim \exp[-\frac{1}{2\sigma}\int_0^t d\tau\, \zeta^2(\tau)]$$

(for the first expression, the time t has been divided into N time steps Δ). The distribution of the function ψ is then given by

$$P[\psi] = P[\zeta]|\det \frac{d\zeta}{d\psi}|, \qquad [2]$$

where $\frac{d\zeta}{d\psi}$ is found by differentiating Eq. [1] with respect to ψ.

In the overdamped limit ($\varepsilon\dot{\psi} + U' = \zeta$), Eq. [2] yields

$$P[\psi] = \left(\frac{\varepsilon^2}{2\pi\sigma\Delta}\right)^{\frac{1}{2}N} \exp(-\frac{\Delta}{2\sigma}\sum_{n=1}^{N}\{[\frac{\varepsilon}{\Delta}(\psi_n-\psi_{n-1}) + U_n']^2\}) \quad , \qquad [3]$$

where $U_n' = U'(\psi_n)$. Letting

$$P(\psi,t) = \int \prod_{n=1}^{N-1} d\psi_n\, P[\psi_0,\psi_1,\ldots,\psi_{N-1},\psi]$$

denote the distribution of ψ at time t given the initial condition ψ_0, the distribution one time step later may be written as

$$P(\psi,t+\Delta) = \left(\frac{\varepsilon^2}{2\pi\sigma\Delta}\right)^{1/2} \int d\eta\, e^{-\frac{\varepsilon^2\eta^2}{2\sigma\Delta}} P(\psi+\eta,t) \exp[\frac{\varepsilon\eta}{\sigma} U'(\psi+\tfrac{1}{2}\eta)$$

$$- \frac{\Delta}{2\sigma} U'^2(\psi + \tfrac{1}{2}\eta) + \frac{\Delta}{2\varepsilon} U''(\psi + \tfrac{1}{2}\eta)] \quad .$$

Expanding this expression in Δ and η finally yields a differential equation for $P(\psi,t)$,

$$\dot{P} = \frac{\sigma}{2\varepsilon^2} P'' + \frac{1}{\varepsilon}U'P' + \frac{1}{\varepsilon}U''P \quad . \qquad [4]$$

Since the equilibrium solution must be $P_0 \sim e^{-\beta U}$, then we obtain the relation $\sigma = \frac{2\varepsilon}{\beta}$, which allows Eq. [3] to be rewritten as $P[\psi] \sim e^{-\beta H}$, where H is the effective Hamiltonian

$$H = \tfrac{1}{2}[U]_0^t + \frac{1}{4\varepsilon} \int_0^t d\tau(\varepsilon^2\dot{\psi}^2 + U'^2 - \tfrac{2}{\beta}U'') \quad , \qquad [5]$$

with $[\,]_0^t$ denoting evaluation at times 0 and t. Furthermore, the solutions of Eq. [4] have the form

$$P(\psi,t) = e^{-\frac{\lambda}{\beta\varepsilon}t}\, e^{-\frac{1}{2}\beta U(\psi)} \rho(\psi) \quad ,$$

with $\rho(\psi)$ satisfying the Schrödinger equation $-\rho'' + V\rho = \lambda\rho$, where the effective potential

$$V = \tfrac{1}{4}\beta^2(U'^2 - \tfrac{2}{\beta}U'')$$

is the same as that appearing in the effective Hamiltonian, Eq. [5]. Thus, in the overdamped limit, the dynamics of Eq. [1] may be studied by examining the statics of this effective Hamiltonian. In particular, the most probable solution ψ will be that which minimizes this Hamiltonian.

In the underdamped case (i.e., the full Eq. [1]), the Jacobian in Eq. [2] is very difficult to evaluate. However, this problem is easily overcome by converting Eq. [1] to two first order equations in ψ and $\phi = \dot{\psi}$. More precisely, we replace Eq. [1] by

$$\dot{\phi} + \varepsilon\phi + U'(\psi) = \zeta \quad ,$$

$$\dot\psi - \phi = \xi \quad ,$$

where ξ is a new (artificial) noise term with the distribution $P[\xi] = \prod_{n=1}^{N} \delta(\xi_n) \sim \delta(\xi)$. $P[\psi]$ is then given by

$$P[\psi] = \int D\phi P[\phi,\psi] \quad ,$$

$$P[\phi,\psi] = P[\zeta]P[\xi]\left|\det \frac{\partial(\zeta,\xi)}{\partial(\phi,\psi)}\right| \quad .$$

This now yields $P[\psi] \sim e^{-\beta H}$ with the new effective Hamiltonian

$$H = \tfrac{1}{2}[\tfrac{1}{2}\dot\psi^2 + U + \tfrac{1}{\varepsilon}\dot\psi U']_0^t + \frac{1}{4\varepsilon}\int_0^t d\tau[\ddot\psi^2 + \dot\psi^2(\varepsilon^2 - 2U'') + U'^2] \quad .$$

However, it is no longer possible to develop a Schrödinger equation for $P(\psi,t)$. Instead, we must be content with a differential equation for the joint distribution $P(\phi=\dot\psi,\psi,t)$,

$$P(\phi,\psi,t) = e^{-\frac{\varepsilon}{\beta}(\lambda-\frac{1}{2}\beta)t}\, e^{-\frac{1}{2}\beta(\frac{1}{2}\phi^2+U+\frac{1}{\varepsilon}\phi U')}\, \rho(\phi,\psi) \quad ,$$

$$-\frac{\partial^2\rho}{\partial\phi^2} + \frac{\beta\phi}{\varepsilon}\frac{\partial\rho}{\partial\psi} + \frac{\beta^2}{4\varepsilon^2}[U'^2 + \phi^2(\varepsilon^2 - 2U'')]\rho = \lambda\rho \quad .$$

Using these results, we can find the effective Hamiltonians for the driven damped pendulum $\ddot\psi + \varepsilon\dot\psi + \sin\psi - F(t) = \zeta$. In the overdamped and underdamped cases, we have, respectively,

$$H_P^{(o)} = -\tfrac{1}{2}[\cos\psi + F\psi]_0^t + \frac{1}{4\varepsilon}\int_0^t d\tau(\varepsilon^2\dot\psi^2 + \sin^2\psi - 2F\sin\psi - \tfrac{2}{\beta}\cos\psi),$$

$$H_P^{(u)} = \tfrac{1}{2}[\tfrac{1}{2}\dot\psi^2 + \tfrac{1}{\varepsilon}\dot\psi\sin\psi - \tfrac{1}{\varepsilon}F\dot\psi - \cos\psi - F\psi]_0^t$$

$$+ \frac{1}{4\varepsilon}\int_0^t d\tau(\ddot\psi^2 + \varepsilon^2\dot\psi^2 - 2\dot\psi^2\cos\psi + \sin^2\psi - 2F\sin\psi).$$

For sinusoidal driving $F(t) = \Gamma\sin\omega_d t$, with ω_d less than, but on the order of, the natural frequency $\omega_0 = 1$ of the pendulum, the term $-2F\sin\psi$ tends to lock the pendulum to the driver in each case, resulting in periodic motion of frequency ω_d. In the overdamped limit, the remaining terms in the Hamiltonian density (we will ignore the boundary terms) tend to damp the motion (since they favor small ψ and $\dot{\psi}$), so that all we see are damped periodic oscillations. However, in the underdamped case, the term $-2\dot{\psi}^2\cos\psi$ favors increased complexity. Since the term $-2F\sin\psi$ keeps the oscillations locked at the frequency ω_d, then the easiest way to increase the complexity (as the driver strength Γ is increased) is to add new frequencies, so that this system follows the quasi-periodicity route to chaos.

We will now extend this to a one-dimensional system of coupled pendula (i.e., a sine-Gordon system),

$$\ddot{\psi} + \varepsilon\dot{\psi} + \psi'' + \sin\psi - F(x;t) = \zeta \quad , \tag{6}$$

$$P[\zeta] = \left(\frac{a\Delta}{2\pi\sigma}\right)^{\frac{1}{2}MN} \exp\left(-\frac{a\Delta}{2\sigma}\sum_{m=1}^{M}\sum_{n=1}^{N}\zeta_{mn}^2\right)$$

$$\sim \exp\left[-\frac{1}{2\sigma}\int_0^t d\tau \int_0^L dx\, \zeta^2(x,\tau)\right],$$

where $\dot{\psi} = \frac{\partial\psi}{\partial t}$ and $\psi' = \frac{\partial\psi}{\partial x}$. As before, in the first expression, we have divided the time t into N time steps Δ and the length L of the system into M steps a, and, as before, we intend to let Δ, $a \to 0$ at the end of the computation. However, this procedure is not well-defined, since it is necessary to know what happens to the ratio $\frac{\Delta}{a}$. Since the true physical system is on a lattice with a fixed interparticle spacing, then the appropriate procedure is to let $\Delta \to 0$ first (at fixed a), and then to let $a \to 0$ (i.e., $\frac{\Delta}{a} \to 0$). Therefore, we will put Eq. [6] on a lattice, yielding

$$\ddot{\psi}_m + \varepsilon\dot{\psi}_m + \frac{\partial U}{\partial\psi_m} = \zeta_m \quad ,$$

$$U = \sum_{m=1}^{M}\left[\frac{1}{a^2}\psi_m(\psi_m - \psi_{m-1}) - \cos\psi_m - F_m\psi_m\right] \quad .$$

Since this has the same form as Eq. [1] (with the scalar function ψ having been replaced by an M-component vector function), then we may proceed as before to yield the effective Hamiltonian for the overdamped and underdamped cases, respectively,

$$H_{SG}^{(o)} = \frac{1}{4a^2}\{[\psi^2]_0^t\}_0^L - \frac{1}{2a}\int_0^L dx[\cos\psi + F\psi]_0^t$$

$$- \frac{1}{2\varepsilon a} \int_0^t d\tau [\psi' \sin\psi - F\psi' + F'\psi]_0^L$$

$$+ \frac{1}{4\varepsilon a} \int_0^t d\tau \int_0^L dx (\varepsilon^2 \dot\psi^2 + \psi''^2 + 2\psi'^2 \cos\psi + \sin^2\psi - \frac{2}{\beta}\cos\psi$$

$$- 2F \sin\psi + 2F''\psi) \quad ,$$

$$H_{SG}^{(u)} = \frac{1}{2a}\{[\frac{1}{2a}\psi^2 - \frac{1}{\varepsilon}\dot\psi\psi']_0^t\}_0^L + \frac{1}{2a}\int_0^L dx [\frac{1}{2}\dot\psi^2 + \frac{1}{\varepsilon}\dot\psi'\psi'$$

$$+ \frac{1}{\varepsilon}\dot\psi \sin\psi - \frac{1}{\varepsilon}F\dot\psi - \cos\psi - F\psi]$$

$$+ \frac{1}{2\varepsilon a}\int_0^t d\tau [\dot\psi'\dot\psi - \psi'\sin\psi + F\psi' - F'\psi]_0^L + \frac{1}{4\varepsilon a}\int_0^t d\tau \int_0^L dx (\ddot\psi^2 + \psi''^2$$

$$- 2\dot\psi'^2 + \varepsilon^2\dot\psi^2 - 2\dot\psi^2 \cos\psi + 2\psi'^2 \cos\psi + \sin^2\psi - 2F\sin\psi + 2F''\psi) \quad .$$

For periodic boundary conditions, the first and third sets of terms vanish in each case.

We will again examine the response of the system to sinusoidal driving $F(x;t) = \Gamma \sin \omega_d t$ with $\omega_d \lesssim \omega_0 = 1$, and we will use periodic boundary conditions (and we will again concentrate on the Hamiltonian density). In the overdamped limit, the pendula lock to the driver, as in the single-pendulum system. The additional terms $\psi''^2 + 2\psi'^2 \cos\psi$ discourage inhomogeneity, resulting in period-1 oscillation, the simplest case allowed by the boundary conditions (the term $2F''\psi$ vanishes for homogeneous driving). Thus, in the overdamped limit, we again see damped, fairly homogeneous oscillation. The only exception occurs for large amplitude oscillation (which can occur for strong driving Γ, or low driving frequency ω_d). When the amplitude becomes large enough for $\cos\psi$ to become negative for appreciable amounts of time, then the term $2\psi'^2 \cos\psi$ begins to allow some inhomogeneity. However, because of the term ψ''^2, this inhomogeneity remains slowly varying.

In the underdamped case, the additional terms $\psi''^2 + 2\psi'^2 \cos\psi$ again discourage inhomogeneity and result in period-1 oscillation (and the term $2F''\psi$ again vanishes); although, the additional term $-2\dot\psi'^2$ now allows a little more inhomogeneity. Thus, as in the overdamped case, the typical motion is to act like a single pendulum with only a slight amount of inhomogeneity, so that the system again follows the quasi-periodicity route to chaos with a period-1 spatial structure.

However, for low frequencies, we can again get large amplitudes, and $\cos\psi$ can again become negative for appreciable amounts of time. In this case, the terms $2\psi'^2 \cos\psi - 2\dot{\psi}'^2$ favor inhomogeneity, while the term $-2\dot{\psi}^2 \cos\psi$ no longer favors increasing the complexity in time. Thus, rather than becoming quasi-periodic in time, the system increases its inhomogeneity. However, the term ψ''^2 still encourages this inhomogeneity to be slowly varying, so that the spatial frequency simply doubles (this is the simplest increase in spatial complexity which is compatible with the periodic boundary conditions). As the driver strength Γ is increased, though, the system must eventually become chaotic, even though it does not do so via quasi-periodicity.

This latter scenario will be encouraged by increasing the inhomogeneity of the system. One way of achieving this is by inhomogeneous driving $F(x;t) = \Gamma(x) \sin \omega_d t$, in which case, the pendula will attempt to lock to the inhomogeneity of the driver (however, the terms $\psi''^2 + \psi'^2 \cos\psi$ will still try to reduce this inhomogeneity for small amplitudes). Another possibility is to drop the periodic boundary conditions. In addition to the fact that the oscillations will no longer be locked into spatially periodic patterns, there are several boundary terms which favor inhomogeneity (even in the overdamped case). The effects of such increased inhomogeneity are currently under investigation.

4. Discussion.

In conclusion, we have emphasized the sense of unity and typicality which is now emerging in studies of space-time complexity (or pattern formation, or coherence and chaos). In view of this and the clear importance of these questions throughout the natural sciences (see §1), it seems fair to claim that elucidating patterns in space-time complexity is the primary direction for dynamical-systems research -- building on the beautiful results and expectations generated from the recent studies of abstract, low-dimensional maps [7]. This invites contributions from many perspectives: dynamical systems, perturbation theory, modulational-instability theory, nonlinear mode-reduction, physical experiments, large-scale numerical simulation, etc. Despite the difference between examples drawn from contexts dominated by diffusion, reaction-diffusion, dispersion, etc., it is essential that the various theoretical and experimental communities stay in close contact.

We have summarized the variety of analytic tools being applied to these problems elsewhere [12]. Likewise, the variety of models and systems studied can be found in the reference list. Here we will merely re-emphasize three points: (i) Well-directed measurements on well-characterized, condensed-matter systems have a major role to play. Happily, such systems and measurements are now expanding rapidly (§1). (ii) The idea of an explicit, nonlinear mode reduction from which a low-dimensional dynamical system can be constructed is tempting; however, carrying through this procedure explicitly will be very difficult in most cases, as was discussed in §2. This is despite elegant and important proofs of bounds on the number of determining modes and the fractal dimension in large classes of equations [18] -- in practice we cannot identify a sufficiently good modal set, and the number of (approximate) modes involved suggests that a statistical approach will be the most plausible. In particular, we typically need to treat both localized-coherent and extended modes self-consistently and with nonequilibrium statistics (§2). Mappings (cf. §3) betwen pde's, cellular automata, and coupled-map lattices are important in understanding classes of space-time complexity. In particular, we expect that the concept of competing interactions will play an increasingly central role, connecting the nonequilibrium problems with spatially-complex systems, which are now somewhat familiar in solid-state physics [9,10].

We are grateful to many past and present colleagues for their advice, particularly K. Fesser, P. S. Lomdahl, and D. W. McLaughlin.

† Permanent Address

Department of Mathematics and Statistics
University of Pittsburgh
Pittsburgh, PA 15261, U.S.

References

[1] For example, Physics in One Dimension, eds. J. Bernasconi and T. Schneider (Springer-Verlag, 1981).

[2] A. Seeger, proceedings of this conference.

[3] J. M. Kosterlitz and D. J. Thouless, Prog. Low Temp. Phys., Vol. VII B, ed. D. F. Brewer (North-Holland, 1978).

[4] J. F. Currie et al., Phys. Rev. B 22, 477 (1980); A. R. Bishop, in Ref. [1].

[5] N. Theodorakopoulos, Z. Physik B 46, 367 (1982).

[6] See articles in Solitons, eds. S. Trullinger and V. Zakharov (North-Holland, in press).

[7] See, for example, Physica D 7 (1983), eds. D. K. Campbell and H. A. Rose.

[8] A. R. Bishop, Proceedings of the Ninth Gwatt Workshop, Gwatt, Switzerland, October 17-19 (1985).

[9] For example, P. Bak, Rep. Prog. Phys. 45, 587 (1982).

[10] S. Aubry, in Ref. [7].

[11] J. Oitmaa and A. R. Bishop, in preparation.

[12] A. R. Bishop, in Dynamical Problems in Soliton Systems, Proceedings of the Seventh Kyoto Summer Institute, ed. S. Takeno (Springer-Verlag, 1985); A. R. Bishop and P. S. Lomdahl, Physica D, in press.

[13] D. R. Grempel, R. E. Prange, and S. Fishman, Phys. Rev. A 29, 1639 (1984).

[14] K. Nakamura, K. Nakahara, and A. R. Bishop, Phys. Rev. Lett. 54, 861 (1985).

[15] E. Abrahams et al., Phys. Rev. Lett. 42, 673 (1979).

[16] Physica D 12 (1984), eds. A. R. Bishop, L. J. Campbell, and P. J. Channell.

[17] For example, E. Ben-Jacob et al., Phys. Rev. A 29, 330 (1984); M. D. Kruskal, C. Oberman, and H. Segur, preprint (1985).

[18] G. I. Sivashinsky, Acta Astronaut. 4, 1177 (1977); J. M. Hyman and B. Nicolaenko, preprint (1985).

[19] M. Lowe and J. P. Gollub, Phys. Rev. A 31, 3893 (1985); P. Coullet, preprint (1985).

[20] For example, R. P. Hall, M. Sherwin, and A. Zettl, Phys. Rev. B 29, 7076 (1984).

[21] For example, G. Gibson and C. Jeffries, Phys. Rev. A 29, 811 (1984); L. P. Levy, Phys. Rev. B 31, 7077 (1985).

[22] For example, M. P. Soerensen et al., Phys. Rev. Lett. 51, 1919 (1983); P. Martinoli et al., preprint (1985).

[23] G. A. Held and C. Jeffries, Phys. Rev. Lett. 55, 887 (1985).

[24] For example, R. Blinc et al., Phys. Rev. B 29, 1508 (1984).

[25] For example, D. W. McLaughlin, J. V. Moloney, and A. C. Newell, Phys. Rev. Lett. 51, 75 (1983).

[26] J. Wu and I. Rudnick, Phys. Rev. Lett. 55, 204 (1985).

[27] J. C. Wheatley et al., preprint (1985); W. Lauterborn, J. Acoust. Soc. Am. 59, 283 (1976).

[28] See Cellular Automata, eds. J. D. Farmer, T. Toffoli, and S. Wolfram (North-Holland, 1984).

[29] For example, K. Kaneko, Prog. Theor. Phys. 72, 480 (1984).

[30] N. G. van Kampen, J. Stat. Phys. 17, 71 (1977); E. Domany, Phys. Rev. Lett. 52, 871 (1984); G. Grinstein et al., preprints (1985); E. Jen, preprint (1985).

[31] For example, T. Schneider, M. Zannetti, and R. Badii, Phys. Rev. B 31, 2941 (1985).

[32] For example, E. Domany and J. E. Gubernatis, Phys. Rev. B 32, 3354 (1985).

[33] U. Frisch, B. Hasslacher, and Y. Pomeau, preprint (1985).

[34] G. Mayer-Kress and H. Haken, preprint (1984).

[35] D. Bennett, A. R. Bishop, and S. E. Trullinger, Z. Physik B 47, 265 (1982); A. R. Bishop et al., Phys. Rev. Lett. 50, 1095 (1983).

[36] E. A. Overman, D. W. McLaughlin, and A. R. Bishop, Physica D, in press; also see K. Fesser et al., Phys. Rev. A 31, 2728 (1985).

[37] O. H. Olsen et al., J. Phys. C 18, L511 (1985).

[38] R. K. Dodd et al., Solitons and Nonlinear Wave Equations (Academic Press, 1982).

[39] A. R. Bishop et al., in preparation.

[40] For example, N. Bekki and K. Nozaki, in Ref. [12].

[41] H. T. Moon and M. V. Goldman, Phys. Rev. Lett. 53, 1821 (1984); G. Pelletier et al., Physica D, in press.

[42] B. Horovitz, et al., preprint (1985); S. Aubry et al., in preparation.

[43] P. Grassberger and I. Procaccia, Phys. Rev. Lett. 50, 346 (1983); also see articles in Dimensions and Entropies in Chaotic Systems: Quantification of Complex Behavior, ed. G. Mayer-Kress (Springer Series in Synergetics, in press).

THE TOPOLOGY OF SEMIDEFECTS AND SOLITONS

H.-R. Trebin
Institut für Theoretische und Angewandte Physik, Universität Stuttgart, Pfaffenwaldring 57, D-7000 Stuttgart 80, W. Germany

Abstract

The concept of semidefects is introduced and topological tools are offered to describe interactions of semidefects, full defects, and topological solitons.

1. Introduction

Ten years ago, during the first Symposium on Trends in Applications of Pure Mathematics to Mechanics, Prof. Dominik Rogula presented a paper entitled "Large deformations of crystals, homotopy, and defects".[1] It contained one of the first applications of homotopy theory to the classification of defects in ordered media. In the years following, the method was used with much success to analyse defect structures in the superfluid helium phases, in liquid crystals and - with some dispute - in solid crystals. New topology-induced phenomena were discovered, like catalysis of defects by line singularities, or topological obstructions to the crossing of defect lines. Method and applications are described in a series of reviews[2-4].

Homotopy theory also allowed to determine the transformation properties of defects in phase transitions. Mostly the order parameter is decorated by additional degrees of freedom at the transition point. A standard example is provided by nematic liquid crystals. In the uniaxial nematic phase, the order parameter is a line segment (director), characterizing the axis of alignment of the constituent elongated molecules. In the transition to the biaxial nematic phase, a second axis of preferred orientation arises, orthogonal to the first one. The resulting order parameter is depicted as cross with bars of different lengths, and the symmetry of the liquid crystal is lowered from $D_{\infty h}$ to D_{2h}. Homotopy theory relates the defects of both phases[4,5]. Out of these considerations of transformation properties of defects the notion of semidefects has grown. In the phase transition from a uniformly aligned uniaxial nematic phase to the biaxial nematic phase only singularities in the secondary axes can show up (fig. 1a). Defects of this kind, whe-

a) b)

Fig. 1 a) Semidefect in a nematic liquid crystal
b) Semidefect (dislocation) in a smectic-A liquid crystal

re only part of the order parameter is singular, are denoted semidefects. An example for a semidefect in a smectic-A liquid crystal is the dislocation depicted in fig. 1b: the long-range order of the molecules' axes is unperturbed, whereas the positional order of the molecules' centers of mass is defected.

The intention of this article is to demonstrate that semidefects are structures which display a rich variety of properties differing from those of ordinary ("full") defects. In particular, they are very similar to topological solitons and also strongly interact with them.

In section 2, some preliminaries from the homotopic defect classification are listed. In section 3, the notion of semidefects is deepened and their relation to full singularities is illuminated. In section 4, the topological tools are applied to describe several interaction processes between semidefects and topological solitons.

2. Topologically stable defects and solitons

Point singularities in two-dimensional space, and line singularities in three-dimensional space are - in essence - labeled by the elements of the fundamental group $\pi_1(V)$ of the (reduced) order parameter space V (or space of degeneracy V). Point singularitites in three-space are labeled by the elements of the second homotopy group $\pi_2(V)$. Planar topological solitons, like Bloch or Neél walls, can be interpreted as unfolded point singularities (in two dimensions) or unfolded line singularities (in three dimensions), and hence the classifying group is $\pi_1(V)$ (fig. 2). Linear topological solitons are unfolded point singularities and are characterized by the elements of $\pi_2(V)$. The labels frequently are denoted "topological charges". In general they are not

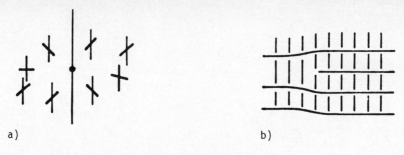

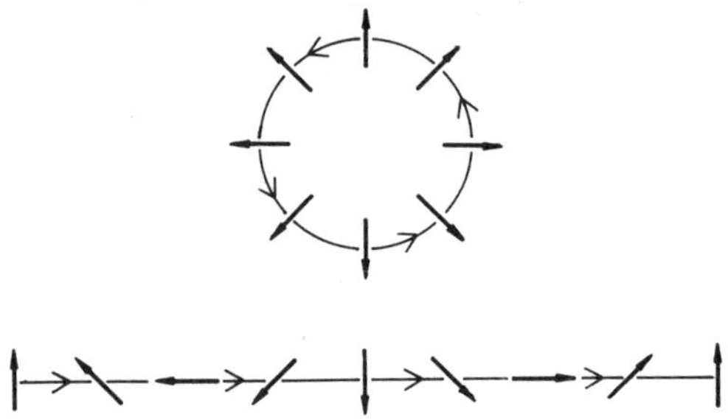

Fig. 2 When encircling a point singularity one meets the same sequence of order parameters as across a planar topological soliton . Such a soliton can therefore be regarded as an unfolded point singularity (in two dimensions).

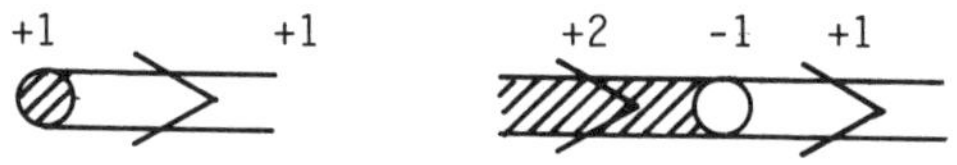

Fig. 3 Point singularities add topological flux to linear solitons or absorb flux from them.

integers, but "matrix charges". A linear topological soliton, if furnished with an orientation, is a tube of "topological flux". A point singularity can absorb flux from the tube or add flux to it (fig. 3). The topological charge of an interfacial point singularity is $\alpha\beta^{-1} \in \pi_2(V)$, where α is the group element for the flux flowing in, β that for the flux flowing out. We will see that similar rules hold for semidefect lines.

3. Semidefects

3.1 Definition and further examples

Semidefects exist in media, whose order parameter can be divided into two (or more) coupled components: into a "rigid" component, which dislikes forming singularities due to energy reasons, and into a "soft" component. A semidefect is the region, where the rigid component forms a continuous field, while the soft component forms a singularity. A field in the rigid component is called partial order, a field in the complete order parameter, consisting of coupled rigid and soft component, is called full order. In case of nematic liquid crystals, the coupling is due to the orthogonality condition of main axis (rigid component) and secondary axis (soft component). In case of the smectic-A liquid crystals the directors (rigid component) stand perpendicular to the layers (soft component). The double-layer system of the smectic-A_2 phase exhibits semidefect walls (fig. 4a), if we view the mass density wave of the layers as partial order, the decoration of the layers as soft component of the order parameter. In the crystal of fig. 4b a semidefect wall is visible, where the superlattice is stacked faultily, whereas the sublattice is regular. The rigid component of the order parameter is the sublattice, the soft component is the superlattice, and both are coupled by a rational relation of the lattice constants.

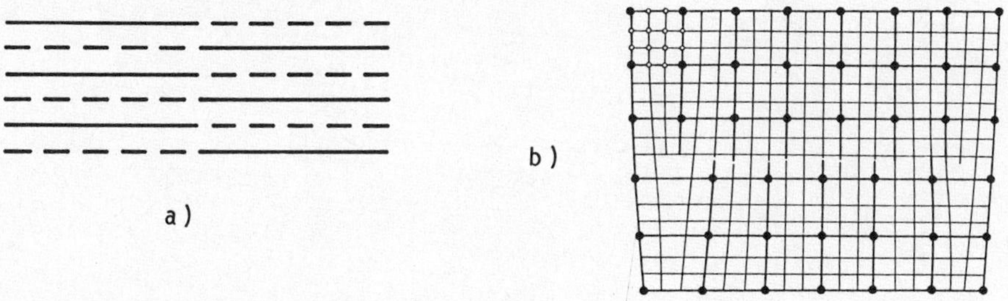

Fig. 4 a) Semidefect wall in the double-layer system of a smectic-A_2 liquid crystal.

b) Semidefect wall in a crystal composed of sublattice and superlattice. The wall terminates at line singularities in the full order, which in this example are dislocations of the sublattice.

3.2 Properties

The example of the superlattice (fig. 4b) exposes an important property of semidefects: they can terminate in the bulk, if they are bounded by a singularity in the full order. Here the bounding singularity is a dislocation of the sublattice (a partial dislocation of the superlattice). With regard to this property semidefects resemble linear topological solitons whose flux is absorbed or emitted by a point charge.

The semidefects in nematic liquid crystals (fig. 1a) are labeled by half-integer winding numbers. Given a source singularity ("hedgehog") of the uniaxial nematic phase, then the secondary axes of the biaxial phase form a tangential line field on each sphere centered on the singular point. Such a field must have singularities of total winding number n = 2. If these singularities condense into one on each sphere, the source emits an n = 2 semidefect line (fig. 5a). If the winding number is shared by two singularities, the source is an interface between two semidefect lines of winding numbers n = - 1 and n = +1 (fig. 5b).

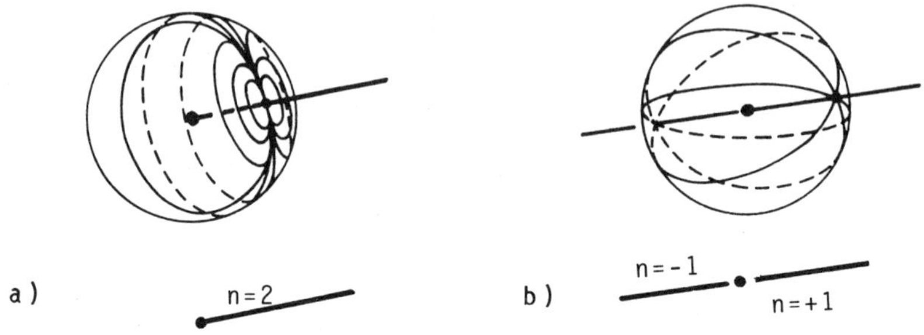

Fig. 5 A source singularity (topological charge 1) of a uniaxial nematic liquid crystal
a) emits a semidefect line of winding number n = 2
b) forms an interface between semidefect lines of winding numbers n = -1 and n = +1

A topologically stable semidefect can relax to the nondefect, but only, if in the course of the decay also the rigid component of the order parameter becomes singular. This condition lends stability to the semidefect.

3.3 Topological description: exact sequence

In the topological description of semidefects three order parameter spaces are involved: the space B of the complete order parameter, the space U of the rigid component, and the space H of the soft component. For nematic liquid crystals, $B = O(3)/D_{2h}$ is the set of positions of a cross in space, $U = O(3)/D_{\infty h} = P^2$ the set of positions of a line (the projective plane P^2), and $H = D_{\infty h}/D_{2h} = P^1$ (the projective line) is the set of positions of the secondary axis, when the main axis is fixed. (For the representation of order parameter spaces as coset spaces see the reviews[2-4]). In n-dimensional space semidefects, defects in the partial order and defects in the full order of dimension d are classified by the elements of the homotopy groups $\pi_r(H)$, $\pi_r(U)$, and $\pi_r(B)$, respectively, $r = n-d-1$.

The three order parameter spaces can be incorporated into a single mathematical structure, viz. a fiber bundle. B is the bundle space, U the base space, and H the fiber. According to Steenrod[6], the homotopy groups of the three spaces are related by a sequence of group homomorphisms, out of which we are going to interpret the following section:

$$\ldots \to \pi_2(B) \overset{j_2}{\to} \pi_2(U) \overset{\partial_2}{\to} \pi_1(H) \overset{i_1}{\to} \pi_1(B) \overset{j_1}{\to} \pi_1(U) \to \ldots \qquad (1)$$

The sequence is exact, which means, that the kernel of each homomorphism equals the image of the preceding homomorphism (fig. 6).

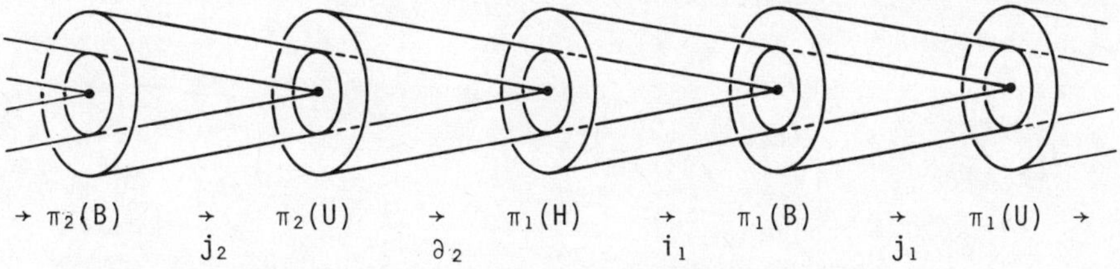

Fig. 6 In this illustration of an exact sequence of group homomorphisms (taken from ref. 3) each disk represents a group. The center is the group identity. Each inscribed disk marks the kernel of the following homomorphism and the image of the one preceding.

i) Homomorphism j_2 tells the fate of a point singularity in the complete order, if in a phase transition the soft component of the order parameter vanishes, yielding a field in the partial order. Defects in $\ker j_2$ become unstable singularities of the partial order, defects outside $\ker j_2$ turn into their image.

ii) Homomorphism ∂_2 relates linear semidefects and point singularities of the partial order. The elements of $\partial_2^{-1}(\rho) \subset \pi_2(U)$ characterize the possible boundaries of a semidefect-line $\rho \in \pi_1(H)$. Elements of $\partial_2^{-1}(\rho\sigma^{-1})$ characterize the possible interfaces between two semidefects of types ρ and $\sigma \in \pi_1(H)$. Point singularitites of $\ker\partial_2$ do not form boundaries of stable semidefect lines. They possess, however, an inverse image under homomorphism j_2, because $\operatorname{im} j_2 = \ker\partial_2$. If in a phase transition the soft component of the order parameter is being added to the rigid component, and the partial order turns into the full order, then point singularitites outside $\ker\partial_2$ break into stable semidefect lines, point singularities inside $\ker\partial_2$ return into an element of their inverse image.

iii) Homomorphism i_1 relates semidefects and full defects of the same dimension. The semidefects of keri_1 are unstable as defects in the complete order, i.e. they can decay if a singularity in the partial order is allowed to appear intermediately. Since the sequence is exact, and hence keri_1 = im∂_2, the semidefects which can relax via a singularity of the partial order are exactly those which can terminate in the bulk.

3.4 Semidefects and solitons in the presence of full line singularities

Whenever we label a singularity or a topological soliton by an element of a homotopy group, it is understood that all testloops, testspheres, and testareas are tied to a base point in physical space as well as in order parameter space. If in the presence of a line singularity $\kappa \in \pi_1(B)$ a semidefect line is found to be of type $\rho \in \pi_1(H)$ by a testloop as in fig. 7a, then a testloop encircling the semidefect line on the other side of κ yields a different group element, denoted $\kappa(\rho)$. The mappings $\rho \to \kappa(\rho)$ constitute a group action of $\pi_1(B)$ on $\pi_1(H)$. Apart from acting on itself and on all higher homotopy groups $\pi_r(B), r > 1$, the group $\pi_1(B)$ also acts on the group $\pi_2(U)$, whose elements label point singularities and linear topological solitons in the partial order. If as in fig. 7b the test of a linear topological soliton by an area on one side of the line singularity results in type $\gamma \in \pi_2(U)$, then the testarea on the other side produces a type denoted $\kappa(\gamma) \in \pi_2(U)$.

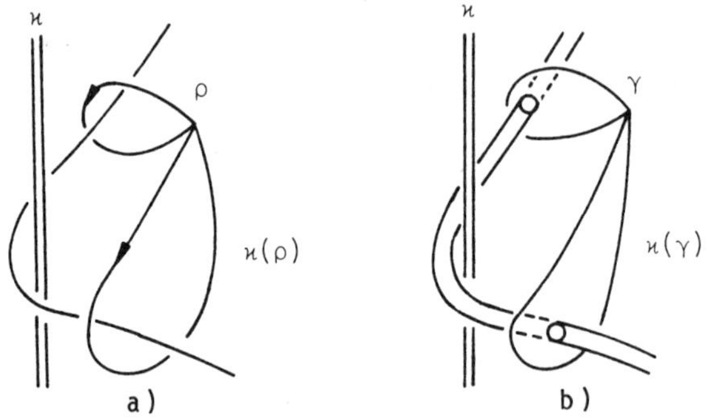

Fig. 7 a) Testloops around a semidefect line yield different homotopy group elements ρ and $\kappa(\rho)$, depending upon which side of the full line singularity κ they have been placed.
b) An area testing the linear topological soliton yields on one side of the full line singularity type γ, on the other side type $\kappa(\gamma)$.

Furthermore the group $\pi_1(B)$ acts on the group $\pi_1(U)$, whose elements characterize line singularities in the partial order. All the group actions are elaborated in detail in ref. 7. Without base points, the true labels of singularities and solitons are the orbits of the various homotopy groups under the action of $\pi_1(B)$, and in the exact sequence (1) each member has to be factorized by $\pi_1(B)$. Since images and kernels are composed of complete orbits, the interpretation of the exact sequence does not change.

4. Semidefects and solitons

4.1 Mutual transformation

From the exact sequence (1) we can derive the following transformation processes:

i) A semidefect ρ, which can terminate in the bulk (i.e. $\rho \in \mathrm{im}\ \partial_2$), may break into a pair of point singularities β^{-1} and $\beta \in \partial_2^{-1}(\rho)$ as in fig. 8a. The two points are connected by a linear topological soliton of type β^{-1}.

ii) A linear topological soliton of type $\beta \in \pi_2(U)$ may break up into a pair of point singularities β and β^{-1} as in fig. 8b. If β lies outside the kernel of ∂_2, a stable semidefect line of type $\rho = \partial_2(\beta^{-1})$ stretches over the space between the points.

iii) A semidefect of type $\rho \in \pi_1(H)$ can convert into any other type σ, if $\rho\sigma^{-1}$ possesses a boundary ($\rho\sigma^{-1} \in \mathrm{im}\ \partial_2$, i.e. ρ and σ are in the same coset of $\mathrm{im}\partial_2$ in $\pi_1(H)$). The conversion is performed by a pair of point singularities β, $\beta^{-1} \in \partial_2^{-1}(\rho\sigma^{-1})$ which, if created on the line ρ and pulled apart, span line σ as in fig. 8c.

4.2 Crossing of a full line singularity and a semidefect line

If a semidefect line is bent around a full line singularity $\kappa \in \pi_1(B)$ as in fig. 9, and if the testloop on one side tells it to be of type $\rho \in \pi_1(H)$, then the testloop on the other side yields type $\kappa(\rho)$. The bridge formed by arriving and departing section of the semidefect carries the label $\kappa(\rho)\rho^{-1}$. Whether the lines are allowed or prohibited to cross depends on the type of the bridge. In contrast to the crossing processes of two full line singularities[8,9], there are three possibilities in

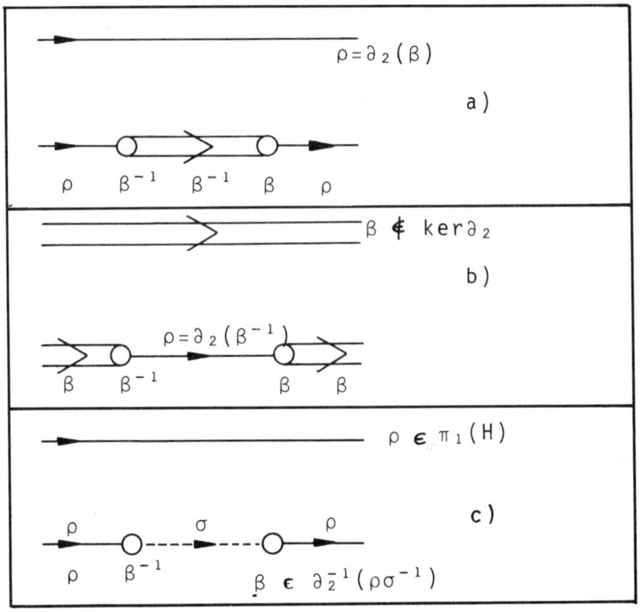

Fig. 8 Transformation processes of linear semidefects and topological solitons

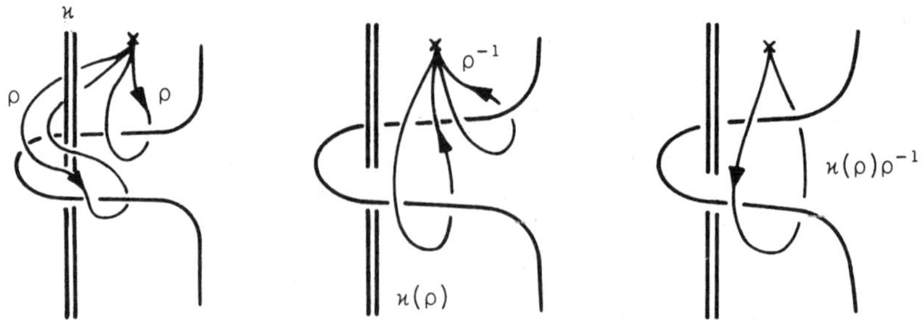

Fig. 9 Crossing of a full line singularity and a semidefect line: the bridge connecting the two singularities is a semidefect line of label $\kappa(\rho)\rho^{-1} \in \pi_1(H)$. If the semidefect is replaced by a linear topological soliton $\gamma \in \pi_2(U)$, the bridge is of type $\kappa(\gamma)\gamma^{-1} \in \pi_2(U)$

our case:

i) $\kappa(\rho)\rho^{-1} = e$ (the identity of $\pi_1(H)$): the bridge can fade away, uninhibited crossing is possible (fig. 10 a).

ii) $\kappa(\rho)\rho^{-1} \in \mathrm{im}\partial_2 \setminus \{e\}$: after crossing, the two lines are connected by a stable semidefect line. This line, however, may break into a pair β, β^{-1} of point singularities in the partial order, $\beta \in \partial_2^{-1}(\kappa(\rho)\rho^{-1})$, which are tied together by a linear topological soliton β^{-1}. Crossing is possible by pair generation (fig. 10b).

The remaining linear semidefect changes its type at the interfacial point singularity β.

iii) $\kappa(\rho)\rho^{-1} \notin \mathrm{im}\partial_2$: the bridge is both a stable semidefect and a stable full line singularity. Crossing is obstructed (fig. 10c).

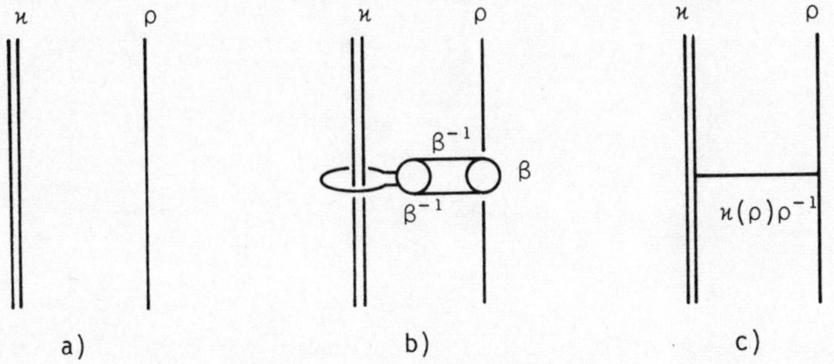

Fig. 10 Three possible results for crossing of a full line singularity and a semidefect line:
a) No obstruction. b) Pair production. c) Topological obstruction.

4.3 Crossing of a full line singularity and a linear topological soliton

Whether a linear topological soliton $\gamma \in \pi_2(U)$ and a full line singularity $\kappa \in \pi_1(B)$ can cross, depends on the type $\kappa(\gamma)\gamma^{-1}$ of the bridge, which is formed by arriving and departing section of the soliton similarly as in fig. 9. Again there are three possibilities:

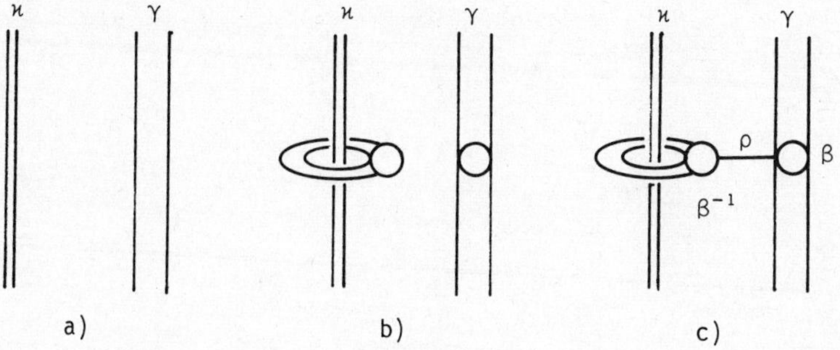

Fig. 11 Three possible results for crossing of a full line singularity and a linear topological soliton
a) No obstruction. b) Pair production. c) Obstruction by a connecting semidefect line.

i) $\kappa(\gamma)\gamma^{-1} = e$ (the identity of $\pi_2(U)$): crossing is possible without obstacle (fig. 11a).

ii) $\kappa(\gamma)\gamma^{-1} \in \ker\partial_2 \setminus \{e\}$: the bridge can break into two point singularities β^{-1}, β. The point $\beta = \kappa(\gamma)\gamma^{-1}$ absorbs topological charge from the remaining soliton and changes its type (fig. 11b).

iii) $\kappa(\gamma)\gamma^{-1} \notin \ker\partial_2$: the bridging soliton may break into two point defects β^{-1}, $\beta = \kappa(\gamma)\gamma^{-1}$. These, however, are linked by a linear semidefect of type $\partial_2(\kappa(\gamma)\gamma^{-1}) \in \pi_1(H)$ (fig. 11c). Crossing is obstructed.

5. Final remarks

The concept of semidefects is not restricted to cases, where the symmetry group H_2 of the full order parameter is a subgroup of the symmetry group H_1 of its rigid component. In ref. 7 semidefects are defined for two further cases:

i) for cholesteric liquid crystals, where the subgroup relation $H_2 < H_1$ is not valid;

ii) for crystals, where stacking faults are identified as semidefects. The set $A = H_1/H_2$ of energetically favored positions within the order parameter space $V = G/H_2$ cannot expressed as coset space in this case.

Section 4 has displayed a rich chemistry of textures and transformation processes for semidefects and solitons. The possible phenomena can only be analysed by topological tools and never are read off the complicated nonlinear differential equations governing the order parameter fields. Homotopy theory provides selection rules for the allowed conversions of textures and thus plays the same presorting role for nonlinear fields as for example group theory does it for the linear fields of quantum mechanics.

Acknowledgment

Most of this work was done in collaboration with Dr. R. Kutka, Fa. Kontron, D-8057 Eching, W. Germany

References

1. D. Rogula, 1976, in: Trends in Applications of Pure Mathematics to Mechanics, edited by G. Fichera (Pitman, New York) pp. 311-331: "Large deformations of crystals, homotopy, and defects".
2. N.D. Mermin, 1979, Rev.Mod.Phys. 51, 591-648: "The topological theory of defects in ordered media".
3. L. Michel, 1980, Rev.Mod.Phys. 52, 617-651: "Symmetry defects and broken symmetry. Configurations. Hidden Symmetry".
4. H.-R. Trebin, 1982, Adv.Phys. 31, 195-254: "The topology of nonuniform media in condensed matter physics".
5. H.-R. Trebin, 1981, MATCH 10, 211-222: "Investigation of defects in ordered media by methods of algebraic topology".
6. N. Steenrod, 1951, Princeton University Press: "The topology of fiber bundles".
7. R. Kutka, H.-R. Trebin, in preparation: "Semidefects".
   A short version of this paper has been published in 1984 by the same authors under the same title in J. Physique Lett. 45, 1119-1123
8. V. Poenaru, G. Toulouse, 1977, J. Physique 38, 887-895: "The crossing of defects in ordered media and the topology of 3-manifolds".
9. K. Jänich, H.-R. Trebin, 1981, in: Physics of Defects, Les Houches, Session XXXV 1980, editors R. Balian, M. Kléman, J.-P. Poirier, North-Holland, pp. 421-429: "Disentanglement of line defects in ordered media".

SOLITONS AND STATISTICAL THERMODYNAMICS*

Alfred Seeger
Max-Planck-Institut für Metallforschung, Institut für Physik,
and
Universität Stuttgart, Institut für theoretische und angewandte Physik,
Postfach 800665, D-7000 STUTTGART-80, Fed. Rep. of GERMANY

After an introductory section dealing with the development of new scientific concepts, the paper discusses the distinction between "solitary waves" and "solitons". It is emphasized that "soliton behaviour" has to be defined in terms of the interactions between solitary waves. In the particle analogue of solitons this corresponds to a multi-particle property.

The bulk of the paper deals with the soliton properties of kinks, a concept derived from dislocation theory. It is shown that for the treatment of kink-pair formation (i.e., the generation of soliton - anti-soliton pairs) it is advantageous to replace the sinusoidal potential connected with the Enneper equation by the so-called Eshelby potential. Kink pairs in unstable equilibrium under a constant applied force and the decay and vibration modes associated with them are discussed in some detail. The results are applied to the calculation of the rate of kink-pair formation in terms of Kramers' theory of thermally activated rate processes. The perfect match of the information furnished by soliton theory and the input required by statistical thermodynamics is emphasized.

1. Introduction

Since its early days, the history of science has been to a large extent a struggle for the "right concepts". This is particularly true of the relationship between mechanics and mathematics. "Right" concept are those which allow us to formulate simple laws of nature that, nevertheless, are general enough to be widely applicable and to describe the "essentials" of many different situations. Finding the right concepts involves very much the art of abstraction. The great masters in this art were the Greek geometers, who invented the notions of the mathematical point, the mathematical line, the simple curves such as the straight line, the circle, the hyperbolae with the concept of asymptotes, etc.

In dealing with physical phenomena (rather than geometry) "Western" science has fully recognized the need for and the power of abstractions since, roughly speaking, the time of Galilei. This approach has remained powerful up to the present day and has spread to other fields including the life sciences, economy, sociology etc. Classical examples from the field of mechanics are the introduction of such concepts as the "mass point", the "rigid body", "frictionless motion", or the "harmonic oscillator". The "soliton" constitutes an important recent addition to this list.

The concept of the soliton has emerged independently in two different areas of mechanics, namely in the theory of plastic deformation of crystals and in fluid dynamics (the theory of water waves in shallow canals). The common denominator is that both fields involve "strongly"

*Dedicated to Professor Frank Nabarro on the occasion of his 70th birthday on March 7th, 1986.

non-linear phenomena (not just small perturbations of linear phenomena) but that, nevertheless, results of striking simplicity and coherency may be obtained.

The emergence of the same concept in fields differing so much suggests wider applicability. This has indeed been borne out by the developments of the last two decades. An incomplete list of additional fields in which the soliton concept has been found useful includes plasma physics, elementary particle physics, ferromagnetism, ferroelectricity, superconductivity, non-linear optics, transmission lines, and molecular crystals. An idea of the wide range of applicability of solitons in materials science may be obtained from a recent review [1].

The name "soliton" appears to have first been given to certain solutions of the so-called Korteweg - de Vries equation in the theory of shallow-water waves [2]; from there its use has spread to the other fields. At the time of the original discovery of soliton behaviour in the field of crystal plasticity [3-5], which preceded that in fluid dynamics by more than a decade, the name was "Eigenbewegungen" (=characteristic motions), emphasizing at the same time the similarity with and the difference from "Eigenschwingungen" (=characteristic vibrations).

Quite often in the history of physics it was found that when a new physical concept was introduced or a new field emerged, the mathematical tools required were already existing, having originated in an seemingly unrelated context. Famous examples are Riemannian geometry in the case of Einstein's general theory of relativity, linear algebra in the case of characteristic vibrations (phonons), matrix calculus in the case of Heisenberg's quantum mechanics, the theory of linear second-order differential equations in the case of Schrödinger's wave mechanics, and the absolute differential calculus in the case of gauge theory. Solitons are no exception. Here the appropriate mathematics had been developed by the mathematicians A. Enneper, V. Bäcklund, S. Lie, L. Bianchi, and G. Darboux (to name only the main contributors) in the context of the differential geometry of pseudospherical surfaces between 1870 and the turn of the century. More about this will be said in Sect. 2.

The two "historical" approaches to solitons, mentioned above, are based on quite different non-linear partial differential equations with two independent variables, namely the Enneper equation[1)] in the case of crystal physics, and the Korteweg - de-Vries equation in the case of shallow-water waves. The present paper deals mainly with the former equation because of the author's personal involvement in revealing and exploiting its soliton properties, because of the tendency in the mathematical literature on solitons to put much stronger emphasis on the Korteweg - de Vries equation in spite of the fact that its soliton properties were discovered by numerical computation [2] and that the mathematical techniques for treating it analytically were developed only later, and, last but not least, because of the intimate relationship of Enneper's equation to other branches of physics.

One of the important criteria for the power and the usefulness of a new concept in science is how well it intertwines with established branches of science. On this criterion solitons in crystals as a model for kinks in dislocation lines (the approach to be followed in this paper)

----

1) The Enneper equation (2.16) is more widely known as "sine-Gordon equation". To the present writer, the name Enneper equation appears preferable for historical (cf. Sect. 2) as well as linguistic reasons; it will be used throughout this paper.

fare very well. Simultaneously with their discovery it was realized [5,6] that they lend themselves to treatments by statistical thermodynamics. It was, in fact, the wish to be able to tackle certain problems in dislocation theory connected with thermal activation, among them the formation of kink pairs in dislocation lines, that led the present writer to the search in the mathematical literature that eventually resulted in the discovery of the soliton properties.

As will be explained in Sect. 5, there is virtually a one-to-one correspondence between the information required as input into statistical thermodynamics and that provided by the soliton theory of kinks. To the present writer the perfect match between soliton theory and statistical thermodynamics is even more striking than the well-known one between Hamiltonian mechanics and statistical thermodynamics, since it involves more subtle features. While he is confident that in due course the close correspondence between solitons and statistical mechanics will bear fruit in other applications as well, at the present time it has definitely been best developed for kinks in dislocations.

As emphasized above, our scientific notions are of necessity based on abstractions and simplifications. This makes it imperative to investigate and understand the nature of their limitations and their limits of applicability. It is characteristic of unifying concepts such as "the soliton" that these limitations may take quite different forms in different areas. Again the field of crystal plasticity is very suitable for studying the problem involved. A quite general model of a dislocated crystal (going back to the author's work of the early 1960's with G. Stenzel and P. Schiller [7]), in which the "embedding" of the soliton aspects in a three-dimensional framework can be studied in detail, has been developed over the years. Space limitations do not allow these important questions to be treated in this paper, however. The reader is referred to a recent review [8].

Successful new concept have a tendency for their names to be misused and to become associated with objects or phenomena that should be clearly distinguished from the original idea. In this regard, solitons are no exception either. A widespread misuse is to employ the name "soliton" for "solitary waves" or "solitary-wave pulses."[2)]

Part of Sect. 2 will be devoted to expounding in some detail the difference between "solitons" on the one hand and "solitary waves" or "solitary-wave pulses" on the other. Again, for doing this the Enneper equation is far better suited than the Korteweg - de Vries equation. Enneper's equation belongs to a class of non-linear wave equation for which the existence of solitary waves of finite amplitude is close to trivial (cf. Sect. 2). Unlike this, the characteristic compensation of non-linearity and dispersion resulting in solitary-wave solutions of the Korteweg - de Vries and related equations is not nearly as easy to recognize.

Sect. 2 also serves to introduce the Bäcklund transformation technique the analytical tool that was essential for the discovery of the soliton properties. A fuller account of this technique in the case of Enneper's

----

2) The above-mentioned confusion on nomenclature, which unfortunately is quite common, may possibly be traced back as far as the name-giving paper [2], in which "solitary wave pulse" and "soliton" are used interchangably. It would in fact be better to speak of "soliton behaviour" or the "soliton properties" of solutions of certain partial differential equations since the essential features of solitons are displayed also by solutions which are usually not called "solitons" (cf. Sect. 3).

equation and of its relationship to the differential geometry of pseudospherical surfaces may be found elsewhere [9].

Equations exhibiting soliton behaviour describe mathematical models of a physical reality that is usually quite complex and in which the soliton behaviour may be hidden. Once the soliton aspects of a physical system have been recognized and isolated, it may be permissable to replace the soliton equations by others that do no longer possess the soliton properties but may be more convenient from a mathematical point of view. A classical example, to be taken up in Sect. 3, is the replacement of the sinusoidal potential leading to Enneper's equation by a polynomial approximation, the so-called Eshelby potential. This replacement destroys the soliton properties but allows us to treat much more easily the effects of external forces.

The most important ingredients to be put into a treatment of a physical system by means of statistical thermodynamics are its energy levels. For the statistical thermodynamics of solitons we have to know the energy levels of their "excited states" (or at least their distribution functions). The mathematical tool for obtaining them, a perturbation technique, was initiated even before the soliton properties were discovered [10]. As a method for solving initial-value problems [10] the perturbation technique may be considered as a fore-runner of the inverse scattering technique (see, e.g. [11-13]). In their practical application the two methods often coincide.

The perturbation technique initiated by Seeger and Kochendörfer [10], which leads to the concept of massive phonons [14], is taken up in Sect. 4. It is carried there only as far as is required for the particular problem in statistical thermodynamics to be treated in Sect. 5, namely the rate of formation of kink pairs on a dislocation line. What can be said on these questions within the available space is far from doing justice to this field, which is in rapid development. The author hopes that, nevertheless, he has succeeded in conveying to the reader a feeling for the intimate relationship between soliton theory and statistical mechanics alluded to above.

In the Appendix a problem is taken up that arises in the perturbation theory of Sect. 4 but that has much broader significance in the quantum-mechanical treatment of double wells. For situations in which explicit solutions of the Schrödinger equation for a single potential well but not for the double-well problem are available, an approximation is developed that makes use of both the WKBJ technique [15] and the knowledge of the single-well solutions.

Two final remarks on the presentation adopted:

(i) In this paper we have chosen kinks in dislocations as the physical system to be considered. Since this is just one example of many systems to which the general ideas and the results may be applied we have employed dimensionless units right from the beginning. On a few occasions, however, we shall switch over to dimensional quantities in the interest of clarity. For a treatment of kinks in dislocation in dimensional units see [8].
(ii) The reader will note that the exposition has a slight "historical touch". The present paper is definitely not intended to give the science history of solitons (this remains to be written), but since unfortunately in this field correct presentations of the historical aspects are the exception rather than the rule, the author thought it important to provide references to the origin of some of the ideas involved. The reader interested in further details may wish to consult brief accounts published elsewhere [9, 16].

2. The Distinction Between Solitons and Solitary Waves - The Bäcklund Transformation.

Let us consider non-linear wave equations for a real field variable Φ in one space (z) and one time (t) dimension,

$$\frac{\partial^2\Phi}{\partial z^2} - \frac{\partial^2\Phi}{\partial t^2} + F(\Phi) = 0 \qquad (2.1)$$

where $F(\Phi)$ denotes a non-linear function of the field variable to be specified later. Eq.(2.1) is Lorentz-invariant, i.e., it admits solutions of the form

$$\Phi = \Phi_0(\zeta) \qquad (2.2a)$$

with

$$\zeta = \frac{z \mp vt}{(1 - v^2)^{1/2}} \quad , \qquad (2.2b)$$

v representing a velocity. Such solutions obey the ordinary first-order differential equation

$$\frac{d^2\Phi_0}{d\zeta^2} + F(\Phi) = 0 . \qquad (2.3)$$

Eq.(2.3) may be integrated to give

$$\frac{1}{2}\left(\frac{d\Phi_0}{d\zeta}\right)^2 = U(\Phi_0) - C_0 \quad , \qquad (2.4)$$

where C_0 denotes a constant of integration and

$$U(\Phi) \equiv -\int_0^{\Phi} F(\Phi')d\Phi' \qquad (2.5)$$

has the meaning of a "potential energy" from which the "force" $F(\Phi)$ may be derived. Further integration leads to

$$\zeta-\zeta_0 = \pm\, 2^{1/2}\int^{\Phi_0(\zeta)}[U(\Phi') - C_0]^{1/2}d\Phi' . \qquad (2.6)$$

Suppose now that $F(\Phi)$ has at least two real zeros at which $dF/d\Phi=0$, and let us specify that two adjacent ones of these occur at $\Phi = 0$ and $\Phi = \hat{\Phi}$ We further assume that $U(0) = U(\hat{\Phi})$, i.e., that the two minima of $U(\Phi)$, at $\Phi = 0$ and $\Phi = \hat{\Phi}$, are of equal height. We may then choose C_0 such that $d\Phi_0/d\zeta = 0$ at $\Phi_0 = 0$ and $\Phi_0 = \hat{\Phi}$. (Because of $U(0) = 0$, following from (2.5), this means $C_0=0$.) Fig. 1 indicates qualitatively the nature of the solutions obtained in this way; for reasons that will become clear later we shall call them <u>kink solutions</u> or just "kinks". (Fig. 1 corresponds to the choice of the + sign in (2.6) and shows a "positive" kink; negative kinks are obtained by reversing the sign of $\zeta-\zeta_0$.)

Eq. (2.1) may be derived from the following Lagrangian density:

$$\mathcal{L}(\Phi, \partial\Phi/\partial t, \partial\Phi/\partial z) = \frac{1}{2}\left(\frac{\partial\Phi}{\partial t}\right)^2 - \frac{1}{2}\left(\frac{\partial\Phi}{\partial z}\right)^2 - U(\Phi) . \qquad (2.7)$$

On the right-hand side of (2.7), the first term denotes the density of kinetic energy. The potential energy density consists of two terms, viz., the "elastic energy" $(\partial\Phi/\partial z)^2/2$ and the "internal energy" $U(\Phi)$.

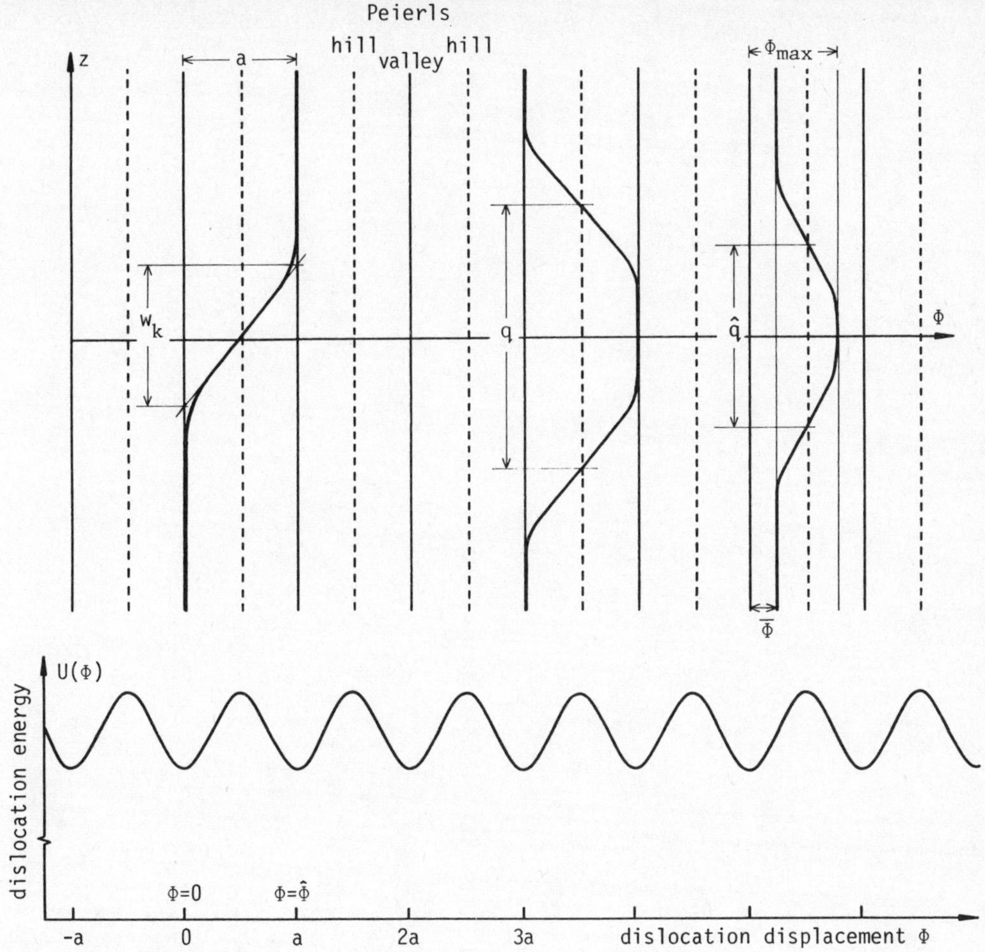

Fig. 1.

Bottom: Potential energy $U(\Phi)$ of a straight dislocation line as a function of its displacement Φ in the glide plane. Top: Various dislocation shapes on a glide plane. The Peierls hills and valleys are indicated. Left: Single positive kink ($d\Phi/dz>0$) of width w_k and height a located at z = 0, widely separated by kink - kink distance q. Right: Kink pair at equilibrium separation $\hat{q}$ in unstable equilibrium under applied stress σ.

The density of the total energy associated with the Lagrangian (2.7) is given by

$$\epsilon_{tot} = \frac{1}{2}\left(\frac{\partial\Phi}{\partial t}\right)^2 + \frac{1}{2}\left(\frac{\partial\Phi}{\partial z}\right)^2 + U(\Phi). \qquad (2.8)$$

Provided ϵ_{tot} tends to zero sufficiently rapidly for $z \to \pm\infty$, we may define the total energy as

$$E_{tot} = \int_{-\infty}^{+\infty} \left\{ \frac{1}{2} \left(\frac{\partial \Phi}{\partial t}\right)^2 + \frac{1}{2} \left(\frac{\partial \Phi}{\partial z}\right)^2 + U(\Phi) \right\} dz \, . \tag{2.9}$$

Insertion of (2.2) into (2.8) gives us

$$\epsilon_{tot} = \frac{1}{2} \frac{1+v^2}{1-v^2} \left(\frac{d\Phi_o}{d\zeta}\right)^2 + U(\Phi_o) . \tag{2.10}$$

For the kink solutions we obtain, making use of (2.4) and $C_o = 0$,

$$\epsilon_{tot} = \frac{1}{1-v^2} \left(\frac{d\Phi_o}{d\zeta}\right)^2 \tag{2.11}$$

and

$$E_{tot} = \frac{E_k}{(1-v^2)^{1/2}} \quad , \tag{2.12a}$$

where

$$E_k = \int_{-\infty}^{+\infty} (d\Phi_o/d\zeta)^2 d\zeta = 2^{1/2} \int_{\Phi=0}^{\Phi=\bar{\Phi}} [U(\Phi)]^{1/2} \, d\Phi \tag{2.12b}$$

is the <u>rest energy</u> of the kink. It may be calculated from $U(\Phi)$ without the need to know the kink solution $\Phi_o(\zeta)$ explicitly.

We note that the energy associated with the kink solution is spatially located in the region in which $d\Phi_o/dz$ or $d\Phi_o/d\zeta$ is significantly different from zero. This "concentration of energy" may travel with constant velocity v without change of shape. Its total energy E_{tot} shows the well-known relativistic velocity dependence (2.12a), with a "limiting velocity" c (the analogue of the velocity of light in the special theory of relativity) that has been put to unity by the choice of the units of space and time. The kink solution of the non-linear wave equation (2.1) shows thus a strong resemblance to the behaviour of a point mass in mechanics. E.g., we may attribute to the kink solutions a kink rest mass

$$m_k = E_k/c^2 \, . \tag{2.13}$$

Solutions of non-linear partial differential equations with spatially concentrated energy and capable of translation with constant velocity without change of shape are now called <u>solitary waves</u>. From the preceding discussion it should be clear that for non-linear wave equations of the type (2.1) the existence of solitary waves is an almost trivial matter, since apart from general regularity conditions the only requirement is that the "potential energy" $U(\Phi)$ shows the behaviour discussed above, i.e., that it possesses at least two minima of equal height.

In the case of non-linear wave equations of the type (2.1) the existence of solitary waves may be understood in physical terms as follows.

The solutions

$$\Phi \equiv 0, \quad \Phi \equiv \bar{\Phi} \tag{2.14}$$

are <u>trivial</u> solutions corresponding to identically vanishing energy density (under the normalization $U(0) = U(\Phi) = 0$ chosen above). Kink-type solutions are possible in precisely those cases in which (2.1) admits a least two such solutions, i.e., in which $U(\Phi)$ exhibits at least two absolute minima. Since the solutions of the type (2.14) correspond to the lowest possible energy of the system (namely zero under the chosen normalization), they are termed "vacuum solutions" or simply <u>vacua</u>. <u>Kinks</u> are solutions of (2.1) which connect one vacuum state (attained, say, at large negative z) with another one (e.g., at large positive z).

The reader will realize that the preceding - admittedly somewhat abstract - characterization of kink solutions describes a situation that is quite common in crystal physics. Consider, e.g., a <u>ferromagnetic crystal</u> with several "directions of easy magnetization". Homogeneous magnetization in any one of these directions represents one of the possible vacuum states. The ferromagnet may exhibit a "domain structure", i.e., its magnetization may take up different directions of easy magnetization in different parts of the crystal. In the "domain walls" (e.g., Bloch or Neél walls) between the domains of homogeneous magnetization the magnetization direction rotates from one direction of easy magnetization to another. This is illustrated in Fig. 2 for the case of a Neel wall in a uniaxial ferromagnet (i.e., a ferromagnet with two [oppositely orientated] directions of easy magnetization). Similar situations may arise in antiferromagnets, ferroelectrics, antiferroelectrics, or ordered alloys. In the latter case the different "vacuum states" are known as "phases", the walls separating them as "anti-phase boundaries".

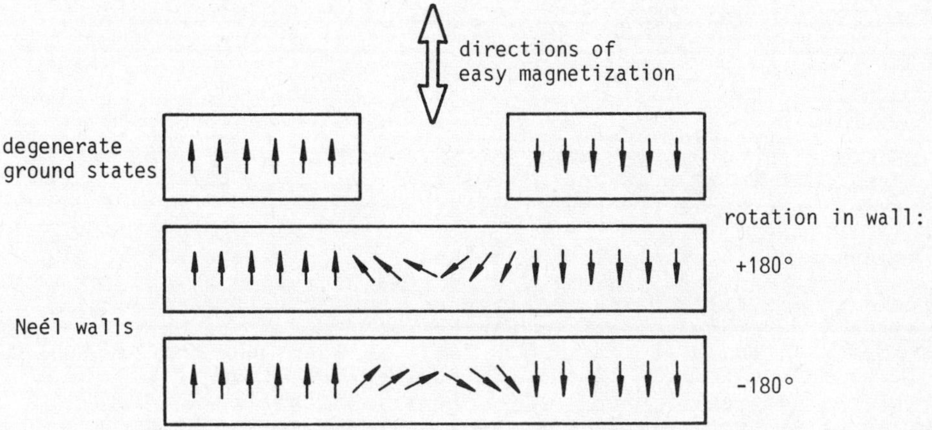

Fig. 2.

Domain structure in a ferromagnet with two directions of easy magnetization. In the Neél walls the magnetization direction rotates from one "ground state" to the other one by ± 180°.

Let us now consider a simple <u>mechanical model</u> which allows us to visualize the physical principles involved in the formation of kinks. It has the additional advantages that it is described rather well by

(2.1) with suitable choices of F(Φ), and that it is of considerable importance for the mechanics of crystalline solids.

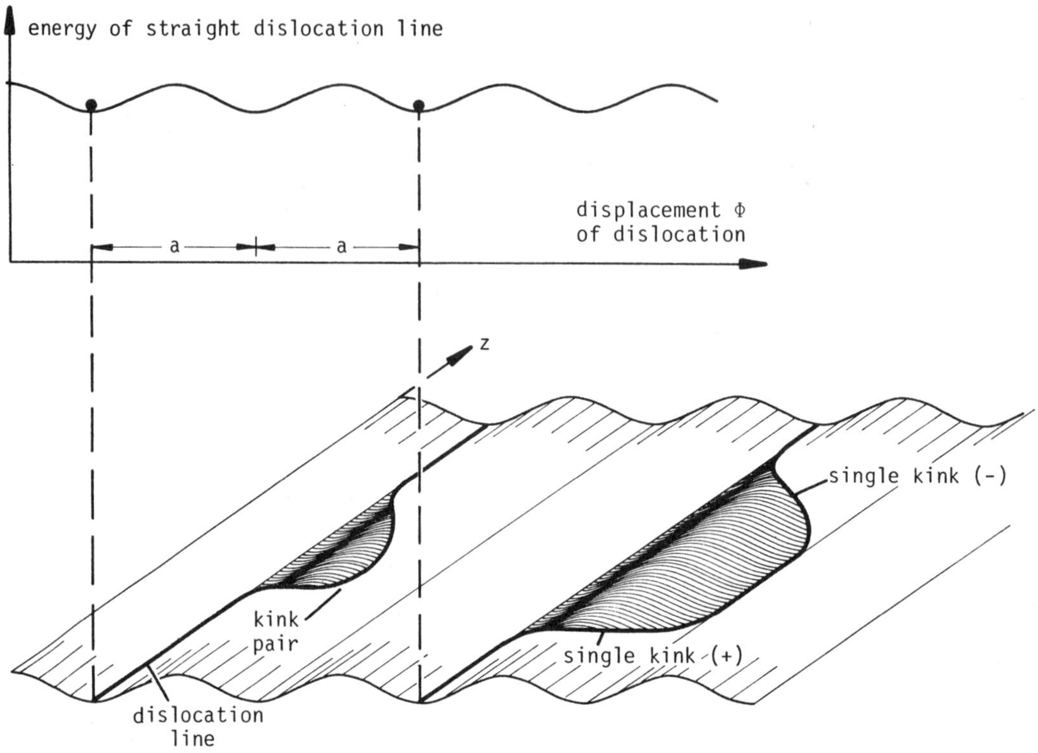

Fig. 3.

The "corrugated sheet" model of flexible dislocation line. The shaded parts of the sheet indicate the areas swept out by the dislocation line when a narrow kink pair or a pair of widely separated kinks of opposite sign are formed.

The energy of a straight dislocation line lying along a crystallographic direction and moving slowly on a crystallographic glide plane is a periodic function of its position coordinate on the glide plane. This may be illustrated by a "corrugated-roof model" as shown in Fig. 3. Since R. Peierls [17] was the first to attempt a quantitative estimate of the energy profile of Fig. 3 (at the suggestion of E. Orowan), its valleys are denoted as <u>Peierls valleys</u>, its hills as <u>Peierls hills</u>. The period a of the "Peierls potential" is related to the periodicity of the crystal structure.

Since dislocation lines cannot end inside a crystal (see, e.g., [5,18]), the straight parts of a dislocation line lying partly in one Peierls valley, partly in an adjacent one, must be connected by a dislocation segment cutting across the Peierls hill separating the two valleys (Fig. 3). Such connecting segments in otherwise straight dislocation lines are known as "kinks" [5,18], following a somewhat different usage of the word by E. Orowan in an related area of the mechanics of crystals. From here the use of the word "kink" appears to have spread to other scientific fields.

We may consider Φ as proportional to the displacement of a dislocation line in its glide plane (for the normalization, see below) and identify the Peierls potential per unit dislocation length with the internal-energy density $U(\Phi)$, normalized to $U = 0$ when the dislocation line lies entirely at the bottom of a Peierls valley. Straight dislocations thus correspond to the vacuum states of the system "crystal plus dislocation". In an infinitely extended crystal (i.e., if we may neglect boundary effects) a straight dislocation line on a given glide plane may take up an (enumerably) infinite number of positions of lowest internal energy at the bottoms of the Peierls valleys. This means that the system "infinite crystal plus dislocation" has an infinitely degenerate vacuum. Kinks in the dislocation constitute "excited states" of this system.

One sees immediately that the formation of a kink requires a finite amount of energy. This means that an energy gap exists between the ground state of the system (the "vacuum") and the kink excitations. The physical origin of the kink energy lies in the fact that the dislocation line has to cross a Peierls hill, i.e., that it must pass through states with non-vanishing internal energy $U(\Phi)$. From the point of view of keeping the internal energy as small as possible, the dislocation line should cross the Peierls hill orthogonally to the direction of the valleys. Such a configuration (called "abrupt kink") would entail a lengthening of the dislocation line by a and thus a considerably expenditure in dislocation line energy, however.

In many situations the tendency of a dislocation to resist an increase of its length may be adequately described by a line-tension S, giving rise to the "elastic energy" in the Lagrangian density (2.7)[3]. It is therefore energetically more favourable for the dislocation line to follow an oblique path across the Peierls hill in order to reach the best possible compromise between elastic and internal energy. The static differential equation (2.3), from which the kink solution [Eq. (2.6) with $C_0 = U(0) = U(\Phi) = 0$] and the kink energy E_k [Eq. (2.12b)] have been derived, is the Euler - Lagrange differential equation of the variational principle demanding that the sum of the internal and the elastic energy should be minimal.

In the model of Fig. 3 the solitary waves discussed above correspond to the motion of kinks along otherwise straight dislocation lines. In order to decide whether they should be called solitons one has to consider the interaction between kinks. In the mass-point analogy we have mentioned, this means that we have to leave the one-particle viewpoint so far adopted.

Consider two kinks of opposite sign on the same dislocation.[4] In the particle picture they correspond to a particle - anti-particle pair. They may annihilate each other, releasing the energy $2E_k$ (if they are at rest) and resulting in a vacuum state (a kink-free dislocation line). By supplying the energy $2E_k$ (or more) to the "vacuum" we may obtain a

----

[3] In the present context the term "elastic energy" is most appropriate, since the physical origin of the line tension does lie in the increase of the energy of the elastic strain field of a dislocation line associated with its lengthening. In the kink-pair problem, to be taken up in Sect. 3, the line tension provides an adequate description of the elastic interaction between kinks on the same dislocation line as long as it is not of "long range" in the sense of footnote 13.

[4] As a convention, we speak of positive or negative kinks depending on whether $d\Phi/dz$ is positive or negative. Occasionally we use the expressions kink or anti-kink instead.

kink pair by pair-creation.[5)]

From the possibility of kink - anti-kink annihilation it follows that an attractive interaction between kinks of opposite sign exists. E.g., we may expect that a positive and a negative kink put at rest at a large distance from each other are accelerated towards one another and annihilate one another. This is a highly non-linear process, and while it is clear that the total energy $2E_K$ of the system will be conserved it is not obvious how it will be distributed over the various excited modes. Physical intuition lets us expect that at least part of the energy is transferred to "radiation", i.e. wave-type solutions of (2.1) transporting energy away from the spatial region in which the annihilation of the kinks takes place. (cf. Sect. 4).

The process just described is indeed what occurs for a general choice of $U(\Phi)$ satisfying the conditions stated above. At first sight a quantitative analytical treatment appears hardly to be possible since it would involve the solution of the non-linear partial differential equation (2.1). In general this requires numerical computations that in the immediate post-war years, when the physical problems discussed above arouse, looked forbiddingly difficult[6)].

However, in 1950 the present author discovered that in the special case of

$$F(\Phi) = -\sin\Phi \qquad (2.15)$$

the mathematical tool required for an analytical solution had existed in the literature on differential geometry for more than half a century under the name of "Bäcklund Transformation".

Together with his collaborator H. Donth [27] the author soon found that not only could exact closed-form solutions of

$$\frac{\partial^2\Phi}{\partial z^2} - \frac{\partial^2\Phi}{\partial t^2} = \sin\Phi \qquad (2.16a)$$

for the kink - kink interaction problem sketched above as well as for many other physically interesting situations be obtained but that these solutions often behaved in a manner which differed strikingly from the "physical intuition" prevailing at the time.

---

5) The creation of kink pairs in dislocation lines under the influence of an applied stress and of thermal fluctuations was treated by the writer and his collaborators in a series of papers [5-8, 19-24]. The name double-kink generation introduced by them has been accepted in the entire dislocation literature. However, the author has recently found it appropriate to reserve the name "double kink" to kinks not extending between nearest but between next-nearest Peierls valleys [23, 25]. He hopes that for the process considered above "kink-pair formation" or "kink-pair generation", which emphasize the analogy to elementary particle physics, will be generally adopted in the dislocation literature.

6) The first successful attempt at the numerical solution of the kink-pair annihilation problem was published in 1962 by Perring and Skyrme [26] for the case of eq.(2.16a) in the context of elementary-particle theory. They rediscovered indeed the solution found a decade earlier by Seeger, Donth, and Kochendörfer [3] by the analytical technique to be described presently. From a historical point of view it is interesting to note that in the case of the Korteweg - de Vries equation, the discovery of the solitons by computation in the mid-1960's preceded the analytical treatment.

Before describing the surprising features of the solutions just mentioned, let us digress briefly to a discussion of the occurrence of (2.16a) in differential geometry.

In 1870 the Göttingen mathematician A. Enneper observed [28] that the angle Φ between the asymptotic lives of a <u>pseudospherical surface</u> (= a surface of constant negative Gaussian curvature, which in the present discussion in taken as unity without loss of generality) obeys

$$\frac{\partial^2 \Phi}{\partial \xi \partial \eta} = \sin \Phi , \tag{2.16b}$$

where η = const. and ξ = const. are the asymptotic lines of the surface. As is well known, (2.16a) and (2.16b) are equivalent to each other. In the theory of hyperbolic partial differential equation (ξ,η) are known as <u>characteristic coordinates</u> related to (z,t) by

$$\xi = (z + t)/2 \quad , \quad \eta = (z - t)/2 \ . \tag{2.16c}$$

In the differential geometry of pseudospherical surfaces the lines z = const. and t = const. are the lines of curvature.

The Gaussian curvature of a surface is an <u>intrinsic</u> property, i.e., it remains invariant under metric-preserving deformations. This has led to a detailed study of the transformations of pseudospherial surfaces into each other, notably by L. Bianchi, A. V. Bäcklund, G. Darboux, and R. Steuerwald.[7)]

Finding such transformations means obtaining a prescription of how to deduce from one pseudospherical surface, characterized by the solution $\Phi = \Phi_0(\xi,\eta) = \Phi_0(z,t)$ of (2.16), another one $\Phi_1(\xi,\eta) = \Phi_1(z,t)$ (or, preferably, a family of such solutions).

In 1882 A. V. Bäcklund showed that from a given solution Φ_0 of (2.16) one may obtain a two-parameter family of solutions Φ_1 by integrating the following system of first-order differential equations:

$$\begin{aligned} \frac{1}{2}\frac{\partial(\Phi_1 - \Phi_0)}{\partial \xi} &= \frac{1 + \sin\sigma}{\cos\sigma} \sin\frac{\Phi_1 + \Phi_0}{2} \\ \frac{1}{2}\frac{\partial(\Phi_1 + \Phi_0)}{\partial \eta} &= \frac{1 - \sin\sigma}{\cos\sigma} \sin\frac{\Phi_1 - \Phi_0}{2} \ . \end{aligned} \tag{2.17}$$

The Φ_1 family is known as the "Bäcklund transform" $B_\sigma\Phi_0$ of Φ_0; the system (2.17) of first-order differential equations is called "Bäcklund transformation". The parameter σ, which may be chosen arbitrarily, is denoted as the <u>parameter</u> of the <u>Bäcklund transformation</u> B_σ. (The second parameter of the Φ_1 family is given by the constant of integration of the system (2.16)).

From (2.16) and (2.17) the use of the names has spread to other partial differential equations. In a generalized usage one understands now by "Bäcklund transformation" a system of lower-order partial differential equations that relates different solutions of (the same or different) higher-order differential equations to each other. The simplest example

----

7) A brief account of these investigations as well as detailed references may be found elsewhere [9].

of a Bäcklund transformation in this generalized sense are the Cauchy - Riemann first-order differential equations relating to each other the real and imaginary parts of an analytic function of a complex variable, each of which obeys Laplace's equation in two dimensions.

The solutions of (2.16) obtained by solving (2.17) may be used as starting functions for further Bäcklund transformations. In 1892 L. Bianchi [29] proved his celebrated "teorema di permutabilità", stating that the successive application of Bäcklund transformations to solutions of (2.16) is commutative, i.e., that

$$B_{\sigma_2} B_{\sigma_1} \Phi_0 = B_{\sigma_1} B_{\sigma_2} \Phi_0 \equiv \Phi_3 \tag{2.18}$$

holds. Furthermore, he showed that the family Φ_3 of solutions resulting from (2.18) obey the relationship

$$\tan \frac{\Phi_3 - \Phi_0}{4} = \frac{\cos \frac{\sigma_1+\sigma_2}{2}}{\sin \frac{\sigma_1-\sigma_2}{2}} \tan \frac{\Phi_1 - \Phi_2}{4} , \tag{2.19}$$

where

$$\Phi_1 \equiv B_{\sigma_1} \Phi_0 , \quad \Phi_2 \equiv B_{\sigma_2} \Phi_0 \tag{2.20}$$

are two different Bäcklund transforms of the same starting function Φ_0.

If in Bäcklund's differential equations (2.17) we put $\Phi_0 = 0 \pmod{2\pi}$, i.e., if we start out from one of the "vacuum solutions" of (2.16), eqs. (2.17) simplify to

$$\frac{\partial(\Phi_1/2)}{\partial\zeta} = \frac{1 + \sin\sigma}{\cos\sigma} \sin(\Phi_1/2)$$
$$\frac{\partial(\Phi_1/2)}{\partial\zeta} = \frac{1 - \sin\sigma}{\cos\sigma} \sin(\Phi_1/2) . \tag{2.21}$$

Since from (2.5) and (2.15) it follows that

$$U(\Phi) = 2 \sin^2(\Phi/2) , \tag{2.22}$$

with $\zeta = (1 + \sin\sigma)\xi/\cos\sigma$ or $\zeta = (1 - \sin\sigma)\eta/\cos\sigma$ eqs.(2.21) become equivalent to the first integral (2.4) in the case of a single kink ($C_0 = 0$). This means that the Bäcklund transforms of the vacuum solutions of (2.16) describe positive (for $-\pi/2 < \sigma < \pi/2$) or negative (for $\pi/2 < \sigma < 3\pi/2$) kinks moving with the velocity $v = -\sin\sigma$. These solutions may be written as

$$\begin{aligned} \Phi_1 &= 4 \arctan \exp \{\frac{1 + \sin\sigma}{\cos\sigma}\xi + \frac{1 - \sin\sigma}{\cos\sigma}\eta\} \\ &= 4 \arctan \exp \{\frac{z + t\sin\sigma}{\cos\sigma}\} \\ &= 4 \arctan \exp \{\frac{z - vt}{(1-v^2)^{1/2}}\} . \end{aligned} \tag{2.23}$$

The single-kink solution (2.23) was first given by Frenkel and Kontorova [30], but as an approximate solution of a dynamic generalization of an atomic static lattice model (discrete in the spatial variable!) originally proposed by Prandtl [31] and Dehlinger [32] rather than as an exact solution of (2.16a).[8]

Let us now return to Bianchi's formula (2.19), which in the present writer's opinion represents a most remarkable result. By replacing the trigonometric functions in (2.16) and (2.19) by their arguments we see that (2.19) constitutes a generalization of the well-known "superposition principle" of linear equations. It states that, notwithstanding the strongly non-linear character of (2.16), two solutions Φ_1 and Φ_2 of (2.16) related in a definite way (namely both being Bäcklund transforms of the same solution Φ_0) can be "superimposed" according to (2.19) to give a fourth solution Φ_3. This process may be continued indefinitely, so that by repeated application of (2.19) an unlimited number of Bäcklund transforms $B_{\sigma k}\Phi_0$ (k = 1,2,3, ...) of a given starting function Φ_0 may be "superimposed".

By applying the procedure just outlined to the starting solution $\Phi_0 = 0$ (mod 2π), we may obtain the solutions corresponding to the superposition of an arbitrary number of kink solutions with arbitrary initial positions and velocities. This was first recognized by Seeger, Donth, and Kochendörfer [3], to whose paper the reader is referred for details.[9].

As has already been indicated, the solution of the problem of the interaction between two kinks of opposite sign, which became possible by the methods just described, contradicted strikingly the "intuitive" physical arguments given above. This may be illustrated in a "world-line" diagram as used in special relativity, which becomes particularly simple in one-dimensional situations such as the present one (Fig. 4).

The world line of a particle at rest at z = 0 is the time axis. The dashed line $z_2 = t/2$ represents the world-line of a non-interacting particle that moves in the direction of positive z with the normalized velocity v = 1/2 and passes through z = 0 at time t = 0. The full lines represent the world lines of two kinks of opposite sign as calculated from the solutions of the Enneper equation discussed above, kink 1 being originally at rest near the origin, kink 2 impinging on it from the left with v = 1/2. Outside the "circle of interaction" shown in Fig. 4 the world lines $z = z_1(t)$ and $z = z_2(t)$ are virtually straight lines. The interaction and partial annihilation of the two kinks inside the circle is described by the solution of Enneper's equation but cannot be represented in a particle picture. Near the origin of the diagram the dotted lines lose their meaning as "particle world lines".

The comparison of the left- and right-hand sides of Fig. 4 shows that the velocities of the kinks before and after the collision are the same. This means that they have not exchanged energy and, in particular, energy has not been transferred to other degrees of freedom. This behaviour, which contradicts physical intuition as described above, is

----

8) As far as the present writer can tell, contrary to widespread belief Frenkel and Kontorova have never explicitly considered the partial differential equation (2.16). In a physical context (a one-dimensional lattice model for dislocations analogous to the Prandtl - Dehlinger model) (2.16a) was apparently written down for the first time independly by the present writer [33] and by Frank and van der Merwe [34]. - We hope that the reader going back to the original papers of the early 1950's will not be confused by the fact that as physical objects they treat "one-dimensional dislocations" rather than "kinks".

9) The present writer [9] has summarized in English some of the results together with additional details contained in H. Donth's Diploma thesis [27].

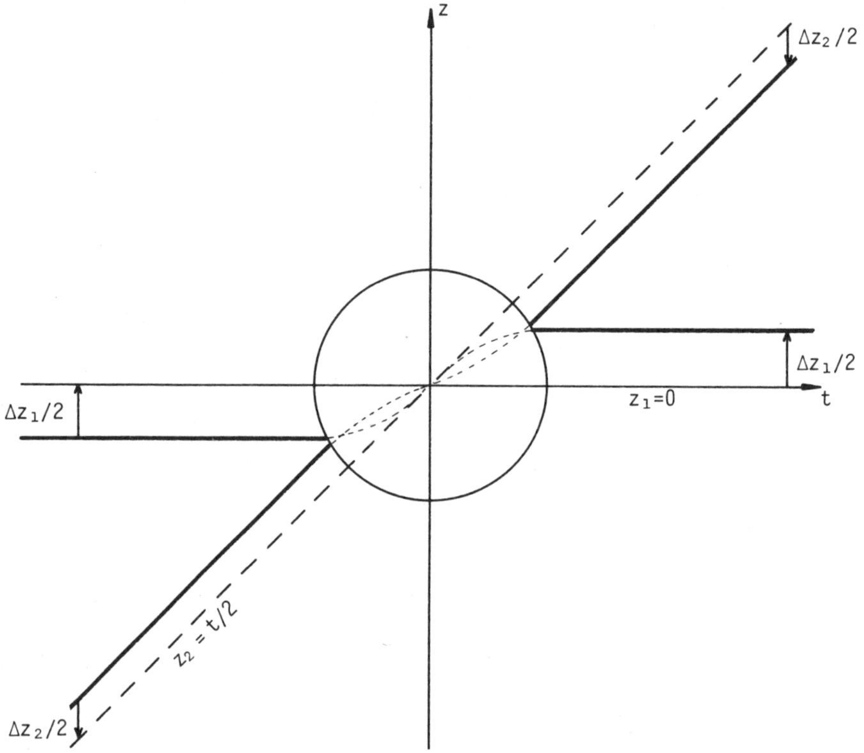

Fig. 4.

World-line diagram of a kink moving in the +z-direction with velocity v = 1/2 and impinging on a kink of opposite sign located at z = $-\Delta z_1/2$. Outside the circle the interaction between the kinks is negligible, hence the world lines of both kinks are straight (full lines). Inside the circle a strong interaction takes place which can no longer be described in a particle picture. The dotted lines are merely intended to guide the eye. The dashed lines and the abszissa represent the world lines of non-interacting kinks.

very characteristic of soliton - soliton interactions. It permits us to describe the translational degrees of freedoms of a soliton ensemble as a one-dimensional gas of non-interacting mass point with suitably chosen mass.

However, it would be wrong to say that "solitons" do not interact with each other, as is occasionally done in the literature. Non-interaction would be incompatible with the non-linear character of (2.1) which, as we have seen, is essential for the existence of kinks. The interaction consists of a shift of the world lines before or after the collision by amount Δz_1 and Δz_2 (cf. Fig.4). These shifts were discovered by Seeger, Donth, and Kochendörfer [3], who called them "Treffstrecke". This concept is closely related to the concept of "time delay" used in the theory of quantum scattering processes [35].

The general expressions for Δz_1 and Δz_2 in the case of collisions between two kinks are given elsewhere [9]. They admit a simple physical

between two kinks are given elsewhere [9]. They admit a simple physical interpretation in terms of the uniform motion of the centre of gravity of the system taking into account the velocity dependence of the kink masses.

When the total energy of a kink - anti - kink pair is less than $2E_k$, none of the two kinks can escape to infinity leaving the other one behind. An oxillatory motion, the "breather", will result [3]. In Sect. 5 we shall make use of the fact that if the centre of gravity of the kink- anti-kink pair is at rest, the oxillation frequency ν is uniquely related to the total energy of the pair [3].

3. Enneper's Equation and Eshelby's Potential

In Sect. 2 we have given an indication of the power of the application of Bäcklund's transformation to the kink solutions of Enneper's equation and of the surprising physics that follows from it. Many more results of a similar nature have been obtained [3,9]. In the case of Enneper's equation the solution of (2.2) and (2.6) for arbitrary C_0 involves elliptic functions and describes sequences of equally spaced kinks of the same ($C_0 < 0$) or of alternating ($C_0 > 0$) sign. From the latter solutions running finite-amplitude wave may be obtained by means of a simple transformation [3]. The Bäcklund transforms of all these solutions, which involve elliptic integrals of the third kind and/or theta functions, could be derived in analytical form [3].

A different set of solutions were obtained by Steuerwald [36] and Seeger [4], independently of each other and by different methods, Steuerwald [36] discussed the significance of these solutions for the differential geometry of pseudospherical surfaces, Seeger [4] and later Seeger and Wesolowski [37] that for crystal physics. The complete family of the Bäcklund transforms of the Steuerwald - Seeger solutions has recently been given in analytical form by Seeger and Wesolowski [38]. They involve elliptic integrals of the third kind and/or theta functions, too.

In a certain sense the solutions obtainable from (2.2) and (2.6) plus the Steuerwald - Seeger solutions constitute a set of basic solutions of Enneper's equation (2.16). They are the only ones that are of the form

$$\Phi = 4 \arctan [Z(z)\cdot T(t)] \quad , \tag{3.1}$$

where $Z(z)$ and $T(t)$ are functions of one independent variable only. Just as (2.19) may be considered to be a generalization of the superposition principle of linear equations, (3.1) may be viewed upon as a generalization of the separation-of-variables ansatz for the solution of linear partial differential equations. As far as the writer knows, all exact solutions of Enneper's equation available in closed-form are either special cases of (3.1) or derivable from them by (possibly repeated) application of Bäcklund's transformation.

As stated above, the Bäcklund transforms of all solutions of (2.16) of the form (3.1) are now known in analytical form. Since the construction of higher-order Bäcklund transforms by means of Bianchi's equation (2.19) involves only algebraic eliminations and (in special limiting cases) differentiations of analytic functions, a vast variety of exact solutions of Enneper's equation (2.16) are known or easily obtainable. Furthermore, by perturbation techniques [9] approximate solutions may be found for all those problems that lie in the neighbourhood of the exact solutions derivable from (2.3) or of their Bäcklund transforms.

For the application of soliton theory to the theory of plastic deformation as well as for other problems it is essential to consider the effects of <u>externally applied shear stresses</u>. A resolved shear stress σ in the glide system of a dislocation of Burgers vector b exerts a force of magnitude $b\sigma$ on the dislocation. This force acts in the glide plane in the direction perpendicular to the dislocation. With suitable normalization, it may be taken into account by adding a work term $-\Phi s$ to the energy density (2.8).

We see immediately that the inclusion of the external forces makes the sinusoidal potential very difficult to treat exactly. The Bäcklund transformation technique is no larger applicable and not even (2.6) may be evaluated in closed form any longer. While it is true that for the so-called <u>static kink-pair problem</u> approximate solutions for the sinusoidal potential may be obtained for small s by means of a perturbation treatment outlined earlier [10] and for large s by a procedure applicable to rather general potentials of $U(\Phi)$ (see below), one would nevertheless like to have quantitative information on the entire range of stresses of applied physical interest.

In 1962 Eshelby [14] pointed out that for certain kink problems the sinusoidal potential (2.22) may be replaced by

$$U_{Esh}(\Phi) = \frac{1}{2}\,\Phi^2(1-\Phi)^2 \quad , \qquad (3.2)$$

where we have normalized U_{Esh} in such a way that for $\Phi \to 0$ it reduces to the same limit as (2.22). Eq.(3.2) satisfies the conditions (cf. Sect. 2) $U(0) = U(\Phi) = 0$ with $\Phi = 1$. The energy barrier between the two minima is $U_{Esh}(1/2) = 1/32$, equal to 1/64 of that of (2.22). For the properties of single kinks in the potential (3.2), first discussed by Eshelby [14], see Table 1.

Apart from the different normalization, the main difference between (2.22) and (3.2) is that the former corresponds to an infinite number of degenerate vacuum states whereas the latter gives rise to only two, the minimum number for the general considerations of Sect. 2 to be applicable. This differences has profound mathematical consequences: For (3.2) an analogue to the Bäcklund transformation does not exist, and the soliton properties as discussed in Sect. 2 do not hold.

Nevertheless, in certain problems dealing with one kink only or with two kinks of opposite sign (i.e., a kink pair) Eq.(3.2) may be an excellent substitute for (2.22), since in these problems the "physics" is confined to the region between two potential walls. Another way of saying this is that only two vacuum states are involved in an essential manner. In Table 1 the results obtained from the two potentials are compared with each other.
The usefuless of considering Eshelby's potential (in the later literature often called Φ^4 potential) becomes obvious when we start treating problems involving a constant applied stress.

From

$$U(\Phi) = U_{Esh} - \Phi\, s \equiv U_s(\Phi) \qquad (3.3)$$

we see that - in constrast to the sinusoidal case discussed above - the analytic form of the radicants in (2.6) and (2.12b) is not radically changed by the inclusion of the $-\Phi s$ term.

With (3.3), Eq.(1.1) takes the form

$$\frac{\partial^2\Phi}{\partial z^2} - \frac{\partial^2\Phi}{\partial t^2} = \Phi\,(1 - 3\,\Phi + 2\,\Phi^2) - s \qquad (3.4)$$

Potential	Sinusoidal	Eshelby
$U(\Phi)$	$2\sin^2(\Phi/2)$	$\frac{1}{2}\Phi^2(1-\Phi)^2$
energy barrier	$U(\pi)=2$	$U(\frac{1}{2})=1/32$
differential equation	$\frac{\partial^2\Phi}{\partial z^2}-\frac{\partial^2\Phi}{\partial t^2}=\sin\Phi$	$\frac{\partial^2\Phi}{\partial z^2}-\frac{\partial^2\Phi}{\partial t^2}=\Phi(1-3\Phi+2\Phi^2)$
dislocation displacement	$a\Phi/2\pi$	$a\Phi$
single kink	$\Phi=4\arctan\exp z$	$\Phi=\frac{1}{2}[1+\tanh(z/2)]$
kink energy	$E_k=8$	$E_k=1/6$
kink width	$w_k=\pi a$	$w_k=2a$

Table 1:

Comparison of the sinusoidal potential and the Eshelby potential (a = separation of Peierls valleys = kink height)

Eq.(3.4) possesses three constant solutions $\Phi = \Phi_K$ (k = 1,2,3), corresponding to the three roots of

$$\Phi_K(1-\Phi_K)(1-2\Phi_K) = s \tag{3.5}$$

They represents straight dislocation lines in stable or unstable equilibrium.

In the following we restrict ourselves to $s \geq 0$ and denote by $\bar{\Phi}$ that solution of (3.5) that reduces to zero if $s \to 0$. We see immediately that

$$\left.\frac{d^2U_{Esh}(\Phi)}{d\Phi^2}\right|_{\Phi=\bar{\Phi}} = -\frac{ds}{d\bar{\Phi}} = (1-6\bar{\Phi}+6\bar{\Phi}^2) \tag{3.6}$$

is positive as long as Φ is less than

$$\Phi_P = \frac{1}{2}(1-3^{-1/2}) \quad . \tag{3.7}$$

The stress corresponding to (3.7),

$$s_P = \left.\frac{dU_{Esh}}{d\Phi}\right|_{\Phi=\Phi_P} = \Phi_P(2-3\Phi_P)/3 = 3^{-3/2}/2 \tag{3.8}$$

is known as the Peierls stress, defined as the maximum resolved shear stress under which a straight dislocation line running parallel to a Peierl valley may be in stable mechanical equilibrium. From the preceding equations one easily finds that

$$s_P - s = (\Phi_P-\bar{\Phi})^2[3^{1/2}+2(\Phi_P-\bar{\Phi})]. \tag{3.9}$$

For the remainder of this section, we concentrate on the static kink-pair problem already mentioned.[10]

----

[10] To some extent the following treatment is a simplified and hopefully more transparent version of earlier work [7]. Nevertheless, the discerning reader will recognize several new results, e.g. the amazingly simple expression (3.15) for the energy of a kink pair.

As discussed in Sect. 1, two kinks of opposite sign on the same dislocation line attract each other with a force that increases with decreasing kink - kink separation. On the other hand, the applied shear stress σ exerts on the kinks constant forces of opposite sign and of absolute magnitude $ab\sigma$, where a is the kink height and b the dislocation strength. For stresses less than the Peierls stress defined above a configuration of static equilibrium must thus exists in which the force due to the applied stress cancels the kink - kink interaction exactly. We denote the solution of (3.4) corresponding to this configuration by $\Phi_0(z)$.

Fig. 1 gives a qualitative sketch of how $\Phi_0(z)$ is expected to look. As has been done in Fig. 1, without loss of generality we may assume that the kink pair is centred at t = 0, i.e., that $\Phi_0(z) = \Phi_0(-z)$. The requirement $\lim(z\to\infty)\Phi_0 = \bar{\Phi}$ means that in (2.6) the constant of integration C_0 has to be chosen in such a way that $U_s(\Phi)-C_0$ has a double root at $\Phi = \Phi$. This allows us to write

$$U_s(\Phi)-C_0 = \frac{1}{2}(\Phi-\bar{\Phi})^2[\Phi^2+2\Phi(\bar{\Phi}-1)+2\bar{\Phi}(\bar{\Phi}-1)+(\bar{\Phi}-1)^2] . \tag{3.10a}$$

The smaller of the zeros of the factor in square brackets,

$$\Phi_{max} = 1 - \bar{\Phi} - [2\bar{\Phi}(1-\bar{\Phi})]^{1/2} , \tag{3.11}$$

is the maximum value that Φ_0 may assume.

With (3.11) we may rewrite (3.10a) as

$$U(\Phi)-C_0 = \frac{1}{2}(\Phi-\bar{\Phi})^2(\Phi-\Phi_{max})[\Phi+\Phi_{max} - 2(1-\bar{\Phi})] . \tag{3.10b}$$

Inserting (3.10) into (2.4) or (2.6) gives us upon integration

$$\Phi_0 = \bar{\Phi} + \frac{4\mu^2}{[2\bar{\Phi}(1-\bar{\Phi})]^{1/2}\,\mathrm{Cosh}(2z\mu) + 1 - 2\bar{\Phi}} \tag{3.12a}$$

$$= \bar{\Phi} + 2(\Phi_{max}-\bar{\Phi})\,\frac{\mathrm{Cosh}(z_0\mu)}{\mathrm{Cosh}(2z\mu) + (\mathrm{Cosh}(2z_0\mu)} \tag{3.12b}$$

$$= \bar{\Phi} + \frac{1}{2}(\Phi_{max}-\bar{\Phi})\,\mathrm{Coth}(z_0\mu)\,\{\mathrm{Tanh}[(z+z_0)\mu] - \mathrm{Tanh}[(z-z_0)\mu]\} \tag{3.12c}$$

with the abbreviations

$$4\mu^2 = \left.\frac{d^2U_{Esh}}{d\Phi^2}\right|_{\Phi=\bar{\Phi}} = (1 - 6\bar{\Phi} + 6\bar{\Phi}^2) = 6(\Phi_P-\bar{\Phi})(1-\Phi_P-\bar{\Phi}) \tag{3.12d}$$

$$\mathrm{Sinh}(2z_0\mu) = \frac{2\mu}{[2\bar{\Phi}(1-\bar{\Phi})]^{1/2}} , \quad \mathrm{Cosh}(2z_0\mu) = \frac{1 - 2\bar{\Phi}}{[2\bar{\Phi}(1-\bar{\Phi})]^{1/2}} . \tag{3.12e}$$

Further useful expressions are

$$2\bar{\Phi}(1-\bar{\Phi}) = \frac{1}{\mathrm{Cosh}^2(2z_0\mu) + 2} \tag{3.13a}$$

$$2\bar{\Phi} = 1 - \frac{\mathrm{Cosh}(2z_0\mu)}{[\mathrm{Cosh}^2(2z_0\mu) + 2]^{1/2}]} \tag{3.13b}$$

$$\Phi_{max} - \bar{\Phi} = \frac{\mathrm{Sinh}^2(z_0\mu)}{[\mathrm{Cosh}^2(2z_0\mu) + 2]^{1/2}} \tag{3.13c}$$

$$s = \frac{\mathrm{Cosh}(2.z_0\mu)}{[\mathrm{Cosh}(2z_0\mu) + 2]^{3/2}} . \tag{3.13d}$$

The energy of a kink pair at rest may be written as

$$E_{kp} = \int_{-\infty}^{+\infty} [\frac{1}{2}(\frac{d\Phi_0}{dz})^2 + U_s(\Phi_0) - U_s(\bar{\Phi})]d$$

$$= 2^{3/2}\int_{\Phi=\bar{\Phi}}^{\Phi=\Phi_{max}} [U_s(\Phi)-C_0]^{1/2}d\Phi \qquad (3.14)$$

$$= \frac{1}{3}\,\mu\,(1 - 6\,s\,z_0).$$

Since in the present model the energy of an isolated kinks is $E_K = 1/6$, we obtain finally, with $z_P \equiv 3^{1/2}$ (see below),

$$E_{kp} = 2E_k\mu(2-12sz_0) = 4E_k\mu[(1-\frac{s}{s_P})-(\frac{z_0}{z_P}-1)+(1-\frac{s}{s_P})(\frac{z_0}{z_P}-1). \qquad (3.15)$$

The preceding formulae provide us with parametric expressions for the dependence of the kink-pair solution $\Phi_0(z)$ and the kink-pair energy E_{KP} on the stress s, which may vary between zero and the Peierls stress s_P. Eq. (3.12c) shows that the kink-pair solution may be looked upon as the superposition of two single kinks of opposite sign but with a modified kink height and kink width.[11)]

The limiting case $s \to 0$ corresponds to $\bar{\Phi} \to 0$, $\Phi_{max} \to 1$, $\mu \to 1/2$, $z_0 \to \infty$, and $E_{KP} \to 2E_K$. In this case $\Phi_0(z)$ reduces to the superposition of two isolated kinks separated by a very large distance z_0.

The opposite limiting case $s \to s_P = 3^{-3/2}/2$ gives us $\bar{\Phi} \to \Phi_P$, $\Phi_{max} \to \Phi_P$, $\mu \to 0$, $z_0 \to [2\Phi_P(1-\Phi_P)]^{1/2} \equiv z_P = 3^{1/2}$. Making use of $\mathrm{Cosh}(2z_0\mu) \to 1$, we see that for s sufficiently close to s_P eq.(3.12) reduces to

$$\Phi_0 = \bar{\Phi} + (\Phi_{max} - \bar{\Phi})\,\mathrm{Sech}^2\mu z. \qquad (3.16)$$

Eq. (3.16) was obtained by several writers, e.g. Nabarro [40], Büttiker and Landauer [41], and Mori and Kato [42], by expanding $F(\Phi_0)$ in powers of Φ_0 and retaining in (2.3) terms up to second order only. These derivations show that in the case of applied shear stresses close to the Peierls stress the solution (3.16) holds for any potential $U(\Phi)$ that is reasonably smooth near its turning point determining s_P. If $U'''(\Phi_P)$ denotes the third derivative of $U(\Phi)$ at that point, the parameter μ in (3.16) is given by [8]

$$\mu^4 = -U'''(\Phi_P)(s_P - s)/8. \qquad (3.17)$$

From (3.17) it follows [40-42] that for s sufficiently close to s_P

$$E_{kp} = \frac{96}{5}\,[-2U'''(\Phi_P)]^{-3/4}\,(s_P - s)^{5/4}\,. \qquad (3.18)$$

For extensive discussions of the other limiting case, $s/s_P << 1$, both for the Eshelby potential [43] and for the general case [8] the reader is referred to the literature. (For the correction of an error in the discussion of E_{KP} in the limit $s \to s_P$ in the Appendix of [43] see [23]).

---

11) This is analogous to Nabarro's dislocation-pair solution of Peierls' integral-differential equation [39].

4. Massive Phonons and Unstable Modes

Eq. (2.1) with $F(\Phi)$ fulfilling the necessary conditions for the existence of kink solutions admits also solutions that are periodic in ζ and which, for $1 < v < \infty$, represent running waves. They may be obtained by choosing in (2.2b)

$$v^2 > 1 \ , \tag{4.1a}$$

resulting in an imaginary ζ coordinate, and

$$C_0 > 0. \tag{4.1b}$$

From (2.4) we see that the maximum amplitude Φ^* of these finite-amplitude oscillations is given by a solution of

$$U(\Phi^*) = C_0. \tag{4.2}$$

v represents their phase velocity. The fact that its modulus is always larger than the limiting velocity $c = 1$ is in accordance with the Lorentz invariance of (2.1) and de Broglie's relationship

$$v\,V = c^2 \tag{4.3}$$

between phase velocity v and group velocity V.

The finite-amplitude oscillations of Enneper's equation (2.16) were first studied by Seeger, Donth, and Kochendörfer [3]. They denoted them as "oszillatorische Eigenbewegungen" in order to distinguish them from the kink solutions (which they called "translatorische Eigenbewegungen") on the one hand and the well-known harmonic "Eigenschwingungen" of linear systems on the other hand. They also showed that, in spite of the non-linear character of the problem, exact solutions for the corresponding wave packets may be obtained by the Bäcklund transformation technique. These wave packets satisfy indeed the relationship (4.3).

The present author demonstrated [4] that the corresponding standing waves of finite amplitude could be obtained in closed form, too. They were found to be of the form (3.1) with $Z(z)$ and $T(t)$ given by elliptic functions of their arguments.

For linear wave equation, e.g. the one-dimensionsal Klein-Gordon equation

$$\frac{\partial^2\Phi}{\partial z^2} - \frac{\partial^2\Phi}{\partial t^2} = \Phi \tag{4.4}$$

obtainable from (2.1) by replacing $F(\Phi)$ by Φ, the relationships between running waves, standing waves, and wave packets are well known. They are based on the "superposition principle" of linear equations mentioned in Sect. 2.

In the case of Enneper's equation the relationships between waves and wave packets of finite amplitudes involve a non-linear superposition principle which mathematically is fully equivalent to that discussed in Sect. 2 in the context of dynamic kink - kink interactions. We could, in fact, have based the definition of soliton behaviour just as well on the properties of the finite-amplitude waves. This demonstrates that it is an unnecessary restriction to associate the soliton properties of certain non-linear partial differential equations only with kink-type or wave-pulse solutions.

A remarkable feature of the wave solutions of (2.16), emphasized by Seeger, Donth, and Kochendörfer [3], is that their frequency does not go to zero as the wavelength goes to infinity. This feature, which they have in common with the optical vibrational modes of crystals, is preserved in the small-amplitude oscillations. Their quantization results in quanta of finite restmass, which Eshelby [14] has termed "heavy phonons" in order to emphasize the distinction from the "light phonons" obtained by quantizing the acoustic modes of solids and resulting in the Debye spectrum. Following a more modern terminology that is in agreement with the nomenclature of elementary-particle physics we shall speak of massive and massless phonons, respectively.

In the dislocation model of Sect. 2 the massive phonons are associated with the vibrations of a dislocation in its Peierls valley. It is obvious that the vibrational modes and the massive-phonon spectrum are modified if the dislocation contains kinks. If the system is described by Enneper's equation, i.e. if we assume $U(\Phi)$ to be sinusoidal, these interactions can be treated exactly by the Bäcklund transformation technique even for finite vibration amplitudes. For many applications, e.g. in statistical mechanics (cf. Sect. 5), it suffices to treat the vibration amplitudes as infinitesimal. This has not only the advantage of mathematical simplicity because we are now allowed to use perturbation theory but permits us to study also non-sinusoidal potentials $U(\Phi)$, for which the Bäcklund transformation technique is not available. In particular, we may employ the Eshelby potential (3.2) in order to investigate the interaction between the massive phonons and kink pairs in unstable equilibrium under an applied stress.

The perturbation-theory technique for studying kink - massive-phonon interactions was initiated by Seeger and Kochendörfer [10] and carried much further by Seeger and Schiller [7,44]. The basic idea is to superimpose on a known solution $\Phi_0(z)$ of (2.1) an infinitesimal one $\phi(z,t)$, i.e., to write

$$\Phi(z,t) = \Phi_0(z) + \phi(z,t) \tag{4.5}$$

and to retain only first-order terms in $\phi(z,t)$.[12]

Insertion of (4.5) into (2.1) gives us

$$\frac{\partial^2 \phi(z,t)}{\partial z^2} - \frac{\partial^2 \phi(z,t)}{\partial t^2} - U''(\Phi_0)\,\phi(z,t) = 0, \tag{4.6}$$

where

$$U''(\Phi_0) \equiv -\frac{dF(\Phi_0)}{d\Phi_0} = \frac{d^2 U(\Phi_0)}{d\Phi_0{}^2}\ . \tag{4.7}$$

By means of the separation ansatz

$$\phi(z,t) = \phi(z)\exp(i\omega t) \tag{4.8}$$

(4.6) may be reduced to the ordinary differential equation

---

12) For the application intended in the present paper, if suffices to take $\Phi_0(z)$ to depend on one independent variable only. In the case of Enneper's equation (2.16) much more general exact solutions $\Phi_0 = \Phi_0(z,t)$ that do not permit the reduction of (4.6) to ordinary differential equations may be taken as zero-order solutions for perturbation theory [9].

$$\frac{d^2\Phi}{dz^2} + [\omega^2 - U''(\Phi_0)]\ \Phi(z,t) = 0\ . \qquad (4.9)$$

Eq.(4.9) has the form of a one-dimensional time-independent Schrödinger equation for the wavefunction $\Phi = \Phi(z)$ if we identify ω^2 with the energy E and $U''(\Phi_0(z))$ with the potential $V(z)$, both in units of $\hbar^2/2m$, where $\hbar$ denotes Planck's constant divided by 2π and m the particle mass. Considering (4.9) as a one-dimensional Schrödinger equation may be very helpful for qualitative considerations.

By introducing $\Phi_0 = \Phi_0(z)$ as the independent variable, (4.9) may be transformed into another useful general form, viz.

$$\frac{d^2\Phi}{d\Phi_0^2} + \frac{d\ln(d\Phi_0/dz)}{d\Phi_0}\frac{d\Phi}{d\Phi_0} + \frac{\omega^2 - U''(\Phi_0)}{(d\Phi_0/dz)^2}\ \Phi = 0 \qquad (4.10a)$$

or, making use of (2.4), into

$$\frac{d^2\Phi}{d\Phi_0^2} + \frac{1}{2}\ \frac{d\ln[U(\Phi_0)-C_0]}{d\Phi_0}\ \frac{d\Phi}{d\Phi_0} + \frac{\omega^2 - U''(\Phi_0)}{2[U(\Phi_0)-C_0]}\ \Phi = 0. \qquad (4.10b)$$

Eq. (4.10b) is particularly useful if $U(\Phi_0)-C_0$ is a polynominal in Φ_0 with first- or second-order zeros only [$U_s(\Phi_0)$ as given by (3.3) provides an example for this]. In this case it falls into the well-studied class of linear second-order ordinary differential equations of the Fuchsian type, which comprises most of the classical second-order differential equations of mathematical physics (see, e.g. [45]).

Insertion of (2.22) and the single-kink solution (2.23) into (4.9) leads to

$$\frac{d^2\Phi}{dz^2} + [\omega^2 - 1 + 2\ \mathrm{Sech}^2 z]\ \Phi = 0 \quad . \qquad (4.11)$$

In the present context (4.11) was first obtained and solved (by transformation to a hypergeometric equation) in the author's thesis (see also Seeger and Kochendörfer [10]). In connection with the kink problem its solutions were later discussed in detail by Seeger and Schiller [44].

Insertion of (3.3) and the kink-pair solution (3.12) into (4.9) gives us

$$\frac{d^2\Phi}{dz^2} + \{\omega^2 - 4\mu^2 - 6(\Phi_0-\bar{\Phi})[\Phi_0-\bar{\Phi}-(1-2\bar{\Phi})]\}\ \Phi = 0 \qquad (4.12a)$$

or

$$\frac{d^2\Phi}{dz^2} + \{\omega^2 - 4\mu^2 + 24\mu^2\ \frac{1 + \mathrm{Cosh}2\mu z_0\ \mathrm{Cosh}2\mu z}{[\mathrm{Cosh}2\mu z_0 + \mathrm{Cosh}2\mu z]^2}\}\ \Phi = 0 \qquad (4.12b)$$

or

$$\frac{d^2\Phi}{dz^2} + \mu^2\{(\frac{\omega}{\mu})^2 - 4 + [\frac{6}{\mathrm{Cosh}2\mu(z+z_0)} + \frac{6}{\mathrm{Cosh}2\mu(z-z_0)}]\}\ \Phi = 0\ . \qquad (4.12c)$$

With the abbreviations

$$x \equiv \mu z, \qquad x_0 \equiv \mu z_0, \qquad (4.13a)$$

eqs. (4.12) may be written as

$$\frac{d^2\Phi}{dx^2} + \{(\frac{\omega}{\mu})^2 - 4 + \frac{24(1 + \mathrm{Cosh}2x_0\ \mathrm{Cosh}2x}{[\mathrm{Cosh}2x_0 + \mathrm{Cosh}2x]^2}\}\ \Phi = 0. \qquad (4.13b)$$

In the limiting case $s \to s_P$, $x_0 \to 0$, eq. (4.13b) becomes

$$\frac{d^2\Phi}{dx^2} - \{\kappa^2 - n(n+1)\,\mathrm{Sech}^2 x\}\,\Phi = 0 \tag{4.14}$$

with

$$n = 3, \quad \kappa^2 = 4 - (\omega/\mu)^2 \quad . \tag{4.15}$$

The case of a single kink (cf. Table 1) in the Eshelby potential is obtained from (4.12) by putting

$$x \equiv \mu(z \pm z_0) \tag{4.16a}$$

and $|z_0| \to \infty$, $\mu \to 1/2$. This limiting process leads to (4.14) with

$$n = 2, \quad \kappa^2 = 4 - \omega^2 \quad . \tag{4.16b}$$

Furthermore, comparison of (4.11) with (4.14) shows that, with

$$x \equiv z \tag{4.17a}$$

and

$$n = 1, \quad \kappa^2 = 1 - \omega^2 \;, \tag{4.17b}$$

eq.(4.14) contains also the case of a single kink in a sinusoidal potential.

From the preceding it is clear that (4.14) with integer n in an important equation in soliton theory. It plays a key rôle in the study of the soliton properties of the Korteweg - de Vries equation, too [11,12]. Darboux [46] realized as early as 1889 that for integer n the general solution of (4.14) may be given in closed form and written as

$$\Phi = \mathrm{Cosh}^{n+1}x \left(\frac{1}{\mathrm{Cosh}x}\frac{d}{dx}\right)^{n+1} [C_1 \exp(\kappa x) + C_2 \exp(-\kappa x)], \tag{4.18}$$

where $C_{1,2}$ denote constants of integration. It appears therefore justified to call (4.14) Darboux's equation, following the usage of von Koppenfels [47].

von Koppenfels [47] studied (4.14) in a wider context, including non-integer n, and established the relationship to the hypergeometric differential equation and the differential equation of the associated Legendre functions. The latter is obtained by means of the substitution

$$\xi = \mathrm{Tanh}\, x, \tag{4.19a}$$

which transforms (4.1) into the associated Legendre equation

$$(1-\xi^2)\frac{d^2\Phi}{d\xi^2} - 2\,\xi\,\frac{d\Phi}{d\xi} + \left[n(n+1) - \frac{\kappa^2}{1-\xi^2}\right]\Phi = 0, \tag{4.19b}$$

or, with

$$\xi = \cos\theta \;; \qquad \theta = \arctan \mathrm{Csch}\xi = 2 \arctan \exp\xi \;, \tag{4.20a}$$

into

$$\frac{d^2\Phi}{d\theta^2} + \cos\theta\,\frac{d\Phi}{d\theta} + \left[n(n+1) - \frac{\kappa^2}{\sin^2\theta}\right]\Phi = 0. \tag{4.20b}$$

Localized solutions of (4.14) have to satisfy the boundary conditions

$$\lim_{|x| \to 0} \Phi(x) = 0 \quad . \tag{4.21}$$

As is easily seen from Darboux's solution (4.18), (4.21) will be satisfied if and only if

$$\kappa^2 = 1,\ 4,\ 9,\ \ldots,\ n^2 \quad . \tag{4.22}$$

The eigenfunctions belonging to the eigenvalues (4.22) are the associated Legendre polynomials $P_n^\kappa(\xi)$, viz. (cf., e.g., [14])

$$n = 1: \quad P_1^1(\xi) = (1-\xi^2)^{1/2} = \frac{1}{\mathrm{Cosh}x} \quad , \tag{4.23}$$

$$n = 2: \quad P_2^2(\xi) = 3(1-\xi^2) = \frac{3}{\mathrm{Cosh}^2 x} \ , \ P_2^1 = 3\xi(1-\xi^2)^{1/2} = \frac{3\ \mathrm{Tanh}x}{\mathrm{Cosh}x} \ , \tag{4.24}$$

$$n = 3: \quad P_3^3(\xi) = 15(1-\xi^2)^{3/2} = \frac{15}{\mathrm{Cosh}^3 x}, \ P_3^2 = 15\ \xi(1-\xi^2) = \frac{15\ \mathrm{Tanh}x}{\mathrm{Cosh}^2 x} \ ,$$

$$P_3^1\ (\xi) = \frac{3}{2}\ (5\xi^2-1)(1-\xi^2)^{1/2} = \frac{3}{2}\ \frac{(5\ \mathrm{Tanh}^2 x - 1)}{\mathrm{Cosh}\ x} \ . \tag{4.25}$$

In discussing the physical meaning of the eigenfunctions, it is expedient to return to the case of a general potential $U(\Phi)$ [eqs. (2.1) and (4.9)]. From the fact that (2.1) does not contain z explicitly it follows that if $\Phi_0(z)$ is a solution of (2.1), $\Phi_0(z+z_0)$ is also one for arbitrary z_0. This means that

$$\Phi = d\Phi_0/dz \equiv 2^{1/2}[U(\Phi_0(z)) - C_0]^{1/2} \tag{4.26}$$

is a solution of (4.9) for the eigenvalue $\omega^2 = 0$, as may be easily verified by inserting (4.26) into (4.9). The physical meaning of this solution is that it corresponds to the infinitesimal translations of $\Phi_0(z)$, i.e., of a single kink in the cases n = 1 and n = 2, and of the kink pair in the case n = 3.

The eigenfunctions of (4.14) associated with $\omega = 0$ are $P_1^1(\xi)$, $P_2^2(\xi)$, and $P_3^2(\xi)$. For n = 1 (Enneper's equation) there are no further eigenfunctions, which means that in this case no massive phonon mode localized at a single kink exists. In the case n = 2 (Eshelby's potential) such a localized mode does exist. It corresponds to $\kappa^2 = 1$, hence $\omega^2 = 3$, and is described by the eigenfunction $\Phi_1 = P_2^1(\xi)$, which possesses a node at the centre of the kink.

For n = 1 and n = 2 the translational modes $\omega = 0$ correspond to the ground-state solutions Φ_0 of the Schrödinger equations (4.14). As a direct consequence of the fact that n = 3 pertains to the kink-pair solution and not to isolated kinks, this is no longer so in that case. We may understand this by the following argument, which holds for any physically acceptable $U(\Phi)$.

By a suitable choice of the z origin, the kink-pair solution $\Phi_0(z)$ of (2.1) can always be chosen to be an even function, so that $d\Phi_0/dz$ is an odd function. Under the general conditions laid down for $U(\Phi)$ in Sect. 2, $\Phi_1 \approx d\Phi_0/dz$ has precisely one node, located at z = 0. Then Sturm - Liouville theory tell us that exactly one eigenfunction with $\omega^2 < 0$ exists and that this is even and nodeless, corresponding hence to the ground-state solution $\Phi_0(z)$ of (4.10).

The physical meaning of the eigenfunction $\Phi_0(z)$ with eigenvalue $\omega_0^2 < 0$ is as follows. As mentioned in Sect. 3, the kink-pair solutions of (2.1) correspond to configurations of unstable equilibrium. Linear combinations of the solutions

$$\Phi(z,t) = \Phi_0(z) \exp(\pm\lambda_0 t) \tag{4.27}$$

of the perturbation equation (4.6) with

$$\lambda_0 = (-\omega_0^2)^{1/2} > 0 \tag{4.28}$$

describe the decay of these unstable configurations. They correspond to the drifting apart of the two kinks if the $\Phi_0 \exp(\lambda_0 t)$ contribution has the same sign as Φ_0, and to the approaching of the kinks in the opposite case.

An upper limit for λ_0 may be obtained from (4.9) by observing that $\omega_0^2-U''(\Phi_0)$ cannot be negative everywhere. This means that $\lambda_0 \leq -U''_{min}$, where U''_{min} denotes the minimum value of $d^2U(\Phi_0)/d\Phi_0^2$. This minimum value is attained at a zero of either $d^3U(\Phi_0)/d\Phi_0^3$ or $dU(\Phi_0)/dz$.

If the applied stress s is sufficiently close to s_P [which, as we have seen, is covered by (4.16) with n = 3], the spatial eigenfunction $\Phi_0(z)$ of the decay mode is the nodeless associated Legendre polynomial $P^3(\xi)$, the decay constant being given by

$$\lambda_0 = 5^{1/2} \mu. \tag{4.29}$$

Let us now return to (4.12), the perturbation equation for kink pairs in the Eshelby potential under constant applied stress $0 < s < s_P$. (For the limiting cases s = 0 and s = s_P exact solutions are known not only for the localized eigenfunctions but, by making κ imaginary, also for the continuous spectrum.)

By means of the substitution

$$\Phi(x) = [\mathrm{Cosh}\ 2x_0 + \mathrm{Cosh}\ 2x]^{-2}\ u(x), \tag{4.30a}$$

or

$$\Phi(x) = [\mathrm{Cosh}\ 2x_0 + \mathrm{Cosh}\ 2x]^{3}\ v(x) \tag{4.30b}$$

eq.(4.13b) may be transformed into

$$[\mathrm{Cosh}2x_0+\mathrm{Cosh}2x]\ u''+12\ \mathrm{Sinh}2x\ v' + (36\ \mathrm{Cosh}2x-\kappa^2[\mathrm{Cosh}2x_0+\mathrm{Cosh}2x])\ u = 0 \tag{4.31a}$$

or

$$[\mathrm{Cosh}2x_0+\mathrm{Cosh}2x]\ v'' - 8\ \mathrm{Sinh}2x\ u' + (16\ \mathrm{Cosh}2x-\kappa^2[\mathrm{Cosh}2x_0+\mathrm{Cosh}2x])\ v = 0 \tag{4.31b}$$

with κ^2 given by (4.15) .

In the terminology of Arscott [48] eqs. (4.31) may be called modified generalized Ince equations. It is well known that insertion of the expansions

$$\left.\begin{array}{l} u(x) \\ v(x) \end{array}\right\} = \sum_{r=0}^{\infty} A_r \text{ Cosh } 2rx \qquad (4.32\text{a})$$

or

$$\left.\begin{array}{l} u(x) \\ v(x) \end{array}\right\} = \sum_{r=1}^{\infty} B_r \text{ Sinh } 2rx \qquad (4.32\text{b})$$

results in three-term recurrence formulae for A_r or B_r, and that these may be treated by the technique of continued fractions [48,49]. Alternatively, the continued-fraction technique may be applied to the three-term recurrence formula obtainable from the "algebraic form" (4.10).

It is easily seen that

$$u(x) = B_1 \text{ Sinh } 2x \qquad (4.33\text{a})$$

is a solution of (4.31 a) for $\kappa^2 = 4$. With

$$B_1 = -\frac{\text{Sinh}^3 2x_0}{[\text{Cosh}2x_0 + 2]^2} \qquad (4.33\text{b})$$

it is equal to $\Phi_1 = d\Phi_0/dz$ already known to be the eigenfunction associated with $\omega^2 = 0$, $\kappa^2 = 4$. An alternative way of writing (4.33) is

$$\Phi(x) = -\frac{B_1}{4 \text{ Sinh}2x_0}\left[\frac{1}{\text{Cosh}^2(x+x_0)} - \frac{1}{\text{Cosh}^2(x-x_0)}\right] . \qquad (4.34)$$

Apart from the second (even) solution for $\omega = 0$, which may be obtained from (4.34) by quadratures but does not satisfy (4.21) and is therefore not very useful in the present context, (4.34) is the only known closed-form solution of (4.13) for general x_0. We have therefore to resort to the numerical evaluation of the three-term recurrence formulae or to various schemes of approximation. This has been done by a group in the author's laboratory comprising Dr. ner.nat. E. Mann, Dipl. Math. E. Mielke, and Dipl. Ing. W. Lay. Some of their results on the discrete eigenvalue spectrum of (4.13) will be summarized below. Details as well as the results on the continous spectrum, which are also very important for the application in statistical mechanics (cf. Sect. 5), will be reported elsewhere.

Considerable insight may be gained by considering (4.13) as the one-dimensional Schrödinger equation for a potential energy consisting of two potential wells [cf. eq. (4.12c)]. In this analogy it is convenient to use for the energy E a scale that differs from that discussed in connection with (4.9) in both its origin and its normalization. We put

$$E = (\omega/\mu)^2 - 4 . \qquad (4.35)$$

This choice ensures that the eigenvalues associated with the <u>discrete</u> part of the eigenvalue spectrum are <u>negative</u>, and that the continuous part of the spectrum belongs to $E \geq 0$.

For the discussion of the <u>negative</u> eigenvalues it is convenient to characterize them by the parameter

$$\kappa = (-E)^{1/2} = [4 - (\omega/\mu)^2]^{1/2} \qquad (4.36)$$

and to consider first two potential wells that are so widely separated that the wavefunctions localized in them do not influence each other. We have then the case n= 2 of (4.14), hence according to (4.22) two pairs of doubly degenerate eigenvalues, $\kappa_0 = 2$ and $\kappa_1 = 1$. The wavefunctions associated with the first pair are symmetric, those associated with the second pair antisymmetric with respect to the centres of the individual wells. The analytical expressions for these wavefunctions and their physical meaning have been discussed above.

When we allow for the interaction between the potential wells, the degenerate eigenvalues split into pairs of non-degenerate ones. The lower eigenvalues of each pair belong to eigenfunctions $\Phi_0(x)$ and $\Phi_2(x)$ that are symmetric with respect to the centre between the two wells (which we choose to be located at $x = 0$), whereas the upper ones are associated with antisymmetric eigenfunctions $\Phi_1(x)$ and $\Phi_3(x)$. If we now let the distance between the potential wells decrease, the symmetry of these wavefunctions does not change. This allows us to determine, from the above discussion of Darboux's equation in the case n = 3, the end points of the energy levels considered as a function of the well separation.

The situation just discussed is illustrated in Fig. 5(a), where the square roots κ_i of the negative energy levels E_i of (4.13) have been plotted as a function of $\eta_0^{-1} \equiv \mathrm{Sech} 2x_0$ [cf. eq. (A.11]. From the discussion of $\Phi_1(z)$ it is known that $\kappa_1 = 2$ for $\eta_0 > 1$. The other curves have been obtained by E. Mielke from the three-term recurrence formulae mentioned above.

Fig. 5b gives the transcription of the results of Fig. 5a into the dependence of the squares of the frequencies of the localized modes, ω_2^2, on the applied stress s, taking into account the s-dependence of μ as discussed in Sect. 3. We see that for the kink pair in the Eshelby potential in addition to the decay mode $\Phi_0(x)$ and the translational mode $\Phi_1(x)$ there are two localized modes $\Phi_2(x)$ and $\Phi_3(x)$. They correspond to even or odd combinations of the localized vibrational modes of the single kinks discussed above.

In Fig. 5a we have indicated the symmetries of the wave functions in the two limiting cases $\eta_0^{-1} = 0$ and $\eta_0^{-1} = 1$. We see that as η_0^{-1} approaches zero, the symmetry of the wavefunctions changes. This means that in that limit the wavefunction cannot be obtained from (4.31) or analogous expressions. If we wish to obtain analytical approximations to the eigenvalues and eigenfunctions in the limit of small s, we have to use different techniques. One such technique, which is based on the WKBJ method [15] but makes use of the fact that the wavefunctions for the single-well problem are known, is developed in the Appendix. It is applicable not only to the discrete part of the eigenvalue spectrum but also to the continuous spectrum.

Also included in Fig. 5a is the upper limit for κ_0 that is obtained by the criterion $\lambda_0^2 \leq -U''_{min}$. In the present case this gives us

$$\kappa_0^2 = \begin{cases} 6\,\mathrm{Coth}^2 2x_0 & (\eta_0 > 2) \\ 12\,\mathrm{Sech}^2 x_0 & (\eta_0 < 2). \end{cases} \qquad (4.37)$$

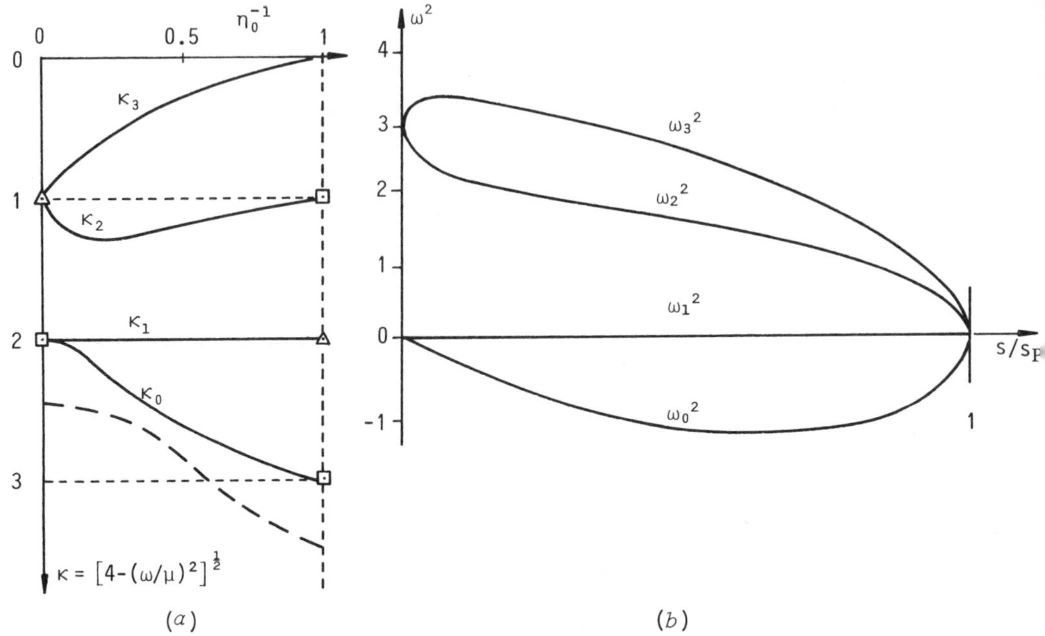

Fig. 5.

The eigenvalues E_i and ω_i^{-2} associated with the localized solutions of eq. (4.13). a): $\kappa_i = (-E_i)^{1/2} = [4 - (\omega_i/\mu)^2]^{1/2}$ plotted as a function of the reciprocal of $\eta_0 = \text{Cosh}2x_0$ [cf. eq. (A.11)]. b) ω_i^2 plotted against the applied stress s divided by the Peierls stress s_P. In both diagrams the left-hand limit describes isolated kinks, whereas the right-hand limit pertains to $s \to s_P$. Even subscript are associated with even eigenfunctions Φ_i, odd subscripts with odd eigenfunctions Φ_i. In the limiting cases squares or triangles indicate eigenvalues associated with symmetric or anti-symmetric eigenfunctions. Also included (dashed line of Fig. 5a) is the upper limit for κ_0 following from (4.37).

5. The Rate of Kink-Pair Formation.

Immediately after the discovery of the soliton solutions of the Enneper equation, considerations involving thermal activation were applied to them [3,5,6]. Specifically, it was proposed [5,21] that the thermally activated generation of kink pairs in dislocation lines was the origin of the so-called Bordoni relaxation [50,51] in the internal friction of metals [cf. footnote5)]. This interpretation has been tested by very detailed experiments and is now generally accepted [52].

Over the years a number of attempts were made to apply to the kink-pair formation problem what is now called soliton theory, starting out with the work of Donth [6] and Seeger, Donth, and Pfaff [21] (see, e.g., [23]). Most experimental investigations of the Bordoni relaxation involve applied stresses that are small compared to the Peierls stress. Recently, considerable interest has arisen in the rôle of kink-pair formation in the plastic deformation of body-centred cubic metals [25,43]. Here the external stresses σ are no longer small compared to the Peierls stress σ_P. Under these circumstances the line-tension model of Sect. 2 should be able to account quite well for the mechanical problems involved [8,43].

Thermally activated processes are usually treated in terms of the Klein - Kramers equation [53-55]. In one space dimension it reads as follows:

$$\frac{\partial \rho}{\partial t} = - F_1(q)\frac{\partial \rho}{\partial p} - \frac{p}{m}\frac{\partial \rho}{\partial q} + \frac{1}{\mu_1 m}\frac{\partial}{\partial p}\{p\rho + mk_B T \frac{\partial \rho}{\partial p}\} \quad . \tag{5.1}$$

In (5.1) $\rho = \rho(p,q,t)$ is the density in phase space (p,q) of an ensemble of particles of mass m and mobility μ_1 (for simplicity both taken as velocity-independent). p and q denote the momentum and space coordinates of the particles, t the time. $F_1 = F_1(q)$ is the force acting on the particles; it is assumed to be independent of p and t. As usual, k_B and T stand for Boltzmann's constant and absolute temperature.

In the application to kink-pair formation [7,8,24] the "reaction coordinate" q describes the separation of the two kinks of a kink pair (cf. Fig. 1). The force $F_1(q)$, which is distinct from the force per unit length $F(\Phi)$ introduced in Sect. 2., results on the one hand from the attraction between these kinks and on the other hand from the external forces $\pm ab\sigma$ (cf. Sect. 3) that tend to pull the kinks apart . At a certain coordinate $q = \hat{q}$ these two forces balance each other, so that we have $F_1(\hat{q}) = 0$. This corresponds to a maximum of the potenti energy associated with $F_1(q)$, hence to a configuration of unstable mechanical equilibrium as discussed in Sect. 4.

A kink-free dislocation is described by q = 0. In thermal equilibrium the phase-space density ρ is high near $\rho = 0$. For the applications indicated above we need to know the rate with which the "particles" concentrated near q = 0 escape over the energy barrier at $q = \hat{q}$. The height of this barrier, which depends on the applied stress σ since both F_1 and $\hat{q}$ are functions of σ (or s), is given by

$$H_{kp}(\sigma) = -\int_0^{\hat{q}} F_1(q)\, dq. \tag{5.2}$$

"Particles" at $q \gg \hat{q}$ correspond to isolated kinks.

At first sight it appears extremely difficult to extend the qualitative picture just described into a quantitative treatment of the kink-pair formation rate based on (5.1). The fact that kinks might be treated as solitons and hence described by a particle picture does not seem to help much since for small kink separations (i.e., nearly annihilated kink pairs) the kink - kink separation coordinate q and hence the function $F_1(q)$ is ill-defined. Even if some plausible assumption for $F_1(q)$ at small q is made[13], it would appear that we still have to solve the difficult task of finding the appropriate solution of a linear partial differential equation with three independent variables and non-constant coefficients.

Fortunately, it turns out that with the assumption

$$H_{kp}(\sigma) \gg k_B T \tag{5.3}$$

----

[13] $F_1(q)$ is known explicitly at large q, since here the kink - kink interaction is dominated by the long-range elastic interactions between kinks [7,8,14,44,56]. This interaction, which is rather easy to deal with (cf., e.g., [8,24]), is not included in the present treatment. For the distinction between "snall" and "large" q, which depends on the applied stress σ, see [8,43].

the difficulties just mentioned can be resolved in a highly satisfactory manner. For the application to crystal plasticity the inequality (5.3) poses virtually no restriction. Under almost all circumstances of practical interest the left-hand side of (5.3) exceeds the right-hand side by a factor ten to twenty. It is true that as σ approaches the Peierls stress σ_P, the left-hand side goes to zero [cf. (3.18)]. However, this is compensated by a corresponding decrease in T. At low temperatures the principal restriction is in fact that the theory as presented here does not allow for the tunnelling underneath the energy barrier at $\hat{q}$.

Kramers [54] shows that provided a "Nachlieferung" condition to be discussed below is fulfilled, under the assumption (5.3) the rate of escape over the potential barrier is given by

$$\Gamma = \nu^0 \; \mathrm{Tanh}\tau \; \exp(-H_{kp}(\sigma)/k_B T) \tag{5.4a}$$

with

$$\mathrm{Sinh}2\tau = 2 \; m \; \lambda \; \mu_1 \quad . \tag{5.4b}$$

In (5.4) two quantities not yet defined appear, viz. ν^0 and λ. In the particle picture used in the derivation of (5.1) ν^0 denotes the vibrational frequency of a particle of mass m at the bottom of the potential well from which we consider escape, whereas λ is proportional to the curvature of the potential energy curve at $q = \hat{q}$ and given by

$$m \; \lambda^2 = \partial F_1(q)/\partial q|_{q=\hat{q},\sigma=\mathrm{const.}} \quad . \tag{5.5}$$

As mentioned above, in the kink picture q denotes the separation between the kinks and anti-kinks of kink pairs. This means that we have to identify m with $m_K/2$, where m_K denotes the kink mass introduced in Sect. 2, and μ_1 with $2\mu_K$, where μ_K denotes the kink mobility[14]. Then (5.5) is applicable directly in the regime of dominant elastic kink - kink interactions, where F(q) is known (cf. footnote[13] and [8]). In all other cases the application of (5.4) to kinks requires more detailed considerations, which in the following can only be sketched.

Closer inspection shows that ν^0 should be identified with the vibrational frequency of a straight dislocation line at the bottom of a Peierls valley. In the normalization adopted in this paper this frequency is given by $(2\pi)^{-1}$ in the absence of an applied stress and by (μ/π) in the Eshelby potential with stress [cf. (3.12d)]. The parameter λ, a measure of the instability of the $\hat{q}$-configuration, is to be identified with the quantity λ_0 introduced through (4.28), characterizing the decay rate of the unstable kink-pair mode.

The preceding discussion may be summarized by stating that the quantities appearing in Kramers' expression (5.4a) do have a simple and clear-cut meaning in terms of soliton theory. We consider as particularly satisfactory the rôle played by the lowest eigenvalue $\omega_0^2 = -\lambda_0^2$ of the perturbation equation (4.9). On physical grounds the kink-pair formation rate is expected to be a <u>functional of the potential</u> $U(\Phi)$. We see that this is realized in the simplest possible manner, namely though the lowest eigenvalue of the linear second-order differential equation (4.9).

---

[14] Note that in the kink picture $(m\mu_1)^{-1}$ is a relaxation time that characterizes the attainment of thermal equilibrium. It may retain its significance even under conditions where μ_K and m_K are difficult to define individually.

In the limit $\mu \to \infty$, Tanh$\tau \to 1$, eqn. (5.4a) reduces to the result of the so-called transition state theory. It was realized already by Kramers [54] that this limit will not be reached because of what he called the "Nachlieferung" problem. In the derivation of (5.4) it is assumed that thermal equilibrium is maintained even at the energy barrier. However, since particles having reached the top of the potential barrier at $q = \hat{q}$ are continually "sucked away" by the applied stress, this assumption will be violated if the coupling of the diffusion particles to the heat bath is too weak. This will certainly be the case for vanishing viscosity, i.e. in the limit $\mu_1 \to \infty$.

The case of extreme underdamping (μ_1 very large) has drawn considerable attention over the years. For recent work on this regime of the Klein - Kramers equation see, e.g., [61-68]. When μ_1 is very large, in the kink picture a kink-pair performs many oscillations of the breather type as described in Sect. 2 before it changes its energy appreciably. Under the analogous conditions in the particle picture the periodic motion inside a potential well may be characterized in terms of the action variable

$$I(E) = \oint p\,dq, \tag{5.6}$$

where the integral is taken around a contour of constant energy E in the (p,q) phase plane. Eq. (5.6) represents the area inside an orbit of constant total energy E in the (p,q) plane.

Kramers [54] introduced the distribution function $\rho(I,t)$, where $\rho(I,t)dI$ is the fraction of the ensemble lying inside a ringshaped area dI of the phase plane, and showed that in the case of extreme underdamping the original Klein - Kramers equation simplifies to

$$\frac{\partial \rho}{\partial t} = \frac{1}{\mu_1 m} \frac{\partial}{\partial I} \left(I\rho + k_B T \frac{\partial \rho}{\partial E}\right) = \frac{1}{\mu_1 m} \frac{\partial}{\partial I} \left(I\rho + k_B T \frac{\partial I}{\partial E} \frac{\partial \rho}{\partial I}\right) . \tag{5.7}$$

The application of (5.7) to the kink-pair formation problem seems to be beset by similar difficulties as discussed above, since it may appear unclear how to evaluate (5.6) when p and q are not defined. However, this evaluation is not necessary [8].

Seeger, Donth, and Kochendörfer [3] obtained the frequency of the breather motion obeying Enneper's equation as

$$\nu = \nu^0 [1 - (E/2E_k)^2]^{1/2} \quad , \tag{5.8}$$

where ν^0 was defined in connection with (5.4) and E_k denotes the energy of an isolated kink. As is well known, the frequency ν of a periodic motion is given by

$$\nu = dE/dI. \tag{5.9}$$

Insertion of (5.8) into (5.9) and integration gives us

$$E = 2E_k \sin (\nu^0 I/2E_k) \tag{5.10}$$

and

$$\nu = \nu^0 \cos (\nu^0 I/2E_k). \tag{5.11}$$

We thus see that soliton theory provides us with precisely that information that is required by Kramers' theory in the extremely underdamped limit.

It is interesting to note that the physical situation covered by (5.7) coincides with that considered by Donth [6] and Seeger, Donth, and Pfaff [21] in the 1950's. However, due to post-war conditions at that time Kramers' paper was not known to the author's group, and the diffusion equation (5.7) was formulated in terms of diffusion along the energy axis rather than along the action-variable axis. The two approaches are equivalent and are both useful (cf., e.g., [8]). In his intuitive derivation of the diffusion equation for kink pairs in energy space Donth made an assumption which restricts the validity of his results, however, as pointed out by Lothe [65].

It is easy to see that in the present case (5.7) does <u>not</u> suffice to calculate the excape rate Γ, since as $E \rightarrow 2E_k$, the oscillation period ν^{-1} of the breather solution goes to zero [cf. (5.8)], so that the basic assumption made in the derivation of (5.7) is violated. In both the early [6,21] and more recent [8] work this difficulty was circumvented by introducing a cut-off energy that depends on the applied stress. A similar problem is encountered in the above-mentioned recent work [57-61] on the case of extreme underdamping. In the present example, in which the I- or E-dependent coefficients of the Smoluchowski-type equation (5.7) are known explicitly over the entire range of these variables, it appears worthwile to replace the Donth - Seeger cut-off procedure by a more sophisticated approach such as that used by Matkowsky et al. [61].

As a final task it remains to relate the escape rate Γ to the <u>net rate of kink-pair formation</u> per unit length of a dislocation line and to experimentally observable quantities. Since this has been dealt with in the literature [8,25], we indicate only the main steps:

a) The <u>translational</u> degrees of freedom of kink pairs may be treated as a gas of non-interacting atoms of mass E_{kp}/c^2. In the limit of low applied stresses this mass equal $2\,m_k$.

b) The <u>vibrational</u> degrees of freedom may be handled by using the perturbation theory approach of Sect. 4 to deduce both the discrete and the continuous part of the vibration spectrum. Each vibrational frequency is associated with an <u>harmonic oscillator</u>. The partition functions may be then be written down exactly and evaluated in closed form in both the high-temperature (classical) and low-temperature limits of quantum statistics [7,23]. In the latter case the zero-point energy of the harmonic oscillators leads to a renormalization of the masses of both isolated kinks and kink pairs.

c) The net rate of kink-pair formation per unit dislocation length is related to the average dislocation velocity v_d and to the rate of plastic strain. These quantities depend on the applied stress mainly through the stress dependence of Γ. In a typical experiment the strain rate is prescribed, and the stress necessary to achieve plastic deformation is observed as a function of strain rate and temperature. This allows us to test the present theory in considerable detail [25,66,67].

APPENDIX

WKBJ-Treatment of Symmetric Double Wells

There is an extensive literature on the treatment of double wells, one of the classical problems in one-particle quantum mechanics, by means of the WKBJ and related techniques (cf., e.g., [68-72]). In this Appendix an approximation procedure appropriate for the kink-pair problem of Sects. 3 and 4 will be outlined.

We consider a one-dimensional Schrödinger equation

$$\frac{d^2\Psi}{dx^2} + [E - V(x)]\,\Psi = 0 \qquad (A.1)$$

with potential energy

$$V(x) = V_1(x) + V_2(x), \qquad (A2.a)$$

where ($x_0 > 0$)

$$V_1(x) = V_0(x_0+x), \quad V_2(x) = V_0(x_0-x)\ . \qquad (A.2b)$$

$V_0(x)$ is assumed to represent a single well and to be non-positive with $\lim(|x|\to\infty)V(x) = 0$.

For negative energies the WKBJ approximation [15] to the even solutions of (A.1) reads

$$\Psi_+ = [V(x)-E]^{-1/4}\ \mathrm{Cosh}\ X(x), \qquad (A.3a)$$

and that to the odd solutions

$$\Psi_- = [V(x)-E]^{-1/4}\ \mathrm{Sinh}\ X(x)\ , \qquad (A.3b)$$

with

$$X = \int_0^{x'=x} [V(x')-E]^{1/2}\,dx'\ . \qquad (A.3c)$$

The functions Ψ_+ and Ψ_- are easily verified to satisfy

$$\left[\frac{d^2}{dx^2} + E - V(x) - W(x;E)\right]\Psi_\pm = 0, \qquad (A.4a)$$

where *)

$$W(x;E) = \frac{5V'^2}{16[V-E]^2} - \frac{V''}{4[V-E]} \qquad (A.4b)$$

$$(V' \equiv dV(x)/dx, \quad V'' \equiv d^2V(x)/dx^2).$$

As may be seen from (A.4b), the expressions (A.3) cease to be good approximations to the exact solutions of (A.1) as $|x|$ approaches $x_E > 0$, which is defined by

$$V(x_E) = E. \qquad (A.5)$$

We denote by $\overline{\Psi}(x)$ those solutions of

$$\frac{d^2\Psi}{dx^2} + [E - V_0(x)]\Psi = 0 \qquad (A.6)$$

that satisfy the boundary condition

$$\lim_{x\to-\infty}\Psi(x) = 0. \tag{A.7}$$

Note that only for special values of E, viz. the eigenvalues of the single-well problem [eq. A.6 with boundary condition $\lim(|x|\to\infty)\Psi(x) = 0$], are the $\bar{\Psi}$ eigenfunctions of the single-well problem. In general, for negative E the functions $\bar{\Psi}$ increase exponentially as $x\to+\infty$.

We now construct approximate solutions of the double-well eigenvalue problem [eq. (A.1) with boundary conditions $\lim\,(|x|\to\infty)\Psi(x) = 0$] in the following way.

For sufficiently large x the single-well solutions $\Psi(x_0\pm x)$ are good approximations to the double-well eigenfunctions as long as the two wells are reasonably separated. On the other hand, for small x (i.e., between the wells), the WKBJ solutions (A.3) should be good approximations as long as V(x) sufficiently exceeds E. The situation is sketched in Fig. 6. The full line gives the total potential V(x), the dashed lines represent the potentials $V_1(x)$ and $V_2(x)$ of the individual wells. The dotted line indicates the potential $V(x) + W(x;E)$ of the Schrödinger equation satisfied by (A.3). Near its singularity at $x = x_E$ it is more attractive than the double-well potential, as may be seen from the wave functions (A.3). On the other hand, in that regime the single-well potential $V_2 = V_0(x_0-x)$, associated with $\bar{\Psi}(x_0-x)$, is more repulsive than the double-well potential V(x), since the attractive contribution from the other well is missing. This suggests that for sufficiently separated wells we should be able to obtain good approximations to the eigenfunctions of (A.1) by matching the logarithmic derivatives of $\bar{\Psi}(x_0-x)$ and of $\Psi_+(x)$ or $\Psi_-(x)$ at a suitably chosen coordinate $x^*<x_E$ and putting $\Psi(x) = \pm\Psi$ for $|x|>x^*$ and $\Psi(x) = \Psi_+(x)$ or $\Psi(x) = \Psi_-(x)$ for $|x|<x^*$. This gives us for the even eigenfunctions

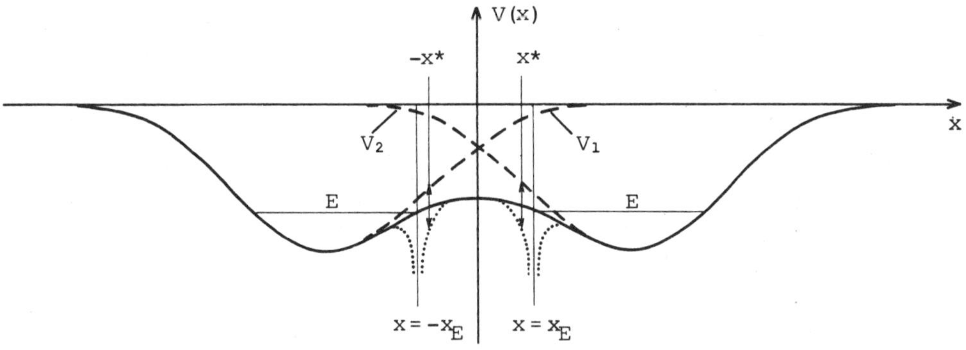

Fig. 6

The double-well potential V(x) of eq. (A.1) (full line). The dashed lines represent the potentials $V_1(x)$ and $V_2(x)$ of the individual wells, the dotted lines indicate the potential corresponding to the WKBJ wavefunction of energy E. As indicated by the perpendicular arrows, at $x=\pm x^*$ the dashed and the dotted lines deviate from the full line by equal amounts.

$$\left.\frac{d\ln\bar{\Psi}(x)}{dx}\right|_{x=x_0-x_E} = \frac{1}{4}\frac{V'(x^*)}{V(x^*)-E} + [V(x^*)-E]^{1/2}\,\mathrm{Tanh}\,X(x^*) \qquad (A.8a)$$

and for the odd eigenfunctions

$$\left.\frac{d\ln\bar{\Psi}(x)}{dx}\right|_{x=x_0-x_E} = \frac{1}{4}\frac{V'(x^*)}{V(x^*)-E} + [V(x^*)-E]^{1/2}\,\mathrm{Coth}\,X(x^*). \qquad (A.8b)$$

Eqs. (A.8) constitute relationships between E and x^*. It is easy to see that for each eigenvalue E in a physically reasonable double-well potential V(x) a solution $x^*<x_E$ of (A.8) can always be found.

Usually we face the reverse problem, i.e., we have to determine, by independent arguments, x^* values for which (A.8) allows us to deduce approximate eigenvalues E that are close to the exact ones. As is indicated in Fig. 6, for $V_0(x)$ not exhibiting humps, $V_2(x) - V(x)$ is non-negative, whereas W(x;E) is non-positive. Since W(x;E) diverges at $x = x_E$, for large enough x_E we can find a value x^* satisfying

$$V_2(x^*) + W(x^*;E) = 0\ . \qquad (A.9)$$

With this choice, at $x = x^*$ the potentials belonging to the approximate solutions $\bar{\Psi}(x_0-x)$ and $\Psi_+(x)$ or $\Psi_-(x)$ deviate from the exact potential V(x) by equal amounts but in opposite directions (cf. Fig.6).

In the case of the lowest eigenvalue of the kink-pair problem an alternative procedure is available. As explained in Sect. 4, together with the eigenvalue $E = \omega^2 = 0$ this eigenvalue results from the splitting of the eigenvalues $\omega^2 = 0$ of the single wells. This suggests that we might use the same x^* for both the even and the odd eigenfunction. For the odd eigenfunction, however, we know that E = 0 for all x_0. We may use this knowledge for determining x^* as a function of x_0 from (A.8b).

In going from the odd to the even eigenfunction, x_E moves away from x^* as determined by the procedure outlined in the preceding paragraph. Since the critical feature of the present method of combining solutions of the single-well Schrödinger equation with the WKBJ approximaton to the double-well wave function is the divergence of the WKBJ wave-function at x^*, the proposed procedure should give good results for the lowest eigenvalue.

Let us now indicate the application of the method to kink pairs in the Eshelby potential. We shall show that it involves only functions that can be handled analytically.

By inserting [cf. (4.13b)]

$$V(x) - E = \kappa^2 - \frac{24(1 + \mathrm{Cosh}2x_0\,\mathrm{Cosh}2x)}{[\mathrm{Cosh}2x_0 + \mathrm{Cosh}2x]^2} \qquad (A.10)$$

into (A.3c) and using the abbreviations

$$\eta \equiv \mathrm{Cosh}2x\ ,\ \eta_0 \equiv \mathrm{Cosh}2x_0 \qquad (A.11)$$

we obtain

$$X(x) = \frac{1}{2}\int_1^{\eta} \frac{24(\eta_0{}^2-1)(\eta+\eta_0)^{-1} + \kappa^2\eta + (\kappa^2 - 24)\eta_0}{[(\eta^2-1)\{\kappa^2\eta^2+2(\kappa^2-12)\eta_0\eta + \kappa^2\eta_0{}^2-24]^{1/2}}\,d\eta. \qquad (A.12)$$

(A.12) may be evaluated in terms of a linear combination of incomplete elliptic integrals of the third kind, defined by [73]

$$\Pi(\Phi,\alpha^2,k) = \int_0^{\Phi} \frac{d\theta}{(1-\alpha^2\sin^2\theta)[1-k^2\sin^2\theta]^{1/2}} \quad . \tag{A.13}$$

The roots of the radicand in (A.12) are given by

$$\eta_{1,2} = \frac{(12-\kappa^2)\eta_0 \pm 2\,[36\,\eta_0^2 - 6\,\kappa^2(\eta_0^2-1)]^{1/2}}{\kappa^2} =$$

$$= \frac{(12-\kappa^2)\mathrm{Cosh}2x_0 \pm 2\,[36\,\mathrm{Cosh}^2 2x_0 - 6\,\kappa^2\mathrm{Sinh}^2 2x_0]^{1/2}}{\kappa^2} \; , \tag{A.14}$$

$$\eta_3 = 1 \quad , \qquad \eta_4 = -1.$$

According to Byrd and Friedman [73] the modulus k, the argument Φ, and the parameters α and α_1 of the elliptic integrals of the third kind are given by

$$k^2 = \frac{(\eta_2-\eta_3)(\eta_1-\eta_4)}{(\eta_1-\eta_3)(\eta_2-\eta_4)} = \frac{(\eta_2-1)(\eta_1+1)}{(\eta_1-1)(\eta_2+1)} =$$

$$= \frac{[\kappa^2\mathrm{Sinh}2x_0 - 24 + 4\,(36\,\mathrm{Cosh}^2 2x_0 - 6\,\kappa^2\mathrm{Sinh}^2 2x_0)^{1/2}]^2}{\mathrm{Sinh}^2 2x_0[\kappa^4\mathrm{Sinh}^2 2x_0 + 48\,(\kappa^2-12)]} \tag{A.15}$$

$$\sin\Phi = \left|\frac{(\eta_2-\eta_4)(\eta-\eta_3)}{(\eta_2-\eta_3)(\eta-\eta_4)}\right|^{1/2} = \left|\frac{\eta_2+1}{\eta_2-1}\right|^{1/2}\mathrm{Tanh}x = \alpha^{-1}\,\mathrm{Tanh}x \tag{A.16}$$

$$\alpha^2 = \frac{\eta_2-\eta_3}{\eta_2-\eta_4} = \frac{\eta_2-1}{\eta_2+1} =$$

$$= \frac{[6\,\mathrm{Cosh}2x_0 + \kappa^2\mathrm{Sinh}^2 x_0 - (36\,\mathrm{Cosh}^2 2x_0 - 6\,\kappa^2\mathrm{Sinh}^2 2x_0)^{1/2}]^2}{\kappa^2\mathrm{Sinh}^2 x_0\,(12 + \kappa^2\mathrm{Sinh}^2 x_0)} \tag{A.17a}$$

$$\alpha_1^2 = \frac{\eta_0-1}{\eta_0+1}\,\frac{\eta_2-\eta_3}{\eta_2-\eta_4} = \alpha^2\,\mathrm{Tanh}x_0 \quad . \tag{A.17b}$$

The limiting case $\sin\Phi = 1$, $\Phi = \pi/2$, means that the elliptic integrals become <u>complete</u>, i.e., that in (A.12) not only the lower limit of integration coincides with one of the roots of the radicant but also the upper one. The integration extends now up to the point where $V(x) - E = 0$, or, because of (A.5), to

$$\eta_2 = \mathrm{Cosh}2x_E \quad . \tag{A.18}$$

Comparison of (A.18) with (A.16), specialized to $\Phi = \pi/2$, or with (A.17a) gives us the relationship

$$\alpha = \mathrm{Tanh}x_E \quad . \tag{A.19}$$

The incomplete elliptic integral of the third kind, being a function of three variables, is not available in tabulated form. It may be represented in terms of the <u>complete</u> elliptic integral of the third kind plus an additional term involving theta functions. These have to be evaluated by means of (well-convergent) infinite series. The <u>complete</u> elliptic integrals of the third kind, $\Pi(\alpha^2,k) = \Pi(\pi/2,\alpha^2,k)$, may be expressed in terms of elliptic integrals of the first and second kinds, which are well tabulated (cf. [73]).

In the present case we expect x^* to be close to x_E, hence $X(x^*) \approx X(x_E)$. This suggests that in (A.8) we represent the leading term by means of elliptic integrals of the first and second kind and handle the remaining terms by Taylor expansions of the hyperbolic functions in (A.8).

The leading term in $X(x^*)$ may be written as the sum of the following three terms:

$$\frac{12(\eta_0^2-1)}{\kappa^2}\int_1^{\eta_2}\frac{d\eta}{(\eta+\eta_0)[(\eta^2-1)(\eta-\eta_1)(\eta-\eta_2)]^{1/2}} =$$

$$= \frac{12(\eta_0-1)g}{\kappa^2}\,\underline{K}\,\{1+\frac{(\alpha_1^2-\alpha^2)\;Z(\beta_1,k)}{[\alpha_1^2(1-\alpha_1^2)(k^2-\alpha_1^2)]^{1/2}}\}, \qquad \text{(A.20a)}$$

$$\frac{1}{2}\int_1^{\mu_2}\frac{\eta\,d\eta}{[(\eta^2-1)(\eta-\eta_1)(\eta-\eta_2)]^{1/2}} = \frac{g}{2}\,\{1+\frac{2\,\alpha^2\;Z(\beta,k)}{[\alpha_2(1-\alpha^2)(k^2-\alpha^2]^{1/2}}\}\;\underline{K}, \qquad \text{(A.20b)}$$

$$\frac{(\kappa^2-24)\eta_0}{2\kappa^2}\int_1^{\eta_2}\frac{d\eta}{[(\eta^2-1)(\eta-\eta_1)(\eta-\eta_2)]^{1/2}} = \frac{(\kappa^2-24)\eta_0}{2\kappa^2}\;g\;\underline{K}(k) \qquad \text{(A.20c)}$$

The quantities k^2, α^2, and α_1^2 are given by (A.15) and (A.17). $\underline{K} = \underline{K}(k)$ denotes the complete elliptic integral of the first kind. Further abbreviations are

$$g \equiv \frac{2}{[(\eta_1-1)(\eta_2+1)]^{1/2}} , \qquad \text{(A.21a)}$$

$$\sin\beta \equiv \alpha/k = [\frac{\eta_1-1}{\eta_1+1}]^{1/2} , \qquad \text{(A.21b)}$$

$$\sin\beta_1 \equiv \alpha_1/k = [\frac{(\eta_0-1)(\eta_1-1)}{(\eta_0+1)(\eta_1+1)}]^{1/2} , \qquad \text{(A.21c)}$$

$$Z(\beta,k) \equiv E(\beta,k) - \frac{\underline{E}}{\underline{K}}\,F(\beta,k) . \qquad \text{(A.21d)}$$

In the expression (A.21d) for the Jacobian zeta function $Z(\beta,k)$

$$F(\beta,k) = \int_0^{\beta}[1 - k^2\sin^2\theta]^{-1/2}\,d\theta \qquad \text{(A.22a)}$$

and

$$E(\beta,k) = \int_0^{\beta}[1 - k^2\sin^2\theta]^{1/2}\,d\theta \qquad \text{(A.22b)}$$

denote the incomplete elliptic integrals of the first or second kind. $\underline{E} > \underline{E}(k) = E(\pi/2,k)$ is the complete elliptic integral of the second kind.

We note that, with (A.17b),

$$\sin\beta_1 = \sin\beta\;\mathrm{Tanh}x_0 \qquad \text{(A.23)}$$

holds. Further useful relationships are

$$\alpha^2(1-\alpha^2)(k^2-\alpha^2) = \frac{4\,\alpha^6}{(\eta_1-1)(\eta_2+1)} \qquad \text{(A.24)}$$

$$\alpha_1^2(1-\alpha_1^2)(k^2-\alpha_1^2) = \frac{4\alpha^6}{(\eta_0+1)^2}\;\frac{\eta_0-1}{\eta_0+1}\;\frac{(\eta_0+\eta_2)(\eta_0+\eta_1)}{(\eta_2-1)(\eta_2+1)} \qquad (A.25)$$

and, with

$$\eta_1+\eta_2 = -2\,(1-12\,\kappa^{-2});\quad \eta_1\eta_2 = \eta_0^2 - 24\,\kappa^{-2}, \qquad (A.26a,b)$$

$$(\eta_0+\eta_1)(\eta_0+\eta_2) = 2(\eta_0-1)(\eta_0 + 12\,\kappa^{-2}), \qquad (A.26c)$$

$$a_1 \equiv \frac{12\;(\eta_0-1)\;g\;(\alpha_1^2-\alpha^2)}{\kappa^2[\alpha_1^2(1-\alpha_1^2)(k^2-\alpha_1^2]^{1/2}} = -\frac{12}{\alpha_1\kappa}\left[\frac{2(\eta_0-1)}{\kappa^2\eta_0+12}\right]^{1/2}. \qquad (A.27)$$

This gives us finally

$$X(x^*)\approx X(x_E) = \frac{1}{\alpha}\,Z(\beta,k) + a_1\,Z(\beta_1,k) + g\,(1-12\,\kappa^{-2})\,\underline{K}(k). \qquad (A.28)$$

The logarithmic derivative of $\Psi(x)$ required in (A.8) is most easily obtained from Darboux's expression (4.18). Putting $n = 2$ and $C_2 = 0$ gives us

$$\frac{d\,\ln\overline{\Psi}(x)}{dx} = \kappa - 2\,\mathrm{Tanh}\,x + 2\,\frac{(\kappa^2+2)\,\mathrm{Sinh}2x - 3\kappa\,\mathrm{Cosh}2x}{(\kappa^2+2)\,\mathrm{Cosh}2x - 3\kappa\,\mathrm{Sinh}2x + \kappa^2}\,. \qquad (A.29)$$

Insertion of (A.28) and (A.29) into (A.8) allows us to proceed as described above.

ACKNOWLEDGEMENTS

It gives the author great pleasure to dedicate this paper to Professor F.R.N. Nabarro of the University of the Witwatersrand, Johannesburg, and to acknowledge on this occasion gratefully the interactions with him in the field of dislocation theory during the past 36 years. The paper was written while the author was the guest of the University of Stellenbosch and of the National Accelerator Centre at Faure, Republic of South Africa; the Appendix is entirely based on work done at these institutions.

The author should like to express his sincere thanks to the staffs of the Physics Department of the University of Stellenbosch and of the National Accelerator Centre for the excellent working conditions they have provided. He is particularly grateful for the technical assistance rendered by Mrs. A.C. Conradie, Mr. J.G.L. Foster, Mr. G.J. Smit, and Mr. D.N. Steenkamp. Thanks are also due to Dr. R. Landauer, Dipl. Ing. W. Lay, Dr. rer. nat. E. Mann, Dipl. Math. E. Mielke, and Dr. Z. Wesolowski for stimulating discussions and to Dr. H. Dekker (Delft) for making ref. [64] available before publication.

REFERENCES:

1. T.Ő. Ogurtani, Ann. Rev. Mater Sci. 13, 67 (1983).

2. N.J. Zabusky and M.D. Kruskal, Phys. Rev. Letters 15, 240 (1965).

3. A. Seeger, H. Donth, and A. Kochendörfer, Z. Physik 134, 173 (1953).

4. A. Seeger, Z. Naturforschung 8a, 47 (1953).

5. A. Seeger, "Theorie der Gitterfehlstellen" in: Encyclopedia of Physics (S. Flügge, ed.) Vol. VII, pt. 2, pp. 383-665, Springer, Berlin etc., 1955.

6. H. Donth, Z. Physik 149, 111 (1957).

7. A. Seeger and P. Schiller, in: Physical Acoustics (W.P. Mason, ed.) Vol. IIIA, p. 351, Academic Press, New York and London 1966.

8. A. Seeger, Dislocations 1984 (P. Veyssière, L. Kubin, and J. Castaing, eds.) p. 141, Editions C.N.R.S., Paris 1984.

9. A. Seeger, "Solitons in Crystals" in: Continuum Models of Discrete Systems (E. Kröner and K.H. Anthony, eds.), p. 253-327, University of Waterloo Press, Waterloo, Ontario, 1980.

10. A. Seeger and A. Kochendörfer, Z. Physik 130, 321 (1951).

11. G.L. Lamb, Jr. "Elements of Soliton Theory", J. Wiley & Sons, New York etc. 1980.

12. G. Eilenberger, Solitons (Springer Series in Solid State Sciences Vol. 19) Springer, Berlin etc. 1983.

13. P.G. Drazin, Solitons (London Mathematical Society Lecture Note Series, Vol. 85) Cambridge University Press, Cambridge etc. 1983.

14. J.D. Eshelby, Proc. Roy. Soc. London, Ser. A 266, 222 (1962).

15. P.M. Morse and H. Feshbach, Methods of Theoretical Physics, pt. II, McGraw-Hill, New York etc. 1953.

16. A. Seeger, in: Forschung in der Bundesrepublik Deutschland (Ch. Schneider, ed.), p. 587, Verlag Chemie, Weinheim 1983.

17. R. Peierls, Proc. Phys. Soc. (London) 52, 34 (1940).

18. F.R.N. Nabarro, Theory of Crystal Dislocations, At the Clarendon Press, Oxford 1967.

19. A. Seeger, in: Deformation and Flow in Solids (R. Grammel, ed.) p. 322, Springer, Berlin, etc. 1956.

20. A. Seeger, Phil Mag. [8] 1, 651 (1956).

21. A. Seeger, H. Donth, and F. Pfaff, Disc. Faraday Soc. 23, 19 (1957).

22. A. Seeger, J. Physique 32, C2-193 (1972).

23. A. Seeger, J. Physique 42, C5-201 (1981).

24. E. Mann, phys. stat. sol. (b) 111, 541 (1982).

25. F. Ackermann, H. Mughrabi, and A. Seeger, Acta metall. 31, 1353 (1983).

26. J. K. Perring and T.H.R. Skyrme, Nuclear Physics 31, 550 (1962).

27. H. Donth, "Zur eindimensionalen Theorie der plastischen Verformung von Metallen," Diplomarbeit T.H. Stuttgart, Stuttgart, 1951.

28. A. Enneper, "Uber asymptotische Linien," Nachr. Königl. Gesellsch. d. Wissenschaften Göttingen, 1870, pp. 493-511.

29. L. Bianchi, "Sulla trasformazione di Bäcklund per le superficie pseudosferiche," Rend. Acc. Naz. Lincei, (5), 1, (2°sem. 1892) pp. 3-12.

30. T. Kontorova and J. Frenkel, J. Phys. Acad. Sci. USSR 1, 137 (1939).

31. L. Prandtl, Z. angew. Math. Mech. 8, 85 (1928).

32. U. Dehlinger, Ann. d. Physik [5] 2, 749 (1929).

33. A. Seeger, "Zur Dynamik von Versetzungen in Gitterreihen mit verschiedenen Gitterkonstanten," Diplomarbeit T.H. Stuttgart, 1948/49.

34. F.C. Frank and J.H. van der Merwe, Proc. Roy. Soc. London, Ser. A, 201, 1950, 261.

35. Ph. A. Martin, Acta Phys. Austriaca, Suppl XXIII, 157 (1981).

36. R. Steuerwald, "Uber Ennepersche Flächen und Bäcklundsche Transformation," Abhandlungen der Bayerischen Akademie der Wissenschaften, Mathem.-naturwissenschaftl. Abtlg, Neue Folge, Heft 40, München, 1936.

37. A. Seeger and Z. Wesolowski, Jnt. J. Eng. Sci. 19, 1535 (1981).

38. A. Seeger and Z. Wesolowski, to be published.

39. F.R.N. Nabarro, Proc. Phys. Soc. (London) 59, 256 (1947).

40. F.R.N. Nabarro, Adv. Phys. 1, 269 (1952).

41. M. Büttiker and R. Landauer, Phys. Rev. Letters 43, 1453 (1979).

42. T. Mori and M. Kato, Phil Mag. A 43, 1315 (1981).

43. A. Seeger, Z. Metallkde. 72, 369 (1981).

44. A. Seeger and P. Schiller, Acta metall. 10, 348 (1962).

45. E.L. Ince, Ordinary Differential Equations, Longmans, Green and Co., London etc. 1927.

46. G. Darboux, Leçons sur la thèorie gènèrale des surfaces et les applications gèometriques du calcul infinitèsimal, pt. II, Chapt. IX Gauthier-Villars, Paris, 1889.

47. W. von Koppenfels, Math. Annalen 112, 24 (1936).

48. F.M. Arscott, Periodic Differential Equations, Pergamon Press, Oxford etc. 1964.

49. A. Erdèlyi (ed.) Higher Transcendental Functions, Vol. III, McGraw-Hill, New York etc. 1955.

50. P.G. Bordoni, Ric. Sci. 19, 851 (1949).

51. P.G. Bordoni, J. Acoust. Soc. Am. 26, 495 (1954).

52. G. Fantozzi, C. Esnouf, W. Benoit, and I.G. Ritchie, Progress in Materials Science 27, Pergamon, New York etc. 1982, p. 396.

53. O. Klein, Arkiv Mat. Astr. Fys. 16, no. 5 (1922).

54. H.A. Kramers, Physica 7, 284 (1940).

55. N.G. van Kampen, Stochastic Processes in Physics and Chemistry, North-Holland, Amsterdam 1981.

56. H.O.K. Kirchner, Acta Physica Austriaca, 48, 111 (1978).

57. M. Büttiker, E.P. Harris, and R. Landauer, Phys. Rev. B 28, 1268 (1983).

58. B. Carmeli and A. Nitzan, Phys. Rev. Letters 51, 233 (1983).

59. M. Büttiker and R. Landauer, Phys. Rev. Letters 52, 1250 (1984).

60. B. Carmeli and A. Nitzan, Phys. Rev. Letters 52, 1251 (1984).

61. B.J. Matkowsky, Z. Schuss, and C. Tier, J. Statist. Physics 35, 443 (1984).

62. M. Büttiker, in: LT-17, U. Eckern, A. Schmid, W. Weber, and H. Wühl (eds.), Elsevier, Amsterdam 1984, p. 1155.

63. H. Dekker, Phys. Letters 112A, 197 (1985).

64. H. Dekker, Physica A, to be published.

65. J. Lothe, Z. Physik 157, 457, 1960.

66. D. Brunner, J. Diehl, and A. Seeger, in: The Structure and Properties of Crystal Defects (V. Paidar and L. Lejcek, eds.), Elsevier, Amsterdam etc. 1984. p. 175.

67. H.-J. Kaufmann, A. Luft, and D. Schulze, Crystal Res. & Technol. 19, 3 (1984).

68. E.M. Harrell, Comm. Math. Phys. 60, 73 (1978).

69. E.M. Harrell, Commun. Math. Phys. 75 239 (1980).

70. H.M.M. Mansour and H.J.W. Müller-Kirsten, J. Math. Phys. 23, 1835 (1982).

71. R. Paulson and N. Frömann, Ann. Phys. (N.Y.) 163, 227 (1985).

72. R. Paulson, Ann. Phys. (N.Y.) 163, 245 (1985).

73. P.F. Byrd and M.D. Friedman, Handbook of Elliptic Integrals for Engineers and Physicists, Springer, Berlin etc. 1954.

TRANSIENT MOTION OF A SOLITARY WAVE IN ELASTIC FERROELECTRICS

by J. POUGET

Laboratoire de Mécanique Théorique associé au C.N.R.S.
Université Pierre et Marie Curie
Tour 66, 4 Place Jussieu, 75230 Paris Cedex 05, FRANCE

Abstract - A nonlinear transient wave motion involving the structure in domains and walls occuring in elastic ferroelectric crystals in the neighborhood of their phase transition is presented. Here the transient motion is caused by an applied electric field. Nonlinearly coupled equations governing the translational and rotational motions of the chain in the continuum approximation were obtained on the basis of a crystalline model consisting of an atomic chain equipped with microscopic electric dipoles. In the absence of applied field the equations obtained possess a solitary-wave solution which represents a moving ferroelectric 180° domain wall. Strains and stresses accompagny this solitary wave being generated through electromechanical couplings. When an electric field is present, a perturbation scheme is developed in order to obtain the characteristic parameters of the evolving wave. The method based on the averaged-Lagrangian of Whitham where the parameters depend only on time (adiabatic perturbation), allows one to deduce the velocity and position (or phase) of the ferroelectric solitary wave. By way of conclusion, numerical illustrations are given for the transient motion from rest of the electromechanical solitary wave.

1. INTRODUCTION

Critical phenomena relating to phase transition in crystals are often the cause of nonlinear effects. It is well known that ferroelectric crystals undergo a phase transition. The reduction of space group symmetry due to this phase transition leads to the splitting of the crystal into domains separated by walls [1, 2]. The problem is particularly devoted to the case of molecular-ferroelectric crystals in which a molecular group can rotate rigidly. For instance, Sodium Nitrite ($Na\ NO_2$) provides a good prototype for such crystals [3]. The problem that we try to solve is the action of a applied electric field on a wall separating two ferroelectric domains. If an electric field is applied to the crystal the phase transition is altered by this field and the latter can favor the growth of a domain at the expense of the other. We have a means to change a structure in two domains into a single domain structure. Ferroelectricity, phase transition and structures in domains and walls lead to two central ideas : (i) as a result of the phase transition the crystal has a nonlinear behavior with respect to the order parameter (polarization) in the neighborhood of the transition point and (ii) the ferroelectric feature of the crystal leads to ferroelectric modes which are dispersive (in the case of molecular-ferroelectric crystals these are rotational modes). It is therefore not surprising that, in such crystals, we have solitary waves [4, 5, 6] . The domain wall is a sort of "defect" localized in the lattice which can be described by a solitary wave. The strong interactions between ferroelectricity and lattice distortion lead inevitably to determine both electric and mechanical (strain and stress) states of the domain wall. With this aim in view, we consider a set of nonlinear equations in the continuum approximation of a microscopic model consisting of a one-dimensional atomic chain. Then we examine the influence of an applied electric field on the solitary wave through a perturbation method based on the averaged-Lagrangian. Finally, numerical illustrations of this transient motion from rest of a wall separating two ferroelectric domains are given.

2. MODEL AND EQUATIONS

In a previous work [7] an exhaustive study of solitary waves in ferroelectrics crystals (without applied electric field) has been considered. In this study a simplified microscopic model which consists of a one-dimensional monoatomic chain equipped with microscopic electric dipoles associated with a molecular group (e.g. NO_2 for Na NO_2) is considered(Fig. 1). The forces acting on the lattice result from (i) the ionic interactions between neighboring ions (short and long range interactions) as in classical ionic crystals ; (ii) the mutual interactions between the microscopic dipoles, and (iii) the electrostatic interaction due to the applied electric field $\vec{E}$. From the energy balance a set of coupled nonlinear equations is deduced. Three coupled equations are obtained for the motions of the chain, which are (i) the transverse motion of the lattice points (the longitudinal displacement is not considered here, since it is practically uncoupled), and (ii) the rigid-body rotational motion of the molecular group. The continuum approximation of the microscopic model leads to a set of nonlinear equations which, with nondimensional quantities, can be read as [7]

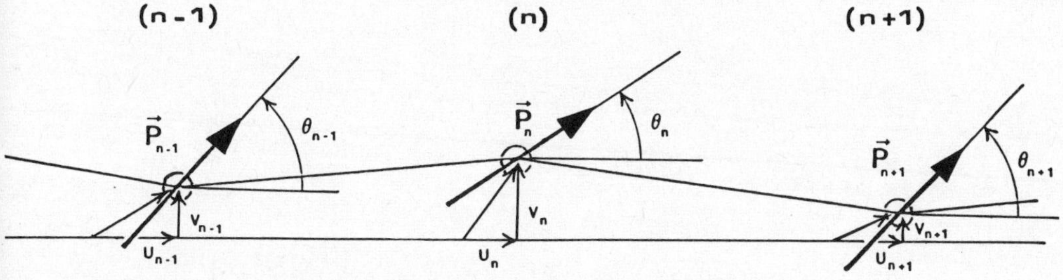

Fig. 1 Monoatomic chain with electric microscopic dipoles

$$(1)\qquad \begin{cases} w_{,t} - V_T^2 v_{,xx} = -\eta(\sin\phi)_{,x} \; , \\ \psi_{,t} - \phi_{,xx} - \sin\phi = \eta v_{,x}\cos\phi + F\cos(\phi/2) \; , \\ v_{,t} - w = 0 \; , \\ \phi_{,t} - \psi = 0 \; . \end{cases}$$

The first equation is the equation of classical elasticity for the transverse displacement v , where V_T is the transverse elastic velocity. The second equation governs the continuum density of dipole rotation ($\phi = 2\theta$). The terms involving η in the right-hand sides are the nonlinear electromechanical couplings where η is the coupling parameter. Eq.$(1)_2$ is a sine-Gordon equation which is nonlinearly coupled to the transverse displacement. This equation, in the absence of coupling, possesses soliton solutions [4, 6] . In addition, eq. $(1)_2$ is modified by a term depending on the applied electric field F. The set of nonlinear equations can be deduce from the following Lagrangian (in the case F = 0) :

$$(2)\qquad L_0 = \frac{1}{2}\Big\{2\phi_{,t}\psi - \psi^2 + 2v_{,t}w - w^2 - \phi_{,x}^2 - V_T^2 v_{,x}^2 - 2(1+\cos\phi) + 2\eta v_{,x}\sin\phi\Big\} .$$

When the applied field is absent, the set of equations (1) possesses a solitary wave solution which can be written as

$$(3)\qquad \phi_0(x,t) = -2\tan^{-1}\left(\sinh\xi/\sqrt{1-2\gamma}\right) ,$$

where we have set

$$(4)\qquad \begin{cases} 2\gamma = \eta^2/(V_T^2 - V^2) \; , \\ \xi = Q(x - X) \; , \end{cases}$$

and the wave number Q ans the velocity V (or Ω = VQ) satisfy the dispersion relation

$$(5)\qquad Q^2(1 - V^2) = 1 - 2\gamma .$$

If F = 0, there is no perturbation and X_t = V = Const., we recover the solution previously studied in Ref. [7] . The deformation $v_{,x}$ and stress can be easily obtained from eqs (3) and $(1)_1$.

3. INFLUENCE OF AN APPLIED FIELD

The problem that we propose to solve is the starting motion of a wall from rest when a field F is suddenly applied (at t_o) to the static solution of eqs(1). We then determine the electric and mechanical states of the domains and walls under the action of the applied electric field. If the latter is applied to a static structure in two domains, the wall does not stay at rest and it begins to move so that the solution of eqs (1), which was initially static, depends now on time t. We in fact suppose that the external field acts on the static solution as a perturbation. The a priori effects of the perturbation are (i) to modulate in time the velocity of the solitary wave, (ii) to alter the thickness of the wall and (iii) to modify the shape of the wall (radiations). With this in mind a perturbation scheme is developed in order to obtain the characteristic parameters of the evolving wave. The method is based on the averaged-Lagrangian of Whitham [8, 9] .

In order to give some idea of the method, we first consider the unperturbed equations (1) (F = 0) corresponding to the Lagrangian density L_o . Eqs (1) can be rewritten in a variational form : $\delta \langle L_o \rangle = \delta \int_S L \, dx = 0$ then $\delta L_o / \delta q = 0$ (where $\langle L_o \rangle$ is the averaged Lagrangian). Let us consider now the perturbed equations (1) where $\varepsilon P(q)$ denotes the perturbation term (here $\varepsilon P = F\cos(\phi/2)$). We assume that eqs (1) can be derived from a Lagrangian L in which the perturbation is present. To find the effect of the perturbation on the solitary wave of the unperturbed equations we assume that the perturbed solution can be well enough approximated by the solution $q = q_o(\alpha_j ; x, t)$; where the α_j 's are the parameters characteristizing the velocity, position, amplitude, etc... of the solitary wave ; moreover, these parameters have a slow time variations (adiabatic time variation hypothesis) [8] . We substitute the solution q_o (with the parameters α_j evolving in time) into the perturbed Lagrangian L and we calculate the global Lagrangian $\langle L \rangle = \int_S L(q_o) dx$ which depends now on the parameters α_j . Up to the first order in ε , the variational principle for $\langle L \rangle$ with respect to the parameters α_j leads to the evolution equations for the parameters

$$\frac{\delta \langle L_o \rangle}{\delta \alpha_i} = \varepsilon \int_S P[q_o] \frac{\partial q_o}{\partial \alpha_i} \, dx . \tag{6}$$

Eq. (6) can be interpreted as an orthogonality condition for the error in eqs (1), where the adiabatic solution $q_o(\alpha_j(t) ; x,t)$ is considered [9] . The advantage of this perturbation method is its simplicity and straightforwardness and this leads to the determination of all the solitary wave parameters without referring to the inverse scattering technique [10, 11, 12] . In addition, the method only requires the knowledge of the solution of the unperturbed system and its Lagrangian and it delivers more informations than the direct energy method does [13] .

But the radiations or the first-order corrective terms in the solitary wave profile cannot be derived from this method and the inverse scattering method must then be used to reach this information [10, 11] .

We apply this perturbation scheme to the present problem. The global Lagrangian $\langle L_0 \rangle$ can be written as

(7) $$\langle L_0\rangle = 8(1-V^2)^{-1/2}\left\{VX_{,t} - 1 + \frac{\eta^2}{18(V_T^2-V^2)}\left[VX_{,t} - 1 + 6V(1-V^2)(X_{,t}-V)(V_T^2-V^2)^{-1}\right] + \mathcal{O}(\eta^4)\right\}.$$

Eq. (7) is, in fact, an expansion of the Lagrangian $\langle L_0\rangle$ with respect to the small parameter η^2 . On accounting for the solution (3) and eqs (4) and (7) (where V et X depend on time) we write the condition (6) in our case and we obtain

(8) $$\begin{cases} \dfrac{d}{dt}\left[V(1-V^2)^{-1/2}\right] = \hat{F}, \\ \dfrac{dX}{dt} - V = 0, \end{cases}$$

where we have set

(9) $$\hat{F} = \frac{1}{2}F\left[1 + \frac{1}{2}(\eta/3V_T)^2\right]^{-1}$$

which is the effective applied field. The solution of eqs (8) is

(10) $$\begin{cases} V(\tau) = \hat{F}\tau(1+\hat{F}^2\tau^2)^{-1/2}, \qquad (\tau = t - t_0) \\ X(\tau) = \hat{F}^{-1}(1+\hat{F}^2\tau^2)^{1/2} + X_0 . \end{cases}$$

We note that the velocity solution $(10)_1$ coincides with that of a relativistic particle of unit rest mass which is uniformly accelerated by $\hat{F}$.

4. CONCLUSION

With the help of an appropriate numerical scheme for nonlinear hyperbolic equations [13] we give some numerical simulations of the transient motion from rest of a wall separating two ferroelectric domains. Fig 2a gives the computed solution in rotation ϕ . Between times $t = 0$ and $t = t_0$ the wall is at rest. An electric field is applied at $t = t_0$. For $t \gtrsim t_0$ the wall begins to move and, progressively, gains a constant velocity which depends on the applied field. Fig. 2b represents the corresponding elastic transverse displacement which characterizes the lattice distortion. Within the first-order range (the solution without the radiations), we recover, in a certain sense, the solitary wave solution without applied field [12] . In addition, we remark that the thickness of the wall becomes nar-

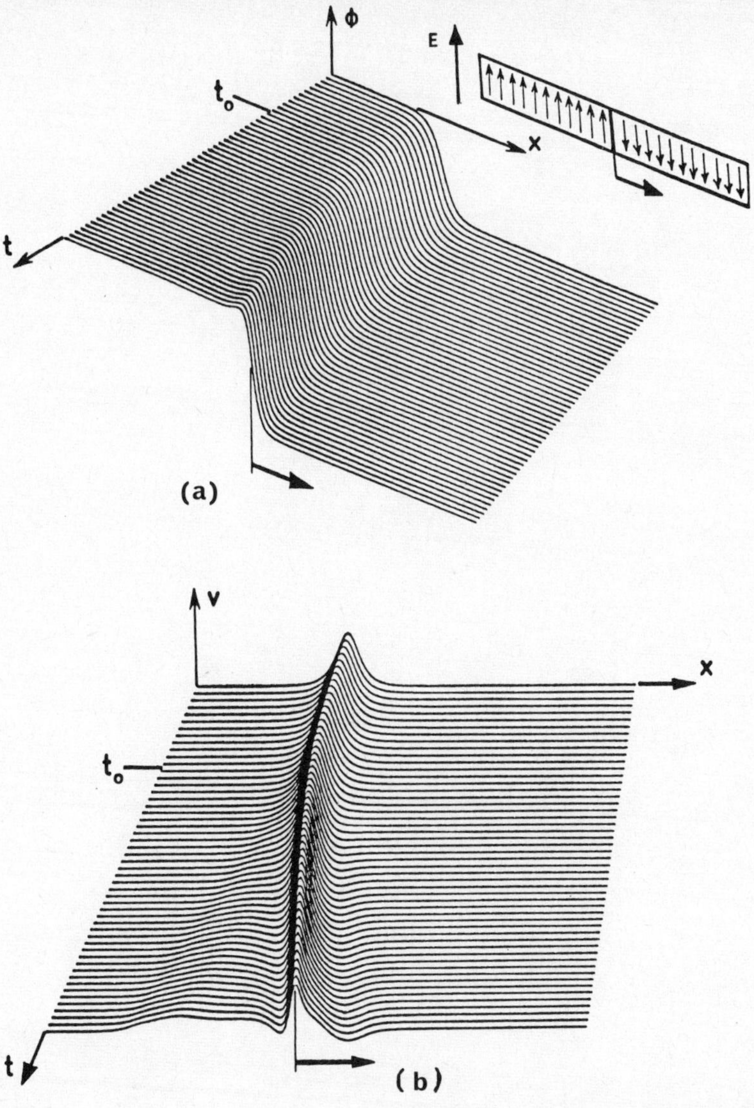

Fig. 2 Transient motion of a ferroelectric wall under the action of an applied electric field : (a) dipole rotation ϕ and (b) corresponding elastic transverse deplacement v.

rower as the velocity increases. From the computation of the solution in rotation it is numerically possible to reach the velocity of the solitary wave as a function of time and to compare this velocity to that obtained by the perturbation method (c.f. eq. (10)). The results are collected in Fig. 3. This ascertains the validity of the perturbation method. Moreover, in Fig. 4 the radiation (i.e. the difference

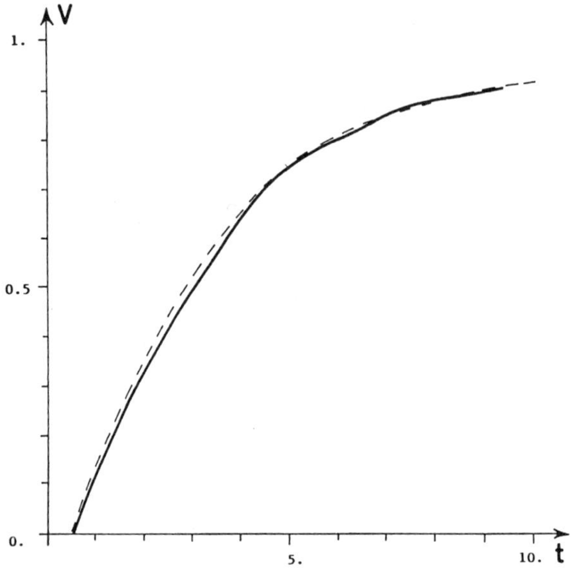

Fig. 3 - Variation in time of the wall velocity ; (——) numerical result, (– – – –) result obtained through the perturbation method.

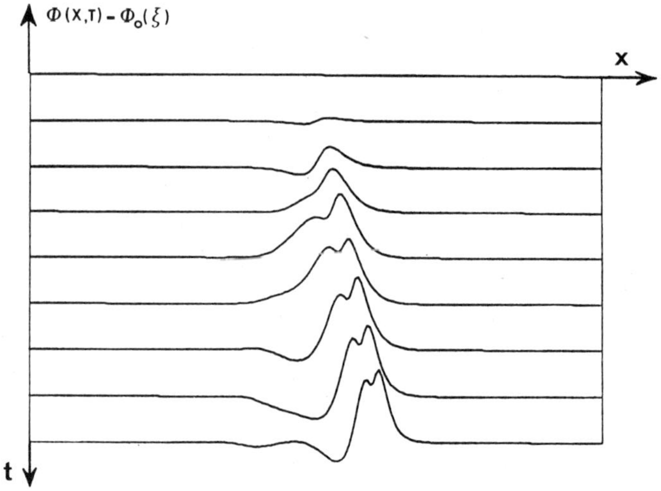

Fig. 4 - Change in the profile of the solitary wave in rotation extracted from Fig. 2.a.

$\phi(x,t) - \phi_0(\xi))$ in rotation is plotted for various values of time with the view of giving the structure of the change in profile of the solitary wave. The main part of the radiations is a sort of hump with trailing decreasing oscillations while in front there are no radiations. This problem of domain wall belongs to a general class of problems dealing with microstructured elastic media in which the propagation of nonlinear waves of the soliton type takes place [14, 15] .

Aknowledgment - The present work was supported by theme "Nonlinear Waves in Electromagnetic Elastic Materials "of A.T.P." Mathématiques Appliquées et Méthodes Numériques Performantes" (M.P.B., C.N.R.S., Paris, France).

Références

[1] I.S. ZHELUDEV, Physics of crystalline Dielectrics (Plenum, New York, I971).

[2] V. JANOVEC, Ferroelectrics, 12, 43, (I976).

[3] S. SUZUKI and M. TAKAGI, J. Phys. Soc. Jpn, 30, 188, (I971).

[4] A.C. SCOTT, F.Y. CHU and D.W. McLAUGHLIN, Proc. IEEE, 61, 1443, (I973).

[5] A.C. NEWELL, J. Applied Mech., 50, 1127, (I983).

[6] R.K. DODD, J.C. EILBECK, J.D. GIBBON and H.C. MORRIS, Solitons and Nonlinear Wave Equations (Academic Press, London, I982).

[7] J. POUGET and G.A. MAUGIN, Phys. Rev., B30, 5306, (I984).

[8] G.B. WHITHAM, Linear and Nonlinear Waves (J. Wiley - Interscience, New York, I974).

[9] V.L. BERDICHEVSKII, PMM U.S.S.R., 46, 244 (I982).

[10] J.P. KEENER and D.W. McLAUGHLIN, Phys. Rev., A16, 777, (I977).

[11] V.I. KARPMAN, Physics Scripta, 20, 462, (I979).

[12] J. POUGET and G.A. MAUGIN, Phys. Rev., B31, 4633, (I985).

[13] J. POUGET and G.A. MAUGIN, Phys. lett., 109A, 389, (I985).

[14] G.A. MAUGIN and J. POUGET. Solitons in microstructured elastic media - Physical and Mechanical Aspects, in Continuum Models of Discrete Systems (5) Ed. A.J.M. Spencer, A.A. Balkema, Amsterdam (I985).

[15] G.A. MAUGIN, Solitons and domain structure in elastic crystals with a microstructure, these proceedings.

DRIVEN KINKS IN SHAPE-MEMORY ALLOYS

F. Falk
Dept. Physics, Universität-GH-Paderborn
D 4790 Paderborn, F.R.G.

Abstract

Based on an one-dimensional model a Ginzburg-Landau theory of the martensitic phase transition in shape-memory alloys is adopted. Order parameter is the shear strain. The free energy density is an even polynomial of the 6th degree in the shear strain and depends quadratically on the strain gradient. The non-linear equation of adiabatic motion shows, in an infinite system, solitary wave solutions of two types. Soliton-like solutions may be interpreted as nuclei of one phase in a matrix of the other one. In this paper the second type of solutions, namely kinks are dealt with. Kink solutions represent domain walls either between high temperature phase and low temperature phase or between different variants of the low temperature phase. Without external forces the kinks occur only at rest. Driven by external forces the domain wall motion obeys a Rankine-Hugoniot equation. In the stress-strain diagram a generalized Maxwell equal area construction applies.

1. Introduction

The pseudoelastic as well as the ferroelastic behaviour of shape memory alloys is due to a first order martensitic phase transition which is connected with a shear deformation parallel to the habit plane [1]. In addition to temperature, external shear stress induces the phase transition from the high temperature austenitic to the low temperature martensitic phase. Furthermore, shear strain causes deformation twinning between different martensite variants. In the temperature range of interest domains of austenite and martensite variants coexist. The phase transformation proceeds by domain wall motion, which is responsible for both, pseudoelastic and ferroelastic deformation behaviour [2]. Based on a one-dimensional nonlinear model, in two previous papers static domain walls without external driving force [3] and solitons representing nuclei [4] are studied. The objective of this paper is to investigate, within the same model, the motion of kink type solitary waves representing domain walls under external shear forces.

2. One-Dimensional Model and Equation of Motion

In the one-dimensional model [3,4] the crystal is built up by stacking atomic planes which are parallel to the shear plane. Therefore the shear displacement $u(x,t)$ is perpendicular to the stacking direction x (Fig. 1). Within each layer the displacement is uniform. As a consequence the mass density ρ remains constant throughout the crystal. The shear strain e is defined by

$$e(x,t) = \frac{\partial u(x,t)}{\partial x}$$

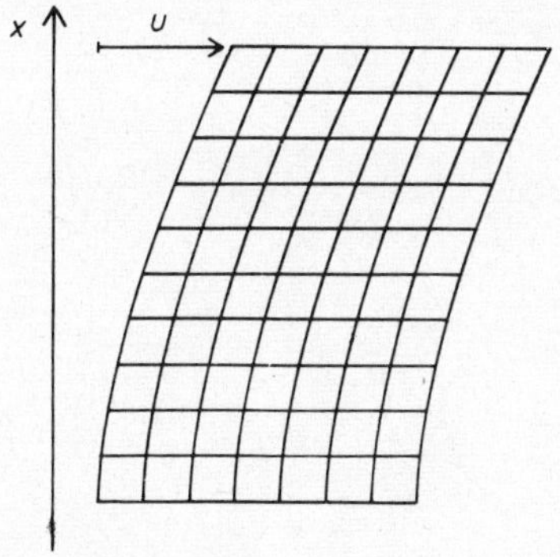

Fig. 1: One-dimensional model. Shear displacement u is perpendicular to stacking direction x.

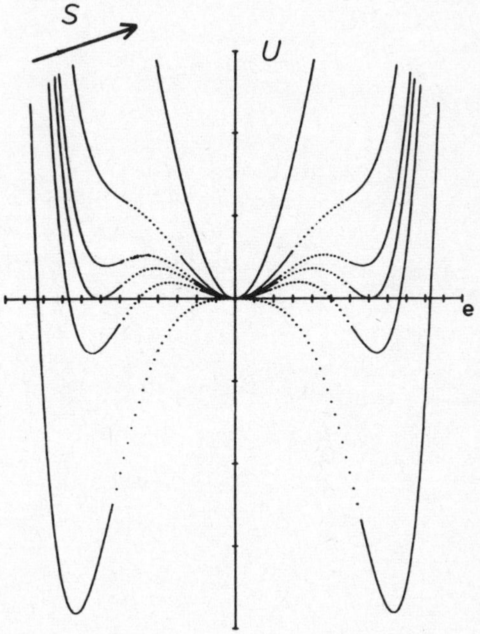

Fig. 2: Internal energy as a function of shear strain for different values of entropy S. U_0 is suppressed.

Depending on e and the entropy density S, the internal energy density U_L (Fig. 2).

$$U_L(e,S) = \alpha\, e^6 - \beta\, e^4 + \gamma(S)e^2 + U_0(S)$$

describes the behaviour of a static homogeneous system [4]. α and β are positive material parameters, $\gamma(S)$ is a monotoneous function changing its sign in the entropy range of interest. $U_0(S)$ is some positive, monotoneous function of S which is of no importance in the present context. The high temperature minimum of U_L at $e = 0$ represents austenite whereas the low temperature symmetric minima correspond to the martensite variants. At intermediate temperature, due to the presence of three minima, three phases may coexist in one crystal.
In order to deal with domain boundaries where the strain varies strongly the internal energy U_L has to be supplemented by a strain gradient term $\delta\, e'^2 = \delta\, (\partial e/\partial x)^2$ Stability requires positive δ. The kinetic energy density is given by $E_{kin} = \rho/2\, \dot{u}^2$ where $\dot{u} = \partial u/\partial t$ means shear velocity. Hence the Hamiltonian density reads

$$H = U_L + \delta\, e'^2 + \rho/2\, \dot{u}^2$$

By rescaling energy, displacement, length, and time [4] we find

$$H = e^6 - e^4 + g(S)e^2 + e'^2 + \frac{1}{2}\dot{u}^2$$

From H the shear stress σ and couple stress μ are defined according to

$$\sigma(e,S) = \frac{\partial H}{\partial e} = \frac{\partial U_L}{\partial e} = 6e^5 - 4e^3 + 2g(S)e \tag{1}$$

$$\mu(e') = \frac{\partial H}{\partial e'} = 2e' \tag{2}$$

The equation of motion reads [4]

$$\ddot{u} - \sigma' + \mu'' = 0 \tag{3}$$

The boundary conditions at infinity are given by

$$\mu = 0 \qquad x \to \pm\infty \tag{4}$$

$$\sigma(+\infty) = \sigma_+ = a_+$$

$$\sigma(-\infty) = \sigma_- = -a_-$$

where $a_\pm$ is the surface force. From the last equations, combined with the stress-strain relation, the values of strain at infinity $e_\pm$ follow, if the phase occuring at infinity and the surface forces are prescribed.

3. Kink Solutions

In the following stationary kink type solitary waves are dealt with. This means that a strain profile $e(x,t) = e(x-vt) \equiv e(z)$ representing a domain wall moves at constant speed v. Integrating the equation of motion (3) once yields

$$\sigma(e) - v^2 e + C_2 = 2e'' = R'(e) \tag{5}$$

with the constant of integration C_2 following from the boundary condition (4)

$$C_2 = v^2 e_\pm - \sigma_\pm \tag{6}$$

This equation implies the Rankine-Hugoniot relation

$$v^2 = \frac{\sigma_+ - \sigma_-}{e_+ - e_-} \geqq 0 \tag{7}$$

by which the velocity of the domain wall is determined. It is in close analogy to a similar relation for shock waves in gas dynamics. Integrating (5) once more yields the inverse function of e(z), namely

$$z(e) = \int \frac{de}{\sqrt{R(e)}}$$

with

$$R(e) = e'^2 = U_L(e) - v^2/2\, e^2 + C_2\, e + C_3 > 0 \qquad (8)$$

An explicit evaluation of the integral is possible only in some cases. Since for $e \to e_\pm$, z should go to infinity the relations

$$0 = R(e_+) = U_L(e_+) - v^2/2\, e_+^2 + C_2\, e_+ + C_3$$

$$0 = R(e_-) = U_L(e_-) - v^2/2\, e_-^2 + C_2\, e_- + C_3$$

hold. Subtracting both equations yields, with (6)

$$U_L(e_+) - U_L(e_-) = \frac{1}{2}\ (\sigma_+ - \sigma_-)(e_+ - e_-)$$

Because of (1) the left hand side is given by integrating σ:

$$\int_{e_-}^{e_+} \sigma\, de = \frac{1}{2}\,(\sigma_+ + \sigma_-)\ (e_+ - e_-) \qquad (9)$$

This equation admits a generalized Maxwell construction. In a stress-strain diagram (Fig. 3) the chord connecting the (σ, e)-values at infinity has to cut the $\sigma(e)$-curve so as to equate the dashed areas. The slope of the chord then represents v^2 (7). From this construction it follows that in order to get adiabatically moving domain walls the values of σ_+ and σ_- cannot been chosen arbitrarily. In addition, the inequalities (7) and (8) pose restrictions on the ranges of temperature and of surface forces a thorough discussion of which is published elsewhere.

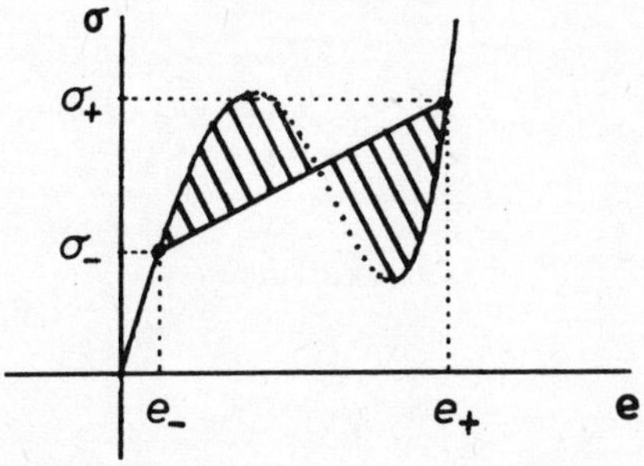

Fig. 3: Generalized Maxwell equal area construction in stress-strain diagram.

References

1. Delaey L., Krishnan R.V., Tas H., Warlimont H.: J. Mater. Sci. 9, 1521 (1974).
2. Christian, J.W.: Metall. Trans. A 13, 509 (1982).
3. Falk, F.: Z. Phys. B - Condensed Matter 51, 177 (1983).
4. Falk, F.: Z. Phys. B - Condensed Matter 54, 159 (1984).
5. Falk, F.: Acta metall. 28, 1773 (1980).

THE LONG-TIME BEHAVIOUR FOR PERTURBED WAVE-EQUATIONS AND RELATED PROBLEMS

Wiktor Eckhaus

Mathematisch Instituut
Rijksuniversiteit Utrecht

Contents*

* *This contribution to the proceedings contains chapters 1,2 and 3. The complete version of the paper appeared as a Preprint of the Mathematical Institute at Utrecht.*

1. INTRODUCTION.

There has been a considerable interest in the recent years in nonlinear evolution equations that arise under various conditions as approximations from the full description of some physical phenomena. Famous examples are: the Korteweg - de Vries equation (originating from the theory of water waves) and the Nonlinear Schrödinger equation (from optics). The derivation of such approximate equations, for some given physical problem, has been the subject of many publications, and also the subject of discussion and controversy (examples of controversial aspects and attempts at clarification are in Meyer (1979), Bona, Pritchard and Scott (1983), Franzen and Keverkian (1984)). Systematic approaches to the process of derivation of approximate equations are in Lamb (1980), Keverkian and Cole (1981), Dodd, Eilbeck, Gibbon and Morris (1982).
Expansions with respect to some small parameters are introduced in these approaches at a very stage of the analysis and the formalism of multiple scaling is often used. The methods require working out the equations for a certain number of terms of the expansions before relevant information on the first term is obtained. For example, in problems of wave-trains modulation, the non linear Schrödinger equation which determines the first term of the expansion, is typically produced by a "non secularity condition" for the third term.

The analysis to be described in the chapters that follow arose from a desire to understand better the meaning and the origin of the approximate model equations. It turns out that the cumbersome machinery of formal expansions is irrelevant for the derivation of these equations. The main steps are simply transformations of variables which produce from some given perturbed equation another perturbed equation. The approximate equations, or model equations, are formal limits of the various perturbed equations. Let us illustrate this by the example of long wave of small amplitude in shallow water.

In a first step, from the physical description, after scaling and recognition of small parameters (Appendix to this paper) one derives a *perturbed wave equation*

$$(1.1) \qquad \left(\frac{\partial^2}{\partial x^2} - \frac{\partial^2}{\partial t^2}\right) \varphi = \varepsilon\, P(\varphi) + O(\varepsilon^2)$$

Here x and t are nondimensional physical space and time variables. The perturbation terms $P(\varphi)$ are computed explicitly (see section 3.1). Higher terms $O(\varepsilon^2)$ can be computed if one wishes to do so. The formal limit equation, i.e. the wave equation, can be expected to produce approximations on compact intervals of x and t.

Next, for long-time behaviour, and taking into account the structure of solutions of the wave equations, consider the transformation

$$(1.2) \qquad \bar{x} = x - t \quad ; \quad \tau = \varepsilon t$$
$$\varphi(x,t) \quad \rightarrow \quad \bar{\varphi}(\bar{x},\tau)$$

using the explicit expression for $P(\varphi)$ one easily finds (section 3.1)

$$(1.3) \qquad \frac{\partial \bar{u}}{\partial \tau} + \frac{3}{2}\bar{u}\frac{\partial \bar{u}}{\partial \bar{x}} + \frac{1}{6}\frac{\partial^3 \bar{u}}{\partial \bar{x}^3} = O(\varepsilon)$$
$$\bar{u} = \frac{\partial \bar{\varphi}}{\partial \bar{x}}$$

The result is a *perturbed K.d.V. equation.*
The formal limit equation, K.d.V. can be expected to produce approximations for τ in compact intervals. If one wishes to study the behaviour on even longer time scales, the perturbation terms $O(\varepsilon)$ must be taken into account (Eckhaus and van Harten (1981), de Kerf (1985)).

The first three chapters that follow are about perturbed wave equations and perturbed hyperbolic systems of first order equations. The main line of thinking is developed in chapter 2. Chapter 3 and 4 are exercises, applications and extensions. Finally, in chapter 5, along a similar line of thinking, some problems for dispersive waves are studied. The main aim of the analysis throughout the paper is to gain better understanding of the phenomena. No attempt has been made to single out "new results".

2. SOME BASIC CONSIDERATIONS ON LONG-TIME BEHAVIOUR.

2.1 An introductory exercise.

We consider in this section a perturbed first order equation

$$(2.1.1) \qquad \left(\frac{\partial}{\partial t} + \frac{\partial}{\partial x}\right)u_\varepsilon = \varepsilon\, P\left(\frac{\partial}{\partial x}, \frac{\partial}{\partial t}; \varepsilon\right) u_\varepsilon$$

The notation in the right-hand side should be understood as follows:

$P\left(\frac{\partial}{\partial x}, \frac{\partial}{\partial t}; \varepsilon\right)$ is a differential operator of some unspecified order in terms of the derivatives with respect to x and t. The perturbation is in general nonlinear, however $P\left(\frac{\partial}{\partial x}, \frac{\partial}{\partial t}; \varepsilon\right)$ has a formal limit as $\varepsilon \downarrow 0$. We have displayed the derivations with respect to x and t explicitly in order to be able to follow in the course of the analysis the effects of changes of variables.

We study solutions $u_\varepsilon(x,t)$, $x \in (-\infty, \infty)$, $t \in [0, T]$ for large T and small ε. We assume at the outset that u_ε and its derivatives to any order needed in the analysis, are uniformly bounded for $\varepsilon \downarrow 0$.

All solutions of the unperturbed equation (i.e. with the right-hand side of (2.1.1) put equal to zero) represent "waves" which move with a constant speed to the right, without changing shape. It is therefore natural to study the landscape of solutions of the perturbed equation in a moving coordinate system and further more to extend the observation over a long period of time. To this purpose we introduce the change of variables

$$\bar{x} = x - t, \quad \tau = \varepsilon t \tag{2.1.2}$$

this induces

$$u_\varepsilon(x,t) \to \bar{u}_\varepsilon(\bar{x},\tau)$$

and from (2.1.1) it follows that

$$\frac{\partial \bar{u}_\varepsilon}{\partial \tau} = P\left(\frac{\partial}{\partial \bar{x}}, -\frac{\partial}{\partial \bar{x}} + \varepsilon\frac{\partial}{\partial \tau}; \varepsilon\right)\bar{u}_\varepsilon \tag{2.1.3}$$

If P is sufficiently neat then the right-hand side can be decomposed (taking the formal limit) and one obtains

$$\frac{\partial \bar{u}_\varepsilon}{\partial \tau} - P\left(\frac{\partial}{\partial \bar{x}}, -\frac{\partial}{\partial \bar{x}}; 0\right)\bar{u}_\varepsilon = \delta(\varepsilon)\, P^{(1)}\left(\frac{\partial}{\partial \bar{x}}, \frac{\partial}{\partial \tau}; \varepsilon\right)\bar{u}_\varepsilon \tag{2.1.4}$$

with $\delta(\varepsilon) = o(1)$.

The equation (2.1.4) is a perturbed *evolution equation for* $\bar{u}_\varepsilon$ on long time scales. To avoid all misunderstanding let us specify explicitly that

$\left(\frac{\partial}{\partial \bar{x}}, -\frac{\partial}{\partial \bar{x}}; \varepsilon\right)$ is obtained from the original operator $P\left(\frac{\partial}{\partial x}, \frac{\partial}{\partial \tau}; \varepsilon\right)$ by taking the formal limit for $\varepsilon \downarrow 0$ and further more replacing each derivative with respect to x by derivative with respect to $\bar{x}$ and each derivative with respect to t by minus the derivative with respect to $\bar{x}$.

$P^{(1)}$ follows by comparison of (2.1.3) and (2.1.4)

As an example consider P to be a quasi-linear third order operator of the form

(2.1.5)
$$P\left(\frac{\partial}{\partial x_1}, \frac{\partial}{\partial \varepsilon} ; \varepsilon\right) u_\varepsilon = \left(a_{10}\frac{\partial u_\varepsilon}{\partial x} + a_{01}\frac{\partial a_\varepsilon}{\partial t}\right) u_\varepsilon + $$
$$+ \left(a_{20}\frac{\partial^2}{\partial x^2} + a_{11}\frac{\partial^2}{\partial x\partial t} + a_{02}\frac{\partial^2}{\partial t^2}\right) u_\varepsilon + $$
$$+ \left(a_{30}\frac{\partial^3}{\partial x^3} + a_{21}\frac{\partial^3}{\partial x^2\partial t} + a_{12}\frac{\partial^3}{\partial x\partial t^2} + a_{03}\frac{\partial^3}{\partial t^3}\right) u_\varepsilon$$

with a_{ij} constants. Then

(2.1.6)
$$P\left(\frac{\partial}{\partial \bar{x}}, -\frac{\partial}{\partial \bar{x}} ; 0\right)\bar{u}_\varepsilon =$$
$$= (a_{10} - a_{01})\, u_\varepsilon \frac{\partial u_\varepsilon}{\partial \bar{x}} + (a_{10} - a_{11} + a_{02})\frac{\partial^2 \bar{u}_\varepsilon}{\partial \bar{x}^2} + (a_{30} - a_{21} + a_{12} - a_{03})\frac{\partial^3 \bar{u}_\varepsilon}{\partial \bar{x}^3}$$

The equation (2.1.4), with (2.1.6), is in general a perturbed *Burgers - Korteweg de Vries equation*. The perturbation term $P^{(1)}$ can easily be computed by performing in (2.1.5) the transformation of variables (2.1.2) in full.

Consider now, as a model equation for long-time behaviour, the unperturbed version of (2.1.4), i.e.

(2.1.7)
$$\frac{\partial \bar{u}_0}{\partial \tau} - P\left(\frac{\partial}{\partial \bar{x}}, -\frac{\partial}{\partial \bar{x}} ; 0\right)\bar{u}_0 = 0$$

The validity of $\bar{u}_0$ as an approximation of $\bar{u}_\varepsilon$ should be expected to be restricted to bounded intervals of the τ-variable. If one wishes to study $\bar{u}_\varepsilon$ on intervals such that $\tau = 0(\frac{1}{\delta})$ then the perturbation term $\delta(\varepsilon)P^{(1)}\bar{u}_\varepsilon$ will in general affect the solution. For the K.d.V. equation this is well-known (see for example Eckhaus and Van Harten 1983 chapter 7). We conclude with another simple but interesting example of (2.1.1): A problem of *hyperbolic singular perturbation* is given by

(2.1.8)
$$\left(\frac{\partial}{\partial t} + \frac{\partial}{\partial x}\right) u_\varepsilon = \varepsilon\left[\frac{\partial^2 u_\varepsilon}{\partial x^2} - \frac{1}{c^2}\frac{\partial^2 u_\varepsilon}{\partial t^2}\right]$$

The problem has been discussed in Cole (1962) and a rigorous theory has been given by Geel (1978). One finds that if the subcharacteristics, that is the lines x-t=constant, are "time-like" (which is the case when $c > 1$), one has the approximation

$$u_\varepsilon - u_0 = 0(\varepsilon)$$

(2.1.9) $$\left(\frac{\partial}{\partial t} + \frac{\partial}{\partial x}\right) u_0=0$$

valid in compacts $t \in [d,T]$, $d > 0$, $T > 0$.

Let us now consider the long-time behaviour. By transformation (2.1.2) we find

(2.1.10) $$\frac{\partial \bar{u}_\varepsilon}{\partial \tau} = (1-\frac{1}{c^2}) \frac{\partial^2 \bar{u}_\varepsilon}{\partial \bar{x}^2} + \frac{\varepsilon}{c^2}\left[\frac{\partial^2 \bar{u}_\varepsilon}{\partial \bar{x}\partial \tau} - \varepsilon\frac{\partial^2 \bar{u}_\varepsilon}{\partial \tau^2}\right]$$

We thus obtain, as model equation (2.1.7)

(2.1.11) $$\frac{\partial \bar{u}_0}{\partial \tau} = (1-\frac{1}{c^2}) \frac{\partial^2 \bar{u}_0}{\partial \bar{x}^2}$$

If $c > 0$ (subcharacteristics of (2.1.8) are "time-like") then the result is a forward diffusion equation.

2.2 The perturbed wave equation.

Let $u_\varepsilon(x,t)$ be defined as a solution of

(2.2.1) $$\left(\frac{\partial^2}{\partial x^2} - \frac{\partial^2}{\partial t^2}\right) u_\varepsilon = \varepsilon\, P\left(\frac{\partial}{\partial x}, \frac{\partial}{\partial t}\,;\, \varepsilon\right)u_\varepsilon$$

with P and u_ε as in the preceding section.

All solutions of the unperturbed equation decompose into waves moving to the right and to the left with constant speed, while each of the two waves preserves its shape for all time. By special choise of initial conditions one can produce as solutions waves which move in only one direction.

Let us in the spirit of the preceding section, first follow the waves moving the right by introducing the transformation

(2.2.2)
$$\bar{x} = x - t \;;\; \tau = \varepsilon t$$
$$u_\varepsilon(x,t) \rightarrow \bar{u}_\varepsilon(\bar{x},t)$$

The equation (2.2.1) transforms to

$$(2.2.3)\qquad 2\frac{\partial^2\bar{u}_\varepsilon}{\partial\tau\partial\bar{x}} = P\left(\frac{\partial}{\partial\bar{x}}, -\frac{\partial}{\partial\bar{x}} + \varepsilon\frac{\partial}{\partial\tau} ; \varepsilon\right)\bar{u}_\varepsilon + \varepsilon\frac{\partial^2\bar{u}_\varepsilon}{\partial\tau^2}$$

For a sufficiently neat operator P one can again decompose the right-hand side and obtain

$$(2.2.4)\qquad 2\frac{\partial^2\bar{u}_\varepsilon}{\partial\tau\partial\bar{x}} - P\left(\frac{\partial}{\partial\bar{x}}, -\frac{\partial}{\partial\bar{x}}; 0\right)\bar{u}_\varepsilon = \delta(\varepsilon)\ P^{(1)}\left(\frac{\partial}{\partial\bar{x}}, \frac{\partial}{\partial\tau}; \varepsilon\right)\bar{u}_\varepsilon + \varepsilon\frac{\partial^2\bar{u}_\varepsilon}{\partial\tau^2}$$

With $\delta(\varepsilon) = o(1)$. For the simplicity of presentation we shall usually take $\delta = \varepsilon$. The unperturbed version of (2.2.4) is the model equation for long time behaviour:

$$(2.2.5)\qquad 2\frac{\partial^2\bar{u}_0}{\partial\tau\partial\bar{x}} - P\left(\frac{\partial}{\partial\bar{x}}, -\frac{\partial}{\partial\bar{x}} ; 0\right)\bar{u}_0 = 0$$

The equation (2.2.5) can further be simplified to a standard version of an evolution equation if P satisfies certain conditions.

Firstly suppose the operator P to be such that

$$P\left(\frac{\partial}{\partial\bar{x}}, -\frac{\partial}{\partial\bar{x}} ; 0\right)\bar{u}_0 = \hat{P}\left(\frac{\partial}{\partial\bar{x}}\right)\frac{\partial\bar{u}_0}{\partial\bar{x}}$$

In other words: $P\left(\frac{\partial}{\partial\bar{x}}, -\frac{\partial}{\partial\bar{x}} ; 0\right)\bar{u}_0$ does not contain the function $\bar{u}_0$ itself, but only its derivatives. In that case, writing

$$(2.2.6)\qquad w_0 = \frac{\partial\bar{u}_0}{\partial\bar{x}}$$

one has

$$(2.2.7)\qquad 2\frac{\partial w_0}{\partial\tau} - \hat{P}\left(\frac{\partial}{\partial\bar{x}}\right)w_0 = 0$$

In a second case P is such that

$$(2.2.8)\qquad P\left(\frac{\partial}{\partial x}, \frac{\partial}{\partial t} ; 0\right) = \frac{\partial}{\partial x}\tilde{P}\left(\frac{\partial}{\partial x}, \frac{\partial}{\partial t}\right)$$

one can then integrate (2.2.5) with respect to $\bar{x}$, and *assuming the solutions and its derivatives to vanish for* $|\bar{x}|\to\infty$ one obtains

(2.2.9) $$2\frac{\partial}{\partial\tau}0 - \tilde{P}\left(\frac{\partial}{\partial\bar{x}}, -\frac{\partial}{\partial\bar{x}}\right)\bar{u}_0 = 0$$

The reader may amuse himself by defining families of operators $P\left(\frac{\partial}{\partial x}, \frac{\partial}{\partial t}, \varepsilon\right)$ which produce in this way interesting model evolution equations, such as K.d.V., Burgers or Burgers-K.d.V. We illustrate here the precedure by the following elementary example. Let $u_\varepsilon(x,t)$ be the solution of the initial value problem

$$(1-\varepsilon)\frac{\partial^2 u_\varepsilon}{\partial x^2} - \frac{\partial^2 u_\varepsilon}{\partial t^2} = 0$$

(2.2.10) $$(u_\varepsilon)_{t=0} = U_0(x) \; ; \; \left(\frac{\partial u_\varepsilon}{\partial t}\right)_{t=0} = - U_0'(x)$$

with $U_0(x)$ some smooth function, decaying repidly for $|x| \to \infty$. It is easy to show that

(2.2.11) $$u_\varepsilon(x,t) = U_0(x - \sqrt{1-\varepsilon}\, t) + 0(\varepsilon)$$

uniformly for $x \in (-\infty,\infty)$, $t \in [0,\infty)$.

Following the procedure of this section we introduce the transformation (2.2.2) and find for the transformed equation (2.2.3)

(2.2.12) $$2\frac{\partial^2 \bar{u}_\varepsilon}{\partial\tau\partial\bar{x}} = \frac{\partial^2 \bar{u}_\varepsilon}{\partial \bar{x}^2} + \varepsilon\frac{\partial^2 u_0}{\partial\tau^2}$$

The model equation reads

(2.2.13) $$\left(2\frac{\partial}{\partial\tau} - \frac{\partial}{\partial\bar{x}}\right)\frac{\partial\bar{u}_0}{\partial\bar{x}} = 0.$$

Hence, after imposing an initial condition, we find as solution

(2.2.14) $$\bar{u}_0(\bar{x},\tau) = U_0(\bar{x} + \tfrac{1}{2}\tau)$$

We compare this result with the exact solution (2.2.11) which, after transformation to $\bar{x},\tau$ variables and a trivial expansion, can be written in the form

(2.2.15) $$\bar{u}_\varepsilon(\bar{x},\tau) = U_0(\bar{x} + \tfrac{1}{2}\tau + 0(\varepsilon\tau)) + 0(\varepsilon)$$

Clearly $\bar{u}_0$ is indeed a nice approximation as long as τ is bounded.

Returning to the general perturbed equation (2.2.1) one must of course remark that an entirely parallel analysis can be given by following the waves to the left. For that purpose one introduces

$$\tilde{x} = x + t \; , \; \tau = \varepsilon t$$

(2.2.16) $$u_\varepsilon(x,t) \to \tilde{u}_\varepsilon(\tilde{x},\tau)$$

One then obtains, as a model equation

(2.2.17) $$2\frac{\partial^2\tilde{u}_0}{\partial\tau\partial\tilde{x}} + P\left(\frac{\partial}{\partial\tilde{x}}, \frac{\partial}{\partial\tilde{x}}; 0\right)\tilde{u}_0 = 0$$

2.3 Questions of validity.

We expect that solutions $\bar{u}_0(\bar{x},t)$ and $\tilde{u}_0(\tilde{x},t)$ of model equations (2.2.5), respectively (2.2.17) are approximations for the long-time behaviour of some (classes of) solutions of the perturbed wave equation (2.2.1). In this section we sketch lines of reasoning that can be used in investigating the question of validity.

Let us consider $\bar{u}_0(\bar{x},t)$. We summarise first the main steps of the preceding section while introducing some shorthand notation.

The function $u_\varepsilon(x,t)$ satisfies the perturbed wave equation

(2.3.1) $$L_\varepsilon u_\varepsilon = \left(\frac{\partial^2}{\partial x^2} - \frac{\partial^2}{\partial t^2}\right)u_\varepsilon - \varepsilon P\left(\frac{\partial}{\partial x}, \frac{\partial}{\partial t}; \varepsilon\right)u_\varepsilon = 0$$

After transformation

$$\bar{x} = x - t \quad , \quad \tau = \varepsilon t$$

(2.3.2) $$u_\varepsilon(x,t) \to \bar{u}_\varepsilon(\bar{x},\tau)$$

the perturbed wave equation becomes

(2.3.3) $$\bar{L}_\varepsilon\bar{u}_\varepsilon = -2\frac{\partial^2\bar{u}_\varepsilon}{\partial\bar{x}\partial\tau} - P\left(\frac{\partial}{\partial\bar{x}}, -\frac{\partial}{\partial\bar{x}} + \varepsilon\frac{\partial}{\partial\tau}; \varepsilon\right)\bar{u}_\varepsilon - \varepsilon\frac{\partial^2\bar{u}_\varepsilon}{\partial\tau^2} = 0$$

The operator P can be decomposed (taking the formal limit) as follows:

$$P\left(\frac{\partial}{\partial\bar{x}}, -\frac{\partial}{\partial\bar{x}} + \varepsilon\frac{\partial}{\partial\tau}; \varepsilon\right)\bar{u}_\varepsilon = P\left(\frac{\partial}{\partial\bar{x}}, -\frac{\partial}{\partial\bar{x}}; 0\right)\bar{u}_\varepsilon$$

(2.3.4) $$+ \varepsilon P^{(1)}\left(\frac{\partial}{\partial\bar{x}}, \frac{\partial}{\partial\tau}; \varepsilon\right)\bar{u}_\varepsilon$$

The function $\bar{u}_0(\bar{x},\tau)$ satisfies the model equation

$$(2.3.5)\quad \bar{L}_0\bar{u}_0 = 2\frac{\partial^2\bar{u}_0}{\partial\bar{x}\partial\tau} - P\left(\frac{\partial}{\partial\bar{x}}, -\frac{\partial}{\partial\bar{x}}\,;\,0\right)\bar{u}_0 = 0$$

The considerations that follow apply also to the problem studied in section 2.1, if the definitions of L_ε , $\bar{L}_\varepsilon$ and $\bar{L}_0$ are modified accordingly.

We would like to establish:

Proposition 1. *Let* $\bar{u}_0(\bar{x},\tau)$ *be a solution of* $\bar{L}_0\bar{u}_0 = 0$.

There exists a solution $\bar{u}_\varepsilon(\bar{x},\tau)$ *of* $\bar{L}_\varepsilon\bar{u}_\varepsilon = 0$

Such that $\bar{u}_\varepsilon - \bar{u}_0 = o(1)$

for $\tau \in [0,\bar{T}]$, $\bar{T} \in \mathbb{R}_+$ *and* $\bar{X} \in I$, *where* I *is either an arbitrary compact interval, or (preferably)* $\bar{X} \in (-\infty, +\infty)$.

A link between the functions $\bar{u}_2$ and $\bar{u}_0$ is provided by the observation that

$$(2.3.6)\qquad \bar{L}_\varepsilon\bar{u}_0 = -\,\varepsilon\, P^{(1)}\left(\frac{\partial}{\partial\bar{x}}, \frac{\partial}{\partial\tau}\,;\,\varepsilon\right)\bar{u}_0 = O(\varepsilon)$$

Hence, functions $\bar{u}_0$ are *formal approximations* to solutions $\bar{u}_\varepsilon$ of (2.3.3). Therefore, closely related to Proposition 1 is the following.

Proposition 2. *Let* $\bar{u}_0$ *be a formal approximation* $\bar{L}_\varepsilon\,\bar{u}_0 = \varepsilon f(\bar{x},\tau;\varepsilon) = O(\varepsilon)$ *(with possibly certain conditions imposed on the behaviour of* f *with* $\bar{x}$ *and* τ*).*

Then there exists a solution $\bar{u}_\varepsilon$ *of* $\bar{L}_\varepsilon\bar{u}_\varepsilon = 0$

such that $\bar{u}_\varepsilon - \bar{u}_0 = 0\ (1)$

for τ *and* $\bar{x}$ *as in Proposition 1.*

In an attempt to demonstrate Proposition 2 one would have to study explicitly the full perturbed wave equation.

On the other hand one could use an alternative approach which starts with the observation that

$$(2.3.7)\qquad \bar{L}_0\bar{u}_\varepsilon = \varepsilon\, P^{(1)}\left(\frac{\partial}{\partial\bar{x}}, \frac{\partial}{\partial\tau}\,;\,\varepsilon\right)\bar{u}_\varepsilon$$

Hence, *functions* $\bar{u}_\varepsilon$ *which are such that*

$$(2.3.8)\qquad P^{(1)}\left(\frac{\partial}{\partial\bar{x}}, \frac{\partial}{\partial\tau}\,;\,\varepsilon\right)\bar{u}_\varepsilon$$

is bounded are formal approximations of solutions $\bar{u}_0$ *of the model equation.* This then leads to:

Proposition 3. *Let* $\bar{u}_\varepsilon$ *be a formal approximation of the model equation*

$$\bar{L}_0\bar{u}_\varepsilon = \varepsilon f\,(\bar{x},\tau;\varepsilon) = O(\varepsilon)$$

(with possibly certain conditions imposed on the behaviour of f *with* $\bar{x}$ *and* τ*).*
Then there exists a solution $\bar{u}_0$ *of* $\bar{L}_0\bar{u}_0 = 0$
such that $\bar{u}_\varepsilon - \bar{u}_0 = o(1)$
for T *and* $\bar{x}$ *as in Proposition 1.*

In order to demonstrate Proposition 3 one must study perturbed model equations. For some well known solution equations, such as K.d.V. or nonlinear Schrödinger, important steps in this direction have already been made (see for example Eckhaus of Van Harten 1983 chapter 7, or De Kerf 1984).

The proposition 3 is of course not sufficient to prove the Proposition 1. To close the reasoning one needs a hypothesis, or preferably a regularity result, asserting that there exist classes of solutions $\bar{u}_\varepsilon$ of (2.3.3) satisfying the condition (2.3.8).

Let us finally (although this may seem almost trivial) examine in what sense $\bar{u}_0$ is a formal approximation for the original (not transformed) perturbed wave equation (2.3.1). The answer is immediate:

$$L_\varepsilon\bar{u}_0 = -\,\varepsilon^2\, P^{(1)}\left(\frac{\partial}{\partial\bar{x}}\,,\,\frac{\partial}{\partial\tau}\,,\,\partial\right)\bar{u}_0 = O(\varepsilon^2) \qquad (2.3.9)$$

The error of order ε^2 should not be surprising, but is rather consistent with the whole of the analysis. Suppose, to the contrary, that for some function $\bar{u}_0^*(\bar{x},\tau)$ one has

$$L_\varepsilon\bar{u}_0^* = \varepsilon\, f(\bar{x},\tau)$$

Then the model equation for $\bar{u}_0^*$ would not be (2.3.5), but would be modified to

$$\bar{L}_0\bar{u}_0^* = f(\bar{x},\tau)$$

We conclude this section with some elementary estimates for solutions of the *inhomogeneus wave equation* which shed some additional light on the preceding discussion.

Let $u(x,t)$, $x\in(-\infty,\infty)$, $t \geqslant 0$ be the solution of

$$\left(\frac{\partial^2}{\partial x^2} - \frac{\partial^2}{\partial t^2}\right) u = -\,\varepsilon^p f(x,t)$$

(2.3.10) $$u(x,0) = \left(\frac{\partial u}{\partial t}\right)_{t=0} = 0$$

By Duhamel's principle u is given by the formula

(2.3.11) $$u(x,t) = \frac{1}{2}\varepsilon^p \int_0^t \int_{x-(t'-t)}^{x+(t'-t)} f(x',t')\, dx'dt'$$

Consider u for a bounded function f:

(2.3.12) $$|\, f(x,t) \,| \leqslant f_0$$

It follows from (2.3.11) that

(2.3.13) $$|\, u(x,t) \,| \leqslant \frac{1}{2}\, \varepsilon^p\, f_0 t^2$$

The estimate is sharp, because for $f=f_0$ $u=\frac{1}{2}\, \varepsilon^p f_0 t^2$ is the solution of (2.3.10). The implications of this estimate are disturbing, because for p=1 $u=0(\frac{1}{\varepsilon})$ for $\tau = \varepsilon t = 0(1)$, which leaves little hope for validity of the model equations. The situation improves very much if we consider integrable functions, and more precisely

(2.3.14) $$\int_{-\infty}^{\infty} |f(x,t)|\, dx \leqslant M$$

for $\tau = \varepsilon t$ on compact intervals. It is not difficult to deduce now, from (2.3.11), that

(2.3.15) $$|u(x,t)| \leqslant \frac{1}{2} M \varepsilon^{p-1} \tau$$

It follows that $0(\varepsilon)$ perturbations of the wave equation (p=1) contribute in 0(1) to the solutions on $\tau=0(1)$ timescales, and this is in accordence with the structure of the model equations. It follows further that $0(\varepsilon^2)$ perturbations of the wave equation contribute $0(\varepsilon)$ to solutions on $\tau=0(1)$ intervals, and are hence still negligable. This last conclusion is what we would like to prove for the *perturbed wave equation*.

2.4. Superposition of waves.

We recall that all solutions of the unperturbed wave equation can be seen as a linear superposition of waves moving to the right and to the left. For the perturbed wave equation we have found a model equation for $\bar{u}_0(\bar{x},\tau)$ representing long-time behaviour of waves moving to the right, and similarly a model equation for $\tilde{u}_0(\tilde{x},\tau)$ representing long-time behaviour of waves moving to the left. It is tempting to investigate the possibility of treating the general case starting with a superposition of $\bar{u}_0$ and $\tilde{u}_0$. Such a superposition is of course a trivial matter if P in the perturbed wave equation is a linear operator. We therefore concentrate on the nonlinear case and take P to be such that for any pair of functions f,g one has

$$P(f+g) = Pf + Pg + H(f,g) \tag{2.4.1}$$

where H is an operator working on pairs of functions f,g. We further have

$$\begin{aligned} &Pf = 0 \qquad \text{if} \qquad f = 0 \\ &H(f,g) = 0 \qquad \text{if} \qquad f = 0 \quad \text{or} \quad g = 0 \end{aligned} \tag{2.4.2}$$

This implies (assuming continuity of the operator) that, for bounded functions $\hat{f},\hat{g}$:

$$\begin{aligned} &P(\varepsilon\hat{f}) = 0(\delta_1(\varepsilon)) = o\ (1) \\ &H(\varepsilon\hat{f},g) = 0(\delta_2(\varepsilon)) = o\ (1) \\ &H(f,\varepsilon\hat{g}) = 0(\delta_3(\varepsilon)) = o\ (1) \end{aligned} \tag{2.4.3}$$

For simplicity of the presentation we shall take

$$\delta_i(\varepsilon) = \varepsilon \qquad i = 1,2,3 \tag{2.4.4}$$

With this preparation we consider the perturbed equation

$$L_\varepsilon u_\varepsilon = \left(\frac{\partial^2}{\partial x^2} - \frac{\partial^2}{\partial t^2}\right)u_\varepsilon - \varepsilon\, P\left(\frac{\partial}{\partial x}, \frac{\partial}{\partial t}; \varepsilon\right) u_\varepsilon \tag{2.4.5}$$

with some initial conditions.

We study, in a formal way, the decomposition

$$u_\varepsilon(x,t) = \bar{u}_\varepsilon(\bar{x},\tau) + \tilde{u}_\varepsilon(\tilde{x},\tau) + \varepsilon\, \psi_\varepsilon(x,t) \tag{2.4.6}$$

with

$$\bar{x} = x-t, \ \tilde{x} = x+t, \ \tau = \varepsilon t \tag{2.4.7}$$

The aim is to have valid approximations

$$\bar{u}_\varepsilon(\bar{x},\tau) - \bar{u}_0(\bar{x},\tau) = o\ (1) \tag{2.4.8}$$

with $\bar{u}_0$ a solution of the model equation

$$\bar{L}_0\bar{u}_0 = 2\,\frac{\partial^2\bar{u}_0}{\partial\bar{x}\partial\tau} - P\left(\frac{\partial}{\partial\bar{x}}, -\frac{\partial}{\partial\bar{x}}; 0\right)\bar{u}_0 = 0 \tag{2.4.9}$$

Similarly

$$\tilde{u}_\varepsilon(\tilde{x},\tau) - \tilde{u}_0(\tilde{x},\tau) = o\ (1) \tag{2.4.10}$$

$$\tilde{L}_0\tilde{u}_0 = 2\,\frac{\partial^2\tilde{u}_0}{\partial\tilde{x}\partial\tau} + P\left(\frac{\partial}{\partial\tilde{x}}, \frac{\partial}{\partial\tilde{x}}; 0\right)\tilde{u}_0 = 0 \tag{2.4.11}$$

The term $\varepsilon\psi_\varepsilon$ represents interactions. The goal of our analysis is to derive conditions under which one can expect that the interaction term will be small.

Let us observe at this point that the formula (2.4.6) features five different variables i.e. t, τ, x, $\bar{x}$, $\tilde{x}$, while in reality there are only two variables: x,t. For long-time behaviour $t=O(\frac{1}{3})$ and $\tau=O(1)$. Suppose that one follows in the x,t plane waves such that $\bar{x} = x-t = O(1)$.
Then $\tilde{x} = x+t = \tilde{x} + 2\frac{\tau}{\varepsilon} = O(\frac{1}{\varepsilon})$. Similarly, if $\tilde{x} = O(1)$ then $\bar{x} = O(\frac{1}{\varepsilon})$. The conclusion is: for a meaningful decomposition (2.4.6), the interaction term $\varepsilon\psi_2(x,t)$ should be small for $x=O(\frac{1}{\varepsilon})$ and $\tau=O(\frac{1}{\varepsilon})$. Furthermore the approximation (2.4.8) should be valid for $\bar{x} = O(\frac{1}{\varepsilon})$ and similarly (2.4.10) should be valid for $\tilde{x}=O(\frac{1}{\varepsilon})$. This implies that $\bar{u}_0(\bar{x},\tau)$ should at least be bounded for $|\bar{x}|\to\infty$, and similarly $\bar{u}_0(\tilde{x},\tau)$ should be bounded for $|\tilde{x}|\to\infty$.

We introduce now the decomposition (2.4.6) in the equation (2.4.5). Working out the formules one gets:

(2.4.12)

$$\bar{L}_\varepsilon\bar{u}_\varepsilon + \tilde{L}_\varepsilon\tilde{u}_\varepsilon + \left(\frac{\partial^2}{\partial x^2} - \frac{\partial^2}{\partial t^2}\right)\psi_\varepsilon - H(\tilde{u}_\varepsilon,\bar{u}_\varepsilon) + \varepsilon\,\{H[(\bar{u}_\varepsilon + \tilde{u}_\varepsilon),\psi_\varepsilon] + P(\psi_\varepsilon)\} = 0$$

Taking the formal limit, while using (2.4.8), (2.4.10), one obtains

$$\left(\frac{\partial^2}{\partial x^2} - \frac{\partial^2}{\partial t^2}\right)\psi_0 = H(\tilde{u}_0,\bar{u}_0) \qquad (2.4.13)$$

In accordance with our discussion we must require that *there exists a function* $\psi_0(x,t)$ *which satisfies* (2.4.13) *and is bounded for* $|x| \leq \frac{1}{\varepsilon}L$, $t \in [0,\frac{1}{\varepsilon}T]$, *with* L *and* T *arbitrary constants.*
In order to study the behaviour of ψ_0 one must know explicitly the right hand side of (2.4.13). This follows from the explicit form of P_1 working out the formula 2.4.1. We shall show later on in applications that the condition formulated above is indeed satisfied in some problems of interest. For example for water waves one thus finds K.d.V. equations for waves moving to the right and to the left with virtually no interactions.

Let us remark that the computation of $\varepsilon\psi_0$ does not provide yet a "next approximation" for u_ε, because using (2.4.8) and (2.4.10) terms of the same order of magnitude are neglected. The study of ψ_0 is neccessary to assure oneself that the decomposition (2.4.6) has any chance of succes. This can also be seen from the study of the functions under consideration as formal approximations, along the line of section 2.3.

We already know that

(2.4.14) $$L_\varepsilon \bar{u}_0 = O(\varepsilon^2)$$

and similarly

(2.4.15) $$L_\varepsilon \tilde{u}_0 = O(\varepsilon^2)$$

One easily finds further that

(2.4.16) $$L_\varepsilon(\bar{u}_0 + \tilde{u}_0) = O(\varepsilon)$$

However

(2.4.17) $$L_\varepsilon(\bar{u}_0 + \tilde{u}_0 + \varepsilon\psi_0) = O(\varepsilon^2)$$

When studying superposition of waves, the inclusion of the small interaction term $\varepsilon\psi_0$ assures that one has consistent formal approximations.

We finally comment briefly on the question how the initial conditions for u_ε should be incorporated in the approximations. It seems natural to require that $\bar{u}_0 + \tilde{u}_0$ should satisfy the original initial condition in the first approximation. This can easily be achieved as follows:

Suppose one has

(2.4.18) $$(u_\varepsilon)_{t=0} = u_0(x) \ , \ \left(\frac{\partial u_\varepsilon}{\partial t}\right)_{t=0} = v_0(x)$$

Let $F(x,t) = F_1(x-t) + F_2(x+t)$ be the solution of

$$\left(\frac{\partial^2}{\partial x^2} - \frac{\partial^2}{\partial t^2}\right) F = 0$$

(2.4.19) $$(F)_{t=0} = u_0 \ ; \ \left(\frac{\partial F}{\partial t}\right)_{t=0} = Q_0$$

This leads to the initial conditions

(2.4.20) $$\bar{u}_0(\bar{x},0) = F_1(\bar{x}) \ ; \ \tilde{u}_0(\tilde{x},0) = F_2(\bar{x})$$

2.5. Perturbed hyperbolic systems

In applications one is often confronted with perturbed hyperbolic systems of first order equations. In this section we consider such a system for a pair of functions $u_\varepsilon(x,t)$, $v_\varepsilon(x,t)$:

(2.5.1) $$\frac{\partial u_\varepsilon}{\partial t} + \frac{\partial v_\varepsilon}{\partial x} = \varepsilon\, P\left(\frac{\partial}{\partial x}, \frac{\partial}{\partial t}; \varepsilon\right)(u_\varepsilon, v_\varepsilon)$$

$$\frac{\partial v_\varepsilon}{\partial t} + \frac{\partial u_\varepsilon}{\partial x} = \varepsilon\, Q(\frac{\partial}{\partial x}, \frac{\partial}{\partial t} ; \varepsilon)\ (u_\varepsilon, v_\varepsilon)$$

The right-hand sides are obvious generalizations of the notation used in the preceding sections: P has the properties defined earlier but works now on pairs of functions, Q is a similar operator.

In steps of the analysis in which little detail about the structure of the right-hand sides is needed, we shall often abbreviate and write

(2.5.2)
$$\frac{\partial u_\varepsilon}{\partial t} + \frac{\partial v_\varepsilon}{\partial x} = \varepsilon\, P$$

$$\frac{\partial v_\varepsilon}{\partial t} + \frac{\partial u_\varepsilon}{\partial x} = \varepsilon\, Q$$

Our analytis is conceptually parallel to the preceding sections. We shall therefore skip some obvious repetitions of reasonings.
Adding and subtracting we obtain from (2.5.2)
The system

(2.5.3)
$$\left(\frac{\partial}{\partial t} + \frac{\partial}{\partial x}\right)(u_\varepsilon + v_\varepsilon) = \varepsilon(P+Q)$$

$$\left(\frac{\partial}{\partial t} - \frac{\partial}{\partial x}\right)(u_\varepsilon - v_\varepsilon) = \varepsilon(P-Q)$$

Let u_0, Q_0 be solutions of the unperturbed system. It is obvious that

(2.5.4)
$$u_0 + v_0 = U(x-t)$$

$$u_0 - v_0 = V(x+t)$$

By special choise of initial conditions the solutions are waves which move in one direction, without changing shape.

As in section 2.2 we follow waves that move to the right by introducing the transformation

(2.5.5)
$$\overline{x} = x-t \quad ; \quad \tau=\varepsilon t$$

$$u_\varepsilon(x,t) \to \overline{u}_\varepsilon(\overline{x},\tau) \ ; \ v_\varepsilon(x,t) \to \overline{v}_\varepsilon(\overline{x},\tau)$$

In the new coordinates the system (2.5.3) transforms into

(2.5.6)
$$\frac{\partial}{\partial\tau}(\overline{u}_\varepsilon + \overline{v}_\varepsilon) = P + Q$$

$$-2\,\frac{\partial}{\partial\overline{x}}\,(\overline{u} - \overline{v}_\varepsilon) = \varepsilon(P-O) - \varepsilon\frac{\partial}{\partial\tau}\,(\overline{u}_\varepsilon - \overline{v}_\varepsilon)$$

Taking the formal limit in eqs.(2.5.6), and denoting by P_0, Q_0, the formal limits of the perturbation terms, lead to the model equations

$$(2.5.7)\qquad \frac{\partial}{\partial\tau}(\bar{u}_0 + \bar{v}_0) = P_0 + Q_0$$

$$\frac{\partial}{\partial\bar{x}}(\bar{u}_0 - \bar{v}_0) = 0$$

It follows that

$$(2.5.9)\qquad \bar{u}_0 - \bar{v}_0 = f_0(\tau)$$

One can put $f_0(\tau) = 0$ by requiring that solutions vanish for $|\bar{x}| \to \infty$. Working out the details of the right-hand side of (2.5.7) one obtains as final form the model equation

$$(2.5.10)\qquad 2\,\frac{\partial\bar{u}_0}{\partial\tau} = P\left(\frac{\partial}{\partial\bar{x}}\,,\,-\frac{\partial}{\partial\bar{x}}\,;\,0\right)(\bar{u}_0,\bar{u}_0)$$

$$+\,Q\left(\frac{\partial}{\partial\bar{x}}\,,\,-\frac{\partial}{\partial\bar{x}}\,;\,0\right)(\bar{u}_0,\bar{u}_0)$$

A very similar analysis can of course be performed for waves moving to the left. One then starts with the transformation

$$\tilde{x} = x+t \quad , \quad \tau = \varepsilon t$$

$$(2.5.11)\qquad u_\varepsilon(x,t) \to \tilde{u}_\varepsilon(\tilde{x},\tau)\ ;\ v_\varepsilon(x,t) \to \tilde{v}_\varepsilon(\tilde{x},\tau)$$

and obtains as model equations

$$(2.5.12)\qquad \tilde{u}_0 + \tilde{v}_0 = 0$$

$$2\frac{\partial\tilde{u}_0}{\partial\tau} = P\left(\frac{\partial}{\partial\tilde{x}}\,,\,\frac{\partial}{\partial\tilde{x}}\,;\,0\right)(\tilde{u}_0,\,-\,\tilde{u}_0)$$

$$(2.5.13)\qquad -\,Q\left(\frac{\partial}{\partial\tilde{x}}\,,\,\frac{\partial}{\partial\tilde{x}}\,;\,0\right)(\tilde{u}_0,\,-\,\tilde{u}_0)$$

Finally one can embark on the analysis of superposition of waves, along the lines of section 2.4. In the general case the formulas become somewhat cumbersome. We shall not pursue the details, but we shall show instead how the analysis can be considerally simplified, for large classes of systems (2.5.1). Such is the case if the right-hand side of (2.5.1) contains derivatives of u_ε, v_ε with respect to x and t, and not the functions u_ε, v_ε themselfs. One of the two functions can then be eliminated in the leading terms of the perturbations. We shall use this procedure in applications later on. In the context of this section the procedure runs as follows:

We observe from (2.5.1) that

$$\frac{\partial v_\varepsilon}{\partial x} = -\frac{\partial u_\varepsilon}{\partial t} + O(\varepsilon) \tag{2.5.14}$$

$$\frac{\partial v_\varepsilon}{\partial t} = -\frac{\partial u_\varepsilon}{\partial x} + O(\varepsilon)$$

Let us denote by

$$\hat{P}\left(\frac{\partial}{\partial x}, \frac{\partial}{\partial t}; \varepsilon\right) u_\varepsilon$$

the expression obtained by formal substitution in the right-hand side of (2.5.1) of the relations

$$\frac{\partial v_\varepsilon}{\partial x} = -\frac{\partial u_\varepsilon}{\partial t}\ ; \ \frac{\partial v_\varepsilon}{\partial t} = -\frac{\partial u_\varepsilon}{\partial x}$$

In a similar way we define

$$\hat{Q}\left(\frac{\partial}{\partial x}, \frac{\partial}{\partial t}; \varepsilon\right) u_\varepsilon$$

Now, substituting in full (2.5.14) in the right-hand side of (2.5.1) produces

$$\frac{\partial u_\varepsilon}{\partial t} + \frac{\partial v_\varepsilon}{\partial x} = \varepsilon\, \hat{P}\left(\frac{\partial}{\partial x}, \frac{\partial}{\partial t}; \varepsilon\right) u_\varepsilon + O(\varepsilon^2) \tag{2.5.15}$$

$$\frac{\partial v_\varepsilon}{\partial t} + \frac{\partial u_\varepsilon}{\partial x} = \varepsilon\, \hat{Q}\left(\frac{\partial}{\partial x}, \frac{\partial}{\partial t}; \varepsilon\right) u_\varepsilon + O(\varepsilon^2)$$

Finally, by cross-differentiation and obvious elimination we obtain the perturbed wave equation

$$\frac{\partial^2 u_\varepsilon}{\partial x^2} - \frac{\partial^2 u_\varepsilon}{\partial t^2} = \varepsilon\left\{\frac{\partial}{\partial x}\hat{Q}\left(\frac{\partial}{\partial x}, \frac{\partial}{\partial t}; \varepsilon\right)u_\varepsilon - \frac{\partial}{\partial t}\hat{P}\left(\frac{\partial}{\partial x}, \frac{\partial}{\partial t}; \varepsilon\right)u_\varepsilon\right\} + O(\varepsilon^2) \tag{2.5.16}$$

Turning to sections 2.2 and 2.4 we observe that the perturbations which are $O(\varepsilon^2)$ do not alter the results. In a good mathematical tradition the problem is thus reduced to the one treated before. Returning to unidirectional waves one can verify that the results of this section coincide with results for eq. 2.5.16, along the lines of section 2.2. The exercise is left to the reader.

3. STRAIGHTFORWARD APPLICATIONS.

3.1. K.d.V. for waterwaves.

We take from the appendix the basic equations for small amplitude waves in shallow water, in the simplest case of one dimensional waves and constant depth.

(3.1.1) $$\frac{\partial\varphi_0}{\partial t} + \eta + \frac{1}{2}\alpha\left(\frac{\partial\varphi_0}{\partial x}\right)^2 - \frac{1}{2}\delta^2\frac{\partial^3\varphi_0}{\partial x^2\partial t} = R_1$$

(3.1.2) $$\frac{\partial\eta}{\partial t} + \frac{\partial^2\varphi_0}{\partial x^2} + \alpha\frac{\partial}{\partial x}\left(\eta\frac{\partial\varphi_0}{\partial x}\right) - \frac{1}{6}\delta^2\frac{\partial^4\varphi_0}{\partial x^4} = R_2$$

(3.1.3) $$R_{1.2} = 0(\alpha\delta^2) + 0(\delta^4)$$

All quantities and variables are dimensionless. φ_0 is the approximation to the velocity potential, η is the elevation of the water surface. The two small parameters α and δ are defined through:

$$\alpha = \frac{\text{wave amplitude}}{\text{depth of fluid}}$$

$$\delta = \frac{\text{depth of fluid}}{\text{characteristic length of wave phenomena.}}$$

Through the scaling performed in the appendix all quantities in (3.1.1), (3.1.2) are of order unity, for x and t of order unity. We consider the case that

(3.1.4) $$\alpha = \delta^2$$

and we write

(3.1.5) $$\varepsilon = \alpha = \delta^2$$

The equations (3.1.1), (3.1.2) can be transformed to a perturbed hyperbolic system

(3.1.6) $$\frac{\partial u}{\partial t} + \frac{\partial\eta}{\partial x} + \varepsilon\left\{u\frac{\partial u}{\partial x} - \frac{1}{2}\frac{\partial^3 u}{\partial x^2\partial t}\right\} = 0(\varepsilon^2)$$

$$\frac{\partial\eta}{\partial t} + \frac{\partial u}{\partial x} + \varepsilon\left\{\frac{\partial}{\partial x}(\eta u) - \frac{1}{6}\frac{\partial^3 u}{\partial x^3}\right\} = 0(\varepsilon^2)$$

where

$$u = \frac{\partial\varphi_0}{\partial x}$$

On the other hand (and dropping now for notational simplicity the index zero for the potential) (3.1.1) states that

(3.1.7) $$\eta = -\frac{\partial\varphi}{\partial t} - \varepsilon\left\{\frac{1}{2}\left(\frac{\partial\varphi}{\partial x}\right)^2 - \frac{1}{2}\frac{\partial^3\varphi}{\partial x^2\partial t}\right\} + 0(\varepsilon^2)$$

Using this in (3.1.2) one obtains the perturbed wave equations

(3.1.8) $$\left(\frac{\partial^2}{\partial x^2} - \frac{\partial^2}{\partial t^2}\right)\varphi - \varepsilon\left[2\frac{\partial\varphi}{\partial x}\frac{\partial^2\varphi}{\partial x\partial t} + \frac{\partial\varphi}{\partial t}\frac{\partial^2\varphi}{\partial x^2}\right] + \varepsilon\left[\frac{1}{2}\frac{\partial^4\varphi}{\partial x^2\partial t^2} - \frac{1}{6}\frac{\partial^4\varphi}{\partial x^4}\right] = 0(\varepsilon^2)$$

One can use either (3.1.8) or (3.1.6) as a starting point of the analysis. Each of the two approaches has some interest of its own. In this section we shall work with the perturbed wave - equation (3.1.8). In the next section the hyperbolic system (3.1.6) will be used.

Proceeding as in section 2.2 we follow waves moving to the right over a long-time, through the transformation

$$(3.1.9) \qquad \bar{x} = x-t \quad ; \quad \tau = \varepsilon t$$

$$\varphi(x,t) \rightarrow \bar{\varphi}(\bar{x},\tau)$$

This produces

$$(3.1.10) \qquad 2\,\frac{\partial^2\bar{\varphi}}{\partial\tau\partial\bar{x}} + 3\,\frac{\partial\bar{\varphi}}{\partial\bar{x}}\,\frac{\partial^2\bar{\varphi}}{\partial\bar{x}^2} + \frac{2}{6}\,\frac{\partial^4\bar{\varphi}}{\partial\bar{x}^4} = O(\varepsilon)$$

Hence the model equation for long-time behavior is the famous Korteweg - de Vries equation

$$(3.1.11) \qquad \frac{\partial\bar{u}_0}{\partial\tau} + \frac{3}{2}\,\bar{u}_0\,\frac{\partial\bar{u}_0}{\partial\bar{x}} + \frac{1}{6}\,\frac{\partial^3\bar{u}_0}{\partial\bar{x}^3} = 0$$

with $\bar{u}_0 = \dfrac{\partial\bar{\varphi}_0}{\partial\bar{x}}$

We note that from (3.1.7), by transformation (3.1.9) and writing

$$\eta(x,t) \rightarrow \bar{\eta}(\bar{x},t)$$

one obtains

$$(3.1.12) \qquad \bar{\eta} = \frac{\partial\bar{\varphi}}{\partial\bar{x}} + O(\varepsilon)$$

This means that one can repeat the reasoning starting from (3.1.10) and obtain as model equation for $\bar{\eta}$ again the K.d.V. equation

$$(3.1.13) \qquad \frac{\partial\bar{\eta}_0}{\partial\tau} + \frac{3}{2}\,\bar{\eta}_0\,\frac{\partial\bar{\eta}_0}{\partial\bar{x}} + \frac{1}{6}\,\frac{\partial^3\bar{\eta}_0}{\partial\bar{x}^3} = 0$$

Of course, in an entirely similar way waves moving to the left can be followed, using the transformation

$$(3.1.14) \qquad \tilde{x} = x+t \quad , \quad \tau = \varepsilon t$$

$$\varphi\,(x,t) \rightarrow \tilde{\varphi}\,(\bar{x},\tau)$$

This produces, as model equation, another version of the K.d.V. equation

$$(3.1.15) \qquad \frac{\partial\tilde{u}_0}{\partial\tau} + \frac{3}{2}\,\tilde{u}_0\,\frac{\partial\tilde{u}_0}{\partial\tilde{x}} - \frac{1}{6}\,\frac{\partial^3\tilde{u}_0}{\partial\tilde{x}^3} = 0$$

with $\tilde{u}_0 = \dfrac{\partial\tilde{\varphi}_0}{\partial\tilde{x}}$

We shall finally consider, along the line of section 2.4, the superposition of waves moving to the right and to the left. We introduce the decomposition

$$(3.1.16) \qquad \varphi(x,t) = \bar{\varphi}(x,\tau) + \tilde{\varphi}\,(\tilde{x},\tau) + \varepsilon\psi(x,t)$$

Working out the equation (3.1.8) one finds

$$(3.1.17)\quad \frac{\partial}{\partial\bar{x}}\left\{\frac{\partial\bar{\varphi}}{\partial\tau}+\frac{3}{4}\left(\frac{\partial\bar{\varphi}}{\partial\bar{x}}\right)^2+\frac{1}{6}\frac{\partial^3\bar{\varphi}}{\partial\bar{x}^3}\right\}$$

$$-\frac{\partial}{\partial\tilde{x}}\left\{\frac{\partial\tilde{\varphi}}{\partial\tau}+\frac{3}{4}\left(\frac{\partial\tilde{\varphi}}{\partial\tilde{x}}\right)^2-\frac{1}{6}\frac{\partial^3\tilde{\varphi}}{\partial\tilde{x}^3}\right\}$$

$$+\frac{1}{2}\left(\frac{\partial^2}{\partial x^2}-\frac{\partial^2}{\partial t^2}\right)\psi+\frac{1}{2}\left\{\frac{\partial\tilde{u}}{\partial\tilde{x}}\frac{\partial^2\bar{u}}{\partial\bar{x}^2}-\frac{\partial\bar{u}}{\partial\bar{x}}\frac{\partial^2\tilde{u}}{\partial\tilde{x}^2}\right\}=0(\varepsilon)$$

As explained in section 2.4, the main question is whether the interaction term $\varepsilon\psi$ can be expected to remain small for $t = 0(\frac{1}{\varepsilon})$ and $\bar{x}\in(-\infty,\infty)$.

Let $\bar{\varphi}_0(\bar{x},\tau)$, $\tilde{\varphi}_0(\tilde{x},\tau)$ be solutions of (3.1.11) resp. (3.1.15) which we assume bounded for $\bar{x}\in(-\infty,\infty)$ $\tilde{x}\in(-\infty,\infty)$, $\tau=0(1)$. We define:

$$(3.1.18)\quad \psi_0=\frac{-1}{4}\left[\tilde{u}_0\frac{\partial\bar{u}_0}{\partial\bar{x}}-\bar{u}_0\frac{\partial\tilde{u}_0}{\partial\tilde{x}}\right]$$

A straight forward computation shows that

$$(3.1.19)\quad \left(\frac{\partial^2}{\partial x^2}-\frac{\partial^2}{\partial t^2}\right)\psi_0+\left\{\frac{\partial\tilde{u}_0}{\partial\tilde{x}}\frac{\partial^2\bar{u}_0}{\partial\bar{x}^2}-\frac{\partial\bar{u}_0}{\partial\bar{x}}\frac{\partial^2\tilde{u}_0}{\partial\tilde{x}^2}\right\}=0(\varepsilon)$$

It follows that the function

$$\bar{\varphi}_0+\tilde{\varphi}_0+\varepsilon\psi_0$$

is a formal approximation for the perturbed wave equation (3.1.8), producing an error on the right hand side that is $0(\varepsilon^2)$.

The function ψ_0 is nice and bounded for $x\in(-\infty,\infty)$, $\tau=0(1)$, indicating that the interaction effects do not accumulate. We refer further to the discussion in sections 2.4 and 2.3 and conclude here that the necessary condition for the interaction term to remain small is satisfied in the problem under consideration. Of course, we do not have a complete theory yet, but it is certainly not inconsistent at this stage to expect that right- and left- moving water waves will evolve each according to its K.d.V. equation with virtually no interaction.

3.2. K.d.V. versus B.B.M.

The Korteweg- de Vries equation has not remained unchallenged as a model equation for water waves. Benjamin, Bona and Mahony (1972) have proposed a slightly different model and the comparison of the two models has been the subject of various publications. We refer in particular to Bona, Pritchard and Scott (1983) and the literature mentioned the there.

The purpose of this section is to rederive the K.d.V. equation, taking as a starting point the hyperbolic system (3.1.6). This derivation has some merits of its own, and has further the advantage that the reasoning stays rather close to the argumentation used by the protagonisets of the B.B.M. equation.

On the basis of section 2.5, and upon some reflection on the particular structure of the hyperbolic system (3.1.6), it appears interesting to introduce a decomposition

$$U = U + \varepsilon G \qquad (3.2.1)$$
$$\eta = U - \varepsilon G$$

This decomposition implies that one wants te study waves moving to the right, because (3.2.1) with $\varepsilon=0$ is true for solutions of the unperturbed system which are waves moving to the right.

Substitution and some trivial manipulations lead to the *decoupled system*

$$\frac{\partial U}{\partial t} + \frac{\partial U}{\partial x} + \varepsilon \frac{1}{2}\left\{3U \frac{\partial U}{\partial x} - \frac{1}{6}\frac{\partial^3 U}{\partial x^3} - \frac{1}{2}\frac{\partial^3 U}{\partial x^2 \partial t}\right\} = O(\varepsilon^2) \qquad (3.2.2)$$

$$\frac{\partial G}{\partial t} - \frac{\partial G}{\partial x} + \frac{1}{2}\left\{-U \frac{\partial U}{\partial x} + \frac{1}{6}\frac{\partial^3 U}{\partial x^3} - \frac{1}{2}\frac{\partial^3 U}{\partial x^2 \partial t}\right\} = O(\varepsilon^2) \qquad (3.2.3)$$

It should be emphasized that the equation (3.2.2) is correct if G is defined as order one solution of (3.2.1).

We note that (3.2.2) also implies

$$\frac{\partial}{\partial t} + \frac{\partial}{\partial x} = O(\varepsilon) \qquad (3.2.4)$$

Using this, (3.2.2) becomes

$$\frac{\partial U}{\partial t} + \frac{\partial U}{\partial x} + \varepsilon \left\{\frac{3}{2} U \frac{\partial U}{\partial x} + \frac{1}{6}\frac{\partial^3 U}{\partial x^3}\right\} = O(\varepsilon^2) \qquad (3.2.5)$$

Let us now define as a model equation

$$\frac{\partial U_0}{\partial t} + \frac{\partial U_0}{\partial x} + \varepsilon \left\{\frac{3}{2} U_0 \frac{\partial U_0}{\partial x} + \frac{1}{6}\frac{\partial^3 U_0}{\partial x^3}\right\} = 0 \qquad (3.2.6)$$

The equation (3.2.6) becomes the K.d.V. equation after the transformation

$$\bar{x} = x - t \quad , \quad \tau = \varepsilon t \qquad (3.2.7)$$
$$U_0(x,t) \rightarrow \bar{U}_0(\bar{x},\tau)$$

On the other hand, on the basis of (3.2.4) one can also write (among other forms)

$$\frac{\partial U}{\partial t} + \frac{\partial U}{\partial x} + \varepsilon \left\{\frac{3}{2} U \frac{\partial U}{\partial x} - \frac{1}{6}\frac{\partial^3 U}{\partial x^2 \partial t}\right\} = O(\varepsilon^2) \qquad (3.2.8)$$

and proclaim as a model equation

$$\frac{\partial U_1}{\partial t} + \frac{\partial U_1}{\partial x} + \varepsilon \left\{\frac{3}{2} U_1 \frac{\partial U_1}{\partial x} - \frac{1}{6}\frac{\partial^3 U_1}{\partial x^2 \partial t}\right\} = 0 \qquad (3.2.9)$$

The equation (3.2.9) is the B.B.M. equation. The relation to the K.d.V. equation becomes clear after transformation (3.2.7) with

$$(3.2.10) \qquad U_1(x,t) \rightarrow \bar{U}_1(\bar{x},\tau)$$

The equation (3.2.9) then becomes

$$(3.2.11) \qquad \frac{\partial \bar{U}_1}{\partial \tau} + \frac{3}{2}\bar{U}_1 \frac{\partial \bar{U}_1}{\partial \bar{x}} + \frac{1}{6}\frac{\partial^3 \bar{U}_1}{\partial \bar{x}^3} = \varepsilon \frac{1}{6}\frac{\partial^3 \bar{U}_1}{\partial \bar{x}^2 \partial \tau}$$

In other words: *the* B.B.M. *equation is a perturbed* K.d.V. *equation.*

On the basis of our discussion in section 2.3 we observe that solutions of the B.B.M. equation are formal approximations to the solutions of the K.d.V. equation, and vice-versa. We expect (and hope) that formal approximations are true approximations, at least on compact τ-intervals. For the K.d.V. versus B.B.M. this has been demonstrated explicitly in Bona, Pritchard & Scott (1983).

What reasons (if any) are there then to prefer the B.B.M. equation? According to protagonists the main reasons are that the B.B.M. equation has a more realistic dispersion relation and is better suited for numerical computations. In this connection the equation (3.2.11) has also been called *a regularized version of* K.d.V.

From our point of vieuw the use of the B.B.M. equation as an alternative model is not illegitimate as long as the relation with the K.d.V. equation, outlined in this section, is properly understood. However, some forms of presentation of these equations encountered in the literature may lead to serious confusion.
This arises as follows:

Let us introduce, in a formal way, the transformations:

$$(3.2.12) \qquad \hat{U}_{0.1} = \varepsilon\, U_{0.1} \;;\; \hat{x} = \frac{x}{\sqrt{\varepsilon}} \;;\; \hat{t} = \frac{t}{\sqrt{\varepsilon}}$$

then (3.2.6) becomes

$$(3.2.13) \qquad \frac{\partial \hat{U}_0}{\partial \hat{t}} + \frac{\partial \hat{U}_0}{\partial \hat{x}} + \frac{1}{2}\hat{U}_0 \frac{\partial \hat{U}_0}{\partial \hat{x}} + \frac{1}{6}\frac{\partial^3 \hat{U}_0}{\partial \hat{x}^3} = 0$$

while (3.2.9) leads to

$$(3.2.14) \qquad \frac{\partial \hat{U}_1}{\partial \hat{t}} + \frac{\partial \hat{U}_1}{\partial \hat{x}} + \frac{1}{2}\hat{U}_1 \frac{\partial \hat{U}_1}{\partial \hat{x}} - \frac{1}{6}\frac{\partial^3 \hat{U}_1}{\partial \hat{x}^2 \partial \hat{t}} = 0$$

The K.d.V. and the B.B.M. equations are thus brought on the same footing. However, the operation is puzzling : it is in contradiction with the basic scaling (see appendix) to consider $\hat{U}_{0.1}$ of order unity, while it does not seem very interesting to consider $\hat{x}$, $\hat{t}$ of order unity.

A serious student of the literature is therefore relieved when he notices (see Bona, Pritchard and Scott) that the equations (3.2.13), (3.2.14) are studied with initial conditions given in the form

(3.2.15) $$(\hat{U}_{0.1})_{\hat{t}=0} = \varepsilon g(\sqrt{\varepsilon}\,\hat{x})$$

and that results are given for $\varepsilon^{3/2}\hat{t} = O(1)$. This means that the analysis remains in the framework of (3.2.6), (3.2.9), (3.2.13), (3.2.11), and that formulas (3.2.12), (3.2.13), (3.2.14) are only an attempt to give the K.d.V. and the B.B.M. equations the same formal status.

3.3 K.d.V.-Burgers for ion acoustic waves.

We take from Franzen and Kevorkian (1984) the following system of equations

(3.3.1)
$$\frac{\partial n}{\partial t} + \frac{\partial}{\partial x}(nu) = 0$$
$$\delta^2 \frac{\partial^2 \Phi}{\partial x^2} = e^{\Phi} - n$$
$$n\left(\frac{\partial u}{\partial t} + u\frac{\partial u}{\partial x}\right) = -n\frac{\partial \Phi}{\partial x} + \mu \frac{\partial^2 u}{\partial x^2}$$

All quantities and variables are dimensionless, δ and μ are small parameters. It is beyond our interests to expound the relation of the system (3.3.1) to the physical problem mentioned in the title of this section. We shall be concerned with mathematical modeling for solutions that are near the equilibrium solution $u=\Phi=0$, $n=1$. For this purpose we write

(3.3.2)
$$n=1+\varepsilon\eta$$
$$\Phi=\varepsilon\varphi$$
$$u=\varepsilon q$$

This produces the system:

(3.3.3) $$\frac{\partial \eta}{\partial t} + \frac{\partial q}{\partial x} + \varepsilon\frac{\partial}{\partial x}(\eta q) = 0$$

(3.3.4) $$\varphi - \eta + \frac{1}{2}\varepsilon\,\varphi^2 - \delta^2\frac{\partial^2\varphi}{\partial x^2} = O(\varepsilon^2)$$

(3.3.5) $$\frac{\partial q}{\partial t} + \frac{\partial \varphi}{\partial x} + \varepsilon\left\{\eta\frac{\partial q}{\partial t} + q\frac{\partial q}{\partial x} + \eta\frac{\partial \eta}{\partial x}\right\} - \mu\frac{\partial^2 q}{\partial x^2} = O(\varepsilon^2)$$

Concerning the three small parameters ε, δ^2, μ we take the case

(3.3.6) $$\varepsilon = \delta^2 = \mu$$

The equation (3.3.4) permits to eliminate η from (3.3.3), (3.3.5). However, one can also (and this requires somewhat less labour) observe from (3.3.4) that

(3.3.7) $$\varphi = \eta + O(\varepsilon)$$

and rewrite (3.3.4) in the form

$$\varphi = \eta - \frac{1}{2}\varepsilon\eta^2 + \varepsilon\frac{\partial^2\eta}{\partial x^2} + O(\varepsilon^2) \tag{3.3.8}$$

This permits to eliminate φ from (3.3.5) which leads to the perturbed hyperbolic system

$$\frac{\partial\eta}{\partial t} + \frac{\partial q}{\partial x} + \varepsilon\frac{\partial}{\partial x}(\eta q) = 0 \tag{3.3.9}$$

$$\frac{\partial q}{\partial t} + \frac{\partial\eta}{\partial x} + \varepsilon\left\{\eta\frac{\partial q}{\partial t} + q\frac{\partial q}{\partial x} - \frac{\partial^2 q}{\partial x^2} + \frac{\partial^3 q}{\partial x^3}\right\} = O(\varepsilon^2) \tag{3.3.10}$$

The analysis of section 2.5 immediately applies to this system. For the waves moving to the right

$$\bar{x} = x-t \quad , \quad \tau = \varepsilon t$$

$$\eta(x,t) \to \bar{\eta}(\bar{x},\tau) \quad , \quad q(x,t) \to \bar{q}(\bar{x},\tau) \tag{3.3.11}$$

one gets as model equations

$$\bar{\eta}_0 - \bar{q}_0 = 0 \tag{3.3.12}$$

$$2\frac{\partial\bar{\eta}_0}{\partial\tau} + 2\bar{\eta}_0\frac{\partial\bar{\eta}_0}{\partial\bar{x}} - \frac{\partial^2\bar{\eta}_0}{\partial\bar{x}^2} + \frac{\partial^3\bar{\eta}_0}{\partial\bar{x}^3} = 0 \tag{3.3.13}$$

For the waves moving to the left

$$\tilde{x} = x+t \quad , \quad \tau = \varepsilon t$$

$$\eta(x,t) \to \tilde{\eta}(\tilde{x},\tau), \quad q(x,t) \to \tilde{q}(\tilde{x},\tau) \tag{3.3.14}$$

one gets as model equations

$$\tilde{\eta}_0 + \tilde{q}_0 = 0 \tag{3.3.15}$$

$$2\frac{\partial\tilde{\eta}_0}{\partial\tau} - 2\tilde{\eta}_0\frac{\partial\tilde{\eta}_0}{\partial\tilde{x}} - \frac{\partial^2\tilde{\eta}_0}{\partial\tilde{x}^2} - \frac{\partial^3\tilde{\eta}_0}{\partial\tilde{x}^3} = 0 \tag{3.3.16}$$

The equations (3.3.13) and (3.3.16) are (versions of) the K.d.V.- Burgers equation.

We finally investigate superposition of waves moving to the right and to the left. One should observe that the system (3.3.9), (3.3.10) cannot be reduced to a single perturbed wave equation, so that the analysis outlined in section 2.5 cannot be directly applied. However, it will appear that some manipulation of (3.3.9), (3.3.10) brings the analysis close to section 2.5 and reduces the computation effort.
To investigate the superposition we write

$$\begin{aligned} \eta(x,t) &= \bar{\eta}(\bar{x},t) + \tilde{\eta}(\tilde{x},T) + \varepsilon\,\Theta(x,t) \\ q(x,t) &= \bar{\eta}(\bar{x},t) - \tilde{\eta}(\tilde{x},T) + \varepsilon\,\psi(x,t) \end{aligned} \tag{3.3.17}$$

straightforward insertion of (3.3.17) into (3.3.9), (3.3.10) requires a considerable bookkeeping. Instead we observe that (3.3.9)(3.3.10) also leads to a system of coupled perturbed wave equations, one of which reads:

$$\left(\frac{\partial^2}{\partial x^2} - \frac{\partial^2}{\partial t^2}\right)q + \varepsilon\left\{\frac{\partial^2}{\partial x^2}(\eta q) - \frac{\partial}{\partial t}\left[q\frac{\partial q}{\partial x} - \eta\frac{\partial \eta}{\partial x}\right]\right.$$

(3.3.18)

$$\left. + \frac{\partial^3 q}{\partial x^2 \partial t} - \frac{\partial^4 \eta}{\partial x^3 \partial t} - \frac{\partial^4 \eta}{\partial x^3 \partial t}\right\} = O(\varepsilon^2)$$

Inserting (3.3.17) into (3.3.18) we find.

$$\frac{\partial}{\partial \bar{x}}\left\{2\frac{\partial \bar{\eta}}{\partial \tau} + 2\bar{\eta}\,\frac{\partial \bar{\eta}}{\partial \bar{x}} - \frac{\partial^2 \bar{\eta}}{\partial \bar{x}^2} + \frac{\partial^3 \bar{\eta}}{\partial \bar{x}^3}\right\} +$$

(3.3.19)

$$+ \frac{\partial}{\partial \tilde{x}}\left\{2\frac{\partial \tilde{\eta}}{\partial \tau} - 2\tilde{\eta}\frac{\partial \tilde{\eta}}{\partial \tilde{x}} - \frac{\partial^2 \tilde{\eta}}{\partial \tilde{x}^2} - \frac{\partial^3 \tilde{\eta}}{\partial \tilde{x}^2}\right\} +$$

$$+ \left(\frac{\partial^2}{\partial x^2} - \frac{\partial^2}{\partial t^2}\right)\psi - 2\left\{\bar{\eta}\,\frac{\partial^2 \tilde{\eta}}{\partial \tilde{x}^2} - \tilde{\eta}\,\frac{\partial^2 \bar{\eta}}{\partial \bar{x}^2}\right\} = O(\varepsilon)$$

Taking into account (3.3.15),(3.3.16) leads to the model equation for the interaction term

(3.3.20) $$\left(\frac{\partial^2}{\partial x^2} - \frac{\partial^2}{\partial t^2}\right)\psi_0 = 2\left\{\bar{\eta}_0\,\frac{\partial^2 \tilde{\eta}_0}{\partial \tilde{x}^2} - \tilde{\eta}_0\,\frac{\partial^2 \bar{\eta}_0}{\partial \bar{x}^2}\right\}$$

One easily verifies that the function

(3.3.21) $$\psi_0 = -\frac{\partial \tilde{\eta}_0}{\partial \tilde{x}}\int^{\bar{x}} \bar{\eta}_0 d\bar{x} + \frac{\partial \bar{\eta}_0}{\partial \bar{x}}\int^{\tilde{x}} \tilde{\eta}_0\, d\tilde{x}$$

satisfies (3.3.20) with an error of $O(\varepsilon)$. Recalling the discussion of section 2.5 we conclude that if $\bar{\eta}_0(\bar{x},\tau)$, $\tilde{\eta}_0(\tilde{x},\tau)$ are integrable with respect to $\bar{x}$, $\tilde{x}$ over the whole real line, then one has indeed good reasons to expect that the interaction term $\varepsilon\psi$ will remain small for $\tau = O(1)$.

Repeating the analysis with, as a starting point, the perturbed wave equation for η one reaches similar conclusions concerning the interaction term $\varepsilon\Theta$.

REFERENCES

T.B.Benjamin, J.L.Bona, J.J.Mahony. Model equations for long waves in nonlinear dispersive systems.
Phil.Trans. Roy. Soc. London A.272 (1972)

J.L.Bona, W.G.Pritchard, L.R.Scott. A comparison of solutions of two model equations for long waves.
Lectures in Applied Mathematics AMS Vol.20 (1983)

J.D.Cole. Perturbation methods in applied mathematics.
Ginn - Blaisdell (1968)

D.David, D.Levi, P.Winternitz. Solitons in straits of varying width and depth.
Université de Montréal, Çentre de Recherches Mathematique Preprint (1985)

R.K.Dodd, J.C.Eilbeek, J.D.Gibbon, H.C.Morris. Solitons and nonlinear wave equations.
Academic Press (1982)

W.Eckhaus, A.van Harten. The inverse scattering transformation and the theory of solitons.
North-Holland (1981)

C.L.Franzen, J.Kevorkian. Uncoupled nonlinear evolution equations by multiple scale and reductive perturbation methods.
Applied Math.Program, University of Washington, Seattle. Preprint (1984)

R.Geel. Singular perturbations of hyperbolic type. Mathematical Center Amsterdam Troef (1978)

B.B.Kadomtsev, V.I.Petviashvili. The stability of solitary waves in weakly dispersive media.
Dokl.Akad. Nauk SSR 192 (1970)

J.Kevorkian, J.D.Cole. Perturbation methods in applied mathematics.
Springer (1981)

F.de Kerf. Perturbation theory for the Korteweg-de Vries equation.
Preprint Mathematical Institute Utrecht (1985)

G.L.Lamb, Jr. Elements of soliton theory.
Wiley-Interscience (1980)

R.E.Meyer. On the meaning of the Boussinesq and Korteweg-de Vries equations.
Bull.Cal. Math.Soc. 71 (1979)

S.Novikov, S.V.Manakov, L.P.Pitaerskii, Y.E.Zakharov. Theory of Solitons.
English translation by Consultant Bureau New York (1984)

G.B.Whitham. Linear and nonlinear waves.
John Wiley (1974)

SOLITONS AND DOMAIN STRUCTURE IN ELASTIC CRYSTALS

WITH A MICROSTRUCTURE:

Mathematical aspects.

G.A.Maugin
Laboratoire de Mécanique Théorique associé au C.N.R.S.
Université Pierre et Marie Curie ,Paris,France .

Abstract

Elastic crystals with a microstructure include ferroelectric crystals, ferromagnetic crystals and crystals with internal mechanical degrees of freedom. In recent works concerning the discrete or continuum modelling of the behavior of such elastic crystals, we have been able to delineate â general descriptive framework in which,using the concepts of solitary waves and solitons, the dynamics of simple structures in domains and walls can be accommodated. To that purpose the notion of Bloch and Néel walls in deformable crystals was introduced in all cases. The general nonlinear mathematical problem obtained concerns a nonlinear hyperbolic dispersive system made of a sine-Gordon equation, or a double sine-Gordon equation, for an internal parameter related to the microstructure, which is nonlinearly coupled to one or two wave equations governing elastic displacements. While exact stable nonlinear solutions of the solitary-wave type can be exhibited in a more or less straightforward manner ,the problem of the interaction of such wave motions (representing then the collision of walls and the coalescence of domains) and that of the transient motion of such waves when acted upon by an external stimulus (then representing the starting motion of walls) can be tackled only by using more sophisticated methods such as singular perturbations, Whitham's averaged Lagrangian method and those methods familiar in soliton theory (Bäcklund transformations, inverse scattering method). This is the concern of the present lecture with an emphasis on perturbations.

1.- INTRODUCTION

The present contribution concerns the following type of hyperbolic systems of equations in the (X,t) plane for the dependent variables (u_α, ϕ) :

$$(1.1)\quad \begin{cases} W_\alpha(u_\alpha) = \eta_\alpha \, f_\alpha(\phi) \quad , \quad \alpha = 1,2 \quad , \\ NL(\phi) = \sum_\alpha \eta_\alpha \, g_\alpha(\phi \, , \, u_\alpha) + \varepsilon \, h(\phi) \quad , \end{cases}$$

where W_α are linear (d'Alembertian) wave operators, NL is a nonlinear dispersive wave operator such as a sine-Gordon or a double sine-Gordon operator, f_α and g_α are known functions or differential operators and the η_α's and ε are possibly infinitesimally small parameters. Usually, $\alpha = 1,2$, and $\alpha=1$ only in the most thoroughly studied case (see below) . Systems of the type (1.1) were obtained by :

(i) Pouget and Maugin [1]-[2] in their nonlinear dynamical study of the domain-wall structure of elastic ferroelectric crystals of the type of sodium nitrite; then $\alpha=1$, NL = SG $\equiv$ sine-Gordon, $u_1=$ v: transverse elastic displacement, ϕ : twice the angle of rotation of the central molecular group , η : electromechanical (piezoelectric) coupling coefficient, ε : applied electric field;

(ii) Maugin and Miled [3] in the study of the motion of magnetoelastic Néel walls in thin ferromagnetic films ; then $\alpha=1,2$, u_1=u: longitudinal elastic displacement, u_2= v: transverse elastic displacement , ϕ : precession angle of the magnetic spins in the plane of the wall, η_1 and η_2 : magnetostriction coefficients, ε : applied magnetic field ;

(iii) Maugin and Miled [4] and Pouget and Maugin [5] while studying the nonlinear one-dimensional dynamics of elastic solids endowed with a rigid microstructure, either in the Eringen formalism of micropolar elastic solids or in the Ericksen "director" formalism for oriented elastic solids ; then the η_α's are the additional "elasticity" coefficients, ε corresponds to an external volume couple and NL may be a double sine-Gordon operator instead of a sine-Gordon one.

In all these examples ϕ is related to some rotational degree of freedom. More on the construction of these models and their physics is to be found in the original papers and a companion"physics" lecture [6]. The ferroelectric case is briefly recalled in Section 2 by way of example.

In the physical examples obtained the nature of the nonlinear operator NL is such that one can expect, through compensation between nonlinear and dispersive effects, a <u>soliton</u> behavior for solutions ϕ whenever $\eta_\alpha = \varepsilon = 0$. Furthermore, for physically admissible limit conditions at $\pm\infty$ and the functions and operators f_α and g_α obtained in the various cases, in the three above cases it was possible to exhibit <u>exact</u> stable solutions of the <u>solitary-wave</u> type for (ϕ, u_α), when $\varepsilon = 0$, through a direct integration (see example below) . It may happen in certain cases that this holds true also for $\varepsilon \neq 0$ when the corresponding additional term is in fact absorbed in NL by a redefinition of the latter without change in its nature [3]. The essential problem that we briefly tackle here is how to obtain a solution, and what are the features of the said solution, when, to reproduce a physical structure in more than two domains (hence at least <u>two</u> walls), we need to consider solutions which a priori depend on several phase factors (e.g., multiple solitons). A singular perturbation method which brings into the picture a <u>modulation</u> of the free parameters of a zeroth order solution has to be used (see Section 3 with example in Section 4). Likewise, such a modulation , of one,or more, free parameters depending on the refinement of the method, has to be introduced when $h(\phi)$ cannot be simply incorporated in NL (Section 5). In all cases, as is often the case for dynamical systems of this type, Hamiltonian and Lagrangian formulations of the nonlinear hyperbolic dispersive system have to be envisaged and this also helps in the numerical simulation of the complex interaction phenomena for which the direct analysis can provide hardly more than asymptotic evaluations of the behavior. The present study partakes in a general study of resonance couplings and symmetry breaking in electromagnetic deformable solids [7]-[8].

2.- PHYSICAL MODEL

While studying a dynamical model for the structure in domains and walls in elastic ferroelectric crystals presenting a central molecular group that carries an electric dipole (e.g., sodium nitrite $NaNO_2$), we have recently shown for a one-dimensional (along X) motion that a large-amplitude internal rotation ϕ, the (infinitesimal) longitudinal elastic displacement u , the (infinitesimal) transverse elastic displacement v and an out-of-plane ,infinitesimally small angular deviation α are governed by the following nonlinear hyperbolic system in the absence of external stimuli (in nondimensional notation; see Figure 1)[1]-[2]:

$$(2.1)_a \qquad \frac{\partial^2 u}{\partial t^2} - V_L^2 \frac{\partial^2 u}{\partial X^2} = \gamma \frac{\partial}{\partial X}(1+\cos\phi) \; ,$$

$$(2.1)_b \qquad \frac{\partial^2 v}{\partial t^2} - V_T^2 \frac{\partial^2 v}{\partial X^2} = -\eta \frac{\partial}{\partial X}(\sin\phi) \; ,$$

$$(2.1)_c \qquad \frac{\partial^2 \phi}{\partial t^2} - \frac{\partial^2 \phi}{\partial X^2} - \sin\phi = \eta \cos\phi \frac{\partial v}{\partial X} + \gamma \sin\phi \frac{\partial u}{\partial X}$$

and

$$(2.2) \qquad \frac{1}{K}\frac{\partial^2 \alpha}{\partial t^2} - V_S^2 \frac{\partial^2 \alpha}{\partial X^2} = \frac{1}{2}\left[(1+\cos\phi)\left(1+\gamma\frac{\partial u}{\partial X}\right) - \frac{1}{2}\eta \sin\phi \frac{\partial v}{\partial X} + (\cos\phi)\, V_S^2 \frac{\partial^2}{\partial X^2}\right]\alpha \; .$$

Here the characteristic speed of the ϕ-equation has been taken as the unit for speeds. The mechanical speeds V_L and V_T are much larger than one.

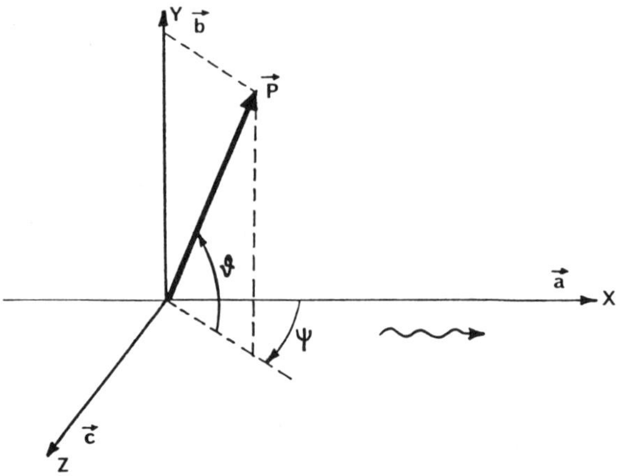

Fig.1.- Notation for eqns.(2.1)-(2.2) ; $\phi = 2\vartheta$, $\alpha = 2\Psi$.

The structure of the above-stated system of dispersive nonlinear hyperbolic equations is of particular interest because it contains simpler systems that we have also constructed in recent studies [3]-[5]. In particular, the first system (2.1) of three equations for the three unknowns (u,v,ϕ) is made of two wave equations for the elastic displacement in the (X,Y) plane of Figure 1 (a situation that recalls the polarization of Rayleigh surface waves in isotropic elasticity) and a sine-Gordon equation for ϕ , each elastic displacement being nonlinearly coupled to ϕ through one of the coupling coefficients γ and η (electromechanical coupling akin to piezoelectricity in the case of $NaNO_2$). As to the remaining angle α, it is governed by a Klein-Gordon equation which is nonlinearly perturbed by the coupled solution of the first system. Clearly, the essential nonlinearity and dispersion are introduced through the angular variable ϕ and both the large excursions of this variable and the dispersive contribution may be related to some phase-transition phenomena. In particular, the relation of the above-recalled model (in a continuum framework) with commensurate-incommensurate phase transitions in ferroelectrics was examined by Pouget and Maugin [1]. The other variables remain small and the problem fully linear with respect to these if it were not for the coupling with ϕ. The interesting system is provided by eqns.(2.1) and eqn.(2.2) may be left aside, eventhough it may have proved essential in the derivation of the whole system (as is the case for ferromagnetic thin films in Ref.3). This system (2.1) in fact appears as a somewhat general system which allows one to study the domain-wall structure in elastic crystals with a microstructure (of which the kinematics is essentially of the rotation type) that these crystals be of the ferroelectric type (case above), of the ferromagnetic type [ϕ then is related to the precession of magnetic spins and γ and η are magnetostriction coefficients; a system analogous to (2.1) was deduced by the author and Miled [3] for the magnetoacoustic dynamics of Néel walls in elastic ferromagnetic thin films of,e.g., yttrium-iron garnet YIG] or of a purely mechanical type such as potassium nitrate KNO_3 whether a nonlinear micropolar description in the manner of Kafadar and Eringen is used (cf. Maugin and Miled [4]) or a continuum description including a field of "directors" in the style of the Leslie-Ericksen theory of liquid crystals is considered (cf. Pouget and Maugin [5] -then ν and η are related to the additional elasticity coefficients of the theory. In the last case, however, the equation that replaces eqn.$(2.1)_c$ may be more involved, being from the start a double-sine-Gordon equation that is nonlinearly coupled to the usual elastic displacements. More on these various models may be found in the review [6]. It is also expected that the above sketched out model applies to the dynamical elasticity of certain long macromolecular chains that display a twist (e.g., D.N.A., polyelectrolytes).

Although most of the following mathematical considerations apply to the whole system (2.1), we shall assume, as shown by Pouget [9] in the ferroelectric case of $NaNO_2$, that $\nu \ll \eta$, so that we are satisfied with the study of the framed subsystem in (2.1),i.e., with $V_T \gg 1$,

$$(2.3)\quad \begin{cases} W(v) = \eta\, f(\phi) , \\ SG(\phi) = \eta\, g(\phi,v) , \end{cases}$$

where

$$\begin{cases} W = \dfrac{\partial^2}{\partial t^2} - V_T^2 \dfrac{\partial^2}{\partial X^2} , \quad SG = \dfrac{\partial^2}{\partial t^2} - \dfrac{\partial^2}{\partial X^2} - \sin , \\ f(\phi) = - \dfrac{\partial}{\partial X}(\sin \phi) , \quad g(\phi,v) = \cos \phi \dfrac{\partial v}{\partial X} . \end{cases}$$

Such a system can formally be rewritten as a <u>nonlinear system of evolution</u> for a four-vector $\vec{U} = (v , \partial v/\partial t, \phi , \partial \phi/\partial t)^T$ as

$$(2.4)\quad \frac{\partial \vec{U}}{\partial t} + N(\vec{U}) = \eta\, \vec{F}(\vec{U}) ,$$

where N is a nonlinear operator and η is the coupling coefficient [e.g., piezoelectricity, magnetostriction, additional (micropolar) elasticity coefficient] which, if necessary <u>only</u>, may be considered as an infinitesimally small parameter.

By direct integration it has been shown by Pouget and the author [1] that for "reasonable" limit conditions at $\pm \infty$, the system (2.3) or (2.4) admits <u>exact</u> propagative solutions in the form of solitary waves for (ϕ,v), which represents ,physically, a 180° Néel wall between two domains with an accompanying acoustic radiation. Such a stable solution is given by

$$(2.5)\quad \begin{cases} \phi = -2\tan^{-1}\left[\dfrac{\sinh \xi'}{\sqrt{1+2\gamma}}\right] , & \xi' = \xi - \frac{1}{2}\ln\left[\dfrac{4(1+2\gamma)}{\delta}\right] , \\ v = \dfrac{2\eta Q}{(\Omega^2-\Omega_T^2)}\left(\dfrac{1+2\gamma}{-2\gamma}\right)^{1/2} \tanh^{-1}\left[\dfrac{\cosh \xi'}{\sqrt{-2\gamma}}\right] + \text{const.} , \end{cases}$$

with

$$(2.6)\quad \xi = QX - \Omega t \ , \quad \Omega_T = V_T Q \ , \qquad \gamma(\Omega, Q) = \frac{1}{2}\eta^2 \frac{Q^2}{(\Omega^2-\Omega_T^2)}$$

on the condition that Q and Ω be related by the pseudo dispersion relation

$$(2.7)\quad D_S(\Omega, Q) = (\Omega^2-\Omega_F^2)(\Omega^2-\Omega_T^2) + \eta^2 Q^2 = 0 \qquad , \Omega_F^2 \equiv Q^2 - 1 \ .$$

The parameter η need not be small in this analysis.
If one considers a crystal structure made of several domains and walls and the whole picture evolves in time because of the action of an external stimulus (electric field, magnetic field or external couple depending on the origin of the microstructure), then the natural question arises of the interaction between solitary waves. In particular, do these solutions behave like true solitons or do they behave in this manner but only approximately ? Both problems of the interactions between such solutions and their evolution in space-time under the influence of external stimuli participate in a unique analytical and numerical framework which is the concern of this lecture. To that purpose it is noticed that the system (2.3) or (2.4) can also be given a Hamiltonian form as also, if necessary, the following nonlinear hyperbolic form in which the left-hand side has the form of a conservative contribution:

$$(2.8)\qquad \frac{\partial \vec{U}}{\partial t} + \frac{\partial}{\partial X}\vec{\mathcal{F}}(\vec{U}) = \vec{G}(\vec{U}) \ .$$

Equations (2.3),(2.4) and (2.8) are but three equivalent formulations of the same system, to which must be added limit conditions at $X = \pm\infty$ and initial conditions at $t = 0$.

3.- MATHEMATICAL FRAMEWORK FOR THE INTERACTIONS OF SOLITONS

In a general manner the system (2.4) has no simple solution such as a single soliton (solitary wave) for rather general initial conditions at (X,0). However, when $\eta = 0$, eqn.$(2.3)_1$ yields a propagative solution, say $v(X-V_T t)$, while eqn.$(2.3)_2$ reduces to an ordinary sine-Gordon equation, which is known to admit multiple-soliton solutions. Whenever $\eta \neq 0$ but small, the additional terms may be considered as perturbations which, in some way, should alter the uncoupled solutions. Equation $(2.3)_1$ remains linear for v even after coupling. Its solution is thus composed of a propagative solution and a particular solution induced by ϕ through η. As to eqn. $(2.3)_2$, it is nonlinear both before and after perturbation by the η term. A special perturbation scheme must be devised for treating the general solution (v,ϕ). In a first step one must determine the "principal" solution for which the effect of perturbations is only to modulate the free parameters (phase and velocity) of the solution. In a second step (it is hopeless to go beyond that analytically), one must determine corrective terms of the first order by constructing a Green function associated with the problem. To that purpose the inverse scattering method must be used. The first-order corrections will represent the radiation of harmonic and/or soliton waves. Such a methodology, already used with success for exploiting the KdV and Schrödinger equations in the presence of perturbations [10]-[12] and the double sine-Gordon equation for interacting solitons [13]-[14] is here generalized to the case of the system (2.3) — also to the case of (1.1) if needed. It can be implemented as follows.

For $\eta = 0$ the system (2.4) admits solutions v_o of the harmonic type for v and ϕ_o of the multiple-soliton type for ϕ (the latter can be constructed by using Bäcklund transformations [15] or Hirota's method [16]). Let $\vec{U}_o$ be this "uncoupled" solution such that

$$\frac{\partial \vec{U}_o}{\partial t} + N(\vec{U}_o) = \vec{0} \tag{3.1}$$

For small $\eta \neq 0$, we look for an asymptotic expansion

$$\vec{U}(X,t) = \vec{U}_o(X,t,\tau) + \eta\, \vec{U}_1(X,t,\tau) + \dots \tag{3.2}$$

such that

$$\lim_{\eta \to 0} \eta\, \vec{U}_1(X,t/\eta) = \vec{0} \quad , \tag{3.3}$$

where $\tau = \eta t$ is a slow time scale. The condition (3.3) is a secularity condition which guarantees that the expansion (3.2) is valid for time intervals of the order of η^{-1} (equivalently, the term $\eta \vec{U}_1$ remains bounded for large time intervals). The first order solution $\vec{U}_1$ satisfies the problem

$$(3.4) \quad \begin{cases} [\hat{N}(\vec{U}_o)]\vec{U}_1 \equiv \dfrac{\partial}{\partial t} \vec{U}_1 + [N^P(\vec{U}_o)]\vec{U}_1 = \vec{\mathcal{F}}(\vec{U}_o) & , \\ \vec{U}_1(X,0) = \vec{0} & , \end{cases}$$

where $\hat{N}(\vec{U}_o)$ is the linearization of eqn. (2.4) about $\vec{U}_o$, N^P is the perturbed operator N and

$$(3.5)\qquad \vec{\mathcal{F}}(\vec{U}_o) = \vec{F}(\vec{U}_o) - \eta^{-1}[\frac{\partial}{\partial t}\vec{U}_o + \vec{N}(\vec{U}_o)] \qquad .$$

This effective source term [not to be mistaken for the one in eqn.(2.8)] accounts for the fact that the condition (3.2) cannot be checked for an $\vec{U}_o$ which satisfies eqn.(3.1) exactly. Accordingly, a certain freedom must be granted to $\vec{U}_o$ which may be modulated on the time scale $\tau = \eta t$, so that this modulated $\vec{U}_o$ solution will satisfy eqn.(3.1) at the order η (and <u>not</u> zero). The first of eqns.(3.4) can be formally solved by representing the inverse operator $[\hat{N}(\vec{U}_o)]^{-1}$ by means of a Green function G :

$$(3.6)\qquad \vec{U}_1(X,t) = \int_0^t (G(X,t|.,t'),\vec{\mathcal{F}}(\vec{U}_o(..,t'))\,dt' \qquad ,$$

where G is a linear operator on a certain Hilbert space $\mathcal{H}$ equipped with the inner product (.,.) involving a spatial integral; G is such that

$$(3.7)\qquad [\hat{N}(\vec{U}_o)]G = 0 , \qquad t > t' \geq 0 , \qquad \lim_{t\to t'} G = 1 \qquad ,$$

or

$$[\hat{N}(\vec{U}_o)]^T G^T = 0 , \qquad t' > t \geq 0 , \qquad \lim_{t'\to t} G^T = 1 \qquad ,$$

where $[\hat{N}(\vec{U}_o)]^T$ is the adjoint of $\hat{N}(\vec{U}_o)$ and 1 is the identity in $\mathcal{H}$.
The efficiency of the above sketched method heavily relies on the ease with which the inverse operator can be found for a given $\vec{U}_o$. In most cases this is a difficult task; the inverse scattering method allows one to find the appropriate Green function in the case of integrable equations of evolution [17]. In the present case we can avoid this complexity since we only need to know the structure of the kernel $K(\hat{N})$ of $\hat{N}(\vec{U}_o)$ or the kernel of $\hat{N}^T$. Indeed, let $\{p_j\}$ denote collectively the set of free parameters ascribed to the U_o solution. The kernel K will consist of two parts, a <u>discrete</u> subspace K_d associated with dispersive waves and a continuum subspace K_c that corresponds to soliton solutions of $\vec{U}_o$. Because $\hat{N}(\vec{U}_o)$ results from the linearization of the original operator about $\vec{U}_o$, the elements of $K_d(\hat{N})$ are simply obtained by differentiating $\vec{U}_o$ with respect to its free parameters. Thus K_d is generated by the finite family of functions

$$(3.8)\qquad \{\, \partial\vec{U}_o / \partial p_j \,\} \qquad ,$$

where j=1,2,..,2N if $\vec{U}_o$ contains a ϕ_o made of N soliton solutions, each with two free parameters (speed and phase). Accordingly, the Green function is decomposed as $G = G_d + G_c$, where G_d admits a representation on the basis (3.8) and G_c is composed of continuous wave trains.

The problem now consists in determining a basis for G_c and the modulation of the free parameters $\{p_j\}$ of $\vec{U}_o$. For the latter, let $\{\vec{b}_j(X,t);j=1,3,..,2N\}$ be a basis that spans $K_d(\hat{N}^T)$. Any part of the effective source (3.5) that is parallel to one of the discrete components will resonate with the Green function and it will produce secular terms. Such disturbing perturbations are eliminated by requiring that $\vec{\mathcal{F}}$ be orthogonal to $K_d(\hat{N}^T)$. That is, on account of (3.5), the following system of ordinary differential equations must hold

$$(3.9)\qquad \sum_{j=1}^{2N} (\vec{b}_k\,,\,\frac{\partial}{\partial p_j}\vec{U}_o)\,\frac{dp_j}{d\tau} = (\vec{b}_k,\,\vec{F}(\vec{U}_o)) \quad ,$$

where τ is the slow time scale and (.,.) denotes the inner product in the space of squared-summable functions. The orthogonality condition (3.9), which is equivalent to the secularity condition (3.3), provides the looked for modulation of the free parameters. Introducing the operator

$$J = \begin{bmatrix} 0 & -1 & 0 & 0 \\ 1 & 0 & 0 & 0 \\ 0 & 0 & 0 & -1 \\ 0 & 0 & 1 & 0 \end{bmatrix} \quad ,$$

we see that the elements of $K_d(\hat{N}^T)$ can be generated by $\{J\,\partial\vec{U}_o/\partial p_j\}$, so that eqn.(3.9) takes on the form

$$(3.10)\qquad \sum_{j=1}^{2N} (\,J\frac{\partial}{\partial p_k}\vec{U}_o\,,\,\frac{\partial}{\partial p_j}\vec{U}_o\,)\,\frac{dp_j}{d\tau} = (\,J\frac{\partial}{\partial p_k}\vec{U}_o\,,\vec{F}(\vec{U}_o))\ ,\ k=1,2,..,2N$$

and thus we only need the zeroth-order solution to proceed to the p_j's as functions of τ.

It remains to find a basis for $K_c(\hat{N})$. The inverse scattering method provides a systematic way for this. It can be shown that [18]

$$(3.11)\qquad G_c(X,t|X',t') = -\frac{i\pi}{4}\int_{-\infty}^{+\infty}\frac{\lambda}{[a(\lambda)]^2}\,\vec{U}_+(X,t;\lambda)\,\vec{U}_-^T(X',t';\lambda)\,d\lambda \ ,$$

wherein

$$(3.12)\qquad \vec{U}_\pm(X,t;\lambda) = \left.\frac{\delta\vec{U}(X,t)}{\delta\rho_\pm(\lambda,0)}\right|_{\rho_\pm=0} \quad ,$$

where $\rho_+(\lambda,0)$ and $\rho_-(\lambda,0)$ are direct and retrograde reflection coefficients (at t=0) that characterize a continuous density of radiation at wave number

$$(3.13)\qquad k(\lambda) = 2\lambda - 1/8\lambda \quad ,$$

and $a(\lambda)$ is the maximum transmission coefficient. It is noticed that $\vec{U}_{\pm}$ are such that

$$(3.14)\quad \begin{cases} [\hat{N}(\vec{U}_o)]\vec{U}_{\pm} = \vec{0}\ , & X \in (-\infty,+\infty)\ , \\ \vec{U}_{\pm}(X,t;\lambda) \simeq \frac{1}{\pi\lambda}\exp\{\mp[k(\lambda)X+\omega(\lambda)t]\} & \text{for } X \rightarrow \pm\infty\ , \end{cases}$$

where $\omega(\lambda) = 2\lambda+(1/8\lambda)$. Once $\vec{U}_{\pm}$ are determined for this problem, then we have all ingredients to construct G_c.

4.- APPLICATION TO THE SOLITON-ANTISOLITON COLLISION

The reader will find in Pouget and Maugin [19] the complete detailed application of the above-given scheme to the system (2.3) or (2.4) when $\emptyset_o$ is made of a soliton and an antisoliton. Then K_d has dimension two and the free parameters p_j are a pseudo wavenumber Q and a phase Δ .Eqns.(3.10)can also be deduced from the Hamiltonian of the zeroth-order solution. Examples of modulation of (Q,Δ) in their plane are given in Figure 2 . The radiations in v and $\emptyset$ at the first order are determined in their

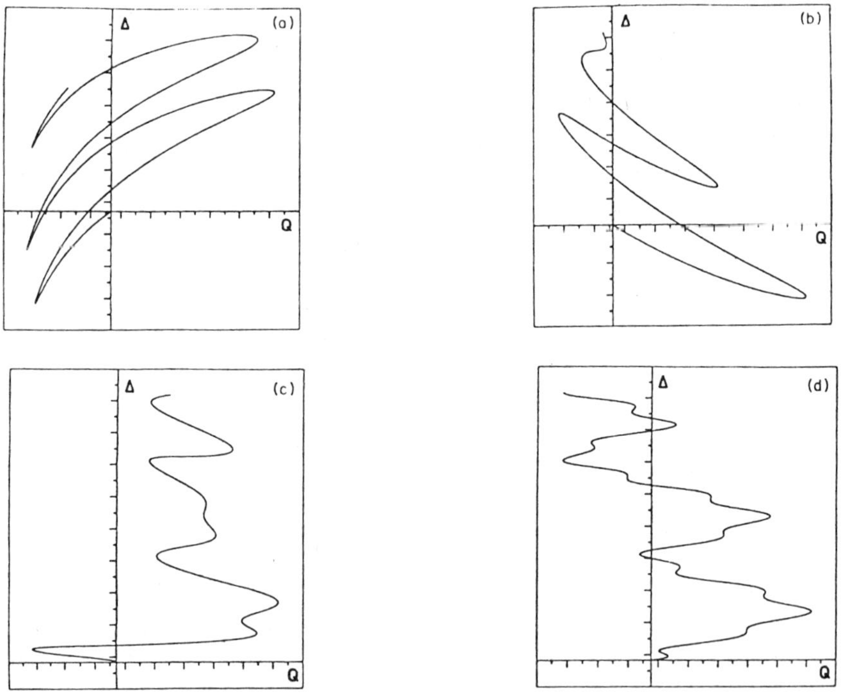

Fig.2.- Modulation of free parameters **Q** and Δ represented in the (Q,Δ) plane for various initial conditions (Q_o, Δ_o);after Pouget and Maugin[19].

asymptotic form by solving problems (3.14) for $\dot{v}_{\pm}$ and $\dot{\phi}_{\pm}$ and using eqn.(3.11). For instance, for v_1, the problem (3.14) reads

$$
(4.1)\quad \begin{cases}
\left(\dfrac{\partial^2}{\partial t^2} - V_T^2 \dfrac{\partial^2}{\partial X^2}\right) \dot{v}_{\pm} = 0 \;, \quad X \in (-\infty,+\infty) \;, \\[2ex]
\dot{v}_{\pm}(X,t;\lambda) \simeq \dfrac{1}{\pi\lambda} \exp\{\mp i[k(\lambda)X + \omega(\lambda)t]\} \;, \quad X \rightarrow \pm\infty \;, \\[2ex]
k(\lambda) = \lambda \;, \quad \omega(\lambda) = V_T\lambda \;,
\end{cases}
$$

while for ϕ_1,

$$
(4.2)\quad \begin{cases}
\left(\dfrac{\partial^2}{\partial t^2} - \dfrac{\partial^2}{\partial X^2} + \cos\phi_o\right) \dot{\phi}_{\pm} = 0 \;, \quad X \in (-\infty,+\infty) \;, \\[2ex]
\dot{\phi}_{\pm}(X,t;\ ) \simeq \dfrac{1}{\pi\lambda} \exp\{\mp i[k(\lambda)X+\omega(\lambda)t]\} \;, \quad X \longrightarrow \pm\infty \;, \\[2ex]
\omega^2(\lambda) = k^2(\lambda) + 1
\end{cases}
$$

After a very long algebra it is found that the radiation v_1 behaves, asymptotically for $X \rightarrow \infty$, like a soliton having two components, one propagating at the speed of the original soliton in ϕ and the other propagating at the speed of elastic waves, while for ϕ_1 the radiation consists in a component propagating at the speed of transverse elastic waves and another at the speed of the ferroelectric mode (mode associated with the internal rotational degree of freedom; it would be a spin wave in ferromagnets and a rotation mode in KNO_3) . This shows that the perturbing term in η plays the role of an electromechanical resonance coupling since we have exchange of the nature of waves at the level of radiations. All the asymptotic results and qualitative interpretations are corroborated by a direct numerical simulation of the system (2.8) by using a "leap-frog" Lax-Wendroff scheme and the border conditions at the end points of the finite spatial interval are given by the asymptotic expressions of the analysis. The detail of this and many graphs are given in Pouget and Maugin [19] for the cases of soliton-soliton, soliton-antisoliton and breather soliton types of interactions.The second of these cases, when ϕ_o is built of a soliton and an antisoliton, is reproduced in Figure 3 which clearly displays the radiations as also the deterioration of the scheme for large time intervals.

5.- EXTERNAL STIMULUS AND PERTURBATIONS

For all cases of microstructure considered, the external stimulus manifests itself through an additional contribution in eqns.$(2.1)_c$ or $(2.3)_2$ that governs the internal degree of freedom subjected to large excursions. For instance, in the ferroelectric

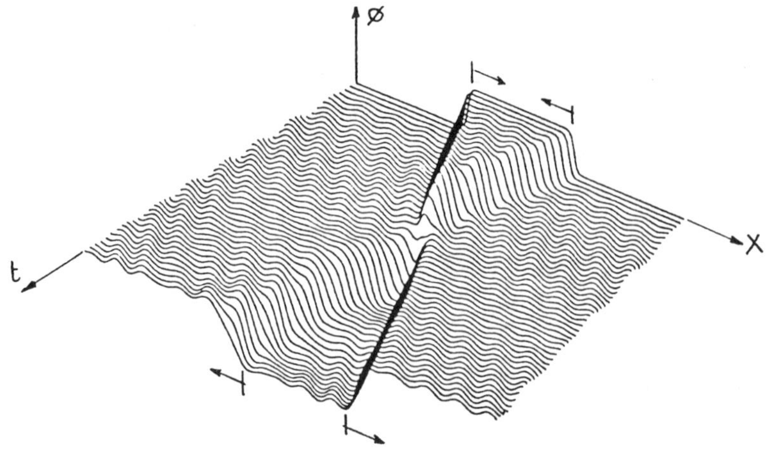

(a)

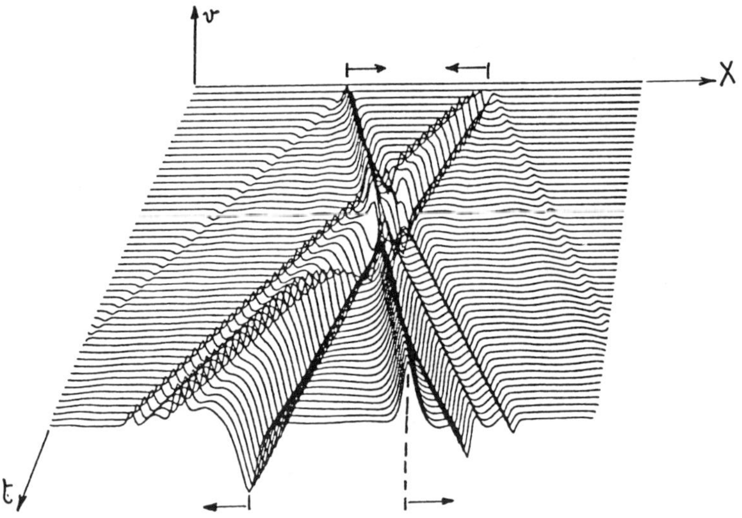

(b)

Figure 3.-Numerical simulation of the soliton-antisoliton collision for the ferroelectric case;(a) rotation $\emptyset(X,t)$ and elastic-wave radiation,(b) Accompanying elastic displacement field with radiation of solitons.

case, a term $F\cos(\phi/2)$ must be added to the right-hand side of eqn. $(2.3)_2$. The system (2.3) therefore reads

$$(5.1)\quad \begin{cases} W(v) = \eta\, f(\phi) , \\ SG(\phi) = \eta\, g(\phi,v) + \varepsilon\, h(\phi) . \end{cases}$$

Equivalently, in lieu of eqn. (2.4) we can write

$$(5.2)\quad \frac{\partial \vec{U}}{\partial t} + N(\vec{U}) = \eta \vec{F}(\vec{U}) + \varepsilon \vec{H}(\vec{U}) .$$

Let

$$(5.3)\quad \begin{cases} \mathcal{H}_o(\vec{U}) = \mathcal{H}(\vec{U};\varepsilon=0) , \\ \mathcal{L}_o(\vec{U}) = \mathcal{L}(\vec{U};\varepsilon=0) , \end{cases}$$

be the Hamiltonian and Lagrangian of the system of evolution (5.2) in the absence of stimulus. It may happen that $\vec{H}$ is not derivable from a potential [the result (5.9) still holds true in this case]. But here we assume that we can introduce the total Hamiltonian

$$(5.4)\quad \mathcal{H}(\vec{U}, \varepsilon\neq 0) = \mathcal{H}_o(\vec{U}) + \mathcal{H}_\varepsilon(\vec{U})$$

where $\mathcal{H}_\varepsilon$ is of order ε . Call $\vec{U}_o$ a solution (possibly static) of the stable solitary-wave type — as given for instance by eqns. (2.5)-(2.7) — of the system (5.1) or (5.3) for $\varepsilon = 0$.

A.- Naïve perturbation of the energy

If the external forcing derives from a potential we can use the definition (5.4). We let one parameter free, say q, and only one , in the $\vec{U}_o$ solution , since there will be only one determining equation, the conservation of energy in the potential associated with the external stimulus, and this parameter q may be modulated in time. The total Hamiltonian at the order ε is written as

$$(5.5)\quad \mathcal{H}(\vec{U}_o(q)) = \mathcal{H}_o(\vec{U}_o(q)) + \mathcal{H}_\varepsilon(\vec{U}_o(q)) ,$$

and the energy conservation

$$(5.6)\quad \frac{d\mathcal{H}}{dt} = 0$$

provides an ordinary differential equation for q , and this may be integrated in time. For the system (5.1) in ferroelectric crystals, q is the velocity of the solitary wave which starts from rest and accelerates in a suddenly applied constant electric field. One simultaneously exploits the smallness of η. Because of the Lorentz invariance of the sine-Gordon equation the result obtained (for $\eta = 0$) is exactly that of a relativistic particle being uniformy accelerated by a constant force. The complete computation, however, is more complex in reason of the η coupling with the wave equation $(5.1)_1$ (see Pouget and Maugin[2]). This perturbation scheme is naîve in that, albeit straightforward, it yields nothing concerning the deformation of the profile (in width and magnitude) of the solitary wave — which physically represents a moving wall separating two domains — as also nothing on the possible radiations accompanying the phenomenon.

B.- <u>Whitham's averaged-Lagrangian method</u>

This method [20] is already more powerful since it is based on the exploitation of variational equations (Euler equations) for each parameter q_j of the $\vec{U}_o$ solution. Consider the evolution system (5.2) and the total Lagrangian

$$\langle \mathcal{L}_o(\vec{U}) \rangle = \langle \mathcal{L}(\vec{U};\, \varepsilon=0) \rangle = \int_{-\infty}^{+\infty} \mathcal{L}(\vec{U};\varepsilon=0)\, dX \tag{5.7}$$

The equation (2.4) is equivalent to $\delta\langle \mathcal{L}_o \rangle = 0$,i.e., $\delta\mathcal{L}_o / \delta\vec{U} = \vec{0}$, with solution $\vec{U}_o$, $\delta/\delta\vec{U}$ denotes the variational derivative. If $\vec{H}$ derives from a potential, the perturbed equations (5.2) can also be deduced from the stationarity of a Lagrangian and the method of Whitham consists in assuming that this latter Lagrangian is nothing but

$$\langle \mathcal{L} \rangle = \int_S \mathcal{L}(\vec{U};\varepsilon\neq 0)\, dX = \int_S \mathcal{L}(\vec{U}_o(X,t;q_j);\varepsilon=0)\,dX + \text{term of order } \varepsilon \text{ involving } \vec{U}_o(X,t;q_j) \tag{5.8}$$

where the parameters q_j are <u>slowly</u> <u>modulated</u> <u>in</u> <u>time</u> (so-called <u>adiabatic</u> time modulation). The variation of $\langle \mathcal{L} \rangle$ with respect to the q_j's then yields the required equations to determine this modulation. More generally, if $\vec{H}$ is not derivable from a potential , then we still consider the variation of $\langle \mathcal{L}(\vec{U}_o(X,t;q_j);\varepsilon=0) \rangle$ with respect to q_j , but using the principle of virtual work we find that the external stimulus contributes an ε term and the variational equation determining the modulation of the q_j's reads

$$\frac{\delta \langle \mathcal{L}_o \rangle}{\delta q_j} = \varepsilon \int_S \vec{H}(\vec{U}_o) . \frac{\partial \vec{U}_o}{\partial q_j}\, dX \quad , \; j=1,2,.. \tag{5.9}$$

and this can be interpreted as the <u>orthogonality</u> <u>condition</u> that must satisfy the error in the system (5.8) when the adiabatically varied solution $\vec{U}_o(X,t,q_j(t))$ is considered instead of the exact, unknown solution. The preliminary work [21] and the more matured contribution of J.Pouget at this Symposium [22] illustrate the application of this perturbation scheme to the precise ferroelectric case where the free parameters q_j ,j=1,2,are the velocity and phase of the solitary wave perturbed by the external stimulus.

C.- Inverse scattering method

The averaged-Lagrangian method does not provide the radiation or first-order corrective terms in the solitary-wave profile. To reach this information one must use the inverse scattering method(see,e.g., Karpman [23]). A direct numerical simulation using the already cited numerical scheme for systems of the type (2.8) provides a solution that corroborates quite well the results of the method B above [22]. The narrowing of the wall and the accompanying radiations are displayed in a clear manner.

6.- CONCLUSION

In this brief review we have emphasized the role of perturbation methods, sometimes allied to some variational formulation, in the study of the interaction phenomena in systems of equations, such as (1.1), which may allow for the existence of soliton solutions for one of the variable inthe absence of couplings. We have left aside the inverse scattering method which, while theoretically powerful, would be rather unmanageable in the present systems. All perturbation methods employed boil down to a modulation of free parameters of a zeroth-order solution and this is the essence of the multiple-scale technique as opposed to the straightforward Poincaré expansion. Here we just want to list a few additional problems to be examined in a near future. The problems reviewed are not the only mathematical aspects of physical systems such as (1.1) or (2.1). In particular, dissipative processes have been left aside. These can be treated in the framework of method B in Section 5 since the method applies equally well for external forcing, or whatever additional term, which does not derive from a potential. Dissipative processes would also present an additional interest in that a very interesting phenomenon could be the existence of strange attractors for systems based on (1.1) and enlarged with such processes. Another problem of interest concerns the solution of the static problem corresponding to the system (1.1) on a finite spatial interval thus allowing the discussion of the domain-wall structure in a specimen of finite extent (see [6] for some numerical periodic solutions for such systems). Finally, the scattering of harmonic waves by a nonlinear solitary-wavelike solution of (1.1) is also of special interest as well as the coupled oscillations (ϕ,v) about a static kink solution of systems (1.1), which has already been studied in the ferromagnetic elastic case by Motogi and Maugin [24].

References

[1] J.Pouget and G.A.Maugin,Solitons and Electroacoustic Interactions in Ferroelectric Crystals-I:Single Solitons and Domain Walls,Phys.Rev.,B30,5306-5325(1984).

[2] J.Pouget and G.A.Maugin,Influence of an Electric Field on the Motion of a Ferroelectric Domain Wall, Phys.Lett., 109A, 389-392(1985).

[3] G.A.Maugin and A.Miled,Solitary Waves in Elastic Ferromagnets, Phys.Rev.B, (submitted for publication in,1985).

[4] G.A.Maugin and A.Miled,Solitary Waves in Micropolar Elastic Crystals, Int.J.Eng. Sci.(submitted for publication in,1985).

[5] J.Pouget and G.A.Maugin, Solitary Waves in Oriented Elastic Crystals (being completed).

[6] G.A.Maugin and J.Pouget,Solitons in Microstructured Elastic Media:Physical and Mechanical Aspects, in Continuum Models of Discrete Systems (5), Ed.A.J.M. Spencer, A.A.Balkema, Amsterdam (1985).

[7] G.A.Maugin,Elastic-Electromagnetic Resonance Couplings in Electromagnetically Ordered Crystals, in Theoretical and Applied Mechanics, Eds.F.P.J.Rimrott and B.Tabarrok,pp.345-355, North-Holland, Amsterdam (1980).

[8] G.A.Maugin,Symmetry Breaking and Electromagnetic-Elastic Couplings, in The Mechanical Behavior of Electromagnetic Solid Continua, Ed.G.A.Maugin, pp.35-46, North Holland, Amsterdam (1984).

[9] J.Pouget, Influence de champs rémament ou initiaux sur les propriétés dynamiques de milieux élastiques polarisables, Doctoral Thesis in Mathematics, Université Pierre-et-Marie Curie, Paris, France (Mars 1984).

[10] V.I.Karpman and E.M.Maslov, Perturbation Theory for solitons, Sov.Phys.J.E.T.P., 46, 281-291(1977).

[11] J.P.Keener and D.W.McLaughlin,Solitons under Perturbations, Phys.Rev., A16, 777-790(1977).

[12] V.I.Karpman and E.M.Maslov, A Perturbation Theory for the Korteweg-deVries Equation, Phys.Lett., 60A, 307-308(1977).

[13] R.K.Bullough, P.J.Caudrey and H.M.Gibbs, in Solitons, Vol.17 of Topics in Current Physics, pp.107-141, Springer-Verlag, Berlin (1980).

[14] A.C.Newell, Synchronized Solitons, J.Math.Phys., 18, 922-926(1977).

[15] A.C.Scott, in Bäcklund Transformations, the Inverse Scattering Method, Solitons and their Applications, Vol515 of Lecture Notes in Mathematics, Ed.R.Miura, pp. 80-105, Springer-Verlag, Berlin (1976).

[16] R.Hirota,Direct Methods in Soliton Theory, in Solitons, Vol.17 of Topics in Current Physics, pp.157-176, Springer-Verlag, Berlin (1980).

[17] D.W.McLaughlin and A.C.Scott, Perturbation Analysis of Fluxon Dynamics, Phys.Rev., A18, 1652-1680(1978).

[18] R.K.Dodd, J.C.Eilbeck, J.D.Gibbon and H.C.Morris, Solitons and Nonlinear Wave Equations, Academic Press, London (1982).

[19] J.Pouget and G.A.Maugin, Solitons and Electroacoustic Interactions in Ferroelectric Crystals-II:Interactions of Solitons and Radiations, Phys.Rev., B31, 4633-4649(1985).

[20] G.B.Whitham, Linear and Nonlinear Waves, J.Wiley-Interscience, New York (1974).

[21] G.A.Maugin and J.Pouget, Transient Motion of a Solitary Wave in Elastic Ferroelectrics, Proc.Intern.Conf.Nonlinear Mechanics, Shanghai, China, Oct.28-31(1985).

[22] J.Pouget, in these proceedings.

[23] V.I.Karpman, Soliton Evolution in the Presence of Perturbation, Physica Scripta, 20, 462-478 (1979).

[24] S.Motogi and G.A.Maugin,Effects of Magnetostriction on Vibrations of Bloch and Néel Walls, Physica statu solidi, a81,519-532(1984).

Acknowledgment.- The present work was supported by Theme "Nonlinear Waves in Electromagnetic Elastic Materials" of A.T.P."Mathématiques Appliquées et Méthodes Numériques Performantes" (M.P.B.,C.N.R.S.,Paris,France).

PHASE DIAGRAM OF ONE-DIMENSIONAL ELECTRON-PHONON AND RELATIVISTIC FIELD THEORY MODELS: RENORMALIZATION-GROUP STUDIES

W. Hanke
Max-Planck-Institut für Festkörperforschung, D-7000 Stuttgart 80, Federal Republic of Germany

Abstract

This paper a) reviews the formal connection between coupled Fermi-Bose models which support soliton excitations in condensed-matter and relativistic-field theory, b) uses renormalization-group (RG) arguments to study their phase diagram at $T=0^{\circ}K$ and, c) presents a RG analysis and comparison with Monte-Carlo data of the effect of Bose quantum fluctuations on the Su-Schrieffer-Heeger (SSH) model for one-dimensional electron-phonon systems like polyacetylene.

I. Introduction

In recent years a fortunate similarity between condensed matter and relativistic field-theory models of coupled Fermi-Bose systems has been revealed. In their study of the properties of quasi-one-dimensional systems like polyacetylene $(CH)_x$, Su, Schrieffer and Heeger (SSH) considered a model in which the phonons interact with electrons by modifying the electron hopping matrix elements. Within the mean-field adiabatic approximation, i.e. treating the phonon degrees of freedom classically, they found a dynamical symmetry breaking of the system: for a half-filled band, the system experiences a Peierls instability. The ground state is dimerized, with the spontaneous breaking of the reflection symmetry resulting in a twofold degeneracy. The low-lying excitations contain, in particular, the soliton states. In the presence of a soliton, there is a c-number solution of the electron field, localized near the soliton, with dispersion-less energy at the center of the gap. As a consequence, one-half a state for each spin orientation is removed from the Fermi sea in the neighborhood of the soliton.[1 2].
In a continuum version[3] this model has a striking similarity to a formal, mathematical model introduced by Jackiw and Rebbi in relastivistic

quantum field theory[4] [5]. Furthermore, when the ionic mass M is put equal to zero, the SSH model of condensed matter physics maps[6] [7] onto another much-studied model of field theory, the N=2 (where N are the spin degrees of freedom) Gross-Neveu model[8]. This equivalence holds also for the M=∞, i.e. adiabatic SSH model, when only the static, semi-classical Gross-Neveu equations are considered[9].

The similarity between the condensed matter and, for example, the Jackiw-Rebbi field-theoretic model is born out both in the ground state and low-lying excitations. In the field theory a one-dimensional, spinless Fermi field is coupled to a broken symmetry Bose field. Again a localized, mid-gap (between positive and negative particles) c-number solution exists to the Dirac equation in the presence of a soliton. Beyond the broken symmetry degeneracy of the ground state, there is a two-fold energy degeneracy of the soliton solution and the two states carry charge $\pm \frac{1}{2}$. Thus, introducing the soliton changes the number of fermions present by a fractional amount, i.e. $\pm \frac{1}{2}$, depending on whether the soliton state is occupied or not. Besides the existence of the mid-gap state also the fermion numbers $\pm \frac{1}{2}$ are common to the condensed matter and field theories. However, in the solid-state model the actual fermion charge is ± 1, due to the spin summation[1].

The aim of this paper is twofold. On the one hand, we want in sections II and III to review some of the ingredients for this fortunate convergence between different branches of theoretical physics. In particular, we emphasize in section II that both the original SSH and the Jackiw-Rebbi model are based in the assumption that the Bose degrees of freedom may be treated classically, neglecting their quantum fluctuations at T=0°K. For the SSH model this mean-field adiabatic approximation (sometimes also called static, i.e. ω=0 appproximation) seems appropriate, since the ionic mass M is very large. For the field-theoretic model this approximation forms for starting point for a weak-coupling expansion[10].

On the other hand, in section III and IV, we use renormalization-group arguments to look at the phase diagram of these one-dimensional Fermi-Bose models at T=0°K. This aims at the central question of whether the ground state is spontaneously broken and thus can have a soliton interpolating between the degenerate vacua if, in fact, the quantum fluctuations of the Bose field are properly taken into account. That this is a crucial issue containing unexpected results becomes evident, when we discuss in section III a renormalization-group analysis[7]. It demonstrates that the low-energy behavior of the N-component SSH model is governed by the zero-mass limit of the theory, the N-component Gross-Neveu model, and not by the much-used adiabatic $M=\infty$ limit. For spinless electrons, it is found that quantum fluctuations destroy the long-range dimerization order in the SSH model for small electron-phonon coupling constants if the ionic mass is finite.

In section IV we discuss recent results[11], we have obtained on the influence of quantum fluctuations in the SSH model for all values of the ionic mass and Fermi-Bose coupling. Starting from a path integral representation for Fermi fields in terms of the Grassmann algebra[12], we integrate out the Fermi variables and obtain an effective potential for the Bose field. Expanding this effective potential in powers of fluctuations around the mean-field adiabatic solution allows for a calculation of the gap in the Fermi spectrum and the phonon-order parameter within a one-loop approximation. Our results are correct to $O(1/N)$ and apply to all values of the ionic mass, from the $M=0$ Gross-Neveu to the $M=\infty$ adiabatic limit. For coupling constants larger than that of polyacetylene the phonon order parameter is found in close agreement with Monte-Carlo simulations of Fradkin and Hirsch[7]. For the set of parameters appropriate of $(CH)_x$, a 25% reduction (15% in the Monte-Carlo data) results in the phonon-order parameter m_p due to fluctuations. For smaller couplings and smaller gaps in the fermionic spectrum the Monte Carlo data substantially derivate from our results due to finite size effects. Here order of magnitude reductions due to the quantum fluctuations are found in m_p compared to the mean-field result.

II. Adiabatic Theory: Symmetry Breaking of the Ground State and Formation of Solitons with Fermion Number ½.

In the following we outline the mean-field theory of the SSH model for $(CH)_x$ and contrast it with the field theory of the Jackiw-Rebbi model. The SSH model is concerned with materials such as polyacetylene (Fig. 1), in which the electrons hop preferentially along a one-dimensional chain. The Hamiltonian is defined as

$$H_{SSH} = -\sum_{n,s} t_{n+1,n}(c^+_{n,s}\, c_{n+1,s} + \text{h.c.}) + \sum_n \left\{\frac{p_n^2}{2M} + \frac{K}{2}(u_n-u_{n+1})^2\right\} \quad (2.1)$$

The first term in (2.1) describes band electrons moving in a one-dimensional lattice (Fig.1) in a tight-binding approximation, while the second term is the harmonic vibrational (phonon) energy.

The hopping matrix element $t_{n+1,n'}$ giving the probability amplitude of an electron hopping from site n to n', depends on the distance between these sites, i.e.

$$t_{n+1,n} = t_0 - \alpha\,(u_{n+1} - u_n). \tag{2.2}$$

This leads to the electron-phonon interaction. The displacements can, in principle, be due to a variety of reasons, e.g. thermal fluctuations, zero-point motions and broken-symmetry effects. Peierls[13] has demonstrated that, in a one-dimensional metal, any non-zero electron-phonon coupling introduces a spontaneous breaking of the reflection symmetry, $u_n \leftrightarrow - u_n$ and a gap 2Δ in the electronic spectrum at the Fermi wave number k_F (Fig. 2).

In $(CH)_x$ there is, on the average, one electron per site, so $k_F = \pi/2a$, where a is the equilibrium lattice spacing. The distortion induces a charge-density wave commensurate with the lattice, in which the nuclear displacements and the electron density oscillate periodically in space, with a wave number $2k_F = \pi/a$. In chemical terminology, double and single bonds alternate on the chain in Fig. 1. This is the insulating ground state, which clearly is two-fold degenerate, due to the invariance of the chain under translations by ± a , ± 2a and so on. In pictorial language, the double bond can be on the left-hand or, equivalently, on the right-hand side of the C-H unit at site M in Fig. 1. A topological soliton then interpolates (as an unsaturated "free radical") between the two ground states.

The basic reason for the symmetry breaking is that, for a small (and n-independent) value of the staggered displacement field

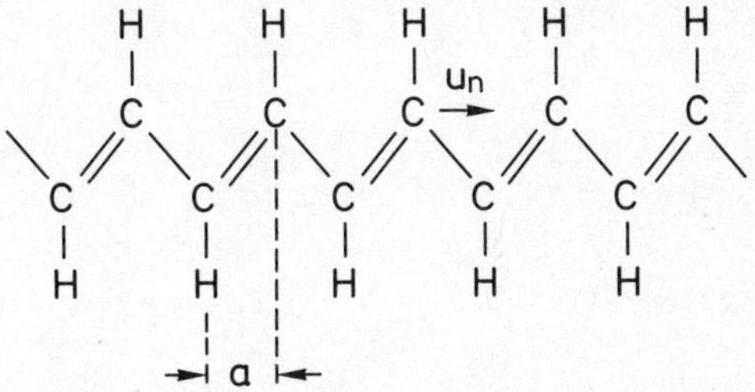

Fig. 2: Electronic band structure due to Peierls metal-insulator transition in polyacetylene.

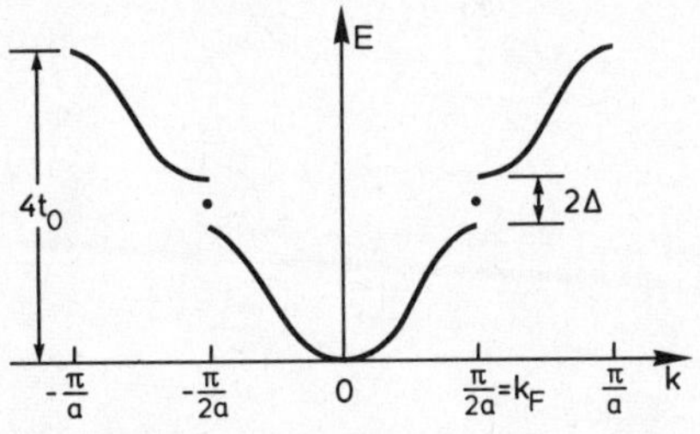

Fig. 1: Perfect dimerized trans-polyacetylene. The coordinate displacement for the n-th C-H group, the phonon field, is termed u_n in accordance with eq. (2.2)

$$\phi_n = (-1)^n u_n, \tag{2.3}$$

the ectronic energy is lowered by an amount proportional to $\phi^2 \ln \phi$, whereas the elastic energy of the lattice increases by an amount proportional to ϕ^2. Within a mean-field adiabatic theory and at $T=0\,^\circ K$ the gain in the electronic energy will thus always be larger than the cost in the elastic energy so that the ground state is distorted. This is true for an arbitrarily small coupling between the electrons and the lattice, and also for an arbitrary number N of spin components of the electrons (N=1 corresponds to spinless electrons, N=2 to electrons with spin ½, etc.).

This Peierls argument is based on the $M=\infty$ adiabatic limit, where the Born-Oppenheimer energy of the electrons is calculated for a given, frozen-in staggered lattice displacement ϕ. Thus, the kinetic energy $p_n^2/2M$ of the lattice vibrations in the Hamiltonian in eq. (2.1) is put equal to zero, and the phonon degrees of freedom are treated classically. Fig. 3 gives a schematic plot of the adiabatic energy per lattice unit((CH)-group) or the double-well potential as a function of the staggered displacement field ϕ, the phonon order parameter. For small values of the coupling constant $g=\alpha(\frac{2}{Kt_0})^{1/2}$ one finds

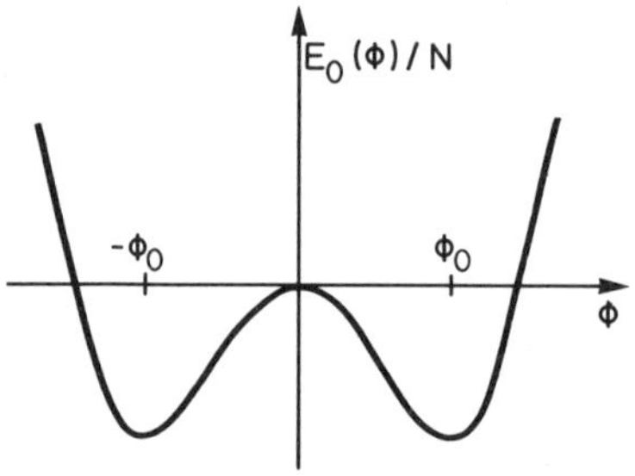

Fig.3: Adiabatic energy per C-H group as a function of the staggered displacement field, (the phonon order parameter $\phi=\phi_n=(-1)^n u_n$ in eq.(2.2)) displaying the degenerate Peierls minima.

$$\phi_0 \approx \text{const} \cdot e^{-\pi/Ng^2} \tag{2.4}$$

and for the gap in the electronic spectrum

$$\Delta_0 \sim g\,\phi_0 \tag{2.5}$$

We will study in section III and IV how these mean-field results are affected when the phonons have a finite frequency ($M<\infty$) and undergo zero-point quantum fluctuations.

The field-theoretic connections of this discrete SSH model become evident if one switches from the SSH model to a continuum field theory, which describes the long-distance behavior. This has been done by Takayama, Lin-Liu and Maki (TLM)[3] in the adiabatic limit. The TLM Lagrangian is

$$\mathcal{L} = \sum_{s=1}^{N} \psi_s^*(x\tau)[\underline{1}\cdot\frac{\partial}{\partial\tau} - i\,\sigma_3\,\frac{\partial}{\partial x} + g\,\phi(x\tau)\sigma_1]\,\psi_s(x\tau)$$

$$+ \frac{M}{2}\,(\frac{\partial}{\partial\tau}\,\phi(x\tau))^2 + \frac{1}{2}\,\phi^2(x\tau) + \frac{\lambda}{2}\,(\frac{\partial}{\partial x}\,\phi(x\tau))^2 \qquad (2.6)$$

Here ψ and ψ^* define anticommuting Grassmann spinor fields consisting of two components:

$$\psi_s(x\tau) = \begin{bmatrix} \psi_{s,\ell}(x\tau) \\ \psi_{s,r}(x\tau) \end{bmatrix}, \qquad (2.7)$$

where ℓ and r stand for the regions around the left and right Fermi points of the zero-dimensional Fermi surface (Fig. 2). The spin index (as in eq. (2.1) for the discrete SSH model) runs from 1 to N. $\phi(x\tau)$ is the phonon field which, for a half-filled band, may be assumed to be real. σ_1 and σ_3 are the Pauli matrices (and γ the Dirac matrices in two dimensions)

$$i\sigma_1 = \gamma_1 = \gamma_0\,\sigma_3 = \begin{bmatrix} 0 & i \\ i & 0 \end{bmatrix}, \qquad \sigma_2 = \gamma_0 \begin{bmatrix} 0 & -i \\ i & 0 \end{bmatrix}$$

$$\sigma_3 = \gamma_5 = \begin{bmatrix} 1 & 0 \\ 0 & -1 \end{bmatrix}, \quad \bar{\psi}_s = \psi_s^*\,\gamma_0 \qquad (2.8)$$

g in eq. (2.6) is the electron-phonon coupling constant and λ determines the bare spatial dispersion of the phonons. We have put the Fermi velocity equal to one in eq. (2.6) for convenience. The parameters in the Lagrangian $\mathcal{L}$ have then the following dimensions: $[\phi] = \kappa$, $[\psi] = \kappa^{1/2}$, $[g] = \kappa^0$, $[M] = [\gamma] = \kappa^{-2}$, where κ has the dimension of a momentum. $\mathcal{L}$ has a discrete global symmetry[7]

$$\psi_s \to \psi_s' = \gamma_5\,\psi_s,\ \bar{\psi}_s \to \bar{\psi}_s' = -\,\bar{\psi}_s\,\gamma_5,\ \phi(x\tau) \to -\,\phi(x\tau) \qquad (2.9)$$

The dimerized state breaks this symmetry, since the electron dimerization order parameter is

$$m_e = \sum_{s=1}^{N} \langle 0|\,\bar{\psi}_s(x\tau)\,\psi_s(x\tau)|0\rangle \qquad (2.10)$$

and

$$\sum_{s=1}^{N} \bar{\psi}_s(x\tau)\,\psi_s(x\tau) \to -\sum_{s=1}^{N} \bar{\psi}_s(x\tau)\,\psi_s(x\tau)$$

under this symmetry. In the $\omega=0$ ($M=\infty$) adiabatic limit m_e is related to the phonon order parameter ϕ by

$$m_e = \frac{2}{g}^{3/2} \sqrt{\frac{K}{t_0}}\, \phi_o \tag{2.11}$$

TLM studied the static ($\omega=0$), adiabatic limit of the Lagrangian in eq. (2.6). In this limit the problem can exactly be mapped onto the Bogoliubov-de Gennes equation[14], which describes the mean-field theory of an inhomogeneous superconductor. The Peierls or uniformly dimerized phase ($\phi(x)=\phi_0$) then just corresponds to the standard solution for the homogeneous superconductor, with order parameter and gap given by eqs. (2.4) and (2.5). The gap separates plus (empty) and minus (occupied) energy states

$$\pm E_k = \pm [k^2+\Delta_0{}^2]^{1/2}. \tag{2.12}$$

Similarly, the soliton phase which interpolates between the degenerate Peierls ground states $\pm\phi_0$, can directly be read off from the superconductivity analogy: if a superconductor is in contact with a normal conductor the gap or order parameter decays into the normal conductor with[15]

$$\Delta(x) = \Delta_0 \tanh (x/\xi) \tag{2.13}$$

where ξ is the correlation length ($\xi = \hbar/\Delta_0$). Thus, the order parameter reduces to the familiar topological kink, with hyperbolic tangent profile, which is appropriate for a double-well potential of the ϕ^4 field theory. Using the symmetries in eq. (2.9), in particular the charge-conjungation (or electron-hole symmetry in (2.1)) symmetry, it can be shown[1-3] that the fermion equation in both cases admits a localized zero-energy (mid-gap) solution, which then implies charge fractionalization: $\frac{1}{2}$ unit of charge corresponds to having the mid-gap state ψ_0 filled or empty. In the above solid-state physics model the fractionalization is hidden behind the spin degeneracy or spin summation. However, it indirectly shows up in unusual charge-spin relations with the neutral soliton having spin $\frac{1}{2}$ and charged solitons being spinless[1,2]. Let us now turn to the relativistic field-theory model of Jackiw and Rebbi[4,5]. Its Lagrangian is of the form

$$\mathcal{L} = i\,\bar{\psi}\,\partial\!\!\!/\psi - g\phi\bar{\psi}\psi + \mathcal{L}_{Bose}(\phi,\pi) \quad , \tag{2.14}$$

where

$$\mathcal{L}_{Bose}(\phi,\pi) = \tfrac{1}{2}\pi^2 + \tfrac{1}{2}(\frac{d}{dx}\phi)^2 + const(\phi_0{}^2 - \phi^2)^2 \qquad (2.15)$$

with π being the momentum conjugate to ϕ. The potential energy density $V(\phi) \sim (\phi_0{}^2 - \phi^2)^2$ in eq. (2.15) is a symmetric, double-well potential of ϕ^4 form. The Dirac spinors in (2.14) have two components, and thus the fermions are spinless.

Again, as in the above condensed-matter example, the adiabatic approximation is assumed. The two models coincide in their fermion parts, but an important difference remains in the boson parts: the SSH-TLM model introduces a symmetry breaking through the dynamical quantum-mechanical coupling between electrons and phonons. In other words, the effective potential in Fig. 3 in which the ions move classically (the Bose field) is generated dynamically. In contrast to this the mathematical field theory of eq. (2.15) puts in the symmetry breaking by assumption, i.e. "by hand". Thus, the ground state of (2.14) is determined by minimizing $V(\phi)$ and solving the free Dirac equation with $\phi = \pm\,\phi_0$ (this is just the Peierls analogue):

$$(\sigma_2(-i)\frac{d}{dx} + \sigma_1 m)\,\psi^{(\pm)} = \pm\, E_k \psi^{(\pm)} \qquad (2.16)$$

where $m_0 = g\phi_0$ and, as in eq. (2.12),

$$E_k = \pm\,[k^2 + m_0{}^2]^{1/2} \qquad (2.17)$$

For the soliton, interpolating between the two ground states, we minimize the potential parts in $\mathcal{L}_{Bose}$

$$-\frac{d^2}{dx^2}\phi(x) + \frac{d}{dx}V(\phi) = 0 \qquad (2.18)$$

which has a hyperbolic tangent solution, precisely as in the condensed matter example in eq. (2.13).

Quantization of the Dirac equations is done as in the ground-state, with the soliton solution $\phi \sim \tanh(x/\xi)$ put in for ϕ in eq. (2.14), i.e.

$$(\sigma_2(-i)\frac{d}{dx} + \sigma_1 m \tanh(x/\xi))u^{(\pm)} = \pm \quad u^{(\pm)} \qquad (2.19)$$

Again, like in the Peierls and superconductivity condensed-matter example, charge-conjugation symmetry paires a positive eigenvalue with a negative one. Furthermore, two degenerate "mid-gap" eigenvalue solutions ψ_0 exist for (2.19), which are charge-conjugation self-conjugate

$$\psi_0^c(x) = \sigma_3\ \psi_0^*(x) = \psi_0(x) \tag{2.20}$$

and which can be shown[5] to carry ½ unit of charge.

III. The Gross-Neveu Model (M=0) and Renormalization-Group Analysis of the Phase Diagram

In this section we closely follow Fradkin and Hirsch[7] in discussing the M=0 limit of the SSH model the Gross-Neveu model and in reviewing scaling arguments concerning the phase diagram for finite mass.
Consider the partition function of the TLM Lagrangian $\mathscr{L}$ in eq. (2.6)

$$Z = \int [D\psi^*][D\psi][D\phi]e^{-S} / \int [D\psi^*][D\psi]e^{-S_0} \tag{3.1}$$

with

$$S = \int_0^\beta d\tau \int_0^L dx\ \mathscr{L}\ , \tag{3.2}$$

where L denotes the length of the system, β is equal to 1/T and we will be interested in the limit $\beta\to\infty$. In eq. (3.1) S_0 is given by eq. (3.2) with $\mathscr{L}$ in eq. (2.6) evaluated for $g = \phi = 0$. $[D\psi]$ is defined as usual

$$[D\psi] = \prod_{s,x,\tau} d\psi_{s,l}(x\tau)\ d\psi_{s,r}(x\tau)\ , \tag{3.3}$$

with the spinor ψ as given in eq. (2.7). $[D\psi^*]$ and $[D\phi]$ are defined by similar expressions.
We can now proceed in two ways. One is to integrate out the fermion Grassmann fields and to work with an effective potential for the phonon field, including quantum fluctuations of the phonons, i.e. keeping M finite in eq. (2.6). This procedure will be followed up in the next section. The second possibility is to integrate out the Bose field, yielding an effective action S_{eff} for the Fermi fields. This possibility was carried out in ref. 7 and is summarized here.
We have

$$\exp(S_{eff}[\bar\psi,\psi]) = \int [D\phi]\ \exp\{-S\} \tag{3.4}$$

with

$$S_{eff} = \sum_{s=1}^{N} [\int d\tau dx \; i \bar{\psi}_s \not{\partial} \psi_s + \frac{g^2}{2} \int d\tau dx \int d\tau' \bar{\psi}_s(x\tau)\psi_s(x\tau) \cdot$$

$$\cdot \; G(\tau-\tau')\bar{\psi}_s(x,\tau')\psi_s(x,\tau')] \qquad (3.5)$$

The phonon Green's function is given by

$$G(\tau-\tau') = i \langle 0| T[\phi(x\tau)\phi(x\tau')]|0\rangle = \frac{\omega}{2} e^{-i|\tau-\tau'|\omega} \qquad (3.6)$$

and, in general is nonlocal in time. However, in the M=0 ($\omega=\infty$) limit G becomes as δ function in the time arguments. The resulting Lagrangian in this zero-mass limit is

$$\mathcal{L} = i \sum_{s=1}^{N} \psi_s \not{\partial}\psi_s + \frac{g^2}{2} [\sum_{s=1}^{N} \bar{\psi}_s\psi_s]^2 \qquad (3.6)$$

It is the Lagrangian of the N-component Gross-Neveu model, which has a continuous SU(N) symmetry, i.e. $\psi'=U\psi$, with U being an SU(N) matrix. This is the original spin rotation symmetry in the N=2 SSH model. The continuous spin rotation symmetry remains unbroken in one dimension (T>0) due to the Mermin-Wagner theorem[16].
However, additionally, the Gross-Neveu model has also a discrete chiral symmetry

$$\psi \rightarrow \psi' = i\gamma_5 \psi, \quad \bar{\psi} \rightarrow \bar{\psi}' = i\bar{\psi}\gamma_5 \qquad (3.7)$$

Since this symmetry is discrete it can be spontaneously broken, thus leading to a dimerized phase with $\langle\bar{\psi}\psi\rangle \neq 0$. This happens for N=2 for all values of g, whereas for N=1 a phase transition to a dimerized state occurs if g is strong enough. This is clearly due to the competition in eq. (3.6) between the kinetic energy term, which favors electron delocalization and the four-fermion interaction term which has a similar phonon mediated origin as the attractive electron interaction in the Cooper pairing of superconductivity. However, here the fermions experience on effective repulsive interaction between nearest neighbors. This becomes evident by integrating out the phonon variables in the discrete SSH version (for N=1 only)

$$H = - t_0 \sum_j (c_j^+ c_{j+1} + h.c.) - \frac{g^2}{4} \sum_j (c_j^+ c_{j+1} + h.c.)^2$$

$$= - t_0 \sum_j (c_j^+ c_{j+1} + h.c.) + \frac{g^2}{2} \sum_j n_j n_{j+1} - \frac{g^2}{2} \sum_j n_j \qquad (3.8)$$

For N=2 eq. (3.8) is modified in that no attractive forward scattering component exists. The Umklapp scattering then always opens up a gap in the electronic spectrum and gives long-range dimerization order. For N=1, the Umklapp scattering is not effective in producing a gap for small g because of the Pauli exclusion principle.

The critical behavior of the M=0 and N=1 Gross-Neveu model can be elegantly analyzed by introducing a boson representation for the fermion fields[17 7]. The renormalization-group properties of the boson theory are determined by the Kosterlitz renormalization group, giving[18] a critical value $g_c^2 = \frac{\pi}{2}$. For all values of N Gross and Neveu found that the effective coupling at short distances is small, whereas at large distances it is large, and the system always iterates to strong coupling in the infrared. In the renormalization group (RG) terminology these features are contained in the β-function which, to one-loop order is

$$\beta(g) = \frac{\partial g}{\partial \ln a_0} = 2^{-3/2} \left[\frac{n-1}{\pi}\right] g^3 \qquad (3.9)$$

with a_0 being the lattice parameter. This result implies, again in RG language, that g becomes relevant due to phonon quantum fluctuations. The gap in the fermionic spectrum behaves like

$$\Lambda_{GN} \sim \Lambda \exp\left[- \frac{\pi}{(n-1)g^2}\right] , \qquad (3.10)$$

where Λ is a large momentum cutoff.

Fradkin and Hirsch[7] presented an RG argument that in fact the low-energy behavior of the SSH model for all M≠0 values is governed by the M=0 limit, and not at all by the adiabatic M=∞ limit.

To see this, we follow their argument, and define an operator $\rho(x,\tau) = \sum_{s=1}^{N} \bar{\psi}_s(x,\tau)\psi_s(x,\tau)$. Then, using eq. (3.5), the effective action gets

$$S_{eff} = \sum_{s=1}^{N} \int dx d\tau \, i \bar{\psi}_s \partial\!\!\!/ \psi_s + \frac{g^2}{2} \int dx \int d\tau d\tau' \rho(x,\tau) G(\tau-\tau') \rho(x,\tau')$$

$$= \sum_{s=1}^{N} \int dx \int d\tau i \bar{\psi}_s \partial\!\!\!/ \psi_s + \frac{g^2}{4} \int dx \int_{-\infty}^{+\infty} d\tau \int^{\infty} ds \omega \rho(x,\tau) [\rho(x,\tau+s) + \rho(x\tau-s)] e^{-\omega s} \qquad (3.11)$$

Expanding $\rho(x,t\pm s)$ in powers of s, the effective Lagrangian becomes

$$\mathcal{L}_{eff} = i \sum_{s=1}^{N} \bar{\psi}_s \partial\!\!\!/ \psi_s + \frac{g^2}{2} \sum_{m=0}^{\infty} \left[\frac{\partial^n}{\partial\tau^n} \rho(x,\tau)\right]^2 \frac{1}{\omega^{2m}}$$

$$= i \sum_{s=1}^{N} \bar{\psi}_s \partial \psi_s + \frac{g^2}{2} \left(\sum_s \psi_s \psi_s\right)^2 + \frac{g^2}{2\omega^2} \left[\sum_s \frac{\partial}{\partial\tau} \bar{\psi}_s \psi_s\right]^2 + .. \qquad (3.12)$$

Therefore, the influence of phonon quantum fluctuations and finite mass ($M>0$) corrections are contained in higher-order time derivatives. Now, we remember that, in our units, the Fermi velocity v_F is put equal to 1; thus time has the dimension of length. So, by straightforward dimensional power counting, near the phase transition, the time derivative terms are irrelevant. For example, the last term in brackets of eq. (3.12) has dimensions $[L^{-4}]$ compared to $[L^{-2}]$ of $(\bar{\psi}\psi)^2$. These operators remain irrelevant for $g^2 \leq \frac{\pi}{2}$ (for $N=1$), and do not contribute to the physics at low energies. Thus, the important and intuitively probably unexpected conclusion is, that the properties of the SSH model near (and below) the dimerization transition (for $N=1$) are, even for finite ionic mass M, always controlled by the behavior at $M=0$. For $N=2$ this implies, that for all $M \neq 0$ the SSH model is dimerized, since again the $M=0$ Gross-Neveu model, which is dimerized for all g, controls the physics. As shown by Fradkin and Hirsch[7], the above naive power counting argument is correct, since the infrared unstable fix point at $g=0$, where the transition to the dimerized state for $N=2$ takes place is asymptotically free. In this case fluctuation effects produce only small logarithmic corrections to free field results. We will make use of some further results of the Fradkin-Hirsch renormalization-group analysis in the next section.

IV. The Effect of Phonon Quantum Fluctuations on the SSH Model for arbitrary Coupling g and ionic mass M

We have learned from the above renormalization-group (RG) arguments that, if we define a dimensionless parameter $\bar{M}$[7]

$$\bar{M} = M \Lambda^2 , \qquad (4.1)$$

where the cutoff $\Lambda \sim 1/a_0$, a_0 being of the order of the lattice spacing, with

$$a_0 \to \infty, \; g \to \infty \text{ also } \bar{M} \to 0, \qquad (4.2)$$

showing that $\bar{M}$ is an irrelevant operator in the RG sense. Conversely, as

$$a_0 \to 0, \quad g \to 0 \quad \bar{M} \to \infty. \tag{4.3}$$

This last line just demonstrates that irrelevant operators have an import effect at short distances, although they drop out from the long-distance physics. In any actual physical situation, however, some amount of this irrelevant operator will always be present. If $g<1$, then the low-lying spectrum will not be affected by the irrelevant operator. This can best be seen from the Gross-Neveu gap in eq. (3.10) which, due to the exponent becomes very small if g is small. On the other hand, for $g\approx 1$, Δ_{GN} is of order Λ, i.e. the cutoff, and there is no justification for dropping the irrelevant operator.

Consequently, in this section, we want to deal with the non-universal corrections and to discuss recent results[11] we have obtained on the influence of quantum fluctuations in the SSH model for all values of M and g. Our approach is based on the path-integral representation in eqs. (3.1) and (3.2) for Fermi fields expressing the SSH-TLM Lagrangian in eq. (2.6) in terms of Grassmann algebra[12]. In contrast to the previous section, this time we integrate out the Fermi variables. We expand the obtained effective Bose potential in powers of fluctuations around the mean-field adiabatic solution and extract the one-point function Γ^1 within a one-loop approximation[12 19]. The resulting analytical expressions for the phonon order parameter and for the gap will then be compared with the Monte-Carlo simulations of Fradkin and Hirsch[7].

We start from the Euclidean version of the SSH model in the continuum approximation as given by eqs. (3.1), (3.2) and (2.6). Additionally, we define Fourier transforms by

$$\phi(x,\tau) = \frac{1}{\sqrt{\beta L}} \sum_{\vec{k}} e^{i(k_1\tau+k_2x)}\phi(\vec{k}) \tag{4.4}$$

$$\phi(\vec{k}) = \frac{1}{\sqrt{\beta L}} \int_0^{\beta} d\tau \int_0^{L} dx \quad e^{-i(k_1\tau+k_2x)}\phi(x\tau) \tag{4.5}$$

with $\vec{k} = (k_1, k_2)$ and the allowed $\vec{k}$-values, $k_1 = \frac{(2n_1+1)\pi}{\beta}$ and $k_2 = \frac{2\pi n_2}{L}$, where n_1 and n_2 are integers.

From eq. (2.6), the electronic part in $\mathcal{L}$ is bilinear in the anticommuting Grassmann fields and, therefore, can be integrated out[12]. The result is, after straightforward manipulations

$$Z = \int [D\phi]\, e^{-\beta L S_{eff}}, \tag{4.6}$$

$$S_{eff} = (2\beta L)^{-1} \sum_{\vec{k}} (1+Mk_1^2+\lambda k_2^2)\phi(\vec{k})\phi(-\vec{k})+U[\phi] \quad (4.7)$$

where

$$U[\phi] = -\frac{N}{\beta L} \ln \{\det[(-ik_1\cdot\underline{1}-\sigma_3 k_2)\delta_{\vec{k},\vec{k}'}+\frac{\sigma_1 g}{\sqrt{\beta L}}\phi(\vec{k}-\vec{k}')]$$

$$/\det(-ik_1\cdot\underline{1}-\sigma_3 k_2)\}. \quad (4.8)$$

Here det stands for the determinant of a matrix which has $\vec{k}$ and the indices of the Pauli matrixes as labels. Next, we split off from the Bose field ϕ a fluctuating contribution $\tilde{\phi}$

$$\phi(\vec{k}) = \sqrt{\beta L}\,\bar{\phi}\delta_{\vec{k},0} + \tilde{\phi}(k) , \quad (4.9)$$

and expand the determinant and thus $U[\phi]$ in eq. (4.8) in powers of fluctuations $\tilde{\phi}$:

$$U[\bar{\phi},\tilde{\phi}] = U^{(0)}[\bar{\phi}] + U^{(1)}[\bar{\phi},\tilde{\phi}] + U^{(2)}[\bar{\phi},\tilde{\phi}] \quad (4.10)$$

The fluctuation-less contribution $U^{(0)}$ is, from eq. (4.8),

$$U^{(0)}[\bar{\phi}] = -\frac{N}{\beta L} \sum_{\vec{k}} \{\ln(\vec{k}^2+g^2\bar{\phi})^2 - \ln(\vec{k}^2)\} \quad (4.11)$$

The first-order contribution $U^{(1)}$ is given by

$$U^{(1)}[\bar{\phi},\tilde{\phi}] = \frac{Ng^2}{2\pi}\,\frac{\bar{\phi}}{\sqrt{\beta L}} \ln\left(\frac{g^2\bar{\phi}^2}{\Lambda^2}\right) \tilde{\phi}(0), \quad (4.12)$$

where we have introduced the cutoff Λ of eq. (4.1) for the $\vec{k}$-integration. The term $U^{(2)}$ second order in the fluctuations is, after some algebra[11], of the form

$$U^{(2)}[\bar{\phi},\tilde{\phi}] = (2\beta L)^{-1} \sum_{\vec{k}} V^{(2)}(\vec{k},\bar{\phi})\tilde{\phi}(\vec{k})\tilde{\phi}(-\vec{k}) \quad (4.13)$$

with

$$V^{(2)}(\vec{k},\bar{\phi}) = -\frac{2Ng^2}{\beta L} \sum_{\vec{k}'} \frac{1}{(\vec{k}'^2+g^2\bar{\phi}^2)} + Ng^2(4g^2\bar{\phi}^2+\vec{k}^2)\; R(\vec{k},g\bar{\phi}) \quad (4.14)$$

and

$$R(\vec{k},g\bar{\phi}) = \sum_{\vec{k}'} (\beta L)^{-1} \frac{1}{(\vec{k}'^2+g^2\bar{\phi}^2)((\vec{k}'^2+\vec{k})^2+g^2\bar{\phi}^2)} \qquad (4.15)$$

Inserting the expansion in fluctuations, eq. (4.10), into eqs. (4.6) to (4.8), we can integrate out the fluctuations $\tilde{\phi}$ for a general value of the constant Bose field $\bar{\phi}$. Performing the limits $L \to \infty$, $\beta \to \infty$. wherever possible, we obtain

$$Z = e^{-\beta L \Gamma[\bar{\phi}]} \qquad (4.16)$$

where the energy density $\Gamma[\bar{\phi}]$ is due to the zero-loop or mean-field contribution $\Gamma_0[\bar{\phi}]$ and due to the one-loop contribution $\Gamma_1[\bar{\phi}]$, i.e.

$$\Gamma[\bar{\phi}] = \Gamma_0[\bar{\phi}] + \Gamma_1[\bar{\phi}] \qquad (4.17)$$

with

$$\Gamma_0[\bar{\phi}] = \tfrac{1}{2}\bar{\phi}^2 - N\int\frac{d^2k}{(2\pi)^2}\{\ln(\vec{k}^2+g^2\bar{\phi})^2 - \ln(\vec{k}^2)\} \qquad (4.18)$$

and

$$\Gamma_1[\bar{\phi}] = \tfrac{1}{2}\int\frac{d^2k}{(2\pi)^2}\ln\{1+Mk_1{}^2+\lambda k_2{}^2+V^{(2)}(\vec{k},\bar{\phi})\} \qquad (4.19)$$

Thus, by forming the derivative with respect to $\bar{\phi}$, we can extract the so-called one-point vertex function[19] $\Gamma^{(1)}[\bar{\phi}]$, the stationary properties of which, i.e. $\Gamma^{(1)}[\bar{\phi}]=0$, determine the electronic gap $\Delta = g\bar{\phi}$:

$$\Gamma^{(1)}[\bar{\phi}] = \delta\Gamma[\bar{\phi}]/\delta\bar{\phi} = 0 \qquad (4.19)$$

This yields the equation for the gap[11]

$$1=2Ng^2\left[\int\frac{d^2k}{(2\pi)^2}\frac{1}{\vec{k}^2+\Delta^2} - \frac{1}{N}\int\frac{d^2k}{(2\pi)^2}\left(4\Delta^2+\vec{k}^2+\frac{Mk_1{}^2+\lambda k_2{}^2}{Ng^2R(\vec{k},\Delta)}\right)^{-1}\right] \qquad (4.20)$$

In the limit $M\to 0$ and neglecting spatial dispersion effects, i.e. $\lambda\to 0$, eq. (4.20) can easily be solved, resulting in the Gross-Neveu gap Δ_{GN} of eq. (3.10).

The general solution for arbitrary M and dispersion λ is detailed in ref. 11. In particular, it makes use of a scaling argument from ref. 7, in that the quantities

$$c_1 = Ng^2 \bar{M} \exp\left\{-\frac{2\pi(1-\frac{g^2 \ln 2}{\pi})}{Ng^2(1-1/N)}\right\} \tag{4.21}$$

and

$$c_2 = Ng^2 \bar{\lambda} \exp\left\{-\frac{2\pi(1-\frac{g^2 \ln 2}{\pi})}{Ng^2(1-1/N)}\right\} \tag{4.22}$$

are invariant with respect to changes in the cutoff Λ within a one-loop approximation. Here the dimensionless quantity $\bar{\lambda}$, like $\bar{M} = M\Lambda^2$, is defined as $\bar{\lambda}=\lambda\Lambda^2$. The final result for the gap is

$$\Delta = \Lambda \exp\left\{-\frac{\pi(1-\frac{g^2 \ln 2}{\pi d_1 d_2})}{Ng^2(1-\frac{1}{Nd_1 d_2})}\right\} , \tag{4.23}$$

where

$$d_1^2 = 1 + 4\pi c_1/(Ng^2)^2 \tag{4.24}$$

and

$$d_2^2 = 1 + 4\pi c_2/(Ng^2)^2 \tag{4.25}$$

Let us discuss various limits contained in eq. (4.23): the adiabatic limit $M \to \infty$, where eq. (4.23) simplifies

$$\Delta_{ad} = \Lambda \exp(-\frac{\pi}{Ng^2}) \tag{4.26}$$

Δ_{ad} is identical with the mean field result in eq. (2.4); b) The limit $\lambda\to\infty$, corresponding to the case where spatial fluctuations become unfavorable. As a result Δ approaches Δ_{ad} again. The number of spin components N enters the expression for the gap in the cases a) and b) only through Ng^2, i.e. via the usual spin factor of the screened electron-phonon interaction (the bubble diagram); c) The limit $M \to 0$, $\lambda\to 0$, corresponding to the Gross-Neveu model. As discussed above here Δ reduces to Δ_{GN} as given by Eq. (3.10) in agreement with Ref. (7). Δ_{GN} contains N in a nontrivial way and reduces to Δ_{ad} only for in the semi-classical limit $N\to\infty$. The general expression eq. (4.23) interpolates between the above limiting cases and allows one to calculate the influence on the gap due to a finite ionic mass and due to a nonvanishing spatial dispersion in the bare phonon part. From Eqs. (3.10), (4.23) and (4.26) follows moreover the inequality $\Delta_{GN} \leq \Delta \leq \Delta_{ad}$.
Figs. 4a and 4b display our results for the phonon order parameter $\phi=m_p$

which, in a one-loop approximation, is still related to the gap Δ through

$$\phi \equiv m_p = \frac{\Delta}{4g} .$$

We have used the units of Fradkin and Hirsch[7] in order to be able to compare our analysis directly with their Monte-Carlo data. In particular, we have rescaled the units of length and energy so that our rescaled parameters for polyacetylene $(CH)_x$, according to SSH[1], are $t_0 = 1$, $M = 928$. $K = 1$ and $g = 0.58$, i.e. $\Delta = 0.295$ with our coupling constant g everywhere being replaced by $2^{1/2}g$.

In Fig. 4a the results for the order parameter m_p are displayed for $M=\infty$ for a 24-site and an infinite lattice (full and dashed lines, respectively)[7]. We note that the finite lattice size introduces substantial effects for $g< 0.6$ and for a correspondingly small gap Δ, which happens roughly when the correlation length $\xi > N_L/4$, N_L being the number of

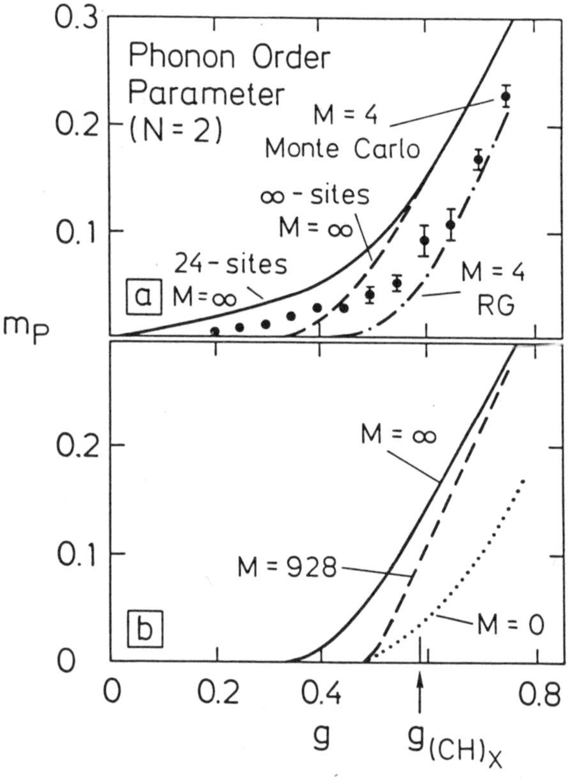

Fig. 4: Phonon order parameter m_p as a function of electron-phonon coupling g for $N = 2$.
a) The solid and dashed lines denote the mean field ($M=\infty$) case in a 24-site and an infinite lattice, respectively. Dots are Monte Carlo data for $M=4$ on a 24-site lattice[7]. The dashed-dotted line gives the result of our theory, eq. (4.23), for $M=4$.
b) The full and dashed lines compare the mean-field ($M=\infty$) case with our result for $M=928$, i.e., the polyacetylene value. The dotted line gives the $M=0$ Gross-Neveu limit from eq. (3.10).

lattice sites and ξ the inverse of the energy gap[7]. Our dash-dotted analytical result for $M = 4$ is essentially within the error bars of the Monte-Carlo results (which are calculated for a 24-site lattice) for coupling strengths $g \geq 0.6$ where finite size effects play no role. Similar as in the case $M=\infty$ our phonon order parameter m_p starts to increase from zero, however, only at significantly larger values of g compared to the finite size Monte Carlo results. The good agreement with the Monte Carlo results for larger coupling strengths is somewhat surprising in view of the fact that the used value of N^{-1}, namely 1/2, cannot be considered as a small expansion parameter.

Figure 4b compares the static $M=\infty$ limit (full line) with the case where M=928 (dashed line) which corresponds to the mass of polyacetylene $(CH)_x$. For the coupling constant g=0.58 of $(CH)_x$ we find a reduction of m_p due to quantum fluctuations from the static mean field value of 0.15 to 0.11. This decrease of 25% is somewhat larger than in the Monte Carlo calculation (15% reduction).[7] The difference is probably due to finite size effects in the numerical calculations. For smaller coupling or gaps the relative influence of fluctuations rapidly increases and the phonon order parameter is by a factor of 10 or more smaller than the mean-field value. The dotted curve represents an upper limit for the influence of fluctuations and is given by the M=0 or Gross-Neveu result of eq. (29). Fig. 4b, finally, displays the above inequality $\Delta_{GN} \leq \Delta \leq \Delta_{ad}$.

With our result for the order parameter in eq. (4.23) it is possible to calculate the influence of dispersion of the bare phonon frequencies on the magnitude of m_p. For $(CH)_x$ we find a negligible (less than 1%) reduction of m_p compared to the dispersionless case.

Acknowledgements The author is grateful for many helpful discussions with Drs. R. Zeyher and D. Schmeltzer.

References:

1. W.P. Su, J.R. Schrieffer and A.J. Heeger, Phys. Rev. Lett. 42 1698 (1979); Phys. Rev. B22, 2099 (1980)
2. M.J. Rice, Phys. Lett. 71A, 152 (1979)
3. H. Takayama, Y.R.Lin-liu and K. Maki, Phys. Rev. B21, 2388 (1980)
4. R. Jackiw and C. Rebbi, Phys. Rev. D13, 3398 (1976)
5. R. Jackiw, Rev. Mod. Phys. 49, 681 (1977)
6. M. Nakahava and K. Maki, Phys. Rev. B25, 7789 (1982)
7. E. Fradkin and J.E. Hirsch, Phys. Rev. B27, 1680 (1983)
8. D. Gross and A. Neveu, Phys. Rev. D10, 3235 (1974)
9. D.K. Campbell and A.R. Bishop, Phys. Rev. B24, 4859 (1981)
10. J. Goldstone and R. Jackiw, Phys. Rev. D11, 1486 (1975)
11. D. Schmeltzer, R. Zeyher and W. Hanke, Phys. Rev. B, to be published
12. L.D. Faddeev and A.A. Slavnov, "Gauge Fields: Introduction to Quantum Theory" (Benjamin, London, 1980), p. 22
13. R.F. Peierls, "Quantum Theory of Solids" (Clarendon Press, Oxford, 1955)
14. P.G. de Gennes, "Superconductivity of Metals and Alloys" (Benjamin, New York, 1966)
15. J. Bar-Sagi and C.G. Kuper, Phys. Rev. Lett. 28, 1556 (1972)
16. N.D. Mermin and H. Wagner, Phys. Rev. Lett. 17, 1133 (1966)
17. M.P.M. den Nijs, Phys. Rev. B23, 6111 (1981)
18. D.J. Amit, Y. Goldschmidt and G. Grinstein, J. Phys. A13, 585 (1980)
19. D.J. Amit, "Field Theory, The Renormalization Group, and Critical Phenomena", (Mc Graw Hill, N.Y., 1978)

Session III:

GAUGE THEORIES IN MECHANICS

THE CROOKED ROAD TO EFFECTIVE STRESS

Dominic G. B. Edelen
and
Dimitris C. Lagoudas
Center for the Application of Mathematics
and
Department of Mechanical Engineering and Mechanics
Lehigh University
Bethlehem, Pennsylvania 18015

Two Lagrangian functions are said to be variationally equivalent if they differ by a null Lagrangian (a Lagrangian whose associated Euler-Lagrange equations are identically satisfied). Kibble [1] noted in his seminal paper of 1961 that variationally equivalent Lagrangians lead to inequivalent gauge field theories, after which this important observation was actively ignored. There is an understandable reason for this situation; variationally equivalent Lagrangian functions are distinguished only by their distinct natural Neumann data, while elementary particle physics rarely if ever considers problems with imposed Neumann data. On the other hand, problems with imposed Neumann data demand inclusion of appropriate null Lagrangians in order that the imposed data be made variationally natural. It is thus clear that gauge theories with imposed Neumann data must make due allowances for the gauge-theoretic inequivalence of variationally equivalent Lagrangians. A specific case in point is the gauge theory of materials with defects that are subjected to imposed tractions on their boundaries.

This body of ideas is the subject of this paper. We first analyze the breaking of variational equivalence of Lagrangians that is induced by the minimal replacement operator of gauge theory in a general context. The results of this study are then particularized to the gauge theory of materials with defects. This allows us to show that the dislocation fields are driven by effective stress rather than the total stress, as has long been known, and that disclination fields are driven by the effective couple stress and the orbital dislocation couples. An interesting by-product of the analysis is the emergence of the Cosserat continuum as the underlying reversible modality.

1. Statement of the problem

Let $\psi^{\alpha}(x^k)$, $\alpha = 1,\ldots,N$ be the components of the state vector of a classical system of fields in space-time E_4 with a Lagrangian function $L_0(x^k, \psi^{\alpha}, \partial_k\psi^{\alpha})$. The field equations that govern the dynamics of the state vector are thus the Euler-Lagrange equations

$$(1.1) \qquad \{E|L\}_{\psi^{\alpha}} = 0 \ , \quad \alpha = 1,\ldots,N \ ,$$

where

(1.2) $\{E|F\}_{\psi^\alpha} := \frac{\partial F}{\partial \psi^\alpha} - \frac{d}{dx^i}\left(\frac{\partial F}{\partial(\partial_i \psi^\alpha)}\right), \quad \partial_i := \frac{\partial}{\partial x^i}$.

We assume for purposes of discussion that the system admits a global internal symmetry group G_0 that acts on the state vector from the left;

$$`\psi^\alpha(x^k) = A^\alpha_\beta\, \psi^\beta(x^k) , \quad A^\alpha_\beta \in G_0 .$$

Since G_0 is global, $dA^\alpha_\beta = 0$ for all $A^\alpha_\beta \in G_0$ and differentiation and the action of G_0 commute;

(1.3) $\partial_i(`\psi^\alpha(x^k)) = A^\alpha_\beta\, \partial_i \psi^\beta(x^k)$.

The internal symmetry group G_0 gives rise to conserved currents via Noether's theorem [2] because the Lagrangian function L_0 is invariant under the action of G_0 ;

(1.4) $L_0(x^k, `\psi^\alpha, \partial_k `\psi^\alpha) = L_0(x^k, A^\alpha_\beta \psi^\beta, A^\alpha_\beta \partial_k \psi^\beta) = L_0(x^k, \psi^\alpha, \partial_k \psi^\alpha)$.

If the internal symmetry group is allowed to act locally, we write $G: E_4 \to G_0$ with $A^\alpha_\beta = A^\alpha_\beta(x^k)$. In this event (1.3) is replaced by $\partial_i(`\psi^\alpha) = A^\alpha_\beta\, \partial_i \psi^\beta + \psi^\beta\, \partial_i A^\alpha_\beta$ and invariance of the Lagrangian is lost. Yang and Mills [3] showed that *minimal replacement*

(1.5) $M : (\psi^\alpha, \partial_k \psi^\alpha) \longmapsto (\psi^\alpha, \partial_k \psi^\alpha + \Gamma^\alpha_{k\beta}\, \psi^\beta)$

gives

(1.6) $\partial_i(`\psi^\alpha) + `\Gamma^\alpha_{i\beta}\, `\psi^\beta = A^\alpha_\beta(\partial_i \psi^\beta + \Gamma^\beta_{i\lambda}\, \psi^\lambda)$,

where the Γ's are the components of Yang-Mills compensating potentials that take their values in the Lie algebra of G_0 and transform under the action of G by

(1.7) $`\Gamma^\alpha_{i\lambda}\, A^\lambda_\beta = A^\alpha_\lambda\, \Gamma^\lambda_{i\beta} - \partial_i A^\alpha_\beta$.

Invariance of the Lagrangian is thus restored provided

(1.8) $L_0 \longmapsto ML_0 , \quad ML_0 := L_0 \circ M$.

The Yang-Mills minimal replacement construct starts with a Lagrangian function L_0 for the system on the classical level, and yet there are many Lagrangian functions with which the construction could start. Let N denote the null class of the Euler Lagrange operator $\{E|\cdot\}_{\psi^\alpha}$:

(1.9) $N = \{\eta \,|\, \{E|\eta\}_{\psi^\alpha} = 0 \quad \text{for all} \quad \psi^\alpha\}$.

Since $\{E|F + H\}_{\psi^\alpha} = \{E|F\}_{\psi^\alpha} + \{E|H\}_{\psi^\alpha}$, the Lagrangians L_0 and $L_0 + \eta$ lead to the same field equations for every $\eta \in N$. Thus,

(1.10) $\quad L_1 = L_0 + \eta , \quad \eta \in N$

is an *equivalence relation* on the collection of all Lagrangian functions. Classically, field theories derived from equivalent Lagrangian functions are equivalent since they have the same field equations.

Equivalent Lagrangian functions need not admit the same global internal symmetry groups (i.e., $\eta \in N$ need not be invariant under the global action of G_0 in (1.10)). Having noted this, we confine our attention to the null class of the Euler-Lagrange operator restricted by G_0.

(1.11) $\quad N_G = \{\eta \in N | \eta(x^k, A^\alpha_\beta \psi^\beta, A^\alpha_\beta \partial_k \psi^\beta) = \eta(x^k, \psi^\alpha, \partial_k \psi^\alpha) \, \forall A^\alpha_\beta \in G_0\}$.

Thus, all Lagrangian functions in the G-restricted equivalence class,

(1.12) $\quad L_1 = L_0 + \eta , \quad \eta \in N_G$,

admit G_0 as a global internal symmetry group. All field theories that derive from Lagrangians in the G-restricted equivalence class are thus G-*equivalent* (i.e., they have the same field equations and the same global internal symmetry group and are thus classically indistinguishable).

The crux of the matter is what happens when G_0 is allowed to act locally by $E_4 \to G_0$. If we start with two G-equivalent Lagrangians L_0 and $L_1 = L_0 + \eta$, $\eta \in N_G$, the required minimal replacement construct gives $L_0 \longmapsto ML_0$ and $L_1 \longmapsto ML_1 = ML_0 + M\eta$. Clearly, ML_0 and ML_1 will be G-equivalent if and only if

(1.13) $\quad M : N_G \to N_G$;

that is, if and only if N_G is closed under minimal replacement. Conversely, *if* N_G *is not closed under minimal replacement, then classically* G-*equivalent theories do not give equivalent theories for the local action of the internal symmetry group.*

2. Characterization of the G-Restricted Null Class

Characterizations of the null class of the Euler-Lagrange operator (variationally trivial Lagrangians) are given in a number of places in the literature [4-7]. A particularly elegant and complete treatment for Lagrangians with dependence on any finite order of partial derivatives is given in reference [8]. For the purposes of this paper, the simplest characterization obtains through use of methods of the exterior calculus, so we follow the discussion given in [9].

We denote the volume element on E_4 by $\mu = dx^1 \wedge dx^2 \wedge dx^3 \wedge dx^4$. Exterior forms of degree k on a space S are denoted by $\Lambda^k(S)$. Let $\{\partial_i \mid 1 \le i \le 4\}$ denote the natural basis for the tangent space, $T(E_4)$, of E_4. We then have

(2.1) $\quad \mu_i = \partial_i \lrcorner \mu , \quad dx^i \wedge \mu_j = \delta^i_j \mu$,

(2.2) $\mu_{ij} = \partial_i \rfloor \mu_j \ , \quad \mu_{ij} = -\mu_{ji} \ , \quad dx^k \wedge \mu_{ij} = \delta^k_i \ \mu_j - \delta^k_j \ \mu_i \ ,$

so that $\{\mu_i \mid 1 \le i \le 4\}$ is a conjugate basis for $\Lambda^3(E_4)$ and $\{\mu_{ij} \mid 1 \le i < j \le 4\}$ is a conjugate basis for $\Lambda^2(E_4)$. Finally, the space $E_4 \times \mathbb{R}_N$ with the global coordinate cover (x^i, q^α) may be viewed as the space of the state since the state can be realized by a map

(2.3) $\Psi : E_4 \longrightarrow E_4 \times \mathbb{R}_N \mid \Psi^* x^i = x^i \ , \quad \Psi^* q^\alpha = \Psi^\alpha(x^k) \ .$

The null class of the Euler-Lagrange operator is given by [8]

(2.4) $N = \{\eta \mid \eta\mu = \Psi^* d\rho \ , \quad \rho \ \varepsilon \ \Lambda^3(E_4 \times \mathbb{R}_N)\} \ .$

Accordingly, since any $\rho \ \varepsilon \ \Lambda^3(E_4 \times \mathbb{R}_N)$ has the form

$$\rho = Q^i \mu_i + \frac{1}{2} Q^{ij}_\alpha \ d\Psi^\alpha \wedge \mu_{ij}$$
$$+ \frac{1}{2} Q_{\alpha\beta i} \ d\Psi^\alpha \wedge d\Psi^\beta \wedge dx^i + \frac{1}{6} Q_{\alpha\beta\gamma} \ d\Psi^\alpha \wedge d\Psi^\beta \wedge d\Psi^\gamma \ ,$$

(2.4) gives the explicit characterization

(2.5) $\eta \ \mu = dQ^i \wedge \mu_i + \frac{1}{2} dQ^{ij}_\alpha \wedge d\Psi^\alpha \wedge \mu_{ij} + \frac{1}{2} dQ_{\alpha\beta i} \wedge d\Psi^\alpha \wedge d\Psi^\beta \wedge dx^i$

$$+ \frac{1}{6} dQ_{\alpha\beta\gamma} \wedge d\Psi^\alpha \wedge d\Psi^\beta \wedge d\Psi^\gamma \ .$$

Here, the Q's are functions of (x^i, Ψ^α) such that

(2.6) $Q^{(ij)}_\alpha = 0 \ , \quad Q_{(\alpha\beta)i} = 0 \ , \quad Q_{(\alpha\beta\gamma)} = 0 \ ,$

and indices enclosed in parentheses denote complete symmetrization [10].

It now remains to obtain the G-restriction, N_G, of the null class. Thus we return to the case where the internal symmetry group acts globally, and hence $dA^\alpha_\beta = 0$ for all $A^\alpha_\beta \ \varepsilon \ G_0$. Under these circumstances $`\Psi^\alpha = A^\alpha_\beta \ \Psi^\beta$ implies $d`\Psi^\alpha = A^\alpha_\beta \ d\Psi^\beta$. It then follows directly from (2.5) that all elements of N_G are of the form (2.5) provided the action of G_0 on the Q's is given by

(2.7) $`Q^i = Q^i \ , \quad `Q^{ij}_\beta \ A^\beta_\alpha = Q^{ij}_\alpha \ , \quad `Q_{\lambda\rho i} \ A^\lambda_\alpha \ A^\rho_\beta = Q_{\alpha\beta i} \ ,$

$$`Q_{\lambda\rho\sigma} \ A^\lambda_\alpha \ A^\rho_\beta \ A^\sigma_\gamma = Q_{\alpha\beta\gamma} \ .$$

When these results are put back into (2.5) we see that every element of N_G is the exterior derivative of an (n-1)-form valued scalar invariant of G_0:

(2.8) $N_G = \{\eta\mu = d\rho \mid \rho = \Psi^* \Lambda^3(E_4 \times \mathbb{R}_N) \ , \quad `\rho = \rho\} \ .$

3. The Image of the G-Restricted Null Class under Minimal Replacement

The local action of the internal symmetry group necessitates the minimal replacement

$$M : (\Psi^{\alpha},\ \partial_k \Psi^{\alpha}) \longmapsto (\Psi^{\alpha},\ \partial_k \Psi^{\alpha} + \Gamma^{\alpha}_{k\beta}\ \Psi^{\beta})$$

where the Γ's are the components of the Yang-Mills compensating potentials. The form of this replacement directly suggests that we introduce a gauge covariant differentiation operation that is defined by

$$D_k \Psi^{\alpha} = \partial_k \Psi^{\alpha} + \Gamma^{\alpha}_{k\beta}\ \Psi^{\beta} \quad . \tag{3.1}$$

Thus, if we define the Yang-Mills connection (gauge connection) 1-forms by

$$\Gamma^{\alpha}_{\beta} = \Gamma^{\alpha}_{k\beta}\ dx^{k} \quad , \tag{3.2}$$

(3.1) gives the gauge covariant exterior derivative

$$D\Psi^{\alpha} = d\Psi^{\alpha} + \Gamma^{\alpha}_{\beta}\ \Psi^{\beta} \quad . \tag{3.3}$$

It is then a simple matter to see that

$$DD\Psi^{\alpha} = \theta^{\alpha}_{\beta}\ \Psi^{\beta} \quad , \tag{3.4}$$

where

$$\theta^{\alpha}_{\beta} = d\Gamma^{\alpha}_{\beta} + \Gamma^{\alpha}_{\lambda} \wedge \Gamma^{\lambda}_{\beta} \tag{3.5}$$

are the curvature 2-forms of the gauge connection. Accordingly, we may write

$$M : (\Psi^{\alpha},\ d\Psi^{\alpha}) \longmapsto (\Psi^{\alpha},\ D\Psi^{\alpha}) \quad , \tag{3.6}$$

which determines the behavior of all the terms involving $d\Psi$'s in (2.5).

What now remains is to determine what happens to the Q's in (2.5) under minimal replacement. We first note that (1.7) and (3.1) through (3.3) imply

$${}^{\backprime}D^{\backprime}\Psi^{\alpha} = A^{\alpha}_{\beta}\ D\Psi^{\beta} \tag{3.7}$$

where ${}^{\backprime}D^{\backprime}\Psi^{\alpha} = d^{\backprime}\Psi^{\alpha} + {}^{\backprime}\Gamma^{\alpha}_{\beta}\ {}^{\backprime}\Psi^{\beta}$ and

$${}^{\backprime}\Gamma^{\alpha}_{\lambda}\ A^{\lambda}_{\beta} = A^{\alpha}_{\lambda}\ \Gamma^{\lambda}_{\beta} - dA^{\alpha}_{\beta} \tag{3.8}$$

follows from (1.7). Now, simply note that ${}^{\backprime}Q_{\lambda}\ A^{\lambda}_{\alpha} = Q_{\alpha}$ implies ${}^{\backprime}Q_{\lambda}\ {}^{\backprime}\Psi^{\lambda} = Q_{\alpha}\ \Psi^{\alpha}$ and hence $d(Q_{\alpha}\Psi^{\alpha}) = D(Q_{\alpha}\Psi^{\alpha})$. An expansion of the right-hand side of this relation by (3.3) and the derivation rule gives $DQ_{\alpha} = dQ_{\alpha} - \Gamma^{\beta}_{\alpha}\ Q_{\beta}$. Accordingly, (2.7) show that

$$\begin{aligned} &DQ^{i} = dQ^{i}\ , \quad DQ^{ij}_{\alpha} = dQ^{ij}_{\alpha} - \Gamma^{\lambda}_{\alpha}\ Q^{ij}_{\lambda} \quad , \quad DQ_{\alpha\beta i} = dQ_{\alpha\beta i} - \Gamma^{\lambda}_{\alpha}\ Q_{\lambda\beta i} - \Gamma^{\lambda}_{\beta}\ Q_{\alpha\lambda i} \quad , \\ &DQ_{\alpha\beta\gamma} = dQ_{\alpha\beta\gamma} - \Gamma^{\lambda}_{\alpha}\ Q_{\lambda\beta\gamma} - \Gamma^{\lambda}_{\beta}\ Q_{\alpha\lambda\gamma} - \Gamma^{\lambda}_{\gamma}\ Q_{\alpha\beta\lambda} \quad . \end{aligned} \tag{3.9}$$

These same results can be obtained by taking the exterior derivatives of (2.7) and then using (3.8) to eliminate the quantities dA^{α}_{β} .

The analysis will be given in detail for the second term in (2.5) since the other terms follow the same pattern. Noting that minimal replacement replaces the operator d by the operator D throughout, we have

$$(3.10)\qquad M(dQ^{ij}_{\alpha} \wedge d\Psi^{\alpha} \wedge \mu_{ij}) = DQ^{ij}_{\alpha} \wedge D\Psi^{\alpha} \wedge \mu_{ij} \quad .$$

However, $Q^{ij}_{\alpha}\, D\Psi^{\alpha} \wedge \mu_{ij}$ is a 3-form valued scalar invariant under the local action of G (see (3.7)) and hence

$$(3.11)\qquad d(Q^{ij}_{\alpha}\, D\Psi^{\alpha} \wedge \mu_{ij}) = D(Q^{ij}_{\alpha}\, D\Psi^{\alpha} \wedge \mu_{ij}) = DQ^{ij}_{\alpha} \wedge D\Psi^{\alpha} \wedge \mu_{ij} + Q^{ij}_{\alpha}\, \theta^{\alpha}_{\lambda}\, \Psi^{\lambda} \wedge \mu_{ij}$$

by (3.4). Elimination of the common terms between (3.10) and (3.11) thus gives

$$M(dQ^{ij}_{\alpha} \wedge d\Psi^{\alpha} \wedge \mu_{ij}) = d(Q^{ij}_{\alpha}\, D\Psi^{\alpha} \wedge \mu_{ij}) - Q^{ij}_{\alpha}\, \theta^{\alpha}_{\lambda}\, \Psi^{\lambda} \wedge \mu_{ij} \quad .$$

When the symmetry relations (2.6) are used, the general result is

$$(3.12)\qquad \begin{aligned} M(\eta\mu) = {} & d(Q^{i}\mu_{i} + \tfrac{1}{2} Q^{ij}_{\alpha}\, D\Psi^{\alpha} \wedge \mu_{ij} + \tfrac{1}{2} Q_{\alpha\beta i}\, D\Psi^{\alpha} \wedge D\Psi^{\beta} \wedge dx^{i} \\ & + \tfrac{1}{6} Q_{\alpha\beta\gamma}\, D\Psi^{\alpha} \wedge D\Psi^{\beta} \wedge D\Psi^{\gamma}) \\ & - \left(\tfrac{1}{2} Q^{ij}_{\alpha}\, \theta^{\alpha}_{\lambda} \wedge \mu_{ij} + Q_{\alpha\beta i}\, \theta^{\alpha}_{\lambda} \wedge D\Psi^{\beta} \wedge dx^{i} + \tfrac{1}{2} Q_{\alpha\beta\gamma}\, \theta^{\alpha}_{\lambda} \wedge D\Psi^{\beta} \wedge D\Psi^{\gamma}\right) \Psi^{\lambda} \; . \end{aligned}$$

Now, it is evident that the exact term in (3.12) can be written

$$\begin{aligned} \Psi^{*} d(Q^{i}\mu_{i} & + \tfrac{1}{2} Q^{ij}_{\alpha}\, Dq^{\alpha} \wedge \mu_{ij} + \tfrac{1}{2} Q_{\alpha\beta i} Dq^{\alpha} \wedge Dq^{\beta} \wedge dx^{i} \\ & + \tfrac{1}{6} Q_{\alpha\beta\gamma}\, Dq^{\alpha} \wedge Dq^{\beta} \wedge Dq^{\gamma}) = \Psi^{*} d\rho \end{aligned}$$

where ρ is a G-invariant 3-form on $E_4 \times \mathbb{R}_N$ for any $\Gamma^{\alpha}_{\beta} \in \Lambda^{1}(E_4)$. Accordingly, the exact term in (3.12) belongs to N_G and we have

$$(3.13)\qquad M(\eta\mu) = \xi - \theta^{\alpha}_{\lambda} \wedge (\tfrac{1}{2} Q^{ij}_{\alpha}\, \mu_{ij} + Q_{\alpha\beta i}\, D\Psi^{\beta} \wedge dx^{i} + \tfrac{1}{2} Q_{\alpha\beta\gamma}\, D\Psi^{\beta} \wedge D\Psi^{\gamma})\Psi^{\lambda}$$

with $\xi = d\rho \in N_G$.

Minimal replacement maps the G-restricted null class into itself if and only if the gauge connection Γ^{α}_{β} *is curvature free.*

This result is stronger than it seems, for $\xi \in N_G$ implies $\{E|\xi\}_{\Psi^{\alpha}} = 0$ for all Ψ^{α} but

$$(3.14)\qquad \{E|\,\xi\}_{\Gamma^{\alpha}_{i\beta}} \neq 0 \quad .$$

Thus, $\xi \in N_G$ *is not variationally trivial with respect to variations of the gauge connection!*

4. Equivalence Breaking

In the interest of simplicity, we confine the discussion to **elements** of

N_G of the form $\eta\mu = \frac{1}{2}\, dQ^{ij}_{\alpha} \wedge d\Psi^{\alpha} \wedge \mu_{ij}$ with $Q^{(ij)}_{\alpha} = 0$, $Q^{ij}_{\lambda} A^{\lambda}_{\alpha} = Q^{ij}_{\alpha}$. When (2.1) and (2.2) are used, we then have $\eta = \partial_i \Psi^{\alpha}\, \partial_j Q^{ij}_{\alpha}$.

Suppose that we start the Yang-Mills construction with two classically equivalent Lagrangians, L_0 and

(4.1) $$L_1 = L_0 + \partial_i \Psi^{\alpha}\, \partial_j Q^{ij}_{\alpha} = L_0 + \eta \quad .$$

After minimal replacement, we have ML_0 and

(4.2) $$M(L_1\mu) = M(L_0\mu) - \theta^{\alpha}_{\lambda} \wedge \frac{1}{2}\, Q^{ij}_{\alpha}\, \mu_{ij} \Psi^{\lambda} + \xi$$

where $\xi \in N_G$.

Let $\{\gamma^{\alpha}_{a\beta}\ ,\ a = 1,\ldots,r\}$ be a basis for the Lie algebra of G_0 . Since Γ^{α}_{β} are Lie algebra valued 1-forms on E_4 , we have

(4.3) $$\Gamma^{\alpha}_{\beta} = W^a \gamma^{\alpha}_{a\beta}\ , \quad W^a = W^a_i\, dx^i \quad ,$$

where $\{W^a \mid 1 \le a \le r\}$ is the set of Yang-Mills potential 1-forms. It then follows [11,12] that

(4.4) $$\theta^{\alpha}_{\beta} = F^a\, \gamma^{\alpha}_{a\beta}\ , \quad F^a = dW^a + \frac{1}{2}\, C_e{}^a{}_f\, W^e \wedge W^f \quad ,$$

where $C_a{}^e{}_f$ are the constants of structure for the group. If we write $F^a = \frac{1}{2} F^a_{ij}\, dx^i \wedge dx^j$, a simple computation and (2.1) show that (4.4) implies

(4.5) $$ML_1 = ML_0 + \frac{1}{2}\, Q^{ij}_{\alpha}\, F^a_{ij}\, \gamma^{\alpha}_{a\lambda}\, \Psi^{\lambda} + \xi$$

with $\{E|\xi\}_{\Psi^{\alpha}} \equiv 0$ for all Ψ^{α} . Following the minimal replacement construct by a minimal coupling construct [3,11], both ML_0 and ML_1 will be augmented by addition of the gauge invariant terms involving F^a_{ij} only. Thus, if we denote these new Lagrangians by $\bar{L}_0$ and $\bar{L}_1$, we have $\bar{L}_0 = L_0(x^j, \Psi^{\alpha}, \partial_i\Psi^{\alpha}, W^a_i, F^a_{ij})$ and $\bar{L}_1 = \bar{L}_0 + \frac{1}{2}\, Q^{ij}_{\alpha}\, F^a_{ij}\, \gamma^{\alpha}_{a\lambda}\, \Psi^{\lambda} + \xi$. The field equations for the Ψ^{α}-fields in the two theories will thus be different if $F^{\alpha}_{ij} \not\equiv 0$:

(4.6) $$\{E|L_0\}_{\Psi^{\alpha}} = 0\ , \quad \{E|L_1\}_{\Psi^{\alpha}} = \{E|L_0\}_{\Psi^{\alpha}} = 0$$

for the global action of G_0 , while

(4.7) $$\{E|\bar{L}_0\}_{\Psi^{\alpha}} = 0\ , \quad \{E|\bar{L}_1\}_{\Psi^{\alpha}} = \{E|\bar{L}_0\}_{\Psi^{\alpha}} + \frac{1}{2}\, Q^{ij}_{\beta}\, F^a_{ij}\, \gamma^{\beta}_{a\alpha} = 0$$

for the local action $G: E_4 \longmapsto G_0$.

Minimal replacement breaks classical field theory equivalence if the gauge connection is not curvature free.

An equivalent and possibly more forceful statement is as follows:

Minimal replacement imbeds any classical field theory in a continuum of inequivalent field theories if the gauge connection is not curvature free.

Of equal importance is the occurrence of the gauge curvature F^a_{ij} on the right-hand side of the second of (4.7). The now standard Yang-Mills construct consists both of the minimal replacement and the minimal coupling arguments. The minimal coupling construct is posited upon the fact that minimal replacement couples the Ψ-fields to the gauge fields only through terms that arise from $Md\Psi^{\alpha} = D\Psi^{\alpha} = d\Psi^{\alpha} + \Gamma^{\alpha}_{\beta}\,\Psi^{\beta}$. This in turn implies that the field equations for the Ψ's involve the Γ's but not the gauge curvature. The above results show that breaking of classical equivalence by minimal replacement also breaks the minimal coupling argument. In fact, if the term $\Psi^{\lambda}\theta^{\alpha}_{\lambda}\,Q_{\alpha\beta i} \wedge D\Psi^{\beta} \wedge dx^{i}$ in (3.13) is used, terms involving the gauge curvature and first derivatives of the Ψ^{α}'s will appear.

Breaking of classical equivalence by minimal replacement also breaks the minimal coupling of the theory.

5. Application to the Gauge Theory of Defects

The homogeneous boundary value problem for an elastic material with dislocations was modified in a previous work [13] by use of appropriate null Lagrangians so as to include specified boundary tractions and initial values for the linear momentum. This class of problems was shown to be well posed in [14]. We now take up the problem of extending these ideas in order to include disclinations as well.

The notation to be used from now on is that established in [15], in order to conform to current practices in continuum theories. The reader should have no difficulty in making the transition from the notation used in the first part of this paper; simply note that the state variables for a material body before minimal replacement are the three coordinate values $\{x^{i} \mid 1 \leq i \leq 3\}$ of the position of a material particle in the current configuration, and that the independent variables are the coordinates of the material particle in the reference configuration, $\{X^{A} \mid 1 \leq A \leq 3\}$, and the time, $T = X^{4}$.

The same null Lagrangian is chosen as in [14]:

$$\eta = d\rho = d(x^{i}dQ_{i}) = dx^{i} \wedge dQ_{i} = \partial_{b}x^{i}\partial_{a}Q^{ab}_{i}\,\mu \quad . \tag{5.1}$$

When it is added to the Lagrangian of elasticity, it does not change the Euler-Lagrange equations, but it does change the natural Neumann data in order to give the appropriate boundary tractions and initial values for linear momentum. Here Q_{i} are 2-forms $Q_{i} = \frac{1}{2}\,Q^{ab}_{i}\,\mu_{ab}$, $Q^{ab}_{i} = -Q^{ba}_{i}$, $dQ_{i} = \partial_{a}Q^{ab}_{i}\,\mu_{b}$. The indices $i, j, \ldots$ and $A, B, C, \ldots$ run from 1 to 3 and $a, b, c, \ldots$ run from 1 to 4. Further, $\mu = dX^{1} \wedge dX^{2} \wedge dX^{3} \wedge dT$ is the 4-dimensional volume element, μ_{a}, $1 \leq a \leq 4$, form a conjugate basis for exterior forms of degree 3 and μ_{ab}, $1 \leq a < b \leq 4$ form a conjugate basis for 2-forms (see [9], pp. 100-104).

The null Lagrangian (5.1) must be invariant under the homogeneous action of the gauge group $G_{0} = SO(3) \triangleright T(3)$ of rigid body rotations and translations. This requires the following transformation law for Q_{i}:

$$\grave{\underset{\sim}{Q}} = \underset{\sim}{Q}\underset{\sim\sim}{R}^{-1} \quad , \tag{5.2}$$

when $\grave{\underset{\sim}{x}} = \underset{\sim\sim}{R}\underset{\sim}{x} + \underset{\sim}{b}$, and we have defined $\underset{\sim}{Q} = [Q_{1}, Q_{2}, Q_{3}]$, $\underset{\sim}{x} = [x^{1}, x^{2}, x^{3}]^{T}$. At this point, the 3×3 orthogonal matrix $\underset{\sim}{R}$ which gives rise to rotations and the vector $\underset{\sim}{b}$ which translates $\underset{\sim}{x}$ do not depend on (X^a) ; they move all of the material particles the same way (homogeneously).

Breaking of the homogeneity of the action of G_0, which comes about when $\underset{\sim}{R}$ and $\underset{\sim}{b}$ depend on (X^a), requires the Yang-Mills minimal replacement construct in order that invariance of the Lagrangian be restored (i.e., in order to restore balance of linear and moment of momentum by Noether's second theorem)

$$dx^i \longmapsto B^i = Dx^i + \phi^i = dx^i + W^\alpha \gamma_{\alpha j}^{\ i} x^j + \phi^i \quad , \tag{5.3}$$

$$dQ_i \longmapsto DQ_i = dQ_i - Q_j \wedge W^\alpha \gamma_{\alpha i}^{\ j} \quad . \tag{5.4}$$

The 1-forms W^α are the compensating potentials for the inhomogeneous rotations, while the 1-forms ϕ^i compensate for the inhomogeneous translations. The matrices $\underset{\sim}{\gamma}_\alpha$ form a basis for the matrix Lie algebra of the representation of the rotation group SO(3) on V_3.

The Lagrangian for an elastic material with dislocations and disclinations is given, after Yang-Mills minimal coupling construct is applied, by

$$L\mu = (ML_o - s_1 L_\phi - s_2 L_W)\mu + B^i \wedge DQ_i \quad . \tag{5.5}$$

Here L_o is the Lagrangian of elasticity and L_ϕ, L_W are the coupling Lagrangians of the compensating fields ϕ^i, W^α (see [15] for explicit evaluations).
The magnitudes of the couplings are determined by the values of the coupling constants s_1 and s_2.

We now form the action integral $A[x^i, \phi^i, W^\alpha] = \int_{B_4} L\mu$. After standard manipulations [9], vanishing of the first variations of A gives the following Euler-Lagrange equations and boundary conditions:

$$D(\underset{\sim}{Z}+D\underset{\sim}{Q}) = -2\underset{\sim}{R} \wedge \underset{\sim}{\Theta} \,, \quad D\underset{\sim}{R} = \tfrac{1}{2}(\underset{\sim}{Z}+D\underset{\sim}{Q}) \,, \quad D\underset{\sim}{G} = \tfrac{1}{2}\bar{\underset{\sim}{J}} \,, \tag{5.6}$$

at all interior points of the body, and

$$\underset{\sim}{x} = \underset{\sim}{f}(X^a) \quad \text{on} \quad \partial B_4^I \,, \tag{5.7}$$

$$\underset{\sim}{Z} + D\underset{\sim}{Q} = 0 \quad \text{on} \quad \partial B_4^{II} \,, \tag{5.8}$$

$$\underset{\sim}{R} \wedge \delta\underset{\sim}{\phi} = 0 \quad \text{on} \quad \partial B_4 \,, \tag{5.9}$$

$$(\underset{\sim}{G}+\underset{\sim}{R}\underset{\sim}{\gamma}\underset{\sim}{x}) \wedge \delta\underset{\sim}{W} = 0 \quad \text{on} \quad \partial B_4 \,. \tag{5.10}$$

Here, ∂B_4^I is the support of the Dirichlet data, ∂B_4^{II} is the support of the Neumann data, and the symbols in the above equations are defined through the constitutive relations

$$Z_i = Z_i^a \mu_a = \frac{\partial ML_o}{\partial B_a^i}\mu_a \,, \quad R_i = \tfrac{1}{2} R_i^{ab}\,\mu_{ab} = \frac{1}{2}\frac{\partial L}{\partial D_{ab}^i}\mu_{ab} \,, \tag{5.11}$$

$$\underset{\sim}{G} = \tfrac{1}{2} G_\alpha^{ab}\mu_{ab}\delta^{\alpha\delta}\underset{\sim}{\gamma}_\delta = -\tfrac{1}{2} s_2 \frac{\partial L}{\partial F_{ab}^\alpha}\mu_{ab}\delta^{\alpha\delta}\underset{\sim}{\gamma}_\delta \,, \tag{5.12}$$

(5.13) $$\underset{\sim}{\bar{J}} = -(2R_i - Q_i)\gamma^i_{\alpha j} \wedge B^j{}_\delta{}^{\alpha\delta}\underset{\sim}{\gamma}_\delta \quad ,$$

where

(5.14) $$F^\alpha_{ab} = \partial_a W^\alpha_b - \partial_b W^\alpha_a + C^\alpha_{\beta\gamma} W^\beta_a W^\gamma_b \quad ,$$

(5.15) $$D^i_{ab} = \partial_a \phi^i_b - \partial_b \phi^i_a + \gamma^i_{\alpha j}(W^\alpha_a \phi^j_b - W^\alpha_b \phi^j_a) + F^\alpha_{ab}\gamma^i_{\alpha j} x^j \quad ,$$

(5.16) $$\underset{\sim}{\Theta} = \frac{1}{2} F^\alpha_{ab}\underset{\sim}{\gamma}_\alpha dX^a \wedge dX^b \quad .$$

It is essential at this point to observe that the integrability conditions for $(5.6)_3$ are non trivial. Using the special way that the Lagrangian depends upon D^i and F^α, these conditions reduce to

(5.17) $$\underset{\sim}{Z} \wedge \underset{\sim}{\gamma}_\alpha \underset{\sim}{B} = \underset{\sim}{Q} \wedge \underset{\sim}{\gamma}_\alpha \underset{\sim}{D} \quad .$$

For the convenience of the reader, we note that (5.6)-(5.10), (5.17) have the following explicit evaluations:

$$\partial_a Z^a_i - Z^a_j W^\alpha_a \gamma^j_{\alpha i} + \frac{1}{2} Q^{ab}_j F^\alpha_{ab}\gamma^j_{\alpha i} = R^{ab}_j F^\alpha_{ab}\gamma^j_{\alpha i} \quad ,$$

$$\partial_a R^{ab}_i - R^{ab}_j W^\alpha_a \gamma^j_{\alpha i} = \frac{1}{2}(Z^b_i + \partial_a Q^{ab}_i - Q^{ab}_j W^\alpha_a \gamma^j_{\alpha i}) \quad ,$$

$$\partial_a G^{ab}_\alpha \delta^{\alpha\delta}\underset{\sim}{\gamma}_\delta - G^{ab}_\beta W^\gamma_\alpha C^\beta_{\gamma\alpha}\delta^{\alpha\delta}\underset{\sim}{\gamma}_\delta = -\frac{1}{2}(2R^{ab}_i - Q^{ab}_i)\gamma^i_{\alpha j} B^j_a \delta^{\alpha\delta}\underset{\sim}{\gamma}_\delta \quad ,$$

$$x^i = f^i(X^a)\Big|_{\partial B^I_4} \quad , \qquad (Z^A_i + \partial_a Q^{aA}_i - Q^{aA}_j W^\alpha_a \gamma^j_{\alpha i}) N_A\Big|_{\partial B^{II}_3} = 0 \quad ,$$

$$(Z^4_i + \partial_A Q^{A4}_i - Q^{A4}_j W^\alpha_A \gamma^j_{\alpha i})\Big|_{T = T_0} = 0 \quad ,$$

$$R^{\alpha A}_i N_A\Big|_{\partial B_3} = 0 \quad , \qquad R^{A4}_i\Big|_{T = T_0} = 0 \quad ,$$

$$(G^{aA}_\alpha + R^{aA}_i \gamma^i_{\alpha j} x^j) N_A\Big|_{\partial B_3} = 0 \quad , \qquad (G^{A4}_\alpha + R^{A4}_i \gamma^i_{\alpha j} x^j)\Big|_{T = T_0} = 0 \quad ,$$

$$Z^a_i B^j_a \gamma^i_{\alpha j} = \frac{1}{2} Q^{ab}_i D^j_{ab}\gamma^i_{\alpha j} \quad .$$

We see from (5.11) that $Z^A_i = -\sigma^A_i$ = -Piola-Kirchhoff stress, $Z^4_i = p_i$ = linear momentum. These relations together with the antisymmetry of $\gamma^i_{\alpha j}$ in i and j reduce (5.17) to the explicit form

(5.18) $$\sigma^A_i B^j_A \gamma^i_{\alpha j} = -\frac{1}{2} Q^{ab}_i D^j_{ab}\gamma^i_{\alpha j} \quad .$$

The presence of the terms on the right-hand side of (5.18) are of particular significance. The interaction of Q_i^{ab} fields with torsion $(\underset{\sim}{D}=\underset{\sim}{\Sigma})$ gives rise to a torque per unit volume. This destroys the symmetry of the Cauchy stress tensor, as expressed by the relations

$$\sigma_i^A B_A^j = \sigma_j^A B_A^i \quad , \tag{5.19}$$

which obtain from (5.18) when the right-hand sides vanish. The above can probably be better understood if we parallel the breaking of the homogeneity of the action of the rotation group SO(3), with the introduction of the deformable directors of Toupin [16]. We should also require that the directors undergo only rigid body rotations, those of a Cosserat continuum, in order for the correspondence to be complete. The introduction of the directors gives rise to the antisymmetric part of the Cauchy stress tensor, and so does the inhomogeneous action of SO(3).

6. Effective Stress, Angular Momentum, and Couple Stress

We will now attempt to give an answer to the question, what drives the dislocation and disclination fields. First we define the "elastic" Piola-Kirchhoff stresses, couple-stresses, linear momenta and angular momenta:

$$S_i^A = \partial_a Q_i^{aA} - Q_j^{aA} W_a^\alpha \gamma_{\alpha i}^j \quad , \quad M_j^{iA} = Q_{[j}^{aA} B_a^{i]} \quad , \tag{6.1}$$

$$P_i = -\partial_A Q_i^{A4} + Q_j^{A4} W_A^\alpha \gamma_{\alpha i}^j \quad , \quad Q_j^{i4} = -Q_{[j}^{A4} B_A^{i]} \quad . \tag{6.2}$$

Equations $(5.6)_2$ and $(5.6)_3$ reduce to

$$\begin{aligned} \partial_a R_i^{\alpha A} - R_j^{aA} W_a^\alpha \gamma_{\alpha i}^j &= -\frac{1}{2}(\sigma_i^A - S_i^A) \quad , \\ \partial_A R_i^{A4} - R_j^{A4} W_A^\alpha \gamma_{\alpha i}^j &= \frac{1}{2}(p_i - P_i) \quad , \end{aligned} \tag{6.3}$$

$$\begin{aligned} \partial_a G^{aA} - G_\beta^{\alpha A} W_\alpha^\gamma C_{\gamma\alpha}^\beta &= -\frac{1}{2}(J_j^{iA} - M_j^{iA})\gamma_{\alpha i}^j \quad , \\ \partial_A G_\alpha^{A4} - G_\beta^{A4} W_\alpha^\gamma C_{\gamma\alpha}^\beta &= \frac{1}{2}(J_j^{i4} - Q_j^{i4})\gamma_{\alpha i}^j \quad , \end{aligned} \tag{6.4}$$

with

$$J_j^{iA} = 2R_{[j}^{aA} B_a^{i]} \quad , \quad J_j^{i4} = -2R_{[j}^{A4} B_A^{i]} \quad . \tag{6.5}$$

It is clear from (6.3) that $(\sigma_i^A - S_i^A)$ is the *effective stress* and $(p_i - P_i)$ is the *effective linear momentum*. We define $(J_j^{iA} - M_j^{iA})$ to be the *effective couple-stress* and $(J_j^{i4} - Q_j^{i4})$ to be the *effective angular momentum*. From (6.3) we see that the effective stresses and effective linear momenta give rise to dislocation fields R_i^{ab}, while from (6.4) we conclude that the effective couple-stresses and effective angular momenta give rise to disclination fields G_α^{ab}.

The "elastic" stresses, couple stresses, linear and angular momenta are solutions to the "elastic" problem, with the same boundary tractions that apply to the true stresses and same initial conditions that apply for the true momenta:

$$(6.6) \qquad \sigma_i^A N_A = S_i^A N_A = T_i(X^B) \quad \text{on} \quad \partial B_3^{II} \,, \quad M_j^{iA} N_A = 0 \quad \text{on} \quad \partial B_3 \,,$$

$$(6.7) \qquad p_i = P_i = p_i^o(X^B) \,, \quad Q_j^{i4} = 0 \quad \text{at} \quad T = T_o \,.$$

They satisfy the field equations

$$(6.8) \qquad DD\tilde{Q} = -\tilde{Q} \wedge \tilde{\Theta} \,, \quad D(\tilde{Q}\tilde{\gamma}_\alpha \wedge \tilde{B}) = D\tilde{Q}\tilde{\gamma}_\alpha \wedge \tilde{B} + \tilde{Q}\tilde{\gamma}_\alpha \wedge \tilde{D} \,.$$

The above relations written out explicitly (using the definitions (6.1), (6.2) as well as the antisymmetry of $\gamma^i_{\alpha j}$ in i and j), take the form

$$(6.9) \qquad \partial_A S_i^A = \partial_4 P_i - \partial_b (Q_j^{ab} W_a^\alpha \gamma_{\alpha i}^j) \,,$$

$$(6.10) \qquad \partial_A M_j^{iA} - S^A_{[j} B^{i]}_A - \frac{1}{2} Q^{ab}_{[j} D^{i]}_{ab} + P_{[j} B^{i]}_4 = \partial_4 Q_j^{i4} \,.$$

The balance of moment of momentum equations for a Cosserat continuum in the current configuration are

$$j m^k_{ij,k} - j t_{[ij]} + j \ell_{[ij]} + P_{[i} \dot{x}_{j]} = \dot{Q}_{[ij]} \,.$$

After being pulled back in the reference configuration, they can be placed into a 1-1 correspondence with (6.10). This justifies the definition of M_j^{iA} as the "elastic" couple-stress and Q_j^{i4} as the "elastic" angular momentum. The interpretation of the term $-\frac{1}{2} Q^{ab}_{[j} D^{i]}_{ab}$ in (5.18) is also correct, because it corresponds to $j\ell_{[ij]}$ which is a couple per unit undeformed volume [16].

Equations (6.9), (6.10) together with the boundary and initial conditions (6.6), (6.7) form the "elastic" problem of a Cosserat continuum. Their solutions will give the "elastic" stresses, couple-stresses, linear and angular momenta, and their difference with the true stresses, couple-stresses, linear and angular momenta will generate the dislocation and disclination fields. Of course equations (6.9), (6.10) cannot be solved independently, since they explicitly contain $\tilde{x}$, $\tilde{\phi}$, $\tilde{W}$ through the B^i's .

The presence of disclinations has the effect of replacing $d\tilde{Q}$ by $d\tilde{Q} - \tilde{Q} \wedge \tilde{\gamma}\tilde{W}$ in the null Lagrangian, hence fixing Q is required in addition to dQ . This is done, however, by the boundary conditions (6.6), (6.7), and for all interior points by the requirement that $-\frac{1}{2} Q^{ab}_{[j} D^{i]}_{ab}$ corresponds to a torque per unit volume applied to the material.

The analogy between the disclinations and the Cosserat elastic continuum serves as a possible explanation as to why it is so difficult to create and sustain disclinations in a material.

A useful insight is finally gained if we look at the equations the other way around. Defining the effective dislocation field as

(6.11) $$\bar{\underset{\sim}{R}} = \underset{\sim}{R} - \frac{1}{2}\underset{\sim}{Q} \quad ,$$

equations (5.6) take the following form:

(6.12) $$D\underset{\sim}{Z} = -\,2\bar{\underset{\sim}{R}} \wedge \underset{\sim}{\Theta} \ , \quad D\bar{\underset{\sim}{R}} = \frac{1}{2}\underset{\sim}{Z} \ , \quad D\underset{\sim}{G} = \frac{1}{2}\bar{\underset{\sim}{J}} \quad ,$$

where $\bar{\underset{\sim}{J}}$ is formed from $\bar{\underset{\sim}{R}}$ in accordance with (5.13). These equations are the same as the corresponding ones that solve the homogeneous boundary value problem, with the only replacement given by eq. (6.11). The true stresses have as a potential the effective dislocation fields and the interaction of the effective dislocation fields with the distortion $\underset{\sim}{B}$ gives rise to disclinations.

References

1. Kibble, T. W. B., J. Math. Phys. 2 (1961), 777.
2. Noether, E., Nachr. Ges. Wiss. Göttingen, Math.-Phys. Kl. 1918, 235.
3. Yang, C. N. and R. L. Mills, Phys. Rev. 95 (1954), 191.
4. Landers, A. W., Jr., "Invariant Multiple Integrals in the Calculus of Variations," in Contributions to the Calculus of Variations, 1938-1941, 175 (Univ. Chicago Press, 1942).
5. Edelen, D. G. B., Arch. Rational Mech. Anal. 11 (1962), 117.
6. Rund, H., The Hamilton-Jacobi theory in the calculus of variations (D. van Nostrand, London, 1966), 250.
7. Dedecker, P., C. R. Acad. Sci. Paris Ser. A-B286 (1978), 547.
8. Anderson, I. M. and T. Duchamp, Am. J. Math. 102 (1980), 781.
9. Edelen, D. G. B., Applied Exterior Calculus (Wiley-Interscience, 1985).
10. Hlavatý, V., Geometry of Einstein's Unified Field Theory (Noordhoff, Groningen, Holland).
11. Actor, A., Rev. Mod. Phys. 51 (1979), 461.
12. Edelen, D. G. B., Ann. Phys. (N.Y.) 133 (1981), 286.
13. Edelen, D. G. B., Lett. Appl. Engng. Sci. 20 (1982), 1049.
14. Edelen, D. G. B., Int. J. Engng. Sci. 21 (1983), 463.
15. Kadić, Aida and D. G. B. Edelen, A Gauge Theory of Dislocations and Disclinations (Lecture Notes in Physics No. 174, Springer, Berlin, 1983).
16. Toupin, R. A., Arch. Rational Mech. Anal. 17 (1964), 85.

GAUGE THEORIES IN MECHANICS

I. A. Kunin
Department of Mechanical Engineering
University of Houston
Houston, TX 77004, U.S.A.

B. I. Kunin
Department of Civil Engineering
Case Western Reserve University
Cleveland, OH 44106, U.S.A.

1. Introduction

In modern physics, gauge theory is considered to be the most powerful method for establishing interactions between fields. The terms gauge invariance (Eichinvarianz) and gauge transformations were introduced by H. Weyl (1918) in his attempt to unify gravitation and electromagnetism by considering the transfer of a length measure (gauge) as a physical process. Despite its mathematical beauty, the physical interpretation of this first gauge theory was untenable. Nevertheless, the related concept of gauge invariance later became one of the most fundamental principles of physics. Electrodynamics (1929), the Yang-Mills theory of nuclear forces (1954), gravitation (1956-   ), electro-weak interactions (1968), theories of elementary particles were understood and developed as gauge field theories. Now there are many introductory and review works on the foundations of gauge theory (see, e.g. [1-4] addressed to physicists and [5,6] to mathematicians).

After this success of gauge theory in fundamental fields, attempts were made to develop a gauge field approach to singularities (defects) in liquid crystals, magnets, spin-glasses, solids, etc. (see e.g. [7-15]).

The question arises: is the gauge theoretical approach limited to rather special macroscopic models or has it more universal meaning? The answer depends on what is understood by a gauge theory. Applications to elementary particles as well as mentioned macroscopic theories of defects are mainly based on the Yang-Mills gauge theory (YM-GT) with definite minimal replacement and coupling prescriptions. A somewhat different gauge approach is related to gravitation theories. At the same time, the gauge principle (of localization of symmetry) itself is not necessarily restricted to those prescriptions and admits alternative realizations.

This paper is an attempt to make a guage approach work in mechanics. The first part of the paper deals with classical mechanics. As a basic example, the Euler equations for a rigid body and their generalizations are shown to admit a description in the scope of a modified gauge theory different from YM-GT.

Some basic models of continuum mechanics are considered in the second part of the paper. The main problem here is an adequate description of translational gauging and gauge transformations. It is shown that continuum mechanics is essentially a translational gauge theory also different from YM-GT.

A reader may ask what are the advantages of a reinterpretation of well-known models as gauge theories. The answer is two-fold. For mechanics, besides a more deep understanding of classical models and their symmetries, this gauge approach may be considered as a first and necessary step before applications to more complicated problems where classical methods and notions often fail (e.g. plasticity, fracture mechanics, turbulence, etc.). For gauge theory, mechanics with its well understood models may be considered as a unique testing ground. We cannot resist here a temptation to give a quotation from the introduction to an excellent book on functional analysis [16]: "When somebody talked to him [the deceased author of the book] about a complicated infinite-dimensional construction, he usually asked: and how does that look in the two-dimensional case? Often enough, that shocking question helped to a better understanding of the mathematical situation". Notice that classical mechanics is one-dimensional...

And, last but not least, this symposium is devoted to "the interaction between pure mathematics and mechanics". The authors are very grateful to the organizers of the symposium for including gauge theory as one of the hot points for such an interaction. Since the language of principal fibre bundles used by mathematicians in gauge theory has not yet become the every-day langauge for physicists, the authors chose an informal free-style type of presentation hoping that this will be equally acceptable for both sides (which might be equivalent to equally unacceptable).

2. The Conventional YM-GT in One-Dimensional Case

Let us start with the basic question: is it possible to consider classical mechanics as a particular case of a gauge theory? Time is the only independent variable in classical mechanics, thus the corresponding gauge theory (if it exists at all) should be a "one-dimensional" one. First, we shall try to apply the standard Yang-Mills gauging procedure to the simplest mechanical system.

Any gauge theory usually starts with equations for the so-called matter fields. Essential to the theory is the assumption that the equations are derived from a variational principle and the corresponding Lagrangian L_0 is invariant with respect to a global (homogeneous) action of a Lie group G_0. The elements $g_0 \in L_0$ are called global gauge transformations and usually are represented by (constant) matrices.

Let us consider free 3-dimensional motion of a non-relativistic particle as the simplest matter field model:

$$L_0(\dot{\underline{x}}) = \frac{1}{2} m (\dot{\underline{x}}, \dot{\underline{x}}) , \qquad m\ddot{\underline{x}} = 0 \tag{2.1}$$

It is clear that L_0 is invariant with respect to the action of the rotation group $SO(3)$ on the position vector $\underline{x}$. We put $G_0 = SO(3)$.

The first step of the gauging is to substitute G_0 by the local gauge group $G = \{g(t)\}$ where $g(t)$ are arbitrary smooth functions with values in G_0, i.e. time-dependent rotations. The gauge principle requires the extension of the G_0-invariance to the much stronger G-invariance (with respect to the infinite-dimensional group G). It is clear that the initial Lagrangian $L_0(\dot{\underline{x}})$ does not satisfy this requirement because now the derivative d/dt does not commute with $g \in G$.

To construct a new G-invariant Lagrangian L_0 the gauge theory prescribes the fundamental minimal replacement recipe: the derivative d/dt is to be replaced by the gauge covariant derivative

$$D_t = \frac{d}{dt} + A(t), \tag{2.2}$$

where a new compensating field $A(t)$ with values in the Lie algebra $SO(3)$ (i.e. skew-symmetric matrices) is introduced. The field $A(t)$, which is also called Yang-Mills potential, transforms under the action of G by

$$g : A \longmapsto A^g = gAg^{-1} - \dot{g}\, g^{-1} . \tag{2.3}$$

Set $D_t^g = d/dt + A^g$. Then D_t obeys the following commutation law with g:

$$D_t^g g = g\, D_t, \tag{2.4}$$

and the Lagrangian $L_0\,(D_t\underline{x})$ is evidently G-invariant.

Thus, replacement $d/dt \to D_t$ restored the invariance of L_0 but the price to be paid is the addition of the new field $A(t)$ to the matter field $\underline{x}(t)$. In order to obtain a closed theory for both interacting fields $\underline{x}(t)$ and $A(t)$, equations involving $A(t)$ are to be added. The second fundamental recipe for constructing such a closed theory postulates:

(a) The complete system of equations is obtained from a G-invariant Lagrangian L.

(b) $L = L(t, \underline{x}, \dot{\underline{x}}, A, \dot{A})$.

(c) L has the form

$$L = L_0\,(D_t\underline{x}) + L_1(F) , \tag{2.5}$$

where $F = D_tA$ is called the Yang-Mills field, and the action of D_t on A is prescribed in a special way.

(d) L_1 is quadratic in F.

The recipe is called the minimal coupling and together with minimal replacement is considered universally as an important integral part of the YM-GT (some deviations are related to (d) only. The definiteness of both recipes is one of the most attractive features of the YM-GT for physicists.

Independently of our emotions on this subject, there is a simple mathematical reality: $F \equiv 0$ in one-dimensional case. Thus, YM-GT does not exist in one-dimensional case and cannot be related to classical mechanics in principle.

On the other hand, some of the YM-GT prescriptions are not necessary consequences of the gauge principles which have both physical and mathematical motivations. Thus, one can try to relax the restrictions. More definitely, we shall completely preserve the minimal replacement construction and only the fundamental postulate (a) of minimal coupling. A simple but important example is considered in the next section. Later this example is used as a model for the more complicated situation in continuum mechanics.

3. Another Gauge Approach

This moment seems appropriate for the first small portion of the gauge philosophy (further portions will be related to continuum mechanics).

Let us look at the gauge theory as a special theory of deformation of (so far, one-dimensional) models invariant with respect to a Lie group G_0. To be more specific, we assume that the equations of a model are derived from a G_0-invariant Lagrangian (whether this latter assumption is really essential will be discussed later on). One may imagine a G_0-box which contains all blocks of the model: state variables (matter fields) ψ, differential operators (polynomials in d/dt), the Lagrangian and a given representation of the group G_0 on all of these.

To deform the model one needs a larger G-box in which the G_0-box may be embedded. The convention is: G consists of G_0-valued functions g(t) (the action on functions $\psi(t)$ is "time-pointwise" $(g\psi)(t) = g(t)\psi(t)$); d/dt is recognized as a covariant derivative, say, D_t^0 which is then deformed within the space of covariant derivatives.

In the one-dimensional case under consideration, the space of all covariant derivatives is a homogeneous space of the group G. For any two derivatives D_t^a and D_t^b, there exists the unique $h \in G$ such that

$$D_t^b = \hat{h}\, D_t^a \doteq h\, D_t^a\, \bar{h} \quad . \tag{3.1}$$

Here $\hat{h}$ is the representation of the element h and is defined ($\doteq$) as the composition of the operators h, D_t^a and $\bar{h} \doteq h^{-1}$. We shall also use the notation

$$D_t^{ha} \doteq \hat{h}\, D_t^a \quad . \tag{3.2}$$

To distinguish the composition $D_t^a g$, $g \in G$ from the action of D_t^a on g, we denote the latter by $(D_t^a g)$. Distinguishing a particular covariant derivative D_t^0 ("fixing a gauge" in physical language) provides us with a (1:1)-correspondence between D's and elements of G: $h \leftrightarrow D_t^h \doteq \hat{h}\, D_t^0$, $h \in G$. In particular, $D_t^e = D_t^0$, where e is the unit element of G.

Let $\mathfrak{g}_0$ and $\mathfrak{g}$ be the Lie algebras of G_0 and G, respectively ($\mathfrak{g}$ consists of $\mathfrak{g}_0$-valued functions of t). Then

$$B^h \doteq D_t^h - D_t^0 = h(D_t^0\, \bar{h}) \tag{3.3}$$

is an element of $\mathfrak{g}$. The identities

$$B^{gh} = gB^h\bar{g} + B^g , \qquad B^e = 0 \tag{3.4}$$

follow from the derivation properties of D_t^0.

We define two different actions $\hat{G}$ and $\tilde{G}$ of the group G on the B-functions:

$$\hat{g}B^k \doteq gB^k\, \bar{g} \quad , \tag{3.5}$$

$$\tilde{h}B^k \doteq B^{hk} \quad , \qquad g,h,k \in G. \tag{3.6}$$

The first $\hat{G}$ is the linear action (the adjoint representation of G) which is identified with the gauge transformations. Two B-functions are gauge-equivalent if they are related by a gauge transformation. Thus, the space of all B's splits into the classes of gauge-equivalent B's (orbits of $\hat{G}$). Evidently, $B^0 = 0$ is the zero-orbit of $\hat{G}$.

The second action $\tilde{G}$ is nonlinear and, in particular, shifts $B^0 = 0$ into $B^h = \tilde{h}B^0$. We shall see that, in contrast to the gauge transformations, $\tilde{G}$-transformations correspond to essential changes of states of a gauge model. The identities (3.4) give a relation between these transformations

$$\hat{g}B^h = \tilde{g}B^h - B^g \quad . \tag{3.7}$$

Notice that it is a specific feature of the one-dimensional case (as well as some of continuum models which will be considered later on) that the groups $\hat{G}$ and $\tilde{G}$ are isomorphic. In the conventional gauge theory, $\hat{G}$ is a subgroup of a much larger group $\tilde{G}$.

The next important block of the G-box is a set of Lagrangians and their transformations under actions of $\hat{G}$ and $\tilde{G}$. This question will be considered in the example of the model we started with in the preceeding section.

To finish with the embedding of the G_0-box into the G-box (i.e. establishing a (1 : 1)-correspondence between conventional and some of the gauge-covariant quantities) we put in a convenient writing the convention stated before

$$\frac{d}{dt} \overset{*}{=} D_t^0 \ . \tag{3.8}$$

The sign $\overset{*}{=}$ will mean that an equality is not gauge-covariant.

As a result, one may say, a model from the G_0-box is embedded into the G-box without changing its real physical meaning. This is analogous to introducing general curvilinear coordinates in an Euclidean space. We will call the preparation of the G-box and the embedding pre-gauging.

We now want to deform the model inside the G-box. The first step is the minimal replacement

$$D_t^0 \to D_t^h = D_t^0 + B^h \overset{*}{=} \frac{d}{dt} + A^h \ , \tag{3.9}$$

where $B^h \overset{*}{=} A^h$ is an arbitrary B-function. The first equality for D_t^h is gauge-covariant while the second is not. The functions B and A coincide under the fixed gauge but have different transformation laws with respect to $\hat{G}$. The inhomogeneous transformation law (2.3) for A is similar to the transformation law for Christoffel symbols Γ, whereas B as well as D_t obeys the homogeneous "tensor" transformation law.

Let us return to the example from the preceeding section and restrict ourselves to the pure gauge model, i.e. the only dynamical variable is B^h. We consider the function $h(t) \in G$ as a potential for B^h and take the Lagrangian L in the form

$$L = L(t, h, D_t^0 h) \overset{*}{=} L(t, h, \dot{h}) \ . \tag{3.10}$$

The simplest one quadratic in $\dot{h}$ is

$$L = L(B^h) = \frac{1}{2} B^T K B, \tag{3.11}$$

where $K = K(t)$ is a (material) tensor of the forth order and the index h for B is omitted. In components

$$L = \frac{1}{2} B_{\alpha\beta} K^{\alpha\beta\lambda\mu} B_{\lambda\mu} \ , \tag{3.12}$$

where K is symmetric in the pairs $\alpha\beta$ and $\lambda\mu$. If $G = SO(3)$, there is additional skew-symmetry inside the pairs.

We require the Lagrangian to be gauge invariant under the action of $\hat{G}$. This defines the transformation law for K

$$\hat{g} : K(t) \longmapsto K_g(t) = \hat{g}(t) K(t) \hat{\bar{g}}(t) \ . \tag{3.13}$$

Let us assume that there is a gauge which makes K time independent. We may assume that this gauge corresponds to D_t^0. In reality, this is a privileged state which defines the choice of D_t^0. Thus, we have a material conservation law for K

$$D_t^0 K_0 \overset{*}{=} \frac{d}{dt} K_0 = 0 \leftrightarrow K_0 \overset{*}{=} \text{const.} \tag{3.14}$$

The Euler equations for the Lagrangian L given by (3.11) can be written in the gauge covariant form

$$\frac{\delta L}{\delta h} = D_t(KB) = (D_t^0 + B)(KB) = 0, \tag{3.15}$$

where the dependence of B on h (3.3) is used.

For G = SO(3), we identify this model with the rotation of a rigid body:

$B \leftrightarrow \underline{\Omega}$ angular velocity with respect to the body,
$K_0 \leftrightarrow$ I inertia tensor with respect to the body,
$K_0 B \leftrightarrow \underline{M}_0$ angular momentum with respect to the body,
$L \leftrightarrow T$ kinetic energy,

$D_t \leftrightarrow \frac{D}{dt} = \frac{d}{dt} - \underline{\Omega}\times$ material derivative,

$\hat{g}B \leftrightarrow g\underline{\Omega}$ rotation of the coordinate system,

$\hat{\bar{h}}B \leftrightarrow \bar{h}\underline{\Omega} = \underline{\omega}$ angular velocity in space,

$$D_t\,(KB) = 0 \leftrightarrow \begin{cases} \frac{D}{dt}\,\underline{M} = 0, \\ \text{or} \\ \frac{d}{dt}\,\underline{M} = \underline{\Omega} \times (I\underline{\Omega}). \end{cases} \tag{3.16}$$

Finally, the energy conservation

$$D_t T \overset{*}{=} \frac{d}{dt} T = 0 \tag{3.17}$$

is equivalent to the conservation law (3.14).

This example permits several conclusions of importance for possible generalizations:

1. The gauge transformations ($\hat{G}$-representation) may be interpreted as a sort of coordinate transformations. They leave L invariant and are symmetries (in the gauge sense). $\tilde{G}$-transformations though preserving the gauge-covariant form of equations essentially change the state of the system (dependence of T and $\underline{\Omega}$ on h).

2. L depends on material tensors K (generally of all orders) which, under the action of $\hat{G}$, tensorially transform simultaneously with state variables leaving L invariant. We write symbolically

$$L = L(h, B^h, K) = L(\hat{g}h, \hat{g}B^h, \hat{\hat{G}}K) . \tag{3.18}$$

3. The gauge theories of fundamental fields use perfectly symmetric Lagrangians where combinations (up to coupling constants) of the space-time and invariant group metrics play the role of K. If one tries to use such an isotropic Lagrangian for the rigid body the Euler equations would degenerate to trivial ones. This underlines the difference between fundamental and phenomenological models.

There are many generalizations of the well-known Euler equations for a rigid body. Let us consider briefly some of them from the gauge theoretical point of view described above.

Pseudo-rigid or affinely-rigid bodies [17-20] are models which in addition to rotational degrees of freedom admit homogeneous deformations. These models are considered as approximations to continuously deformed bodies. A gauge approach to pseudo-rigid bodies requires adding of a potential energy term to the Lagrangian (3.11). This would be an interesting bridge to gauge theories of continuum mechanics.

Another direction of generalization is preserving the basic structure of the Euler equations but with different finite-dimensional Lie groups [21, 22]. This approach is based on a symplectic structure of co-adjoint representations and uses essentially the Hamiltonian formalism. Examples of application: motion of a rigid body (possibly electrically charged or magnetized) in a fluid (in the presence of electric or magnetic fields), dynamics of a rigid body with ellipsoidal cavities filled with a (possibly magnetic) fluid. Development of these models in the scope of the gauge approach should clarify an important interrelation between the Hamiltonian and gauge formalisms.

The most revolutionary generalizations were initiated by Arnold in his pioneering work [23] developed further in [24-26, 21] and many others. In this approach the generalized Euler equations are considered as equations for geodesics on a group (SO(3)-case corresponds to the usual Euler equations). It was shown that the motion of a perfect incompressible fluid is a geodesic curve on an infinite dimensional group SDiff of volume preserving diffeomorphisms of a manifold. Thus, the fluid is considered as a dynamical system (time t is the only independent variable) with the infinite-dimensional configuration space SDiff. Methods of modern global functional analysis used in this approach led to new important results including some existance and uniqueness theorems. One may notice, however, that this global approach develops in the direction which, in a sense, is opposite to the gauge ideology of localization. The dynamical variables of the gauge theory are fields, i.e. functions on space-time, and the main tendency is a space-time localization of a group action. It seems that deep links should exist between these two approaches and it would be interesting to reveal them.

Now, after obtaining some experience in dealing with one-dimensional models, we proceed to continuum mechanics.

4. What Is Gauge Theory?

We need to put this question because, as is discussed below, the standard YM-GT does not work in continuum mechanics. Let us indicate some of the reasons for this:

1. Internal variables like charge, spin, isospin, etc. play an important role in YM-GT. In fact, the gauge group acts in internal spaces related to these variables. To the contrary, the main state variables of continuum mechanics are deformations described by displacements, velocities, etc. The action of the gauge group should be related to space-time rather than to an internal space. The non-linearity of this action (translations!) requires a generalization of the standard minimal replacement.

2. Transformations in such basic models as elasticity and hydrodynamics are curvature-free. As a result, the standard minimal coupling does not work (just as in the one-dimensional case).

3. Equations of phenomenological models (like plasticity) generally are not derived from a Lagrangian.

The question arises: is gauge theory applicable to such models at all? The answer depends on what is understood by gauge theory (for example, one of the creators of YM-GT Yang himself did not consider the general relativity as a "true" gauge theory). Let us consider two "definitions" of gauge theory.

1. Physical (Hehl, et al [27]):
"A gauge theory represents a heuristic framework for deriving the field coupled to a conserved current of a matter field".
2. Mathematical (Trautman [28]):
"A gauge theory is any physical theory of a dynamical variable which, at a classical level, may be identified with a connection on a principal bundle".

These two definitions are far from being identical though they do not contradict each other. After accepting the mathematical criterium, one may ask: what is a connection on a principal bundle?

Definition: connections on a principal bundle are guys living in an abstract space and using a strange language. Ordinary physicists cannot speak with them without translators.

Fortunately: Some of these guys have their fully authorized ambassadors which are called covariant derivatives. In contrast to their bosses, the ambassadors are very friendly towards physicists and even provide vehicles for (parallel) transportation.

Unfortunately: Some of the important guys have no ambassadors.

As an illustration, let us compare the well-known for physicists covariant derivatives in classical differential geometry and YM-GT. Let $\lambda,\mu,\ldots$ and $a,b,\ldots$ be space-time and an internal vector space indices, respectively.

Differential geometry	YM-GT
Covariant derivatives	
$(\nabla_\mu w)^\lambda = \partial_\mu w^\lambda + \Gamma^\lambda_{\mu\nu} w^\nu$	$(D_\mu \psi)^a_\lambda = \partial_{[\mu}\psi^a_{\lambda]} + A^a_{[\mu|b|}\psi^b_{\lambda]}$
∇w	$D \wedge \psi$
$\nabla_u w = u.\nabla$	
Gauge groups	
$g_0 \varepsilon G_0 = GL(n)$ (or its subgroup)	$g_0 \;\varepsilon\; G_0$ internal symmetries
$g(x) \;\varepsilon\; G$	$g(x) \;\varepsilon\; G$
Connections (on space-time)	
$\nabla \leftrightarrow \Gamma$ linear	$D \leftrightarrow A$ G_0-connection
$\nabla = \nabla^\Gamma = \nabla(\Gamma)$	$D = D^A = D(A)$
Curvatures	
$R(u,v) = [\nabla_u,\nabla_v] - \nabla_{[u,v]}$	$F = D \wedge A$
$S(u,v) = \nabla_u v - \nabla_v u - [u,v]$ torsion of ∇	

Notice an essential difference with the one-dimensional case. The covariant derivatives are labeled now by connections rather than elements of the local gauge groups G.

The fields A and Γ (Christoffel's symbols) are identified as gauge potentials in YM-GT and general relativity type gauge theories, respectively. The minimal replacement and coupling recipes are similar for the theories of both types though not identical.

Our main point of interest now is the Galilean group and the corresponding covariant derivatives. The ambassador's problem here is not so simple as in the case of the linear group GL(n).

5. The Galilean and Affine Spaces

Models of non-relativistic mechanics live in the Galilean space-time E_G which has an interesting geometry different from the Euclidean geometry (see e.g. [21, 30-32]). It is convenient for the purposes of this paper to view E_G as a 4-dimensional affine space A_4 with additional structures.

A_4 is a homogeneous space for the affine group A(4) which is a semi-direct product of the linear group GL(4) and the translation group T_4

$$A(4) = GL(4) \triangleright T_4 . \tag{5.1}$$

Let $E_A = E_1 \times E_3$ be the product of two Euclidean spaces which are identified with time and 3-space, respectively. E_A is called Aristotelian (or absolute) space-time. One may consider E_A as A_4 with an additional structure of a trivialized fibre-bundle: the base is time and the fibre is the 3-space. The Galilean space E_G is E_A modulo all shifts of fibres (pure Galilean transformations or boosts) leaving the base (time) invariant. More precisely, E_G is a (trivializable) fibre bundle with the total space A_4, base E_1, fibre E_3, and the projection being an affine map. Later the Galilean transformations will be considered as special gauge transformations and E_A will be interpreted as E_G with a fixed gauge (trivialization).

The spaces A_4, E_G, and E_A are endowed with canonical linear connections with zero curvatures and torsions (flat connections). The parallel transport with respect to these connections is absolute (does not depend on a path). Geodesics coincide with straight lines and define privileged rectilinear coordinate systems (inertial for E_G).

The same spaces considered as a material medium are endowed with a second (material) linear connection which generally is not flat. The existence of two geometries is a characteristic feature of continuum mechanics.

As we know, a linear connection ∇ is related to only one (linear) part of the affine and Galilean groups. On the other hand, as we shall see, the translational component plays a dominant role in continuum mechanics. How does this component contribute to a total gauge derivative?

6. Gauging the Translation Group

The problem of an appropriate gauging of the Galilean group in continuum mechanics has much in common (though not identical) with the well-known problem of gauging of the Poincaré group in theories of gravitation. The main question is how to modify the gauge formalism for a natural incorporation of the translation group. Many different approaches were suggested for solution of this problem (see e.g.

[33-42]). There are even different points of view on the Einstein's general relativity: is it a pure Lorentz or translational gauge theory. Much attention was given to theories with torsion which is often interpreted as a translational gauge field. On the whole, along with optimistic statements on complete solution of the Poincaré (or translation) gauging problem, there are more cautious conclusions [42]: "So far nobody has managed to gauge the translation group in a satisfactory manner".

Our approach to this problem is addressed mainly to the needs of continuum mechanics. Speculations on possible applications to gravitation theory are not discussed here.

We start from the concept of a (generalized) affine connection in the bundle of affine frames over a manifold M given in [43]. We want to realize these connections as the corresponding covariant derivatives. Strictly speaking, a completely adequate geometric structure would be an associated bundle of tangent affine spaces with a nonlinear connection. Such an approach is too cumbersome (Cf. [36]) for continuum mechanics which is based on the language of tensor fields in linear tangent spaces. This motivates our definition of an affine covariant derivative.

It is known [43] that there is a (1 : 1)-correspondence between the set of affine connections on M and the set of pairs (Γ, h) where Γ is a linear connection on M and h a tensor field of type (1,1) which is related to the translational component T_n of $A(n)$ ($n = \dim M$). A connection (Γ, I) where I is the unit tensor will be identified with Γ. We also assume that $\det h \neq 0$.

Let F, X and $T^{(p,q)}$ be spaces of smooth functions, vector and (p,q)-tensor fields on M, respectively. A tensor $h \in T^{(1,1)}$ with $\det h \neq 0$ is uniquely extended to such an automorphism $\hat{h}$ of the tensor algebra T on M, that $\hat{h}$ commutes with contractions and

$$\hat{h}\big|_F = \text{identity}, \qquad \hat{h}\big|_X = h \ ,$$
$$\hat{h}(t \otimes s) = \hat{h}(t) \otimes \hat{h}(s), \qquad t, s \in T \ . \tag{6.1}$$

In particular, if $\langle\theta,w\rangle$ is the standard pairing (contraction) of $\theta \in T^{(0,1)}$ and $w \in X = T^{(1,0)}$, then

$$\langle\theta,w\rangle = \hat{h}\langle\theta,w\rangle = \langle\hat{h}\theta,\hat{h}w\rangle = \langle\hat{h}\theta,hw\rangle \tag{6.2}$$

and thus $\hat{h}^T\theta = \bar{h}^T\theta$ ($\bar{h} = h^{-1}$). More generally, $\hat{h}^T = \hat{h}^{-1}$ with respect to the extension of $\langle,\rangle$ to T.

We define now an affine covariant derivative $D = D(\Gamma,h)$ as a ∇-valued function which is in a (1 : 1)-correspondence with a connection (Γ,h). We put

$$D_u(\Gamma,h) \doteq \hat{h}\nabla^{\Gamma}_{hu}\,\hat{h}^{-1} \doteq u \cdot D(\Gamma,h) \quad . \tag{6.3}$$

It is readily seen that D_u as well as ∇_u is a derivation on T, i.e. it satisfies the Leibniz rule with respect to $\otimes$.

Let us drop the index Γ and define a new linear covariant derivative

$$'\nabla \doteq \hat{h}\,\nabla\,\hat{h}^{-1} \tag{6.4}$$

Then

$$D_u = {}'\nabla_{'u} \quad , \quad 'u \doteq hu \ . \tag{6.5}$$

Thus D_u and $'\nabla_u$ are in (1 : 1)-correspondence and may be viewed as two representations of D. They will be both used depending on convenience.

In continuum mechanics, h will be identified with a transformation (deformation, motion) of a material medium. We shall restrict ourselves to the case when h completely defines a change of a state, i.e. h is the only dynamical variable identified with a translation connection (or with its transformation). Thus, this will be the case of a pure translation gauge theory. A linear connection $\nabla \leftrightarrow \Gamma$ is considered as responsible for internal degrees of freedom which are frozen with respect to a moving medium. The transformation $\nabla \to {}'\nabla$ given by (6.4) is completely induced by the translational part h. In other words, in a pure translational gauge theory, the linear connection though important is not an independent dynamical variable and plays a passive role. The same is valid for a metric tensor.

The new curvature $'R$ and torsion $'S$ corresponding to $'\nabla \leftrightarrow {}'\Gamma$ are given by

$$'R(u,v) = h\ R(u,v)\ \bar{h} \ , \tag{6.6}$$

$$'S(u,v) = S(u,v) + \Omega(u,v) \ , \tag{6.7}$$

where

$$\Omega(u,v) \doteq h[(\nabla_u\bar{h})v - (\nabla_v\bar{h})u] \ . \tag{6.8}$$

Thus, the curvature undergoes coordinate-like transformation and remains gauge equivalent to the initial one.

The torsion increment Ω is a translational component of $'S$ (depending on ∇). A dislocation interpretation of S and Ω will be briefly discussed below.

Let $\tilde{k}$ be a derivation on T that commutes with contractions and is induced by a tensor $k \in T^{(1,1)}$:

$$\tilde{k}\big|_F = 0 \quad , \quad \tilde{k}\big|_X = k \quad ,$$

$$\tilde{k}(t \otimes s) = \tilde{k}t \otimes s + t \otimes \tilde{k}s \ , \quad t, s \in T \tag{6.9}$$

For example, the curvature R may be defined by

$$\tilde{R}(u,v) = [\nabla_u,\nabla_v] - \nabla_{[u,v]} \quad . \tag{6.10}$$

Another example is a relation between ∇_u and the Lie derivative L_u [43]

$$\nabla_u = L_u + \tilde{\nabla u} \ . \tag{6.11}$$

We have the following representations for the covariant derivatives D_u and $'\nabla_u$:

$$D_u = \hat{h}D_u^0 \hat{h}^{-1} = D_u^0 + \tilde{B}_u \ , \tag{6.12}$$

$$'\nabla_u = \hat{h}\nabla_u\hat{h}^{-} = \nabla_u + '\tilde{B}_u \ , \tag{6.13}$$

where $D_u^0 \doteq \nabla_{'u}$, and (1,1)-tensors B_u, $'B_u$ are given by

$$B_u = h(D_u^0 \bar{h}) = -(D_u^0 h)\bar{h} \quad , \tag{6.14}$$

$$'B_u = h(\nabla_u \bar{h}) = -(\nabla_u h)\bar{h} \quad . \tag{6.15}$$

The (1,2)-tensors B and $'B$ defined by

$$B_u = u \cdot B, \quad 'B_u = u \cdot 'B \tag{6.16}$$

play the role of Christoffel symbols for D_u and $'\nabla_u$, respectively.

Notice the relations

$$'B_{'u} = B_u \ , \tag{6.17}$$

$$\Omega(u,v) = 'B_u v - 'B_v u \quad . \tag{6.18}$$

The tensors h form a group H whose action described above will be called "H-transformations". Let G be another copy of H whose action on D_u is given by

$$G \ni g : \ D_u \to D_u^g \doteq \hat{g}\, D_u\, \hat{g}^{-1} \quad . \tag{6.19}$$

This action will be called (translation) gauge transformation, or G-transformation.

These two representations of the same group are analogous to the representations $\tilde{G}$ and $\hat{G}$ from the one-dimensional case.

7. Holonomic Transformations

An element $h \in H$ and the corresponding transformation $\hat{h}$ will be called ∇-holonomic iff $\Omega = 0$ (cf. (6.7), (6.8)). With respect to a composition law, such transformations form what we will call the (∇-) holonomic subgroup of H. This group appears to be very relevant for continuum mechanics.

Now it is assumed that M is a flat space (R=0, S=0) and H will be restricted to holonomic transformations only. We put $\nabla_\mu = \partial_\mu$ with respect to a rectilinear coordinate system and denote

$$'\partial_\mu \doteq h^{\nu}_{.\mu}\partial_\nu \quad , \qquad j \doteq \det h. \tag{7.1}$$

The following are equivalent: h is ∇-holonimic $\leftrightarrow$ $'S = 0 \leftrightarrow {'B}_u v = {'B}_v u$, or in components

$$\leftrightarrow \; {'B}^{\lambda}_{[\mu\nu]} = 0 \;\leftrightarrow\; B^{\lambda}_{\mu\nu} = -\partial_\nu h^{\lambda}_{.\mu} \quad , \tag{7.2}$$

$$\leftrightarrow \; \bar{h}^{\lambda}_{.\mu} = \delta^{\lambda}_{\mu} + \partial_\mu Y^\lambda \leftrightarrow h^{\lambda}_{.\mu} = \delta^{\lambda}_{\mu} + {'\partial}_\mu X^\lambda \; , \tag{7.3}$$

where vector fields X,Y ("displacements") are potentials for h.

Notice also useful formulas

$$\partial_\lambda(\bar{j}\, h^{\lambda}_{.\mu}) = 0, \qquad {'\partial}_\lambda(j\, \bar{h}^{\lambda}_{.\mu}) = 0 \quad , \tag{7.4}$$

$$\partial_\lambda \ln j = -\, {'\partial}_\nu\, \bar{h}^{\nu}_{.\lambda} \; , \quad {'\partial}_\lambda \ln j = \partial_\nu\, h^{\nu}_{.\lambda} \tag{7.5}$$

We say that q is a scalar m-density with respect to H if for $h \in H$

$$\hat{h}q = j^{-m}q \qquad (j = \det h) \; . \tag{7.6}$$

A tensor m-density is a product of q with a tensor.

Let M be endowed with a volume form μ which is a scalar (-1)-density, and p be a vector (+1)-density. The following identities are consequences of (7.2-7.5):

$$\int_V (\mathrm{div}\, h\, p)\mu = \int_V (\mathrm{Div}^0 j\, p)\, \bar{j}\mu \quad , \tag{7.7}$$

$$\leftrightarrow \quad j\, \mathrm{div}\, h\, p = \mathrm{Div}^0\, j\, p \; , \tag{7.8}$$

where V is an arbitrary domain in M, div and Div^0 are divergences relative to ∂_μ and $'\partial_\mu$, respectively.

Let

$$L = \rho_0 L(x,h) + \rho_0 \langle\beta, X\rangle \tag{7.9}$$

be a scalar (+1)-density considered as a Lagrangian with X and h related by (7.3). A scalar (+1)-density ρ_0 and a one-form β are given. The corresponding Euler equations are

$$\mathrm{Div}^0\, P = \rho_0\beta \quad , \qquad P \doteq \rho_0 \frac{\partial L}{\partial h} \; . \tag{7.10}$$

Let η_0 be a metric on M. If we assume that L depends on h through $\hat{h}^{-1}\eta_0 = h^T\eta_0 h$

only, then (7.10) takes the form (making use of (7.8))

$$\mathrm{div}T = \bar{j}\rho_0 b, \qquad T \doteq 2\bar{j}\rho_0 \frac{\partial L}{\partial \eta_0} \tag{7.11}$$

with relations

$$hP = j\,T\,\eta_0, \qquad \beta = \eta_0 b\ . \tag{7.12}$$

The symmetric tensor (+1)-density T will be identified with a stress-momentum tensor.

The Lagrangian (7.9) and equations (7.10) or (7.11) describe a (holonomic) translation gauge model where the only dynamic variable is a translation connection h. However, the equations are written in the form which is not H-covariant. To give them such a form let us introduce two more divergences.

$$\mathrm{Div}\,p \doteq D_\mu p^\mu\ , \qquad {}'\mathrm{div}\,p \doteq {}'\nabla_\mu p^\mu \tag{7.13}$$

where the dependence of D_μ and ${}'\nabla_\mu$ on h follows from (6.3) and (6.4). We have the relations

$$\mathrm{Div}\ p = \hat{h}\ \mathrm{Div}^0\ \hat{h}^{-1}\ p = \mathrm{Div}^0\ p + \tilde{B}_\mu p^\mu \quad , \tag{7.14}$$

$${}'\mathrm{div}\ p = \hat{h}\ \mathrm{div}\ \hat{h}^{-1}\ p = \mathrm{div}\ p + {}'\tilde{B}_\mu p^\mu \quad , \tag{7.15}$$

where $\tilde{B}$ and ${}'\tilde{B}$ are defined by (6.12-6.15). This permits one to rewrite (7.10) and (7.11) in an H-covariant form (cf. with (3.15)).

To complete the gauge description of the model we introduce the (holonomic) translation gauge group G as a second copy of H, define gauge transformations of covariant derivatives by (6.19) and require G-invariance of the Lagrangian. Basically, this is similar to the considered one-dimensional case and we omit details.

8. Motions of Material Medium

Let, as before, E_G be the Galilean space-time and V (dim V = 4) be the associated vector space. Vector and tensor fields on E_G are considered as having their values in a standard tensor algebra $\mathcal{T}$ on V. Let E_0 = time x 3-space be a fixed trivialization of E_G . We think of this as being an "observer". The results of measurements corresponding to different "observers" are related by Galilean transformations.

For $x \in E_G$ and $w \in V$ we have trivialization-dependent representations $x = (x^0 = t, \underline{x})$ and $w = (w^0, \underline{w})^T$, and similar for tensors. Being the product of two Euclidean spaces, E_0 may be endowed with a linear connection $\overset{0}{\nabla}$ and a (trivialization-dependent) metric which is the sum of two metrics

$$\eta_0(w,w) \doteq (w,w)_0 = \lambda_0(w^0)^2 + (\underline{w},\underline{w})_0 \tag{8.1}$$

where $\lambda_0 > 0$ is a parameter having dimension $[\text{velocity}]^2$ and $(\underline{w},\underline{w})_0$ is the standard 3-space scalar product.

With respect to a fixed E_0, orientation-preserving homogeneous Galilean transformations have the form

$$h_0 = \begin{pmatrix} 1 & \underline{0}^T \\ \underline{c} & R \end{pmatrix}, \quad \underline{c} \in V, \quad R \in SO(3) . \tag{8.2}$$

They induce transformations of η_0 as a (0,2)-tensor.

Let ϕ_t be a motion of a body in the 3-space. We refer to the book [32] for the definition of ϕ_t as well as basic notions of the push-forward ϕ_* and pull-back ϕ^* (geometrically invariant description of deformation gradients). We consider regular mappings ϕ only and denote $\bar{\phi} \doteq \phi^{-1}$ (so that $\phi^* = (\phi_*)^{-1} = \bar{\phi}_*$). In space-time language, a motion ϕ_t corresponds to a diffeomorphism $\phi : E_0 \to E_0$ (for simplicity, ϕ is extended to the total E_0). We write

$$\phi : \bar{x} \to x = \phi(\bar{x}) \; ; \qquad \bar{\phi} : x \to \bar{x} = \bar{\phi}(x) . \tag{8.3}$$

The transformations ϕ form a group Diff E_0 with respect to the composition law (E_0 will be dropped). The induced transformations ϕ_* and $\bar{\phi}_*$ of tensor fields on E_0 (functions including) may be considered as belonging to a representation Diff_* of Diff . One may say that ϕ and ϕ_* transport a point $\bar{x}$ and physical events in an infinitesimal neighborhood of $\bar{x}$ (in reference state) "far away" to a point x and its neighborhood (in a current state). This situation is not appropriate for a gauge approach which requires that everything "happens" near a given point. We can use an additional structure of E_0 (or E_G) to overcome this obstacle.

The absolute parallelism in E_0 allows one to define one more representation Diff_0 of Diff. An element $\phi_0 \in \text{Diff}_0$ shifts a tensor at x to the point $\phi(x)$. We define a composition transformation $\hat{h}(\phi)$ by

$$\hat{h} = \phi_* \bar{\phi}_0 \tag{8.4}$$

which is an automorphism of the tensor algebra respecting the space-time structure (in particular, $\hat{h}(dt) = dt$). It leaves invariant points, and functions, and defines (or is defined by) a (1,1)-tensor $h = \hat{h}|_X$. Notice that h and ϕ are (1 : 1) modulo constant translations which permits to identify h with a motion.

It can be verified using criteria (7.2-7.3) that h as defined by (8.4) is a $\overset{0}{\nabla}$-holonomic transformation and thus $H^0 \subset H$ but they do not coincide. The difference has the following origin. Motions as global diffeomorphisms are characteristic for pure elastic bodies only. Even a laminar flow (in a physical sense) of a fluid generally is a local rather than global diffeomorphism (to say nothing about turbulence and plasticity). Notice in this connection that a wide-spread notion of a material manifold as well as equivalence of Lagrange and Euler pictures generally have local rather than global meaning (cf. with the so-called local groups of local transformations [43] related, in a sense, to H). This is a motivation to associate H-transformations with space-time rather than with material picture and extend H^0 to H.

We assume that h = I corresponds to a state at rest. Comparison of (8.4) with usual definitions of kinematics (see, e.g. [32]) leads to the following identification:

$$h \overset{*}{=} \begin{pmatrix} 1 & \underline{0}^T \\ \underline{v} & F \end{pmatrix} , \tag{8.5}$$

where $\underline{v}$ is the <u>spatial velocity</u> and F the <u>spatial deformation gradient</u>, considered as functions of $x = (t,\underline{x})$. The space-time vector $v = (v^0 = 1, \underline{v})$ is referred to as the <u>velocity</u>. Notice that (8.2) is a special case of (8.5).

The requirement for h to be $\overset{0}{\nabla}$-holonomic leads to a number of equivalent compatibility equations for $\underline{v}$ and F which are obtained from (7.2-7.3), in particular, (i, j, k are space indices)

$$(\partial_t + \underline{v} \cdot \underline{\nabla})F^k_{\cdot i} = {}'\partial_i v^k \qquad ({}'\partial_i = F^j_{\cdot i}\partial_j) \quad . \tag{8.6}$$

As was indicated above, h admits potentials X and Y, however they are not considered as measurable quantities in contrast to $\underline{v}$ and F.

Let us classify fields with respect to H-transformations.

1. <u>H-scalars</u>. Space-time quantities such as metric η_0, volume form μ_0, etc. All functions including material ones.

2. <u>H-scalar densities</u>. Material fields such as a mass (+1)-density ρ, a material volume (-1)-density μ, Lagrangian (+1)-densities, etc. Notice that a mass form $m = \rho\mu$ is an H-scalar.

3. <u>H-tensors</u> and <u>H-tensor densities</u>. Material metric η, stress, etc.

4. <u>Translation connections</u>. $h \in H$ and their components $\underline{v}$, F. Recall that the transformation laws for connections (or covariant derivatives) were defined above. In particular, $\underline{v}$ and F are not H-tensors.

The action of the gauge group G is basically the same as that of H with the exception of connections (see (6.19)), Lagrangians and stress tensors. G-invariance of Lagrangians results in the action of G on a stress value only leaving its dependence on h invariant. In contrast to this, H acts on both value and argument h of stress.

9. Conservation of Mass and Kinetic Energy

Let $v_0 \doteq (1,\underline{0}^T)^T$. It follows from (8.5) that $v = hv_0$, and this permits one to identify v_0 with the velocity "at rest" (velocity in the time-direction) and D_{v_0} with the covariant time derivative D_t. A material field P is covariantly constant in time iff $D_t P = 0$.

One can define in a natural way an H-Lie derivative for H-tensors and H-densities. In particular, for an H-scalar n-density q, the derivative in the direction of the above v is given by the formula

$$D_t q \equiv L_v q = (\partial_t + \underline{v} \cdot \underline{\nabla})q + nq \operatorname{div} \underline{v} \quad . \tag{9.1}$$

The proof of (9.1) is based on the identities (7.2) - (7.5) and the representation (8.5) for h. Notice that for n=1

$$D_t q = \operatorname{div}(q\, v) \quad , \tag{9.2}$$

and for n=0 (H-scalars) D_t coincides with what is called the material derivative in continuum mechanics.

The identification of h with a transformation from a state at rest implies that the mass (+1)-density ρ, material volume (-1)-density μ and scalar mass form $m \doteq \rho\mu$ are time-independent for $h = h_0 = I$. Together with the definition of an H-density, this leads to conservation laws

$$D_t\rho = 0, \quad D_t\mu = 0, \quad D_t m = 0 \quad . \tag{9.3}$$

Let us consider now a kinetic energy densities ρK and K (the latter being per unit mass) as functions of h. We define K by (see (8.1))

$$2K \doteq (v,v)_0 = v^T \eta_0 v = \lambda_0 + (\underline{v},\underline{v})_0 \quad , \tag{9.4}$$

the constant λ_0 being identified with $2K_0$.

If we introduce the tensors

$$V_0 \doteq v_0 \otimes v_0 = v_0 v_0^T \text{ (matrix)} \quad , \tag{9.5}$$

$$V \doteq v \otimes v = \hat{h}\, V_0 = v\, v^T \text{ (matrix)} \quad , \tag{9.6}$$

and the material metric

$$\eta \doteq \hat{h}^{-1}\, \eta_0 = h^T\, \eta_0 h \text{ (matrix)} \quad , \tag{9.7}$$

then K admits alternative representations

$$2K = \langle \eta_0, V \rangle = \langle \eta, V_0 \rangle \ . \tag{9.8}$$

Considering ρK as a component of a Lagrangian let us find the contributions P_K and T_K to the tensors P and T defined by (7.10) and (7.11). From (9.7), (9.8) with identities (7.5) and $\rho = \bar{j}\rho_0$, we obtain

$$P_K \doteq \rho_0 \frac{\partial K}{\partial h} = \rho_0 \bar{h} V \eta_0, \tag{9.9}$$

$$T_K = 2\rho \frac{\partial K}{\partial \eta_0} = \rho V \quad , \tag{9.10}$$

the relations (7.12) between P and T being satisfied.

We are prepared now to consider briefly the gauge description of two basic models of continuum mechanics.

10. Ideal Compressible Fluid and Elasticity

These two models are described by Lagrangians of the type (7.9) where

$$L = K - U \quad , \tag{10.1}$$

U being a potential (internal) energy which depends on h through $\underline{\eta} = F^T \underline{\eta}_0 F$ where F is the 3-dim. block (deformation gradient) of h given by (8.5). In addition, we assume that, in the case of fluid, $U = U(\xi)$ where $\xi = \ln j$, $j = \det h = \det F$.

To have equations in 4-dim. forms (7.10) or (7.11), it is convenient to introduce $\underline{\eta}^0 \doteq \begin{pmatrix} 0 & \underline{0}^T \\ \underline{0} & \underline{\eta}_0^{-1} \end{pmatrix}$. We set $T = T_K + T_U$ where

$$T_U = - \frac{dU}{d\xi} \underline{\eta}^0 \doteq p\underline{\eta}^0 \tag{10.2}$$

is the contribution to T due to -U, and p is identified with pressure. Taking $\rho = \rho_0 \exp(-\xi)$ as a new variable and putting $E(\rho) = U(\xi)$ we obtain the standard relation

$$p = \rho^2 \frac{dE}{d\rho} \tag{10.3}$$

The equations of motion (7.11) (with b=0) now take the form

$$\operatorname{div} (T_K + T_U) = 0 \tag{10.4}$$

or, after substituting (9.10) and (10.2),

$$\operatorname{div}(\rho V + p\underline{\eta}^0) = 0 \quad . \tag{10.5}$$

Taking into account that $v^0 = 1$ and using (9.2) we obtain the 3-dim. Euler equations for ideal compressible fluid

$$D_t\rho = 0, \quad \rho D_t\underline{v} + \underline{\nabla} p = 0, \tag{10.6}$$

to which the energy conservation equation may be added

$$\rho D_t(K + W) + \mathrm{div}(p\underline{v}) = 0 . \qquad (10.7)$$

Recall that D_t is the covariant derivative given by (9.1-2) and different from the usual material derivative.

To obtain a completely gauge covariant form of the equations the gradient $\underline{\nabla}$ should be transformed into a covariant gradient $\underline{D}$ using the formulas given above.

Notice an interesting consequence of the gauge approach. The mass conservation is an integral part of the Euler equations of motion rather than an independent postulate.

In the case of elasticity, U is an arbitrary function of $\underline{\eta}$. Equations of motion can be written in two equivalent forms (7.10), (7.11). As before, we put $P = P_K + P_U$ and $T = T_K + T_U$.

It can be shown that 4-dimensional tensors P_U and T_U correspond to 3-dimensional Piola-Kirchhoff and Cauchy stress tensors, respectively (for their definition see [32]). As in hydrodynamics, the mass conservation is incorporated into the Euler equations.

The material metric η in elasticity is not necessarily Euclidean (internal stresses). In this case, η_0 in (9.7) denotes an arbitrary metric in a state of rest rather than a space-time metric. A more general case corresponds to an arbitrary material linear connection ∇. Thus, there exist two geometries: space-time and material, the latter , in general, being non-flat. A crucial criterium which distinguishes elasticity from, say, plasticity is holonomic translational gauge covariance of material geometry: ∇ and η, in the case of elasticity, are not independent dynamical variables. In physical terms, sources of internal stress (dislocations, point defects, etc.) are frozen in a medium. From a gauge theoretical point of view, elasticity is a pure holonomic translation gauge theory.

11. Generalizations

The main attention here was paid to pure translational gauge theories that, in addition, were holonomic. The latter is the simplest case that (paradoxically) was excluded from a usual gauge approach. Generalizations may develop in several directions.

1. Including matter fields. One may start from a process in a medium at rest (diffusion, electromagnetic interactions, etc.). Switching on motion leads to an interaction between gauge translational fields h and matter fields (e.g. magneto-hydrodynamics). Notice that matter fields may be Yang-Mills fields themselves. This would be a combined translation-YM-GT. In relation to this, the authors should confess that taking the example (2.1) of a material point as a matter field had a compositional reason only. It is more natural to consider

points and rigid bodies on an equal footing in the scope of a gauge model based on the entire group of motions ISO(3).

2. Removing of holonomicity restrictions (second gauging) leads to a general translational gauge theory (e.g. plasticity). In our opinion, this should clarify such not well defined notions as elasto-plastic deformations. Notice that elastic (better to say holonomic) transformations should be considered as a part of a translational connection rather than matter fields. An important role will be played by the translational torsion Ω which may be interpreted as a density-flux of micro- or macro-dislocations (e.g. in a Bravais lattice) in contrast to the linear torsion S related to dislocations with internal degrees of freedom (e.g. spin, lattice with a basis). Though it looks as a pure speculation authors expect that the translational gauging may be relevant to turbulence.

3. Switching on of internal degrees of freedom will result in an affine gauge theory where both linear and translation connections are dynamical variables. This third gauging may also include more general connections (geometries) as well as interactions with Yang-Mills fields.

4. The Lagrangian formalism cannot play the same role in phenomenological models of continuum mechanics as in fundamental fields. For non-Lagrangian models, a gauge covariance of equations should substitute the more restrictive requirement of Lagrangian invariance.

Notice that generalized gauge models should incorporate translational, and even holonomic, connections (identified here with motions) as an integral part. There is no continuum mechanics without motions.

12. Conclusions

We have shown that some basic models of classical and continuum mechanics admit a gauge field approach though different from conventional ones. In particular, ideal compressible fluid and elasticity are pure translational gauge theories.

Our optimistic attitude towards applications of gauge theory (GT) to mechanics is based on the following:

1. The usual dynamical variables in mechanics are tensor fields which transform in a specified way under the action of a finite-dimensional group, e.g. SO(3). GT essentially extends the transformation group to an infinite dimensional gauge group and correspondingly adds new fields: connections, or gauge covariant derivatives.

2. This leads to more general notions of gauge symmetry and covariance which are far-reaching generalizations of the usual coordinate invariance. This is especially important for models where classical approach is not completely adequate.

3. GT gives heuristic methods for constructing new models. In a sense, GT is a special theory of model deformations.

4. GT establishes a bridge over a widening gap between mechanics and modern field theories.

5. GT is a very aesthetic theory which combines methods of modern geometry with physical invariance principles. Simply speaking, it is just interesting.

In connection with this, we express deep gratitude to E. Kröner, D. Edelen, A. Kadić, M. Zorawski, R. Rivlin and F. Hehl for hot discussions on GT and on a crucial question: why GT?

References

1. C.N. Yang, in Proc. Hawaii Topical Conf. in Particle Physics 6, ed. by P.N. Dodson, Unif. Press of Hawaii, Honolulu, 1975, 489-561.

2. W. Drechsler, in Fiber bundle techniques in gauge theories, Springer-Verlag, Berlin, 1977.

3. G. Mack, Fortschritte der Physik 29, 135-185, (1981).

4. I. J. R. Aitchison. An informal introduction to gauge field theories. Cambridge Univ. Press, 1982.

5. D. Bleecker, Gauge theory and variational principles. Addison-Wesley, London, 1981, 145-248.

6. Yu. I. Manin, Gauge fields and complex geometry (in Russian). Nauka, Moscow, 1984.

7. J.A. Herz, Phys. Rev. 18B, 4875 (1978).

8. B. Julia, G. Toulouse, J. Physique-Lett. 40, L395-398 (1979).

9. I.E. Dzyaloshinskii, G.E. Volovik, Ann. of Phys. 125, 67 (1980).

10. B. Gairola, in Continuum models of discrete systems - 4, O. Brulin and R.K.T. Hsieh, eds., North-Holland, Amsterdam, 1981, 55-66.

11. I.E. Dzyaloshinskii, in Les Houches - Physics of defects, R. Bahan et al., eds., North-Holland, 1981, 320-360.

12. E. Kröner, ed. Gauge field theories of defects in solids. Max-Plank-Institut, Stuttgart, 1982.

13. N. Rivier, D.M. Duffu, J. Physique, 43, 293 (1982).

14. A. Kadić, D.G.B. Edelen. A gauge theory of dislocations and disclinations. Springer-Verlag, Berlin, Heidelberg, New York, 1983.

15. I.A. Kunin, in The mechanics of dislocations, E.C. Aifantis and J.P. Hirth, eds., Proc. Int. Symp., American Soc. of Metals, Metals, Part, OH, 1985, 69-76.

16. I.M. Glazman, Ju. I. Ljubič, Finite-dimensional linear analysis, MIT Press, Cambridge, MA 1974.

17. H. Cohen, Utilitas Mathematica, 20, 221-247 (1981).

18. H. Cohen, R.G. Muncaster, J. of Elasticity, 14, 127-154 (1984).

19. J.J. Slawianowski, Rep. Math. Phys. 10, 219 (1976).

20. H.P. Berg, Acta Phys. Austr., 54, 191-209 (1982).

21. V.I. Arnold. Mathematical methods of classical mechanics, Springer, Berlin, Heidelberg, New York 1979.

22. O.I. Bogoyavlensky, Commun. Math. Phys. 95, 307-315 (1984).

23. V.I. Arnold, Ann. Inst. Grenoble 16, 319-361 (1966).

24. V.I. Arnold, Usp. Mat. Nauk 24, 225-226 (1969).

25. D.G. Ebin, J.E. Marsden, Ann. of Math. 92, 102-163 (1970).

26. J.E. Marsden, A. Weinstein, Physica 7D, 305-323 (1983).

27. F.M. Hehl, J. Nitsch, P. von der Heyde, in General relativity and gravitation, vol. 1, Plenum, New York, 1980, 329-355.

28. A. Trautman, in General relativity and graviation, vol. 1, Plenum, New York, 1980, 287-307.

29. H. Weyl. Space-time-matter. Dover Publ., New York, 1952.

30. I.M. Yaglom. A simple non-Euclidean geometry and its physical basis. Springer-Verlag, New York 1979.

31. R. Penrose, in Battelle Rencontres, C.M. DeWitt and J.A. Wheeler, eds., Benjamin, New York, 1968, 121-235.

32. J.E. Marsden, T.J.R. Hughes. Mathematical foundaitons of elasticity. Prentice-Hall, Englewood Cliffs, N.J., 1983.

33. F.W. Hehl, P. von der Heyde, G.D. Kerliok, J.M. Nester, Rev. Mod. Phys. 48, 393-416 (1976).

34. Y.M. Cho, Phys. Rev. 14D, 2521-25; 3335-40 (1976).

35. C.P. Luehr, M. Rosenbaum, J. Math. Phys. 21, 1432-38 (1980), 25, 380 (1984).

36. W. Drechsler, Ann. Inst. Poincare, 37A, 155-184 (1982).

37. E.A. Ivanov, J. Niederle, Phys. Rev. 25D, 976-987; 988-994 (1982).

38. R.P. Wallner, Acta Phys. Austr. 54, 165-189 (1982).

39. D.G.B. Edelen, Int. J. Theor. Phys. 23, 949-985 (1984); 24, 659-673 (1985).

40. R. Aldrovandi, E. Stédile, Int. J. Theor. Phys. 23, 301-323 (1984).

41. F.W. Hehl, Found. Phys. 15, 451-471 (1985).

42. M.A. Schweizer, in Cosmology and gravitation: spin, torsion, rotation, and supergravity, P.G. Bergmann and V. DeSabbata, eds., Plenum, NY, 1980, 117-124.

43. S. Kobayashi, K. Nomizu. Foundations of differential geometry, vol. I, Interscience, New York, 1963.

ON THE ROLE OF NOETHER'S THEOREM IN THE GAUGE THEORY OF CRYSTAL DEFECTS

B.K.D. Gairola
Institut für Theoretische und Angewandte Physik, Universität Stuttgart, Pfaffenwaldring 57, D-7000 Stuttgart 80, W. Germany

INTRODUCTION:

In recent years gauge theories have attracted much attention due to the success of the unified gauge theory of electromagnetism and weak interactions and of the theory of strong interactions. Much of the interest is also due to the fundamental character of the gauge formalisms. It is now believed that "gauge invariance" plays a key role in building models of physical theories.

The formalism of gauge theory so successful in describing the interactions of elementary particles has not met with the same success in describing the interaction of crystal defects. Various authors[1-3] have formulated such gauge theories but their versions differ from each other and many questions remain open. Even the question as to what constitutes a gauge theory can be answered in different ways. There is the standard Yang-Mills theory which is modeled on gauge kinematics but with dynamics structured after electromagnetism as Yang and Mills did for the gauge group SU(2) and there is the Einstein theory which is a gauge theory but not of Yang-Mills type. But a gauge theory does not have to be of either type. In our opinion a gauge theory is any theory which has kinematic based on a local gauge group, gauge potential and gauge field strength. Only if correct physical results can be derived from the theory it is viable.

In this paper we investigate the role of Noether's theorem in the gauge theory of crystal defects which break the translational symmetry. In particular we utilize the identitites derived from this theorem to demonstrate that our gauge theory leads to results well known in the continuum theory of defects. We also clarify the nature of the Yang-Mills term which represents the interactions between the defects.

NOETHER'S SECOND THEOREM AND THE BIANCHI-TYPE IDENTITIES

A characteristic feature of any gauge theory is that the number of fields that appear in the Lagrangian is larger than the number of effective degrees of freedom of the theory. Let us recall the usual way a gauge theory is constructed. One starts from geometrical equations which express the action of the given gauge symmetry on a certain set of fields. Then one constructs a Lagrangian which is a gauge invariant local function of the fields. As soon as this function is specified one can identify some of the fields as nonphysical or null fields of the Lagrangian when expressed in the usual Yang-Mills (quadratic) form.

This feature of gauge theories was clearly recognized by Rosenfeld[4,5]. However, much earlier Hilbert[6] had pointed out that the equations of motion of such a theory cannot all be independent of one another, but must satisfy a number of identitites that are in general equal in number to the number of abitrary functions that define an element of the group.

Such identities can be derived from Noether's second theorem and are sometimes called Bianchi-type identities, after the identities of this type that occur in general relativity. Their importance can be realized from the fact that Einstein hesitated a long time to accept the validity of his field equations because of these identities[7].

Noether's second theorem can be stated in the following way: If the action is invariant under infinitesimal transformations of an infinite continuous group parametrized by r arbitrary functions there exist r independent differential identities for the Euler derivatives of the Lagrange function.

These identities were previously derived by the author in the context of spinor fields[8]. We rederive them here in a form given by Bergmann[9].

Consider a field theory whose Lagrangian is a function of tensor field variables and their derivatives collectively denoted by $Y^A(x)$ and $\nabla Y^A(x)$ $(A = 1,2...n)$. In the following we will, generally, suppress the collective index A. The field equations are derivable by variation of the action

$$S = \int_V L[\underline{x}, Y(\underline{x}), \nabla Y(\underline{x})]dv \qquad (1)$$

Since we will be dealing with the gauge theory of crystal defects we are primarily interested in the translation group. Therefore, consider the diffeomorphism

$$\underline{x} \to \underline{x}' = \underline{x} + \delta\underline{x} \tag{2}$$

The total variation of field variables is then given by

$$\delta Y(\underline{x}) = Y'(\underline{x}') - Y(\underline{x}') + Y(\underline{x}') - Y(\underline{x}) = \delta_f Y(\underline{x}) + \delta\underline{x}\cdot\nabla Y(\underline{x}) \tag{3}$$

where $\delta_f Y$ is called the form variation. It commutes with ∇. Similarly

$$\delta\nabla Y(\underline{x}) = \delta_f \nabla Y(\underline{x}) + \delta\underline{x}\cdot\nabla\nabla Y(\underline{x}) \tag{4}$$

In the following it suffices to consider only the linear transformations

$$\delta\underline{x} = \underline{t}(\underline{x})\tau(\underline{x}) \tag{5}$$

and

$$\delta Y(\underline{x}) = \underline{t}(\underline{x})\Phi(x) + (\nabla\underline{t}(\underline{x}))^T\Pi(\underline{x}) \tag{6}$$

where $(\nabla\underline{t})^T$ is the transpose of $\nabla\underline{t}$. It is understood that τ, Φ and Π are tensors of appropriate rank and $t\tau$, $t\Phi$ etc. mean scalar multiplication of $\underline{t}$ with τ, Φ etc. In this paper we denote tensors of second rank by lower case greek letters and upper case letters denote tensors of higher rank. It follows that the form variation is given by

$$\delta_f Y = \underline{t}(\Phi - \tau\nabla Y) + (\nabla\underline{t})^T\Pi = \underline{t}\Psi + (\nabla\underline{t})^T\Pi \tag{7}$$

The diffeomorphism is an invariant transformation if

$$\delta S = \int_{V+\delta V} L(\underline{x}', Y'(\underline{x}'), \nabla' Y'(\underline{x}'))dv - \int_V L(\underline{x}, Y(\underline{x}), \nabla(\underline{x}))dv$$

$$= \int_V [JL(\underline{x}', Y'(\underline{x}'), \nabla' Y'(\underline{x}')) - L(\underline{x}, Y(\underline{x}), \nabla Y(\underline{x}))]dv$$

$$= \int_V \delta L \, dv = 0 \tag{8}$$

This is possible if δL is a divergence i.e. if $\delta L = \nabla\cdot\underline{m}$. Using (4) and the relation

$$J = \det(1 + \nabla\cdot\delta\underline{x}) \qquad (9)$$

we can put (8) in the form

$$\int(\frac{\delta L}{\delta Y}\,\delta_f Y + \nabla\cdot\underline{n})dv = 0 \qquad (10)$$

where $\frac{\delta L}{\delta L}$ is the Euler derivative of L defined by

$$\frac{\delta L}{\delta Y} = \frac{\partial L}{\partial Y} - \nabla\cdot\frac{\partial L}{\partial\nabla Y} \qquad (11)$$

and

$$\underline{n} = L\delta\underline{x} + \frac{\partial L}{\partial\nabla Y}\,\delta_f Y - \delta\underline{m} \qquad (12)$$

Substituting (7) in (10) and assuming that $\underline{t}(\underline{x})$ vanishes on the boundary of integration region we get

$$\int\underline{t}\ [\ \frac{\delta L}{\delta Y} - \nabla\cdot\ (\Pi\,\frac{\delta L}{\delta Y}\ ]dv = 0 \qquad (13)$$

Since the functions $\underline{t}(\underline{x})$ are arbitrary we get finally the Bianchi-type identity satisfied by the Euler derivative

$$\Psi\,\frac{\delta L}{\delta Y} - \nabla\cdot(\Pi\,\frac{\delta L}{\delta Y}) \equiv 0 \qquad (14)$$

Using the relation

$$\nabla\cdot\frac{\partial L}{\partial\nabla Y} = \frac{\partial^2 L}{(\partial Y)\partial\nabla Y}\,\nabla Y + \frac{\partial^2 L}{(\partial\nabla Y)\partial\nabla Y}\,\nabla\nabla Y \qquad (15)$$

and putting

$$\frac{\partial^2 L}{(\partial \nabla Y)\partial \nabla Y} = M \tag{16}$$

in the above identity we find that the term containing the highest derivative is of the form

$$M_{AB}^{jk}\ \Pi_m^{iA}\ \partial_i \partial_j \partial_k Y^B = 0 \tag{17}$$

and this term must vanish by itself. Since this holds for any $\nabla\nabla\nabla Y$ we obtain

$$M_{AB}^{jk}\ \Pi_m^{iA} = 0 \tag{18}$$

In this equation we have to take care to include only that part which corresponds to the symmetry of $\nabla\nabla\nabla Y$. Thus we find that Π represents the null vector of M. In other words the tensor operator M is not of maximal rank. In such cases the corresponding Lagrangian is called singular. This reflects the fact that field Y contains unphysical parts which have to be eliminated to obtain a unique solution.

The above derivation can be generalized in several ways. First of all it can be extended easily to the field theories involving higher order derivatives. Moreover, these derivatives of Y need not be of the form ∇Y, $\nabla\nabla Y$ etc. Suppose such a derivative occurs in the Lagrangian in the form of a differential operator $D(\nabla)$ we find that the Euler derivative is given by

$$\frac{\delta L}{\delta Y} = \frac{\partial L}{\partial Y} + D^+(\nabla)\ \frac{\partial L}{\partial D^+(\nabla) Y} \tag{19}$$

where $D^+(\nabla)$ is the adjoint of $D(\nabla)$. Corresponding changes have to be made in other equations too.

A more general formulation can be given to include the nonlocal theories. We can think of action as a functional of a functional i.e. the Lagrangian itself is a functional. Such cases occur frequently in physics. However, it is easier to work directly with the action functional itself. It can be expanded in a functional Taylor series

$$S[Y + \delta Y] = S[Y] + \sum_{n=1}^{\infty} \frac{1}{n!} \int \dots \int \frac{\delta^n S}{\delta Y(\underline{x}_1) \dots \delta Y(\underline{x}_n)} \delta Y(\underline{x}_1) \dots \delta Y(\underline{x}_n) \cdot$$

$$\cdot \, dv_1 \dots dv_n \tag{20}$$

where $\frac{\delta S}{\delta Y}$ is the Fréchet derivative of S. For linear and bilinear or quadratic terms it is convenient to use the Dirac notation

$$S[Y + \delta Y] = S[Y] + \langle \frac{\delta S}{\delta Y} | \delta Y \rangle + \langle \frac{\delta S}{\delta Y} | B | \frac{\delta S}{\delta Y} \rangle + \dots . \tag{21}$$

where $B = \frac{\delta^2 S}{\delta Y \delta Y}$ is now interpreted as an integral operator.

In this notation equation (18) would be written as $M|\Pi\rangle = 0$ and M is interpreted as an integral operator acting on the vector Π.

CONSEQUENCES FOR THE GAUGE THEORY OF TRANSLATIONAL DEFECTS IN CRYSTALS

We briefly recapitulate the way we have formulated a gauge theory of crystal defects in our previous work. This formulation is based on the fact that a crystal is not a structureless continuum but rather a medium with a lattice structure and therefore, a proper gauge theory should take into account this structure. For this reason we introduce a set of vectors $\underline{e}_a$ defining a preferred coordinate system in the internal space. The crystal is considered as a collection of neighborhoods called internal spaces. Alternatively in the bundle picture we consider internal spaces as fibres or tangent spaces of a base manifold.

Initially the Lagrangian is given by

$$L(Y,\ \underline{e}_a\cdot\nabla Y) = L(Y,\ e_a^i\partial_i Y) \tag{22}$$

where index a refers to the internal space, index i to the base space and y behaves as a tensor with respect to the internal space transformations. The vectors $\underline{e}^a$ reciprocal to $\underline{e}_a$ can be defined by the relation $e_i^a = \partial_i x^a$ where x^a are crystallographic coordinate planes. The Lagrangian is supposed to be invariant under the global translations and rotations. Acording to the minimal prescription invariance against local translation is achieved by the substitution

$$\underline{e}_a\cdot\nabla Y \rightarrow \underline{g}_a\cdot\nabla Y = (\underline{e}_a + \underline{h}_a)\cdot\nabla Y \tag{23}$$

and multiplying the Lagrangian by $\det(\underline{g}_a^i)$ to make it a proper density. To this Lagrangian one must also add a piece depending on a gauge invariant field constructed from the gauge potential $\underline{h}_a$. The gauge invariant field or field strength in this case is called torsion. It is defined by the relation

$$\tau_{ij}^a = \partial_i e_j^a - \partial_j e_i^a \tag{24}$$

and is directly related to the dislocation density. The transformation behaviour of $\underline{h}_a$ is given by

$$\delta h_a^i = h_a^i\ \partial_j t^i + \delta_a^i\ \partial_j t^i \tag{25}$$

The minimal substituion used here is quite different from the usual one in other gauge theories of Yang-Mill's type. Here the gauge potential operates on the partial differential operator.

In our previous work we identified the Y variables with X. That means $\partial_i X^a$ are not the deformation tensors as in [2] but are just the reciprocal vectors e_i^a. In this case the field and the frame with respect to which the field is referred can be merged into a single entity (the triad $\underline{e}_a$). Thus initial Lagrangian does not contain elastic energy but cohesive energy. This corresponds well with the physical picture; no defects and no external sources means no elasticity. An alternative interpretation of $\partial_i X^a$ could be as a plastic deformation. But this does not change our physical picture. The only difference is that in the final Lagrangian the total deformation consists of two factors the elastic deformation and the plastic deformation.

We now see that the total Lagrangian is a function of $\underline{e}_a$ and its derivative which is exactly the kind of Lagrangian we considered in the previous section. However, to simplify further discussion we consider only the linear and static situation. On the other hand we shall use a more general functional or nonlocal formulation. The local theory is then just a special case. The action functional now is simply minus the energy functional. In the linearized version the role of gauge potential is played by distorsion and that of a gauge field strength by the dislocation density. On the other hand from a more macroscopic point of view i.e. when we can not perceive the rotation of triads due to the translation energy functional takes the form

$$E = - E_i[\varepsilon^t] + E_f[\mathrm{inc}\,\varepsilon] \tag{26}$$

where $\varepsilon^t = \varepsilon^p + \varepsilon$. Due to the linear approximation the product of plastic deformation and elastic deformation is replaced by a sum of plastic and elastic strains. The elastic part ε is the gauge potential which transforms under local translation as follows

$$\varepsilon' = \varepsilon + \mathrm{def}\,\delta\underline{x} \tag{27}$$

In equations (26) and (27) there occur the operators def and inc. These are the well known deformation and incompatibility operators first introduced by Kröner[10]. They are defined by

$$(\mathrm{def})_{ijk} = \delta_{ik}\partial_j + \delta_{jk}\partial_i \tag{28}$$

$$(\mathrm{inc})_{ijkl} = \varepsilon_{imk}\varepsilon_{jne}\partial_m\partial_n \tag{29}$$

They satisfy the operator identity inc def = 0 which clearly demonstrates the gauge invariance of inc ε. On the other hand ε^p transforms as

$$\varepsilon'^p = \varepsilon^p - \mathrm{def}\,\delta\underline{x} \tag{30}$$

which follows from the transformation behaviour of reciprocal triads. Thus both $E_i[\varepsilon^p + \varepsilon]$ and $E_f[\mathrm{inc}\ \varepsilon]$ are gauge invariant. The first term E_i actually represents the interaction of the defect with the medium whereas the second term E_f represents the elastic energy of the medium.

Using the notation of equation (20) we can write $\delta E = 0$ in the form

$$- \langle \sigma | \delta\varepsilon \rangle + \langle \chi | \text{inc } \delta\varepsilon \rangle = 0 \tag{31}$$

where

$$\sigma = \frac{\delta E_i}{\delta\varepsilon} \quad \text{and} \quad \chi = \frac{\delta E_f}{\delta \text{inc}\,\varepsilon} \tag{32}$$

Since inc is a selfadjoint operator we obtain from (31) the field equation

$$\sigma = \text{inc } \chi \tag{33}$$

from which it follows that $\text{div}\sigma = 0$. This result is compatible with (33). Equations (26) and (33) show us that the gauge potential term provides the source for the field χ.

Let us now consider the pure elastic term E_f. It is not only gauge invariant but also invariant with respect to global rotations. Therefore, it can be written as a bilinear functional

$$E[\eta] = \langle \eta | J | \eta \rangle \tag{34}$$

where

$$\eta = \text{inc } \varepsilon \quad \text{and} \quad J = \frac{\delta^2 E_f}{\delta\eta\delta\eta} \tag{35}$$

We then have the following relation

$$|\sigma\rangle = \text{inc} | J \eta\rangle \quad \text{or} \quad |\chi\rangle = J|\eta\rangle \tag{36}$$

A more conventional equation can be derived in the following way. From the rotational invariance argument is follows that the first term E_i in (22) can also be written as a bilinear functional

$$E_i[\varepsilon] = \langle \varepsilon | C | \varepsilon \rangle \tag{37}$$

where

$$C = \frac{\delta^2 E_i}{\delta\varepsilon\delta\varepsilon} \tag{38}$$

From the Noether's identity it follows that whereas E_i is a regular functional E_f is not because J is not of maximal rank. Since E_i is a regular functional we can apply the Legendre transformation and write the first term as σ S σ. But we can not apply the Legendre transform to the second term E_f because such a mapping would be singular. Using the relation (33) we then obtain the equation

$$\text{inc S inc } |\chi\rangle = |\eta\rangle \tag{39}$$

This equation for the local case is well known. We now recognize that J would be the Green operator of the integrodifferential operator inc S inc if it could be calculated. However, this is not the case here. We can not solve the integrodifferential equation (39) in a unique manner. The nonuniqueness of the Green operator J was pointed out previously by Kunin[11]. This nonuniqueness is due to the extra freedom of local translations. To achieve a unique solution we have to eliminate this extra freedom. This can be done either by the method of Lagrange multipliers or by inspection. In the case of local theory this latter method was used by Kröner[10] and independently by Marguerre[13]. Their condition is div χ = 0 which is very similar to the constraint conditions in other gauge theories.

Two important conclusion can be drawn from the above discussion. First, the elastic energy functional of defects E is always singular. Second, this functional can have a nonlocal form even in the local case.

Here it may be noted that a Legendre transformation is a mapping from the tangent bundle to the cotangent bundle. We had identified the internal space with a tangent space. We could have also identified it with the cotangent space. This would have led to the so called stress space formulation in which χ play the role of gauge potential. This is just an equivalent formulation.

Acknowledgment

The author would like to thank Prof. E. Kröner for reading the manuscript and critical comments.

References

1. Turski, L., Bull. Acad. Polon. 7 Sér. Sci. Tech 14, 289 (1966)
2. Kadic, A. and Edelen, D.G.B., in Continuum Models of Discrete Systems 4, Brulin, O. and Hsieh, R.K.T. eds., North-Holland, pp. 67-74, Amsterdam (1981)
3. Gairola, B.K.D., in Continuum Models of Discrete Systems 4, Brulin, O. and Hsieh, R.K.T. eds., North-Holland, pp. 55-65, Amsterdam (1981)
4. Rosenfeld, L., Ann. d. Physik 5, 113 (1930)
5. Rosenfeld, L., Ann. Inst. Henri Poincaré 2, 25 (1932)
6. Hilbert, D., Nachr. Ges. Wiss. Göttingen, 395 (1915)
7. Funk, P., "Variationsrechnung und ihre Anwendung in Physik und Technik", Springer, Berlin 1970
8. Datta, B.K., Nuovo Cimento B 6, 1 (1971)
9. Bergmann, P.G., Phys. Rev. 75, 680 (1949)
10. Kröner, E., Z. Physik 139, 175 (1954)
11. Kunin, I.A., "Elastic Media with Microstructure, vol. 2", Springer Berlin (1983)
12. Marguerre, K., Z. Ang. Math. Mech. (ZAMM) 35, 242 (1955)

ON GAUGE THEORY IN DEFECT MECHANICS*

E. Kröner

Institut für Theoretische und Angewandte Physik, Universität Stuttgart
und Max-Planck-Institut für Metallforschung, Stuttgart, F.R. Germany

Abstract

The field theory of the continuized Bravais crystal with crystallographic defects is treated by means of a gauge approach. Starting with the general linear group as gauge group yields the affine connexion Γ of a flat space. The torsion tensor of this space represents the dislocation density. An independent metric tensor g is introduced which describes length measurement in crystals with intrinsic point defects. In this way Γ obtains a part Q (a tensor) which makes it nonmetric with respect to g. This part specifies the numbers and types of the point defects presented in the crystal. This is discussed from the standpoint of internal and external observer. Some open problems for which the help of advanced mathematics is highly appreciated, are indicated.

1. Introduction

In view of the great successes of the gauge theories within the fundamental physical field theories also applications of gauge concepts to theories of defects in ordered structures have gained some popularity (see e.g. refs. [1-3] and the papers of this session). A fundamental problem is always that of the gauge group to be selected. Here it is common to distinguish between groups describing internal symmetries and those describing external symmetries or, more specifically, space-time symmetries. The gauge approach was particularly successful when dealing with internal symmetries. In this case use is made of the so-called minimal replacement, according to which ordinary differentiation, here symbolized by ∂, is replaced by covariant differentiation: $\partial \to \nabla = \partial + \Gamma$,

---

*Dedicated to Professor F.R.N. Nabarro on his 70th birthday

where Γ exhibits the transformation properties of a connexion in differential geometry.

The fact that we use ∂ rather than some transformed ∂ in ∇ means that we take the same coordinate system before and after the minimal replacement, which implies that the considered physical system is always in the same physical space. This is the case when only internal symmetries are considered. If we include also space-time symmetries, then, as a result of the gauge procedure, the system might go from a euclidean space, for instance, to a curved space, so that, after the gauging, the euclidean coordinates must appear replaced by curved-space coordinates.

In this note we are interested in defects in crystals, in particular in the so-called Bravais crystals, because these represent the simplest group of crystals. Only the internal, or intrinsic, crystallographic defects will be considered. Intrinsic (crystallographic) defects imply certain configurations which involve only regular atoms of the considered Bravais crystal. All defects which are formed by atoms external to the crystal are classified as extrinsic. There are three basic types of intrinsic defects, namely point defects (vacancies and self-interstitials), line defects (dislocations) and the more complex interface defects. Only the first two types of defects will be considered here. The restriction to intrinsic defects and the exclusion of external influences acting on the crystal implies that no curvature occurs so that we can apply the gauge theory with internal symmetries.

We shall be content to discuss the geometrical theory of defects because a good physical field theory of defects in crystals, also called the many-defect theory, or defect dynamics, does not exist. Something could be said about the statics of defects, which is well described in the frame of a stress space, where, as in the strain space considered here, methods of differential geometry are best suited (see e.g. Kröner[4,5]). The stress space, however, will not be touched in this note, nor will be the dynamics.

There have been other interesting approaches to the theory of defects in Bravais crystals. Rogula[6] based his investigations on homotopy group theory and introduced the concept of a Bravais lattice space. The relations between this type of theory and the gauge approach need further exploration and will not be discussed here.

2. The concept of continuized crystal

For many purposes the lattice constants can be considered small compared to all other lengths of interest. In such cases the notion of continuized crystal proves to be adequate because it permits to work with continuous functions. As we shall see, the problem then falls within the frame of differential geometry.

The continuized crystal results from a limiting process in which all primitive lattice constants go to zero such that their ratios as well as the local density and defect content remain unchanged. This implies that, in the limit, the mass points of the lattice have vanishing mass. It is clear how the point defects are reduced. The conservation of local dislocation content requires that the number of disloactions intersecting a fixed area element increases such that the sum of the Burgers vectors remains the same. It can easily be seen that, under this condition, the distance between neighbouring dislocations goes to zero when measured from outside, but goes to infinity when measured stepwise in units of the lattice constants, which is the natural way to measure distances in crystals. This means that dislocation is essentially a discrete concept.

The preceding discussion shows that the continuized crystal is not a continuum in the sense of continuum theories. Although its lattice points are infinitely close, they remain countable and therefore form a crystal. To describe the properties of a crystal with continuous functions, one has to take as a basis the continuized crystal and not a continuum.

A physical argument may support the last statement. According to the theory of Peierls[7] and Nabarro[8] a straight dislocation line which moves along its glide plane, experiences a lattice periodic potential. Assuming this as sinusoidal, these authors were able to calculate the stress necessary to move the dislocation along the glide plane, i.e. to lift it over the "Peierls walls". This stress depends on the ratio r of the wave length of the potential over the mutual distance of the two atomic planes adjacent to the glide plane. Since r is kept constant during the limiting process, it follows that the stress which moves a dislocation is the same for the real crystal and the continuized crystal which therefore is a realistic model for the real crystal.

3. The concept of gauging

It is common in gauge theories to start with a homogeneous state and to find those global transformations which do not intervene in the physics of this state. Often this is done by requiring that the Lagrangean remains constant under these transformations. Since not all physical situations, however, are well described by a Lagrangean, we shall often refer to "the physics" or "the physical equations" of the states under global transformation. In our case, the initially considered state is, of course, the ideal (continuized, Bravais) crystal.

In this note we exclude external actions on the crystal, i.e. we assume the standpoint of an internal observer (lattice flea) who can perceive the main crystallographic directions (three at each point) and can jump from one atom to the next along these directions. By counting his steps he can identify atoms respective to some reference atom. We suppose that the internal observer has no organ to measure the length of a single step. That means, the step length has a meaning only in some external geometry, but not in the geometry of the internal observer, i.e. the intrinsic crystal geometry.

Obviously, the internal observer, as defined above, will see no difference between the various Bravais crystals - he is a geometric hero but a physical washout. Expressed differently, we ourselves as external observers say that from the standpoint of the internal observer the ideal Bravais crystal is physically unchanged under any global affine transformation. It is clear that from this starting point we cannot derive the physics of the Bravais crystal, but we can derive its internal geometry which includes, as we shall see, both the point defects and line defects.

For simplicity consider the primitive cubic crystal which, at each point x, has three mutually orthogonal base vectors of (externally and internally) equal length. If these vectors are made the base vectors of a coordinate system, say x^k, then the x^k are called crystallographic coordinates. Let $\{v^k_\alpha(x)\}$ be the set of base vectors at all points. Here the superscripts k (= 1,2,3) indicate the cartesian vector components, and the three base vectors at a point are distinguished by α (= 1,2,3). In the ideal state all base vectors of a given α are parallel and of equal length, so that

$$\partial_m v^k_\alpha = 0 \quad \text{(ideal crystal, } x^k \text{ cartesian)} \quad . \tag{1}$$

If now the crystal is inhomogeneously deformed and the vectors v are dragged along so that they now are the base vectors of the deformed crystal, then

$$\partial_m v^k_\alpha \neq 0 \quad \text{(deformed crystal, } x^k \text{ cartesian)} \; . \tag{2}$$

If, however, we introduce the crystallographic coordinates of the deformed crystal, say $x^{k'}$, which are the coordinates dragged along with the deformation, then, with $\partial_{m'} \equiv \partial/\partial x^{m'}$, we obtain

$$\partial_{m'} v^{k'}_\alpha = 0 \quad \text{(deformed crystal, } x^{k'} \text{ crystallographic)}, \tag{3}$$

obviously valid for the base vectors of any deformed crystal.

Note that the crystallographic coordinates are distinguished in that they describe the <u>configuration</u> of the crystal. Therefore, the calculation of these coordinates is an essential part of the physical problem.

Suppose now that the states of the crystal are geometrically specified by the vectors v and their derivatives ∂v. v and ∂v are considered independent in the sense of general mechanics (e.g. Lagrange formalism).

Imagine the following operation: Subject each lattice triad to a general linear transformation of the form

$$v_\alpha \rightarrow {}'v_\alpha = A_\alpha^{\ \beta} v_\beta \tag{4}$$

where the (3x3) matrix A is constant throughout the crystal. We imagine that the vectors v describe a new ideal crystal which still belongs to the group of Bravais crystals and, therefore, is considered to be in essentially the same state by the internal observer.

The sketched operation can also be described as a coordinate transformation if both initial and final coordinates are chosen crystallographic. Then

$$dx^k \rightarrow dx^{k'} = A_k^{k'} dx^k \tag{5}$$

with the inverse transformation

$$dx^{k'} \rightarrow dx^{k} = A^{k}_{k'}dx^{k'} \quad , \tag{6}$$

where

$$A^{k'}_{k}A^{l}_{k'} = \delta^{l}_{k} \ , \quad A^{k'}_{k}A^{k}_{l'} = \delta^{k'}_{l'} \tag{7}$$

As long as the matrix A is constant, the transformation is called global. The totality of these transformations forms the (global) general linear group, say G, which is now used as the gauge group.

Recall that the choice of this group has been made according to the needs of the internal observer. Following the general gauge concept we now require that the physical equations be invariant under local gauge transformations, i.e. when we apply G locally, which means that G becomes G(x).

The gauge concept signifies that the transformations (4) and (6) with both position-independent and -dependent A correspond to real physical processes. Intuitively it is clear, that such processes can be achieved by dislocations. At the moment, this is not important, however. It is of course erroneous to use the Lagrangean, or some physical equations, of the global problem also for the local problem, in which a much larger manifold of states is considered. This, however, can be reconciled by introducing so-called compensating fields which restore the original correctness.

To understand this better, note that the original Lagrangean, (or physical equations) usually contain the field gradients, in our case the ∂v's. The transformation (4) implies

$$\partial_m v_\alpha \rightarrow \partial_m{}' v_\alpha = \partial_m (A^{\beta}_{\alpha} v_\beta) = A^{\beta}_{\alpha}\partial_m v_\beta + (\partial_m A^{\beta}_{\alpha}) v_\beta \quad . \tag{8}$$

In the global situation the A-matrix is constant so that the last term in eq.(8) vanishes. The transformation of the Lagrangean then leads to

$$L(v_\alpha, \partial_m v_\alpha) \rightarrow L(A^{\beta}_{\alpha} v_\beta, A^{\beta}_{\alpha}\partial_m v_\beta) = L('v_\alpha, \partial_m{}'v_\alpha) = L(v_\alpha, \partial_m v_\alpha), \tag{9}$$

the latter equation because it was assumed from the beginning that the "global" Lagrangean is invariant under the global transformations of the general linear group G. The sequence (9) is no longer true in the local situation where A depends on x. Introducing a new quantity Γ it is possible, however, to write

$$A^{\beta}_{\alpha}\partial_m v_{\beta} + (\partial_m A_{\alpha}^{\ \beta})v_{\beta} = \nabla_m{}' v_{\alpha} \equiv \partial_m{}' v_{\alpha} + \Gamma_{m\alpha}{}^{\beta\prime} v_{\beta} \tag{10}$$

which implies that

$$\nabla_m{}' v_{\alpha} = \nabla_m(A_{\alpha}^{\ \beta} v_{\beta}) = A^{\beta}_{\alpha}\nabla_m v_{\beta} \; . \tag{11}$$

A routine calculation shows that $\Gamma_{ml}{}^{k} \equiv \Gamma_{m\alpha}{}^{\beta}\delta^{\alpha}_{l}\delta^{k}_{\beta}$ transforms like an affine connexion and has here the particular form

$$\Gamma_{ml}{}^{k} = A^{k}_{k'}\partial_m A_{l}^{k'} \tag{12}$$

which is the most general form of an affine connexion in a flat space. Because of eq.(11) the sequence (9) is also true if ∂ there is replaced by ∇ (minimal replacement). It follows from eq.(10) that ∇ degenerates to ∂ in the global situation.

Let us now decompose the connexion in a symmetric and an antisymmetric part according to

$$\Gamma_{ml}{}^{k} = A^{k}_{k'}(\partial_m A_{l}^{k'} + \partial_l A_{m}^{k'})/2 + A^{k}_{k'}(\partial_m A_{l}^{k'} - \partial_l A_{m}^{k'})/2 \; . \tag{13}$$

The part antisymmetric in m, l contains six functional degrees of freedom and is known as Cartan's torsion, a tensor of 3rd rank. Since the matrix A has nine functional degrees of freedom, we conclude that the part symmetric in m, l of eq.(13) contains three functional degrees of freedom.

As is well-known, the torsion tensor has been identified as dislocation density by Kondo[9] and independently by Bilby, Bullough and Smith[10]. We return later to the identification of the symmetric part of the connexion.

4. The concept of metrics (and nonmetrics)

The foregoing investigations have led to the conclusion that the geometrical states (the configurations) of a continuized crystal which is defected by dislocations are well described by the affine connexion of a flat space (which incidentally is a space with teleparallelism). However, the affinely connected spaces of which we speak here, are not completely defined only if in addition to the affine connexion something is determined about the metric. It is clear that the metric, if there is any, must have something to do with the concept of step counting in our crystal. We therefore expect, that a metric law of the form

$$ds^2 = g_{kl}\,dx^k dx^l \tag{14}$$

will be significant in our theory.

It is important to note that the concept of metrics, or metricity, is independent of the concept of affinity. This does not exclude that in certain physical situations metricity and affinity might be related. For instance, if we (can) measure lengths by eq.(14) and if, say in cartesian coordinates x^k,

$$g_{kl} = A_k^{k'} A_l^{l'} \delta_{k'l'} \tag{15}$$

then we say that the connexion (12) is metric "with respect to the metric tensor (15)". The part in quotation marks is often omitted in such a statement. It is nevertheless fundamental and should be kept in mind.

If lengths are measured according to eq.(14) but without validity of (15), then the connexion (12) is called nonmetric (with respect to g). In principle, g can be any symmetric tensor, so that the geometry of the crystal is now described by 9 + 6 = 15 spatial functions. This leads us to a further identification problem (see below).

The introduction of a metric tensor endows the considered space with an internal product. It is then possible to define reciprocal base vectors and an inverse metric tensor g^{lm} by

$$g^{lm} g_{kl} = \delta_k^m \; . \tag{16}$$

g_{kl} and g^{lm} are used for raising and lowering indices and forming invariants.

To obtain more clarity consider the well-known identity (Schouten[11],p. 132), valid for any choice of connexion Γ and metric g:

$$\Gamma_{ml}{}^{k} = \{{}^{k}_{ml}\}_g - \{S_{ml}{}^{k}\}_g + \{Q_{ml}{}^{k}\}_g \quad , \tag{17}$$

where on the right hand side we have the Christoffel symbol (no tensor) with respect to g

$$\{{}^{k}_{ml}\}_g \equiv g^{ks}(\partial_m g_{sl} - \partial_s g_{lm} + \partial_l g_{ms})/2 \quad , \tag{17'}$$

the so-called contortion tensor

$$\{S_{ml}{}^{k}\}_g \equiv g^{ks}(S_{msl} - S_{slm} + S_{lms}) \quad , \tag{17''}$$

and the tensor of nonmetricity

$$\{Q_{ml}{}^{k}\}_g \equiv g^{ks}(Q_{msl} - Q_{slm} + Q_{lms})/2 \quad , \tag{17'''}$$

where

$$Q_{mlk} \equiv - \nabla_m g_{lk} \quad .$$

If g obeys eq.(15), then Q = 0: the connexion is metric. In absence of dislocations we have from (13) and (17)

$$\Gamma_{ml}{}^{k} = A^{k}_{k'}(\partial_m A_l{}^{k'} + \partial_l A_m{}^{k'})/2 = \{{}^{k}_{ml}\}_g - \{Q_{ml}{}^{k}\}_g \quad . \tag{18}$$

If this is to be a connexion in a flat space, then Q cannot be arbitrary. Because the rigorous nonlinear analysis is rather involved, though possible, let us see what can be obtained in a linear approximation. We try the ansatz

$$Q_{mls} = - \partial_m q_{ls} \tag{19}$$

with some tensor q and obtain

$$\Gamma_{ml}{}^{k} = g^{ks}(\partial_m h_{sl} - \partial_s h_{lm} + \partial_l h_{ms})/2 \ , \qquad h_{kl} = g_{kl} - q_{kl} \quad . \tag{20}$$

In the course of the linearization we set $g^{ks} = \delta^{ks} + \varepsilon^{ks}$, h^{ks} $(\equiv g^{kl}g^{rs}h_{lr}) = \delta^{ks} + \eta^{ks}$ and take ε and η very small. Then we can replace g^{ks} by h^{ks} in (20) and obtain in this approximation

$$\Gamma_{ml}{}^{k} = \{{}^{k}_{ml}\}_h \quad , \tag{21}$$

i.e. the Christoffel symbol taken with h instead of g. On the other hand Γ has the form (12). It follows that

$$h_{kl} = A_k^{k'} A_l^{l'} \delta_{k'l'} \tag{22}$$

which means that Γ is metric with respect to h, and not with respect to g, the tensor used to raise and lower indices.

5. The nonmetric identification problem

We now have to identify the tensors g, h and q with quantities occuring in the physics of crystals. Among these three tensors, g is distinguished in that it is used to raise and lower indices and to form inner products. Therefore it must have to do directly with the crystal lattice. At this point, we recall that the intrinsic point defects vacancy and self-interstitial are described geometrically as so-called displacement dipoles lim Su where u is the relative displacement of the two faces of an area element S and the limiting process implies $S \to 0$, $u \to \infty$ such that the product of the two remains finite (Kröner[12,13,14]). A density of such dipoles has the dimension and meaning of a strain. This is the strain which the isolated infinitesimal crystal element suffers when atoms in a certain density are either removed from regular lattice sites (case of vacancies) or added on irregular sites (case of interstitials). This strain, sometimes called extra strain, should not be confused with the strain which would occur if the same operation would take place in a nonisolated crystal element. In fact, the strain in such an element would contain a contribution due to the constraint from its neighborhood.

Obviously, the presence of intrinsic point defects has an important effect on the metricity, i.e. the step counting in crystals. Each time the internal observer arrives at such a defect, he does not know how to continue. Should he count his steps as if the defect did not exist? Let us investigate the consequences of such a prescription. The step counting defines a metric and can therefore be described by the law (14). We should be aware, however, that the crystal described by the metric g is not the real crystal, which contains point defects, but a fictive nondefected crystal. A different way of describing this situation is by saying that the crystal with point defects is not a crystal in the strict sense. It does contain, however, some characteristics of a crystal and may, therefore, be called "pseudocrystal". The pertaining coordinates are then pseudocrystallographic.

Whereas the geometry of the fictive crystal is completely specified by the metric tensor g and the Christoffel symbol derived herefrom, this is not the case with the real crystal (the pseudocrystal). Here the point defects, or the nonmetricity caused by them, must be introduced. This is done by the tensor q as described above. For our whole concept it is important that the internal observer who counts along the pseudocrystallographic coordinate lines, can detect the point defects. Since not all three crystallographic directions are well defined at point defects, the internal observer notices these defects and can record their numbers and types. Thus the tensor q is accessible to the internal observer.

6. Internal vs. external observer

In the preceding sections we have described the (Bravais) crystal, defected by dislocations and intrinsic point defects, in the language of affine nonmetric differential geometry of a flat space. To this end we needed the (pseudo-) metric tensor g and the (flat) affine connexion Γ, eq.(12). g and Γ together are composed of 15 independent functions. Three of these form the symmetric part of Γ and transform like a connexion, i.e. not like a tensor. The other 12 functions are components of tensor fields. They represent the tensor field g (six functions) and the tensor field S (torsion, also six functions) which forms the antisymmetric part of the connexion. Another representation of the 12 tensorial functions is in terms of S and q, where

$$q_{kl} = g_{kl} - h_{kl} \, , \; h_{kl} \equiv A_k^{k'} A_l^{l'} \delta_{k'l'} \quad . \tag{23}$$

Both S and q correspond to crystallographic anomalies (≡ defects) and can be detected and measured by the internal observer on his wandering through the crystal lattice.

The internal observer knows nothing of the external world into which his crystalline world is imbedded. Therefore he has no organ for the cartesian or any other coordinate system by which positions in the external world are specified. On the other hand, our basic geometric quantities g and Γ are functions given in an external, e.g. cartesian, coordinate system. The question then arises how to connect the observations in the two worlds.

The results of the internal observer are given in crystallographic coordinates, or, when point defects are present, in pseudocrystallographic coordinates. In these coordinates, however, $g_{kl} = \delta_{kl}$, from the definition of the internal observer. The crystallographic, or pseudocrystallographic, coordinates have also a meaning in the external world, where they can be reached, e.g. by transformation from cartesian coordinates. This transformation which so-to-speak positions the internal into the external world, is "worth" three functions of positions. Taking these three functions together with the 12 tensor component functions leads to the 15 functions which constitute the quantities g and Γ.

Conclusion

The differential-geometric theory of intrinsic line defects (dislocations) in crystals has been given by Kondo[9] and by Bilby, Bullough and Smith[10] in the early fifties. The corresponding theory of intrinsic point defects (vacancies and self-interstitials) was added only recently by Kröner[13,14], with contributions by Günther and Zorawski[15] and by Gairola[16]. In this note we have described an alternative theory, namely one based on gauge concepts, which reproduces the already obtained results. We have restricted ourselves to the gauge theory related to internal symmetries rather than to external (= space-time) symmetries. The derived geometry therefore always refers to a flat space, i.e. to a space with zero curvature (tensor). Beside this fundamental concept we have used three further basic concepts. The first of these is the concept of the continuized crystal which describes a manifold of points which are countably dense. The second basic concept is that of the (precisely defined) external and internal observer (or world). The perceptive faculty of the external observer is larger in that he knows how to position the crystal world into the real physical world. As a consequence the external observer has more functions of position (15), than the internal observer (12), at his disposal to describe the crystal with defects.

Using the general linear group as gauge group and applying the principle of minimal replacement we find that the connexion always occurring in this approach is the affine connexion of a flat space. The antisymmetric part of this connexion is identified as density of dislocations. To obtain a description of intrinsic point defects, the third basic concept, namely the concept of metrics has to be introduced in addition. This concept is not contained in the conventional gauge approach. Its introduction appears to be indispensable, however, if a nonmetric part of the connexion, which then represents the density of point defects, is to be extracted from the affine connexion.

Intrinsic interface defects were excluded from this discussion. They correspond to a nonconnective situation because paths of parallel displacement cannot be carried through such defects. It is not clear how the gauge concept can be applied to this case. Also its application to external (= space-time) symmetries seems to offer considerable difficulties. All these extensions concern basic mechanics and physics which however require advanced mathematics. The help of mathematicians will therefore be highly appreciated.

Acknowledgement

Thorough discussions with Drs. F. Hehl, B.K.D. Gairola, I.A. Kunin, B. Orlowska, D. Rogula, H.-R. Trebin and M. Zorawski are gratefully acknowledged.

References

1. Kadic, A. and Edelen, D.G.B., A Gauge Theory of Dislocations and Disclinations, Lecture Notes in Physics 174, Springer, Heidelberg, 1983 (290 pp.).

2. Discussion Meeting on Gauge Field Theories of Defects in Solids, Stuttgart 1982, E. Kröner (ed.), avail. at Max-Planck-Institut für Metallforschung, Postfach 800665, D 7000 Stuttgart, F.R. Germany (53 pp.).
3. Kleinert, H., Double Gauge Theory of Stresses and Defects, Phys. Lett. 97 A, 51-54 (1983).
4. Kröner, E., The continuized crystal - a bridge between micro- and macromechanics?, to appear in Z. Angew. Math. Mech. 1986
5. Kröner, E., Continuum Theory of Defects, in: Les Houches, Session 35 (1980) - Physics of Defects, R. Balian et al. (eds.), North-Holland, Amsterdam 1981, 217-315.
6. Rogula, D., Large Deformations of Crystals, Homotopy, and Defects, in: Trends in Applications of Pure Mathematics to Mechanics, G. Fichera (ed.), Pitman, London 1976, pp. 311-331.
7. Peierls, R.E., The Size of a Dislocation, Proc. Phys. Soc. 52, 34-37 (1940).
8. Nabarro, F.R.N., Dislocations in a Simple Cubic Lattice, Proc. Phys. Soc. 59, 256-272 (1947).
9. Kondo, K., On the Geometrical and Physical Foundations of the Theory of Yielding, in: Proc. 2nd Japan Nat. Congr. Appl. Mech., Tokyo, 1952, pp. 41-47.
10. Bilby, B.A., Bullough, R. and Smith, E., Continuous Distributions of Dislocations: a New Application of the Methods of Non-Riemannian Geometry, Proc. Roy. Soc. A 231, 263-273 (1955).
11. Schouten, J.A., Ricci Calculus, Springer, Heidelberg, 1954.
12. Kröner, E., Die Versetzung als Elementare Eigenspannungsquelle, Z. Naturforschung 11a, 969-985 (1956).
13. Kröner, E., Field Theory of Defects in Crystals: Present State, Merits and Open Questions, in: The Mechanics of Dislocations, E.C. Aifantis and J.P. Hirth (eds.), Amer. Soc. for Metals, Metals Park, Ohio, 1985.
14. Kröner, E., Field Theory of Defects in Bravais Crystals, in: Dislocations and Properties of Real Materials, Book No. 323, The Institute of Metals, London, 1985.
15. Günther, H. and Zorawski, M., On Geometry of Point Defects and Dislocations, Ann. d. Physik, VII 42, 41-46 (1985).
16. Gairola, B.K.D., Gauge Invariant Formulation of Continuum Theory of Defects, in: Continuum Models of Discrete Systems 4, O. Brulin and R.K.T. Hsieh (eds.), North-Holland, Amsterdam, 1981.

Session IV:

HYDRODYNAMIC STABILITY

RECENT PROGRESSES IN THE COUETTE-TAYLOR PROBLEM

G.IOOSS
U.A. 168, I.M.S.P.
Université de Nice
Parc Valrose, F-06034 NICE Cedex

I. INTRODUCTION

Experiments of the last years on fluid flows between concentric rotating cylinders [6][17][1][2], showed a large variety of new types of structures. For corotating cylinders, the first bifurcation leads to the well-known Taylor-vortex flow (T.V.), which is steady, axisymmetric and periodic in the axis direction. The secondary bifurcation may lead either to wavy Taylor vortices (W.T.V.), or the so called Twisted vortices (Tw), or wavy outflow (or inflow) boundaries (W.O.B. and W.I.B.). All these flows are periodic in time and have the form of rotating waves, i.e. an observer might choose a suitable rotating frame (constant angular velocity) and just see a steady flow. Further bifurcations mainly lead to quasi-periodic flows, for instance the Modulated Wavy Vortices (M.W.V.) or the wavelets.
The experiments seem to strongly indicate that these quasi-periodic flows are pure superpositions of two rotating waves, with different azimuthal wave numbers !

For counter-rotating cylinders (the angular velocity of the outer cylinder being not too small), the first observed bifurcation lead to spiral vortices which have the structure of a travelling wave in both axial and azimuthal directions. Further bifurcations may lead to very complicated flows, difficult to analyse, except one : the interpenetrating spirals (I.S.). It looks like a pure addition of two travelling waves of spiral types, with different azimuthal wave numbers.

Here, we wish
i) to give general ideas on mathematical basis for explaining the occurence of such flows,
ii) to show that other types of flows, not yet observed, might be obtained in choosing suitable values for the parameters,
iii) to prove the occurence of pure superpositions of travelling waves for some quasi-periodic flows such that (I.S.) and a special kind of (M.W.V.), and
iv) to give for the other quasi-periodic flows, which are observed, the degree of approximation of the truth in saying that we "see" a pure superposition of rotating waves.

II. SYMMETRIES AND HOW TO USE THEM

II.1 The ideal system

The viscous incompressible fluid lies between two concentric cylinders of radii $R_1 < R_2$, and angular velocities Ω_1, Ω_2. Let us replace the (singular) natural boundary conditions on the top and bottom of cylinders by a periodicity condition. It is well-known that, due to the independence of the laws of mechanics with respect to the reference frame, the Navier-Stokes equations are invariant under a suitable representation of the group of rigid motions.

Boundary conditions here restrict this invariance to the action of the subgroups of translations along the z-axis, reflexions $z \mapsto -z$, and rotations about the z-axis. In cylindrical coordinates, these actions on the velocity field $U=(u_r,u_\theta,u_z)$ are defined by :

$$(1)\quad \begin{cases} [\tau_s U](r,\theta,z) = (u_r(r,\theta,z+s),\ u_\theta(r,\theta,z+s),\ u_z(r,\theta,z+s)) \\ [S U](r,\theta,z) = (u_r(r,\theta,-z),\ u_\theta(r,\theta,-z),\ -u_z(r,\theta,-z)) \\ [R_\varphi U](r,\theta,z) = (u_r(r,\theta+\varphi,z),\ u_\theta(r,\theta+\varphi,z), u_z(r,\theta+\varphi,z)) \end{cases}$$

II.2 Linear stability analysis of the Couette flow

The Couette flow, observed for $|\Omega_1|$ (inner cylinder) small enough, possesses all the symmetries of the system : it only has a non-zero azimuthal component $v_\theta^\circ(r)$.

The perturbation U satisfies a functional equation of the form

$$(2)\qquad \frac{dU}{dt} = \mathcal{F}(\mu, U)$$

in a suitable Hilbert space (see for instance [12]) where μ is the set of parameters : $\mathcal{R}$ Reynolds number, Ω_2/Ω_1, R_2/R_1, h height of cylinders. Equation (2) can be understood as an ordinary differential equation, where we only allow $t \geqslant 0$. The steady solution U=0 corresponds to the basic Couette flow. Moreover $\mathcal{F}(\mu,\cdot)$ commutes with the operators τ_s, S, R_φ. These are the equivariance properties of (2).

The stability analysis of the 0 solution of (2), using the h-axial periodicity of the system, leads to a discrete set of eigenvalues σ, and eigenvectors of the form

$$\hat{U}(r)\ e^{i(\alpha z+m\theta)}$$

where $\alpha = 2\pi k/h$, and k and $m \in \mathbb{Z}$.

The fundamental fact is that, for $\mathcal{R}$ small enough, all the eigenvalues σ are such that Re $\sigma<0$, so the Couette flow is stable. There is a critical Reynold number $\mathcal{R}_c$ which depends on Ω_2/Ω_1, R_2/R_1, h (non dimensionalized) where at least one eigenvalue σ_o crosses the imaginary axis.

The numerical results lead to the following, for the optimal wave number α_c (which gives the smallest $\mathcal{R}_c$) : for R_2/R_1 fixed, and $\Omega_2/\Omega_1 > (\Omega_2/\Omega_1)_c(<0)$, then at $\mathcal{R}_c, \sigma_o=0$ and the corresponding eigenvector is axisymmetric (m=0). If $\Omega_2/\Omega_1 < (\Omega_2/\Omega_1)_c(<0)$, then at $\mathcal{R}_c$, $\sigma_o = \pm i\eta_o$ and the corresponding eigenvectors are no longer axisymmetric ($m \neq 0$).

Now, for not necessarily optimal wave numbers α , h is large enough to let us consider that α is close to α_c and the same results are still true qualitatively. Moreover, in playing with the other parameter Ω_2/Ω_1, we can obtain physical situations where modes with different m's become unstable simultaneously. These types of interactions are studied in III and in IV: In III, we consider m=0 (steady mode) and m≠0 (oscillatory mode), while in IV we consider m and m'≠0 (two oscillatory modes).

II.3 System on the center manifold

Another fundamental property here, due to the symmetry S, is that all the previous eigenvalues σ are double (at least). In fact, denoting an eigenvector

$$\zeta = \hat{U}(r)\, e^{i(\alpha z + m\theta)}$$

then

$$S\zeta = (S\hat{U}(r))\, e^{i(-\alpha z + m\theta)} \quad \text{(change } \hat{u}_z(r) \text{ in } -\hat{u}_z(r)\text{)}$$

is an eigenvector too, for the same eigenvalue.

Now assume that for a value of the set of parameters, say $\mu=0$, we have some eigenvalues $\sigma_o, \sigma_1, \ldots, \sigma_N$ on the imaginary axis, other eigenvalues being of negative real part. The functional space may be decomposed as follows [12] :

$$\text{(3)} \qquad U = X + Y$$

where X is a linear combination of eigenvectors (or generalized ones) belonging to the "critical" eigenvalues, and Y lies in a supplementary space. It is a classical result that all the dynamic of (2) is attracted locally on a so called center manifold [15], [11], of equation

$$\text{(4)} \qquad Y = \Phi(\mu, X) = O(|\mu|\,\|X\| + \|X\|^2) \;.$$

The trace of the system (2) on the center manifold (4) satisfies an ordinary differential equation :

$$\text{(5)} \qquad \frac{dX}{dt} = F(\mu, X) \;.$$

Moreover, $F(\mu,\cdot)$ keeps equivariance properties, depending on the actions of operators (1) on the eigenvectors [5], and has a C^k regularity (for any fixed k) in a neighborhood of 0.

Remark : It is remarkable that this technique was systematically used a long time ago, without knowing the mathematical justifications, by J.T.STUART and J.WATSON in the 1960's (see references in [18]).

II.4 Simple primary bifurcating flows

To illustrate simply (5), let us consider the primary bifurcations of the Couette flow. As we noticed in §II.2 there are two different cases.

II.4.1 One steady mode (see the classical method in [14]).

Here we assume that the critical eigenvalue is $\sigma_o=0$ and that the eigenvector satisfies m=0. So, let us denote by

$$\text{(6)} \qquad \zeta_o = \hat{U}(r)\, e^{i\alpha z} \quad , \quad S\zeta_o = \overline{\zeta}_o$$

the two eigenvectors belonging to the 0 eigenvalue. We have obviously :

(7) $R_\varphi \zeta_o = \zeta_o, \quad \tau_s\zeta_o = e^{i\alpha s}\zeta_o, \quad S\zeta_o = \overline{\zeta}_o .$

Then writing $X = x_o\zeta_o + \overline{x}_o\overline{\zeta}_o$, the system (5) becomes

(8) $$\frac{dx_o}{dt} = f_o(\mu, x_o, \overline{x}_o) \quad \text{in } \mathbb{C} ,$$

and (7), associated with the equivariance of (5), leads to :

(9) $$\begin{cases} f_o(\mu, e^{i\alpha s} x_o, e^{-i\alpha s}\overline{x}_o) = e^{i\alpha s} f_o(\mu, x_o, \overline{x}_o) \\ f_o(\mu, \overline{x}_o, x_o) = \overline{f_o(\mu, x_o, \overline{x}_o)} . \end{cases}$$

This gives immediately the structure of (8) :

(10) $$\frac{dx_o}{dt} = x_o(\sigma(\mu) + a|x_o|^2 + \ldots) = x_o\, g(\mu, |x_o|)$$

where g is <u>real</u> and <u>even</u> in $|x_o|$. Puting (10) in polar coordinates leads to the bifurcated axisymmetric <u>Taylor vortex flow</u> (T.V.) (supercritical if $a<0$, subcritical if $a>0$). Note that $|x_o| \sim (\frac{\sigma(\mu)}{-a})^{\frac{1}{2}}$ and that a change in the phase of x_o is the same as applying τ_s to U, i.e. a translation along the axis. The choice of a real x_o leads to U such that SU = U. This property and the invariance under $\tau_{2\pi/\alpha}$ leads to $u_z|_{z=k\pi/\alpha} = 0$, i.e. we have horizontal cells of height π/α.

II.4.2 One oscillatory mode (for the first study, see [9])

Let us assume now that the critical eigenvalues are $\pm i\eta_o$, and that the eigenvectors belonging to them are of the form :

(11) $$\zeta_1 = \hat{U}(r)\, e^{i(\alpha z + m\theta)}, \quad \zeta_2 = S\zeta_1 = (S\,\hat{U}(r))\, e^{i(-\alpha z + m\theta)} \quad \text{for } i\eta_o$$
and $\overline{\zeta}_1$, $\overline{S\zeta_1}$ for $-i\eta_o$.

Here we have

(12) $$X = x_1\zeta_1 + x_2\zeta_2 + \overline{x}_1\overline{\zeta}_1 + \overline{x}_2\overline{\zeta}_2,$$

and (5) may be written

(13) $$\begin{cases} \frac{dx_1}{dt} = f_1(\mu, x_1, \overline{x}_1, x_2, \overline{x}_2) \\ \frac{dx_2}{dt} = f_2(\mu, x_1, \overline{x}_1, x_2, \overline{x}_2) \end{cases}$$

Now, it is easy to check that

(14) $R_\varphi\zeta_1 = e^{im\varphi}\zeta_1, \quad \tau_s\zeta_1 = e^{i\alpha s}\zeta_1, \quad S\zeta_1 = \zeta_2,$

hence the propagation of equivariance leads to

(15) $$\begin{cases} f_1(\mu, e^{im\varphi}x_1, e^{-im\varphi}\overline{x}_1, e^{im\varphi}x_2, e^{-im\varphi}\overline{x}_2) = e^{im\varphi} f_1(\mu, x_1, \overline{x}_1, x_2, \overline{x}_2) \\ f_1(\mu, e^{i\alpha s}x_1, e^{-i\alpha s}\overline{x}_1, e^{-i\alpha s}x_2, e^{i\alpha s}\overline{x}_2) = e^{i\alpha s} f_1(\mu, x_1, \overline{x}_1, x_2, \overline{x}_2) \\ f_2(\mu, x_1, \overline{x}_1, x_2, \overline{x}_2) = f_1(\mu, x_2, \overline{x}_2, x_1, \overline{x}_1) , \end{cases}$$

and it is easy to give the general form of f_1 and f_2. We have

$$(16)\quad \begin{cases} \dfrac{dx_1}{dt} = x_1(\sigma(\mu) + b|x_1|^2 + c|x_2|^2 + \dots) \\ \dfrac{dx_2}{dt} = x_2(\sigma(\mu) + c|x_1|^2 + b|x_2|^2 + \dots) \end{cases},$$

where $\sigma(\mu) = i\eta_o + O(\mu)$.

It is easy to study the phase diagrams of (16), depending on the values of the coefficients b_r, c_r [5][7]. For $b_r < 0$, $c_r < b_r$ we have stable periodic solutions :

$$(17)\qquad x_1 = r_1\, e^{i\Omega_0 t + \varphi_0}, x_2 = 0$$

and the symmetric ones (apply S). The corresponding solution U of (2) has the form

$$(18)\qquad U = U(r, \alpha z + m\theta + \Omega_o t),$$

in cylindrical coordinates, where U is 2π-periodic in its second argument. These solutions are the so-called spiral cells (S_m).

For $c_r > b_r$ and $b_r + c_r < 0$, we have another type of stable periodic solutions :

$$(19)\qquad x_1 = r_1\, e^{i(\Omega_1 t + \varphi_1)},\ x_2 = r_1\, e^{i(\Omega_1 t + \varphi_2)}\ .$$

We can choose the phase in such a way that the corresponding solution U is invariant under S, i.e. we have here horizontal cells of size π/α. These are the so-called ribbon-cells (R_m), not yet observed up to now. Note that these bifurcating periodic solutions have a simple spatial structure since we can write then into the form of a rotating wave :

$$(20)\quad U(t) = R_{\Omega t/m}U(0)\quad ,\quad \text{with } \Omega = \Omega_o \text{ for (18)},\ \Omega_1 \text{ for (19)}.$$

The proof of (20) and of the form (18) is just a trivial consequence of the propagation of equivariances with respect to R_φ and τ_s.

III. INTERACTION OF A STEADY MODE AND AN OSCILLATORY MODE

This situation was studied numerically by Di Prima et al.[8], not using symmetry arguments. Results were first shown to me in September 1983 by J.Sijbrand, even though to my knowledge these are not yet published. Golubitsky and Stewart [10] used symmetry arguments to study all the periodic solutions occuring in this case.

Their work is based on Liapunov-Schmidt analysis, so it has the inconvenient not to give the dynamical behavior of the solutions of the initial data problem. Here we present briefly this case, using the center manifold technique, which keeps all the dynamics of the problem (note that it is sufficient to have a C^k center manifold with k large enough to make the following analysis).

III.1 System on the 6-dimensional center manifold

As indicated in §II.2, in playing with the two parameters $\mathcal{R}$ and Ω_2/Ω_1, we consider here a case where critical eigenvalues are 0 and $\pm i\eta_o$, the 6 eigenvectors belonging to them are

$$(21)\begin{cases}\zeta_o = \hat{U}(r)\, e^{i\alpha z}, \quad \bar{\zeta}_o = S\,\zeta_o \quad \text{for the 0 eigenvalue,}\\ \zeta_1 = \hat{V}(r)\, e^{i(\alpha z + m\theta)}, \quad \zeta_2 = S\,\zeta_1 = (S\,\hat{V}(r))\, e^{i(-\alpha z + m\theta)} \quad \text{for } +i\eta_o,\\ \bar{\zeta}_1, \bar{\zeta}_2 \quad \text{for } -i\eta_o .\end{cases}$$

Here we pose

$$(22)\quad X = x_o\zeta_o + \bar{x}_o\bar{\zeta}_o + x_1\zeta_1 + \bar{x}_1\bar{\zeta}_1 + x_2\zeta_2 + \bar{x}_2\bar{\zeta}_2,$$

and (5) may be written (parameters (μ,ν)) :

$$(23)\quad \frac{dx_j}{dt} = f_j(\mu,\nu,x_o,\bar{x}_o,x_1,\bar{x}_1,x_2,\bar{x}_2), \quad j=0,1,2$$

Propagating the equivariance properties, in the same way as before (see (15)), we can easily prove the following :

Lemma 1 : The 6-dim.-system on the center manifold takes the following form :

$$(24)\quad \frac{dx_j}{dt} = e^{i\psi_j}\, g_j(r_o,r_1,r_2,\theta), \quad j=0,1,2$$

where $x_j = r_j\, e^{i\psi_j}$, g_j are 2π-periodic in $\theta = 2\psi_o - \psi_1 + \psi_2$, and g_o is odd in r_o, even in (r_1,r_2), while g_1 and g_2 are even in r_o, odd in (r_1,r_2). Moreover

$$g_o(r_o,r_2,r_1,-\theta) = \bar{g}_o(r_o,r_1,r_2,\theta)$$
$$g_2(r_o,r_1,r_2,-\theta) = g_1(r_o,r_2,r_1,-\theta).$$

More precisely, the lower order terms in (24) are the following :

$$(25)\begin{cases}\dfrac{dx_0}{dt} = x_o\, P_o(r_o^2,r_1^2,r_2^2) + \beta_o\,\bar{x}_o\, x_1\,\bar{x}_2 + \text{order } 5\\ \dfrac{dx_1}{dt} = x_1\, P_1(r_o^2,r_1^2,r_2^2) + \beta_1\, x_o^2\, x_2 + \text{order } 5\\ \dfrac{dx_2}{dt} = x_2\, P_1(r_o^2,r_1^2,r_2^2) + \beta_1\,\bar{x}_o^2\, x_1 + \text{order } 5\end{cases}$$

with

$$P_o(r_o^2,r_1^2,r_2^2) = a_o\mu + b_o\nu + c_o r_o^2 + d_o r_1^2 + \bar{d}_o r_2^2, \quad a_o,b_o,c_o,\beta_o \in \mathbb{R},$$
$$P_1(r_o^2,r_1^2,r_2^2) = i\eta_o + a_1\mu + b_1\nu + c_1 r_o^2 + d_1 r_1^2 + e_1 r_2^2 .$$

Remark : The system (25) was already written by J.T.Stuart in 1964 (see the review paper [18] for references)!

III.2 Periodic and quasi-periodic solutions

After recombination with (22), (3) and (4), we obtain the following primary bifurcating solutions ($r_o,r_1,r_2,\Omega_o,\Omega_1$ are functions of μ,ν) :

$$(26)\begin{cases}x_o = r_o\, e^{i\varphi_0}, \quad x_1 = x_2 = 0 \quad : \text{Taylor vortices T.V. ;}\\ x_o = 0,\ x_1 = r_1\, e^{i\Omega_o t+\varphi_1}, x_2 = 0 \quad : \text{m-spirals } S_m\\ x_o = 0,\ x_1 = 0,\ x_2 = r_1\, e^{i\Omega_o t+\varphi_2} \quad : \text{symmetric spirals } S(S_m),\\ x_o = 0,\ x_1 = r_2\, e^{i\Omega_1 t+\varphi_1},\ x_2 = r_2\, e^{i\Omega_1 t+\varphi_2} \quad : \text{m-ribbons } R_m .\end{cases}$$

The structures of these flows are described in §II.4.

The study of secondary bifurcating flows leads first to two Hopf-bifurcations from the steady T.V. branch and two bifurcations from the R_m branch leading to periodic (rotating wave) solutions. These secondary branches may join or not the T.V. branch to the R_m branch depending on coefficients. They correspond to the following solutions of (25) :

$$\textbf{(27)}\quad x_o = r_o e^{i\varphi_0},\quad x_1 = r_1 e^{i(\Omega_1 t+\varphi_1)},\quad x_2 = r_1 e^{i(\Omega_1 t+\varphi_2)},$$

where $\theta = 2\varphi_o - \varphi_1 + \varphi_2 = 0$ or π, and r_o, r_1, Ω_1 functions of μ,ν.

If we come back to the flow U, using (22),(3),(4), then for $\theta=0$ we obtain, up to phase shifts in z,θ,t :

$$\textbf{(28)}\quad U = r_o\hat{U}(r)e^{i\alpha z} + r_1\Big[\hat{V}(r)e^{i(\alpha z+m\theta+\Omega_1 t)} + (S\,\hat{V}(r))e^{i(-\alpha z+m\theta+\Omega_1 t)}\Big] + \text{c.c.} + \text{h.o.t.}\quad .$$

This solution is a <u>symmetric</u> rotating wave, so, as seen in §II,4.1, we have horizontal cells of size π/α. These are the <u>twisted vortices</u> [10], also founded in another way in [5] directly from the bifurcation of the T.V. flow. If we are close to the T.V. branch, r_1 is small with respect to r_o, while when we are close to the R_m branch, r_o is small with respect to r_1. At the first orders it looks like a superposition of the T.V. flow and the R_m flow, with exact spatial correspondence for cells.

For $\theta=\pi$, we obtain, up to phase shifts in z,θ,t :

$$\textbf{(29)}\quad U = r_o\hat{U}(r)e^{i\alpha z} + r_1\Big[\hat{V}(r)e^{i(\alpha z+m\theta+\Omega_1 t)} - (S\,\hat{V}(r))e^{i(-\alpha z+m\theta+\Omega_1 t)}\Big] + \text{c.c.} + \text{h.o.t.}\quad .$$

The symmetries of this flow are described in [10] : to change θ in $\theta+\pi/m$ is the same as applying S to U (symmetric solution). This is the <u>wavy-vortex flow</u>. At the first orders it looks like a superposition of the T.V. flow and the R_m flow with shifted cells of $\pi/2\alpha$, as remarked in [18].

Due to the rotationnal symmetry, and the rotating wave structure of each periodic bifurcating solution, the study of the stability reduces to an <u>autonomous</u> stability analysis (we just choose a suitable rotating frame). In this way it is easy to find a secondary bifurcation from the m-spirals which leads to a quasi-periodic flow with principal part

$$\textbf{(30)}\quad U = r_o\hat{U}(r)e^{i(\alpha z+\Omega_0 t)} + r_1\hat{V}(r)e^{i(\alpha z+m\theta+\Omega_1 t)} + \text{c.c.} + \text{h.o.t.}\ ,$$

where $\Omega_o = O(r_1^2)$ is <u>small</u>, and r_o is small with respect to r_1.

This flow is denoted in [10] as <u>Wavy Spirals</u>, and is not yet observed in experiments. Note that in the X part of U, the x_2 composent is $O(r_1\, r_o^2)$ and that $\theta = 2\psi_o(t) - \psi_1(t) + \psi_2(t)$ is constant. Hence $2\Omega_o - \Omega_1 + \Omega_2 = 0$, and we show below that this implies that the flow is superposition of two travelling waves.

Other quasi-periodic solutions may bifurcate from the twisted vortices (no standard denomination), and from the wavy vortices (modulated wavy vortices). They can be obtained in looking for fixed points of the 4-dimensional system in r_o, r_1, r_2, θ , bifurcating from solutions such as (28) or (29) where $r_1 = r_2$ and $\theta = 0$ or π. We then obtain quasi-periodic flows of following form :

(31) $$U = r_o \hat{U}(r)\, e^{i(\alpha z + \Omega_0 t + \varphi_0)} + r_1 \hat{V}(r)\, e^{i(\alpha z + m\theta + \Omega_1 t + \varphi_1)} + r_2 (S\, \hat{V}(r))\, e^{i(-\alpha z + m\theta + \Omega_2 t + \varphi_2)} + \text{c.c.} + \text{h.o.t.}\ .$$

where $2\Omega_o - \Omega_1 + \Omega_2 = 0$, and φ_j, $j=0,1,2$ are constant. We note (the same is true for (30)) that the X part of U takes the form

(32) $$X = X(r, \alpha z + \Omega_o t,\ m\theta + (\Omega_1 - \Omega_o t)$$

in cylindrical coordinates, where X is 2π-periodic in its last arguments.

This shows that

(33) $$R_{(\Omega_0 - \Omega_1)t/m} \circ \tau_{-\Omega_0 t/\alpha} X = X(r, \alpha z, m\theta)$$

is steady. Since the equation (4) of the center manifold is equivariant under R_φ and τ_s, this proves that the velocity field U itself satisfies (33), i.e. in cylindrical coordinates

(34) $$U = U(r,\ \alpha z + \Omega_o t,\ m\theta + (\Omega_1 - \Omega_o)t).$$

This can be visualized as follows : if we choose a suitable rotating frame of angular velocity $(\Omega_o - \Omega_1)/m$, then we observe a travelling wave in the axis direction of velocity$-\Omega_o/\alpha$.

We obtain here a special type of pure superposition of travelling waves. This fact illustrates a general result shown in [3] when a system is invariant under two commuting representations of the group of plane rotations. Moreover this striking property was clearly observed for modulated wavy vortices [17], and (34) corresponds to one of the possible structures of these quasi-periodic flows (no azimuthal symmetry breaking).

Remark : To be able to study the dynamics in the 6 dimensional space of the solutions of (25), we have at least to compute the 4 real coefficients a_o, b_o, c_o, β_o and the 7 complex coefficients $d_o, a_1, b_1, c_1, d_1, e_1, \beta_1$.

IV. INTERACTION OF TWO OSCILLATORY MODES

IV.1 System on the 8-dimensional center manifold

As indicated in §II.2, playing with the two parameters $\mathscr{R}$ and $\Omega_2/\Omega_1(<0)$, we consider here a case where critical eigenvalues are $\pm i\eta_1$ and $\pm i\eta_2$. The 8 eigenvectors belonging to them are noted

(35)
$$\begin{aligned}\zeta_o &= \hat{U}(r)\, e^{i(\alpha z+m\theta)}, \quad \zeta_1=(S\,\hat{U}(r))\, e^{i(-\alpha z+m\theta)} \quad \text{for } i\eta_1\\ \zeta_2 &= \hat{V}(r)\, e^{i(\alpha z+m'\theta)}, \quad \zeta_3=(S\,\hat{V}(r))\, e^{i(-\alpha z+m'\theta)} \quad \text{for } i\eta_2\end{aligned}$$

and $\overline{\zeta}_o,\overline{\zeta}_1,\overline{\zeta}_2,\overline{\zeta}_3$ for $-i\eta_1$ and $-i\eta_2$.

We assume below that m and m' have no common divisor (this simplifies the exposition). Here we pose

(36)
$$X = x_o\zeta_o + x_1\zeta_1 + x_2\zeta_2 + x_3\zeta_3 + \text{c.c.}$$

and (5) may be written into the form (23), j=0,1,2,3.

Using all equivariances as for (15), we can easily prove the following [4] :

Lemma 2 : The 8-dim. system on the center manifold takes the following form

(37)
$$\frac{dx_j}{dt} = e^{i\psi_j}\, g_j(r_o,r_1,r_2r_3,\theta_1,\theta_2) \quad , \quad j=0,1,2,3,$$

where $x_j = r_j\, e^{i\psi_j}$, g_j are 2π-periodic in $\theta_1=\psi_1-\psi_o+\psi_2-\psi_3$ and $\theta_2 = m'(\psi_o+\psi_1) - m(\psi_2+\psi_3)$, g_o and g_1 are even in r_2,r_3 odd in r_o,r_1 and g_2 and g_3 are even in r_o,r_1, odd in r_2,r_3.

Moreover for (j,k)=(0,1) and (2,3), we have :

$$g_j(r_o,r_1,r_2,r_3,\theta_1,\theta_2) = g_k(r_1,r_o,r_3,r_2,-\theta_1,\theta_2) .$$

Remark **1.** The dependency in θ_2 for g_j arises at the order $2(m+m')-1$ in (r_o,r_1,r_2,r_3).

Remark **2.** For $x_o=x_1=0$, we have $g_o=g_1=0$, the same is valid for $x_2=x_3=0$, $x_o=x_3=0$, $x_1=x_2=0$ in changing the indices.
In all these cases the two remaining complex equations have the form

(38)
$$\frac{dx_j}{dt} = x_j\, \tilde{g}_j(r_j,r_k) \quad , \quad \tilde{g}_j \text{ even in } r_j,r_k .$$

Remark **3.** For $x_o=x_2=0$, we have g_o and g_2 of order $2(m+m')-1$ in (r_1,r_3), even though $\frac{dx_1}{dt}$ and $\frac{dx_3}{dt}$ have the form (38), and the analogous result is true for $x_1=x_3=0$.

IV.2 Periodic and quasi-periodic solutions

Let us first precise a little more the system (37), in writing the lower order terms :

$(39)_1$
$$\begin{cases}\dfrac{dx_0}{dt} = x_o\, P_o(r_o^2,r_1^2,r_2^2r_3^2) + f_o x_1 x_2\overline{x}_3 + \text{order } 5\\[2mm] \dfrac{dx_1}{dt} = x_1\, P_o(r_1^2,r_o^2,r_3^2,r_2^2) + f_o x_o\overline{x}_2 x_3 + \text{order } 5\end{cases}$$

$$(39)_2 \quad \begin{cases} \dfrac{dx_2}{dt} = x_2\, P_2(r_o^2, r_1^2, r_2^2, r_2^2) + f_2 x_o \bar{x}_1 x_3 + \text{order } 5 \\ \dfrac{dx_3}{dt} = x_3\, P_2(r_1^2, r_o^2, r_3^2, r_2^2) + f_2 \bar{x}_o x_1 x_2 + \text{order } 5 \end{cases}$$

with

$$P_o(r_o^2, r_1^2, r_2^2, r_3^2) = i\eta_1 + \alpha_o \mu + \beta_o \nu + b_o r_o^2 + c_o r_1^2 + d_o r_2^2 + e_o r_3^2$$

$$P_2(r_o^2, r_1^2, r_2^2, r_3^2) = i\eta_2 + \alpha_2 \mu + \beta_2 \nu + b_2 r_o^2 + c_2 r_1^2 + d_2 r^2 + e_2 r_3^2$$

After recombination with (36), (3) and (4), we obtain the following primary bifurcating solutions, where $r_o, r_1, r_2, r_3, \Omega_o, \Omega_1, \Omega_2, \Omega_3$ are functions of μ, ν :

$$(40) \quad \begin{cases} x_o = r_o\, e^{i\Omega_0 t}, \quad x_1 = x_2 = x_3 = 0 & : \text{Spirals } S_{m_0} \\ x_1 = r_o\, e^{i\Omega_0 t}, \quad x_o = x_2 = x_3 = 0 & : \text{Spirals } S(S_{m_0}) = S_{m_1} \\ x_2 = r_2\, e^{i\Omega_2 t}, \quad x_o = x_1 = x_3 = 0 & : \text{Spirals } S_{m_2'} \\ x_3 = r_2\, e^{i\Omega_2 t}, \quad x_o = x_1 = x_2 = 0 & : \text{Spirals } S(S_{m_2'}) = S_{m_3'} \\ x_o = r_1\, e^{i\Omega_1 t}, \quad x_1 = r_1\, e^{i(\Omega_1 t + \varphi_1)}, \quad x_2 = x_3 = 0 & : \text{Ribbons } R_m \\ x_2 = r_3\, e^{i\Omega_3 t}, \quad x_3 = r_3\, e^{i(\Omega_3 t + \varphi_3)}, \quad x_o = x_1 = 0 & : \text{Ribbons } R_{m'} \end{cases}$$

All these flows are "rotating waves", as described in §II.4.2, bifurcating from the Couette flow.

Since all these flows are rotating waves, the study of their stability is easy and secondary bifurcations lead to quasiperiodic flows : **i)** there are two Hopf bifurcations on each Spiral branch and on each ribbon branch, **ii)** let us call these branches $I.S_{0,3}$, $I.S_{1,2}\left[= S(I.S_{0,3})\right]$, $I.S_{0,2}$, $I.S_{1,3}\left[= S(I.S_{0,2})\right]$, $S.R_o$, $S.R_\pi$.

The branch of quasi-periodic solutions $I.S_{0,3}$ may join S_{m_0} with $S_{m_3'}$, and there are analogous facts for other $I.S$'s. The branches $S.R_o$ and $S.R_\pi$ may join R_m with $R_{m'}$ (these properties depend on the 14 coefficients of the truncated system (39)). Let us precise the spatial structure of these flows.

i) $I.S_{0,3}$ **has the following remarkable form :**

$$(41) \qquad x_o = r_o\, e^{i\Omega_0 t}, \quad x_3 = r_3\, e^{i(\Omega_3 t + \varphi_3)}, \quad x_1 = x_2 = 0 ,$$

where $r_o, r_3, \Omega_o, \Omega_3$ are functions of μ, ν. It is easy to prove the following :

<u>Theorem</u> : The flow $I.S_{0,3}$ is such that the velocity field U, may be written in cylindrical coordinates as :

$$U = \widetilde{U}_o(r, \alpha z + m\theta + \Omega_o t) + \widetilde{U}_3(r, -\alpha z + m'\theta + \Omega_3 t) ,$$

where $\widetilde{U}_o$ and $\widetilde{U}_3$ are 2π-periodic in their 2nd argument.

The analogous result is true for the flow $I.S_{1,2}$.

This property of pure superposition of two spiral waves is due to the fact that the part X of U satisfies this property, as it is easy to see from (41), (36), (35). This was noticed also in [3].
Now to show this for U, we use the fact that the center manifold commutes with $\tau_s - R_{(\alpha s/m)}$, $\tau_s - R_{-(\alpha s/m')}$, $R_\varphi - \sigma_{(m\varphi/\Omega_0)}$ where σ_s is the operator which consists in shifting the time by s.
This flow is called "interpenetrating spirals" for obvious reasons. Moreover, since after computations (see for instance [7]) $\Omega_3\, \Omega_o > 0$, the two spirals of different azimuthal waves numbers m and m' are moving in opposite axial directions. This flow was observed in [2].

ii) **$I.S_{0,2}$ is not so clean as $I.S_{0,3}$:**
In fact if we suppress terms of order 2(m+m')-1 in the system (39), then we obtain a solution analogous to (41) : (not the same r_o, Ω_o as in (41)).

(42) $$x_o = r_o\, e^{i\Omega_0 t}\ ,\ x_2 = r_2\, e^{i(\Omega_2 t + \varphi_2)}\ ,\ x_1 = x_3 = 0$$

This has to be modified at the order 2(m+m')-1 by quasi-periodic terms with the ordinary complications (higher harmonics,...). Nevertheless, up to the order 2(m+m')-1 in the amplitudes, we can say that we have interpenetrating spirals, here moving in the same axial direction. Physically this means that if m and m' are large, the observer see here too a true superposition of spirals of different azimuthal wave numbers. The analogous result holds for the flow $I.S_{1,3} = S(I.S_{0,2})$.

iii) **The flows $S.R_o$ and $S.R_\pi$ have the structure of the superposition of ribbons of the two different wave numbers m and m' :**
If we truncate the system (37) at the order 2(m+m')-1, then $S.R_o$ is given by

(43) $$\begin{cases} x_o = r_o\, e^{i(\Omega_0 t + \varphi_0)}\ , & x_1 = r_o\, e^{i(\Omega_0 t + \varphi_1)} \\ x_2 = r_2\, e^{i(\Omega_2 t + \varphi_2)}\ , & x_3 = r_2\, e^{i(\Omega_2 t + \varphi_3)} \end{cases}$$

where $r_o, r_2, \Omega_o, \Omega_2$ are functions of μ, ν (different from the previous ones) and where $\varphi_1 - \varphi_o + \varphi_2 - \varphi_3 = 0$. The solution $S.R_\pi$ is given by (43) too, but $\varphi_1 - \varphi_o + \varphi_2 - \varphi_3 = \pi$. It is easy to show that, in shifting, z, θ, t we can arrange $S.R_o$ to be invariant under S, i.e. it is a flow in horizontal cells of height π/α. This corresponds to the superposition of the two families of ribbons with the same cells. On the contrary, for $S.R_\pi$ the two families of ribbons are shifted by $\pi/2\alpha$, so this gives wavy cells. Now, the corresponding flow U is not a pure addition of the form

(44) $$U = \tilde{U}_o(r, \alpha z, m\,\theta + \Omega_o\, t) + \tilde{U}_2(r, \alpha z, m'\,\theta + \Omega_2\, t)$$

but this is true up to the order 2(m+m')-1 (hence the highest are m and m', the "cleanest" is this superposition) <u>provided that</u> $m'\Omega_o \neq m\Omega_2$ [4]. It seems that both flows $S.R_o$ and $S.R_\pi$ were not yet observed.

Quasi-periodic flows with 3 frequencies may bifurcate [4] from all these secondary branches, the third frequency being close to 0. This fact is analogous to what happens in the interaction of two simple Hopf bifurcations with no resonance [13]. Moreover, depending on the parameters, we can obtain chaotic behaviors too. To study the dynamics in the 8 dimensional space of the solutions of (37), we have first to compute the 14 complex coefficients precised in (39); this is done in [4].

V. WAVY INFLOW (OR OUTFLOW) BOUNDARIES

We saw in §III.2 that, from the Taylor vortex flow, if we keep the same periodicity $2\pi/\alpha$ in z, then the simplest regimes which may bifurcate are wavy vortices or twisted vortices. This can be obtained by a general analysis starting with the Taylor flow [5]. In fact, experiments described in [1], also show other types of flows, bifurcating from the T.V. flow.

We wish to prove hereafter that these flows result from a symmetry breaking in the number of cells which is in fact divided by 2 (even though an observer might believe to see the same number of "cells").

Instead of considering the Couette flow as the basic solution, we take the Taylor flow. Then the problem is no more invariant under all translations in z, but just under $\tau_{2\pi/\alpha}$. For the analysis we choose a T.V. flows U_o invariant under S (x_o real in (10)), in noting that there are two flows of this type : U_o and $\tau_{\pi/\alpha}U_o$, the difference being that the plane z=0 is an "outflow boundary" for U_o, and an "inflow boundary" for $\tau_{\pi/\alpha}U_o$.

Assume, in what follows, that the T.V. flow becomes unstable in an oscillatory way, i.e. at criticality we have eigenvalues $\pm i\omega_o$, and eigenvectors

(45) $$\zeta_o = \hat{U}(r,z)e^{im\theta} \ , \ \zeta_1 = S\,\zeta_o \quad \text{for } i\omega_o \ ,$$

and $\overline{\zeta}_o, \overline{\zeta}_1$ for $-i\omega_o$.

We have $R_\varphi\,\zeta_j = e^{im\varphi}\zeta_j$, j=0,1, moreover due to

(46) $$\tau_s\, S = S\,\tau_{-s}$$

it is easy to show the following properties [5] :

(47) $$\tau_{2\pi/\alpha}\,\zeta_o = e^{2i\pi\,\ell/n}\,\zeta_o \ , \ \tau_{2\pi/\alpha}\,\zeta_1 = e^{-2i\pi\,\ell/n}\,\zeta_1$$

where n is the number of axial periods in the height h.

The cases when $\ell=0$ correspond to no axial symmetry breaking (like in §III.2). One of the most interesting cases is when $\ell/n = 1/2$. In this case, the bifurcating flows will have twice less cells, since the flow will be invariant under $\tau_{4\pi/\alpha}$.

One of the basic facts here is that 0 is always an eigenvalue for the linearized problem about the T.V. flow, due to the translational invariance of the system. We use the following fundamental decomposition of the space (more convenient than the one used in [5]) :

$$U = \tau_\eta(U_o + X + Y) \tag{48}$$

with
$$X = x_+ \zeta_+ + \bar{x}_+ \bar{\zeta}_+ + x_- \zeta_- + \bar{x}_- \bar{\zeta}_- ,$$

Y in a suitable space of codim 5,

$$\zeta_\pm = \zeta_o \pm \zeta_1 , \quad Y = \Phi(\mu, X) = O[|\mu| \|X\| + \|X\|^2],$$

where Φ is the equation of a center manifold, of dimension 5.

The differential system on the center manifold may be written as follows :

$$(49)\quad \begin{cases} \dfrac{d\eta}{dt} = h(\mu, x_+, \bar{x}_+, x_-, \bar{x}_-) = \alpha\, x_+ \bar{x}_- + \bar{\alpha}\, \bar{x}_+ x_- + \text{order } 4 \\ \dfrac{dx_+}{dt} = f_+(\mu, x_+, \bar{x}_+, x_-, \bar{x}_-) = x_+(i\omega_o + a\mu + b|x_+|^2 + c|x_-|^2) + d\, \bar{x}_+ x_-^2 + \text{order } 5 \\ \dfrac{dx_-}{dt} = f_-(\mu, x_+, \bar{x}_+, x_-, \bar{x}_-) = x_-(i\omega_o + a\mu + b'|x_+|^2 + c'|x_-|^2) + d' x_+^2 \bar{x}_- + \text{order } 5 \end{cases}$$

where all equivariances resulting from (45), (47) are used. For instance, the system (49) has no η in the right hand side, and the invariant manifold Φ is independent of η too.

Obvious solutions of (49) are

i) $x_- = 0$, $x_+ = r_+ e^{i\Omega_+ t}$, $\eta =$ const.

ii) $x_+ = 0$, $x_- = r_- e^{i\Omega_- t}$, $\eta =$ const.

These solutions lead to rotating waves, the solution i) being symmetric for $\eta=0$. We have, for the velocity field U :

$$(50)\quad \begin{cases} \text{(i)}\quad U = U_o(r,z) + [r_+(\hat{U}(r,z) + S\,\hat{U}(r,z))\, e^{i(m\theta+\Omega_+ t)} + \text{c.c.}] + \text{h.o.t.} \\ \text{(ii)}\quad U = U_o(r,z) + [r_-(\hat{U}(r,z) - S\,\hat{U}(r,z))\, e^{i(m\theta+\Omega_- t)} + \text{c.c.}] + \text{h.o.t.} \end{cases}$$

Now, if we start with $U_o' = \tau_{\pi/\alpha} U_o$ instead of the Taylor flow U_o, we can make the same analysis, using eigenvectors

$\zeta_\pm' = \tau_{\pi/\alpha} \zeta_\pm$ Hence we have

$$S\, \zeta_+' = -\zeta_+' \quad \text{and} \quad S\, \zeta_-' = \zeta_-'$$

as it can be checked, using (46). It is then easy to see that we recover the solutions (i) and (ii) shifted by π/α:

$$\tau_{\pi/\alpha}(i):\ U = U_o'(r,z)+\{r_+[\hat{U}(r,z+\tfrac{\pi}{\alpha})+(S\hat{U})(r,z+\tfrac{\pi}{\alpha})]e^{i(m\theta+\Omega_+t)}+c.c.\}+h.o.t.$$
$$\tau_{\pi/\alpha}(ii):\ U = U_o'(r,z)+\{r_-[\hat{U}(r,z+\tfrac{\pi}{\alpha})+(S\hat{U})(r,z+\tfrac{\pi}{\alpha})]e^{i(m\theta+\Omega_-t)}+c.c.\}+h.o.t.$$

But now we remark that $S\tau_{\pi/\alpha}(ii) = \tau_{\pi/\alpha}(ii)$, while we had $S(i)=(i)$. This property, together with the $(4\pi/\alpha)$-periodicity of these flows, show that :

1. for the flow (i), on planes $z = 0+k\frac{2\pi}{\alpha}$ we have $u_z=0$, and the originally flat planes $z = \frac{\pi}{\alpha} + k\frac{2\pi}{\alpha}$ look now wavy.

This is the "wavy inflow boundaries"(W.I.B.).

2. for the flow (ii), it is the converse.

This is the "wavy outflow boundaries" (W.O.B.).

VI. OTHER FLOWS

There are many other typical observed flows [2] in this famous problem, and a way to study them theoretically would be to go on further in the same spirit as here. For instance, the interaction of two oscillating modes $(\pm i\omega_1, \pm i\omega_2)$ starting from the Taylor flow, leads to a dynamic on a 9 dimensional center manifold. This is similar to the problem of §IV, but the system is not so "clean" since we have no longer the invariance under τ_s for any s (specially, when $\ell = 0$ in (47) $\tau_{2\pi/\alpha}$ gives nothing). Nevertheless, the two azimuthal wave numbers m and m′ lead to two families of twisted vortices, and to two families of wavy vortices, which might interact.

Further bifurcations, then give quasi-periodic flows such as the so-called "modulated wavy vortices" and "wavelets" [1]. Notice that the general symmetry breaking arguments of D.Rand in [16] gave nice indications on the structure of modulated wavy vortices.

The striking fact here again, is that we can prove (as in §IV.2) that up to the order $m+m'-1$ (if m is prime with m′), these flows are pure superposition of rotating waves, as conjectured in the experiments [17].

BIBLIOGRAPHY

[1] **Andereck,C.D., R.Dickman, H.L.Swinney,** New flows in a circular Couette system with corotating cylinders, Phys.Fluids 26,(6),1395-1401, 1983

[2] **Andereck,C.D., S.S.Liu, H.L.Swinney,** Flows regimes in a circular Couette system with independently rotating cylinders (preprint)

[3] **Chossat,P.** Bifurcation d'ondes rotatives superposées, C.R. Acad. Sci. Paris, 300, I,7, 209, 1985 and : Interaction d'ondes rotatives dans le problème de Couette-Taylor, C.R. Acad.Sci. Paris, 300, I,8, 251, 1985

[4] **Chossat,P., Y.Demay, G.Iooss,** Interaction de modes azimutaux dans le problème de Couette-Taylor, (en préparation)

[5] **Chossat,P., G.Iooss,** Primary and Secondary bifurcations in the Couette-Taylor problem, Japan J.Appl. Math. 2, 37-68, 1985

[6] **Cognet,G., A.Bouabdallah, A.Aitaider,** Laminar-Turbulent transition, R.Eppler, H.Fasel Ed., 330, Springer 1982

[7] **Demay,Y., G.Iooss,** Calcul des solutions bifurquées pour le problème de Couette-Taylor avec les deux cylindres en rotation, J.Méca. Th. Appl., Vol. spécial, "Bifurcations et comportements chaotiques", 1985

[8] **Diprima, R.C., P.M.Eagles, J.Sijbrand,** Interaction of axisymmetric and non-axisymmetric disturbances in the flow between concentric counter-rotating cylinders, (in preparation), 1984

[9] **Diprima,R.C., R.C.Grannick,** A nonlinear investigation of the stability of flow between counter-rotating cylinders, IUTAM Symp. on Instability of Continuous Systems, H.Leipholz ed., Springer, 1971

[10] **Golubitsky,M., I.Stewart,** Symmetry and Stability in Taylor-Couette Flow, (preprint)

[11] **Iooss,G.,** Bifurcation of maps and applications, North Holland Math. Studies, 36, 1979

[12] **Iooss,G.,** Bifurcation and transition to turbulence in hydrodynamics, C.I.M.E. session on "Bifurcation Theory and Applications", L.Salvadori ed., Lect.Notes in Math. 1057, 152-201, Springer 1984

[13] **Iooss,G.,** Bifurcations élémentaires - successions et interactions, in "Nonlinear Phenomena in Chemical Dynamics", Vidal, Pacault ed., 71-78, Springer 1981

[14] **Kirchgassner,K., P.Sorger,** Branching analysis for the Taylor problem, Quart. J.Mech. Appl. Math. 22, 183, 1969

[15] **Marsden,J.E., M.Mc Cracken,** The Hopf bifurcation and its applications, Applied Math. Sci. 19, Springer 1976

[16] **Rand,D.,** Dynamics and Symmetry : predictions for modulated waves in rotating fluids, Arch. Rat. Mech. Anal. 79, 1, 1-38, 1982

[17] **Shaw,R.S., C.D.Andereck, L.A.Reith, H.L.Swinney,** Superposition of travelling waves in the circular Couette system, Phys. Rev. Let. 47, 17, 1172-1175, 1982

[18] **Stuart,J.T.,** Nonlinear stability theory, Annual Rev. Fluid Mech. 3, 347-370, 1971.

ON PROPAGATION OF THE TRANSITION LAYERS IN SOLUTIONS TO NONLINEAR PARTIAL DIFFERENTIAL EQUATIONS

Zbigniew Peradzyński
Institute of Fundamental Technological Research
Swiętokrzyska 21, 00-049 Warsaw, Poland

Introduction

In this paper we are concerned with the role which is played by the boundary and transition layers in the analysis and in the construction of asymptotic solutions of the initial boundary value problem for the semilinear parabolic equation

$$u_t + \bar{\nabla}(t,x)\cdot\nabla u = Lu + f(u,t,x), \qquad x\in\Omega\subset R^n, \tag{1}$$

with slowly varying coefficients. L denotes here a linear second order uniformly elliptic operator.

Dealing with solutions of differential equations one usually is not concerned with exact numbers, one is rather interested in the approximate "shapes" of the solutions and the deformations they undergo when certain external "physical" parameters are varied. For this reason instead of a single equation it is more natural to consider a family of equations depending on some parameters.

In our case of slowly varying coefficients by applying proper scaling one arrives at the family of equations

$$\varepsilon(u_t + \bar{\nabla}(t,x)\cdot\nabla u) = \varepsilon^2 Lu + f(u,t,x) \tag{2}$$

involving a small parameter ε which can be understood as the ratio of the thickness of the transition or boundary layer to the characteristic length $\frac{f}{f_{,x}}$. Roughly speaking in our attempt we demonstrate that asymptotic solutions of Eq.(2) for $\varepsilon\to 0$ can be constructed by gluing up the local equilibria $u_\alpha(t,x)$ defined by $f(u,t,x) = 0$ with the help of transition layers and boundary layers.

A considerable part of this paper is devoted to the analysis of stationary solutions to Eq.(2) and their stability. The special case of Eq.(2) was considered in connection with plasma maintained by laser radiation [1,2,3]. In these papers we have assumed f to be $I(x)\cdot h(u - u_o) - u$, where h is the Heaviside step function. Due to the fact that in this special example the nonlinearity is concentrated at $u = u_o$ it was possible to obtain at some stages explicit expressions, thus providing us with understanding and experience necessary for further generalizations.

Another reason for which we confine our attention in this paper mainly to the Eq.(1) and its elliptic counterpart is that in these cases the well known techniques based on maximum principle can be applied in order to justify rigorously the asymptotics. These are for instance the monotone iteration scheme, the Serrin's sweeping principle [4,5] both exploiting the notion of super and subsolutions.

As far as more general parabolic or elliptic equations, for instance fully nonlinear or systems of equations, are of interest, still in many cases similar techniques are available (see for instance [6,7,8]). An extension of some of the results of this paper for the case of a system of equations is also possible by applying the implicit function theorem [9].

Finally all results concerning the formal asymptotics included in this paper can be extended without any difficulty also for hyperbolic dissipative equations as for example Eq.(2) with L being hyperbolic $\Delta - a^2\partial_{tt}$. In the following for brevity we assume that the coefficients of Eq.(2) and f are smooth functions of t, x and u.

2. Transition and boundary layers

Let us consider for a moment an equation with coefficients depending on some parameters $\mu \in R^m$, but independent of t and x. In this case without loss of generality we may assume $L = \Delta$

$$u_t + \bar{v}(\mu)\cdot\nabla u = \Delta u + f(u,\mu). \tag{3}$$

We assume also that $\Omega = \mathbf{R}^n$ and that the forcing term f has a number of equilibrium states, say $u_1(\mu),\ldots,u_k(\mu)$, that is $f(u_\alpha(\mu),\mu) \equiv 0$ for $\alpha = 1,\ldots,k$.

We say that $u_\alpha(\mu)$ is a stable equilibrium if $f(u,\mu)$ is decreasing in u in a certain neighbourhood of $u(\mu)$. We will impose, however, a little stronger stability condition, requiring that

$$f_{,u}(u_\alpha(\mu),\mu) < 0 \tag{4}$$

for all μ from a certain open set $Q \subset \mathbf{R}^m$. Having such stable equilibria one may search for solutions representing transitions from one stable equilibrium to another, say from u_α to u_β. Assuming the solution in the form of the travelling wave $u = \phi(\bar{n}\cdot\bar{x} - ct)$, $\bar{n}^2 = 1$ one arrives at the ordinary differential equation

$$(\bar{v}\cdot\bar{n} - c)\phi' = \phi'' + f(\phi,\mu) \tag{5}$$

with boundary conditions $\phi(-\infty) = u_\alpha$, $\Phi(+\infty) = u_\beta$. We assume here that the unit vector $\bar{n}$ is directed in such a way that ϕ is an increasing function of $\bar{n}\cdot\bar{x}$, thus avoiding the ambiguity in the sign of c. According to this convention u_β should be larger than u_α. From the stability condition (3) it follows that the derivative ϕ' is vanishing at $\pm\infty$. Introducing $\eta := \phi'$ as the new dependent variable and taking ϕ as the independent variable one arrives at the first order equation

$$(\bar{v}\cdot\bar{n} - c)\eta = \eta\eta_{,\phi} + f(\phi,\mu) \tag{6}$$

for a non-negative function $\eta(\phi)$.

As $\phi = u_\alpha$ and $\phi' = 0$ at $-\infty$ and $\phi = u_\beta$ and $\phi' = 0$ at $+\infty$ it follows that η should satisfy two boundary conditions $\eta(u_\alpha) = \eta(u_\beta) = 0$. There are exactly two integral curves starting from any stable equilibrium, one in the positive and the other in the negative direction, whose slopes are equal respectively

$$\tfrac{1}{2}\left(\bar{v}\cdot\bar{n} - c + \sqrt{(\bar{v}\cdot\bar{n} - c)^2 - 4f_{,u}}\right), \quad \tfrac{1}{2}\left(\bar{v}\cdot\bar{n} - c - \sqrt{(\bar{v}\cdot\bar{n} - c)^2 - 4f_{,u}}\right).$$

Therefore, in general, it is not possible to satisfy both boundary conditions. Indeed, let the integral curve starting from u_α (i.e. $\eta(u_\alpha) = 0$) in the direction of u_β be of the form $\eta(\phi,\mu,q)$ where $q = \bar{v}\bar{n}-c$ (we know it is unique). If it satisfies also the boundary condition at u_β, then by integrating Eq.(6) with respect to ϕ we obtain

$$q \int_{u_\alpha(\mu)}^{u_\beta(\mu)} \eta(\phi,\mu,q)\,d\phi = \int_{u_\alpha(\mu)}^{u_\beta(\mu)} f(\phi,\mu)\,d\phi. \tag{7}$$

Therefore, in general, it is not possible to satisfy both boundary conditions unless the parameters appearing in Eq.(6) i.e. $q := \bar{v}\cdot\bar{n} - c$ and μ satisfy additional relation (7) which can be written in the form

$$H(\bar{v}\cdot\bar{n}-c,\mu) = 0. \tag{8}$$

In principle relation (8) can be solved for $\bar{v}\cdot\bar{n} - c$ to obtain

$$\bar{v}\cdot\bar{n} - c = F(\mu). \tag{9}$$

Eq.(8) will play important role in the analysis of the asymptotic solutions of Eq.(2).

Examples.

1° For $f = I\cdot h(u - u_o) - u$ the solubility condition (8) is given by

$$c - \bar{v}\cdot\bar{n} = 2\beta\Big/\sqrt{1-\beta^2}, \qquad \text{where } \beta = \frac{2}{I}u_o - 1. \tag{10}$$

This is the case considered in [1,2].

2° Take $f = A \sin ku - B \sin 2ku$, i.e. the form which may represent the two leading terms in the Fourier expansion of a general function f restricted to the interval included between the two successive stable equilibria. In this case we have

$$\bar{v}\cdot\bar{n} - A\left(\frac{k}{2B}\right)^{\frac{1}{2}} = c.$$

DEFINITIONS. The center of the transition layer is defined to be the surface on which $\phi'' = 0$.

Let $\phi''(\xi_o) = 0$. *Similarly we define* $\varepsilon = (u_\beta - u_\alpha)/\phi'(\xi_o)$ *to be the thickness of the layer.*

Suppose now that on a certain hyperplane in $R^1\times R^n$ e.g. $\bar{n}\cdot(\bar{x} - \bar{x}_o) =$

$= \lambda t$ a constant boundary condition is prescribed, say $u_o = 0$. In such a case a boundary layer connecting u_o with some equilibrium state may appear.

The boundary layer is again a solution of Eq.(4) satisfying $u = u_o$ at the surface, and $u = u_\alpha$ at infinite distances from the surface. Again the first order form (6) of Eq.(4) can be used, but now only one boundary condition $\eta(u_\alpha) = 0$ is obtained. One requires, however, that the solution exists in the whole interval (u_o, u_α) and this imposes certain inequality of the form

$$\{\bar{v}\cdot\bar{n} - c - F_\alpha(\mu, u_o)\}\operatorname{sgn}(u_o - u_\alpha) \geqq 0. \tag{11}$$

This can be seen when investigating the equation for the variation $\zeta := \frac{\delta\omega}{\delta q}$, where $\omega = \frac{1}{2}\eta^2$ and $q = \bar{v}\cdot\bar{n} - c$. It is of the following form

$$\zeta_{,\phi} = \sqrt{2\omega} + \frac{q}{\sqrt{2\omega}}\zeta \tag{12}$$

and it implies that $\eta(\phi)$ increases with q for $\phi > u_\alpha$ and $\eta(\phi)$ decreases as function of q for $\phi < u_\alpha$. Therefore, if $u_\beta > u_\alpha$ and if $q_o = \bar{v}_o\cdot\bar{n} - c_o$ satisfies Eq.(9) then for $q < q_o$ not all values of u_o from the interval (u_α, u_β) are available as the boundary condition for the layer connecting u_α and u_o. Indeed, the integral curve of Eq.(6) starting from u_α (i.e. $\eta(u_\alpha) = 0$) hits the u axis before it reaches u_β. This shows also that the critical value of $\bar{v}\cdot\bar{n} - c$ in equality (11), i.e. $F_\alpha(\mu, u_o)$ is given by this value of q for which there exists a solution of the equation

$$\eta\eta_{,\phi} + f = q\eta$$

satisfying $\eta(u_\alpha) = \eta(u_o) = 0$. Clearly, other types of boundary conditions leading to the boundary layers can be investigated in the similar way.

3. Geometrical optics of the transition layers

In the case of slowly varying coefficients, i.e. when parameters μ

depends weakly on t and x it still makes sense to speak of weakly curved transition layers connecting two stable local equilibria $u_\alpha(t,x)$, $u_\beta(t,x)$. If the ratio of the layer thickness to the curvature radius is of order $O(\varepsilon)$ then the error which we make taking μ and $\bar{n}$ not being constant is of order $O(\varepsilon)$ too. Again locally the solubility conditions (8) or (9) must be satisfied.

Suppose now that the position of the center of the layer is defined by $\psi(t,x) = 0$. Expressing $\bar{v}\cdot\bar{n}$ and c in terms of derivatives of ψ

$$\bar{v}\cdot\bar{n} = \bar{v}\cdot\frac{\nabla\psi}{|\nabla\psi|}, \qquad c = \frac{\psi_t}{|\nabla\psi|}$$

and inserting these relations into (8) one arrives at the Hamilton-Jacobi type of equation

$$H\left(\frac{1}{|\nabla\psi|}\cdot(\bar{v}\cdot\nabla\psi + \psi_t),t,x\right) = 0. \tag{13}$$

In virtue of its homogeneity the function ψ is constant on trajectories of the Hamiltonian system

$$\frac{dx^\alpha}{ds} = \frac{\partial H}{\partial p^\alpha}, \qquad \frac{dp^\alpha}{ds} = -\frac{\partial H}{\partial x^\alpha}, \qquad \alpha = 0,\dots,n, \tag{14}$$

where $x^0 = t$ and $H = H\left(\frac{1}{|p|}\cdot(\bar{v}\cdot\bar{p}+p^0),x^0,x\right)$. Therefore the position of the layer at later times can be reconstructed by sending rays of the Hamiltonian system from each point of its initial position. Suppose that it is given by $\psi_0(x) = 0$ then, for any x_0 satisfying $\psi_0(x_0) = 0$, the initial condition for Eqs (14) is given by $t = 0$, $x = x_0$, $p = \nabla\psi_0(x_0)$ and p_0 can be computed by equating the Hamiltonian to zero.

We prefer at the moment to work in the coordinates in which the equation with slowly varying coefficients takes form

$$u_t + \bar{v}(\varepsilon t,\varepsilon x)\cdot\nabla u = Lu + f(u,\varepsilon t,\varepsilon x), \tag{15}$$

where $Lu = a^{\mu\nu}(\varepsilon t,\varepsilon x)u_{,\mu\nu}$. According to section 2 the transition layer solution connecting the states $u_\alpha(\mu)$, $u_\beta(\mu)$ can be expressed as

$$u = \phi(\bar{n}\cdot\bar{x} - ct,\mu), \tag{16}$$

where by (8) $\bar{n}$, c and μ are related. Since (16) is translation invariant we may assume that $\phi''(0,\mu) = 0$ and this makes ϕ to be uniquely determined by $\bar{n}$ and μ. Thus the general form (16) depends on two sets of parameters $\bar{n}$, with $\bar{n}^2 = 1$, and μ. In the case of equation (15) the role of μ is played by εt and εx.

Let us assume now that the states of local equilibrium $u_\alpha(\varepsilon t,\varepsilon x)$, $u_\beta(\varepsilon t,\varepsilon x)$ are defined in $\mathbb{R}^+\times\mathbb{R}^n$, and suppose that a C^∞ function $\psi(\varepsilon t,\varepsilon x)$ satisfies appropriate Hamilton-Jacobi equation in $\mathbb{R}^+\times\mathbb{R}^n$, and let for every $t \geqq 0$, $\psi(\varepsilon t,\varepsilon x) = 0$ defines a smooth surface without the boundary (i.e. closed or infinite). If this is the case then we define

$$\tilde{u}(t,x) = \phi\left(\frac{1}{|\nabla\psi|}\psi(\varepsilon t,\varepsilon x),\varepsilon t,\varepsilon x\right), \tag{17}$$

where ϕ is the general planar solution (16) for the constant coefficients case. One easily verifies that:

LEMMA 1. $\tilde{u}(t,x)$ *satisfies Eq.(15) up to terms of order* $O(\varepsilon)$. *This estimate is, however, not uniform in* t, x *and it is valid only for those points* t, x *for which* $\psi(\varepsilon t,\varepsilon x) \sim O(\varepsilon)$. *On the other hand if* $\psi \sim (\varepsilon^{-\frac{1}{2}})$ *then Eq.(15) is satisfied up to terms of order* $O(\varepsilon^{\frac{1}{2}})$.

In order to construct a uniform approximation we take a $C_0^\infty(\mathbb{R}^1)$ function $\chi(s)$ such that for some convenient positive constant A

$$\chi(s) = 1 \quad \text{for} \quad |s| < A \qquad \text{and} \qquad \chi(s) = 0 \quad \text{for} \quad |s| > 2A.$$

Let now $\chi_\varepsilon = \chi(\varepsilon^{\frac{1}{2}}\psi)$, then we define

$$u_A(t,x) = \begin{cases} \tilde{u}(t,x)\chi_\varepsilon + u_\alpha(\varepsilon t,\varepsilon x)(1-\chi_\varepsilon) & \text{if} \quad \psi \leqq 0, \\ \tilde{u}(t,x)\chi_\varepsilon + u_\beta(\varepsilon t,\varepsilon x)(1-\chi_\varepsilon) & \text{if} \quad \psi \geqq 0. \end{cases} \tag{18}$$

THEOREM 1. If $u_\alpha(\tau,y)$, $u_\beta(\tau,y)$ *are bounded in* $\mathbb{R}^+\times\mathbb{R}^n$ *together with their first and second derivatives then* $u_A(t,x)$ *satisfies Eq.(15) up to terms of order* $O(\varepsilon^{\frac{1}{2}})$ *and this estimate is uniformly valid in* $\mathbb{R}^+\times\mathbb{R}^n$.

This rather easy theorem we leave without the proof.

COROLLARY. It is worthy to note that the notion of the transition layer is still working even if the curvature of the surface $\psi = 0$ *is not of order* $O(\varepsilon)$ *but of order* $O(\varepsilon^p)$ *for* $p < 1$. *Clearly, the estimates in Theorem 1 is then much worse.*

If instead of the Hamilton-Jacobi equation one prefers to work with the Hamiltonian system (14), then having a smooth surface $S \subset \mathbf{R}^+ \times \mathbf{R}^n$ made of trajectories of this system one can easily define a function $\psi(\varepsilon t, \varepsilon x)$ in a $\frac{1}{\varepsilon}$ neighbourhood of S and such that it produces also a good results when applied in (17). One can take for example

$$\frac{\psi}{|\nabla\psi|} = \mathrm{dist}(x, \{S \cap (\{t\} \times \mathbf{R}^n)\}).$$

To close this section let us note that a similar construction as (18) can be done in the case of a curved boundary layer, producing thus uniformly valid $O(\varepsilon^{\frac{1}{2}})$ approximate solutions.

4. The initial value problem

After rescaling the variables t and x the Theorem 1 can be applied to Eq. (2). Then the layer thickness is not of order $O(1)$ but of order $O(\varepsilon)$. The layer solution is of the form

$$u(t,x) = \phi\left(\frac{1}{\varepsilon|\nabla\psi|}\psi, t, x\right) \quad \text{where} \quad \psi = \psi(t,x),$$

and the layer thickness is of order $O(\varepsilon)$; similarly $\chi_\varepsilon = \chi(\varepsilon^{-\frac{1}{2}}\psi)$. Again the Eq. (15) is satisfied in such a case also up to term of order $O(\varepsilon^{\frac{1}{2}})$. Finally the curvature of the layer in the Corollary can be of order $O(\varepsilon^{-p})$.

Let us assume now that a $C^\infty(\mathbf{R}^n)$ initial function $u_o(x)$ is given and such that for almost every x its value $u_o(x)$ belongs to the bassin of attraction of a one of the stable equilibria evaluated at $t = 0$. In other words, we assume that the solution $U(t,x)$ of the initial layer equation

$$u_t^\varepsilon = \frac{1}{\varepsilon} f(u,0,x), \qquad u^\varepsilon(0,x) = u_o(x) \tag{19}$$

is finite for $t \to \infty$.

The limit $U(0,x)$ of $u^{\varepsilon}(t,x)$ for $\frac{t}{\varepsilon} \to \infty$ will follow stable equilibria jumping eventually at some points from one stable state to another (Fig.1). Thus in $O(\varepsilon)$ time scale the transition layers are formed

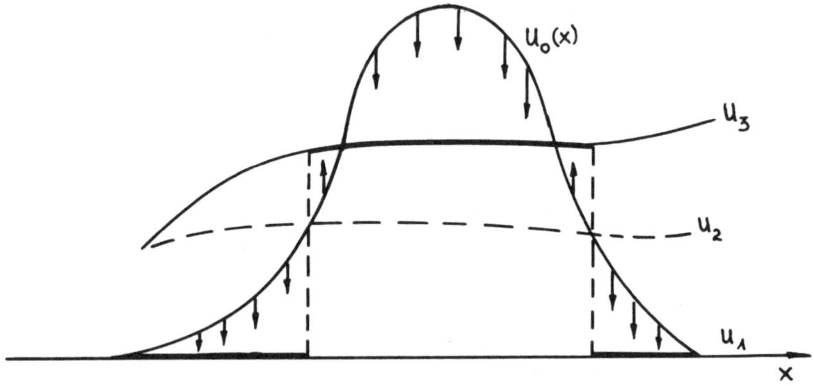

Fig.1 The solution of Eq.(19) is attracted to the stable equilibria.

in the vicinities of jump surfaces. The surfaces on which $U(0,x)$ jumps define the initial positions of the layer when the evolution in the $O(1)$ scale is considered. If these are smooth surfaces the positions of the layers at later times can be determined by solving Eq.(13) or Eqs (14) and then applying the prescription (18), the asymptotic solution in principle can be found. In general, however, the surfaces determined from Hamilton-Jacobi equation cease to be smooth for time large enough. Although in such a case prescription (18) is not working anymore, one may still consider a zero-order approximation which can be identified with the limit of a genuine solution $u(\varepsilon,t,x)$ of Eq.(15) for $\varepsilon \to 0$.

Suppose that for $0 < \varepsilon < \varepsilon_o$, $u(\varepsilon,t,x)$ is a family of solutions of Eq.(15) in $R^+ \times R^n$ satisfying the initial condition $u(\varepsilon,0,x) = u_o(x)$. If in some neighbourhood of (t_o,x_o), $t_o > 0$, the family $u(\varepsilon,t,x)$ is uniformly convergent for $\varepsilon \downarrow 0$, then under the assumptions we made the limit $u(0,t_o,x_o)$ is equal to one of the values $u_\alpha(t_o,x_o)$, $\alpha = 1,..,k$.

This can be seen by using the Green function of the operator of the left hand side of the equation $(\varepsilon v\cdot\nabla - \varepsilon^2 Lu + 1)u = f + u$. For interior points it converges to a δ-distribution when $\varepsilon \downarrow 0$.

Out of all indices $\{1,\ldots,k\}$ we select those which correspond to stable equilibria. We denote this set by $S\{1,\ldots,k\}$. Let us consider a domain $\Omega \subset R^n$ with a smooth boundary $\partial\Omega$ and let $G = (0,T)\times\Omega$. Let us assume also that:

1° $G = \bigcup_{i=1}^{N} \bar{G}_i$, where $\bar{G}_i$ are open with piecewise smooth boundaries ∂G_i. The singularities of ∂G_i are of the corner type, i.e. two smooth pieces of boundary make a dihedral angle.

2° There is a mapping $\pi\colon \{1,\ldots,N\} \to S\{1,\ldots,k\}$ such that if $\partial G_i \cap \partial G_j$ is a piece of a smooth hypersurface then $\pi(i) \neq \pi(j)$.

3° If $\pi(i) \neq \pi(j)$ then $\partial G_i \cap \partial G_j$ satisfies (at all its interior points) the Hamilton-Jacobi Eq.(8) for the transition layer connecting the stable equilibria $u_{\pi(i)}$ and $u_{\pi(j)}$.

We define the function $U(t,x)$ in $\bigcup_{i=1}^{N} G_i$

$$U(t,x) := \sum_{i=1}^{N} \chi[G_i]u_{\pi(i)}(t,x), \tag{20}$$

where $\chi[G_i]$ is the characteristic function of the set G_i.

The question arises as to whether $U(t,x)$ can be considered as the limit for $\varepsilon \downarrow 0$ of a family of genuine solutions $u(\varepsilon,t,x)$ of Eq.(2) which satisfy certain (say independent of ε) initial and boundary conditions. Clearly, in such a case the appearance of the boundary layer at $\partial\Omega$ should be expected. In subsequent sections we tray to answer this question in the case of steady solutions to Eq.(15) assuming of course, that f is t independent. In that case we take $G = \{0\}\times\Omega \sim \Omega$, $U = U(x)$ and the initial condition is not present in the formulation of our problem.

5. Stability of steady layers

From now on we assume that f, u and the coefficients of Eq.(15)

are independent of t, thus arriving at the elliptic problem in Ω

$$\varepsilon v(x)\cdot\nabla u = \varepsilon^2 Lu + f(u,x), \qquad u_{|\partial\Omega} = u_o(x). \tag{21}$$

As u is independent of t the constraint $c = 0$ appears in the Hamiltonian (8)

$$H(\bar{v}\cdot\bar{n},x) = 0. \tag{22}$$

When, however, the layer stability will be considered we will recall the t derivative in Eq.(21) and consequently c in Eq.(22).

In this section stability of formal solutions, as given by (18) or (20), will be considered within the framework of formal theory (i.e. based on Eq.(8)). It turns out, however, that even on the basis of the formal theory, stability arguments can be used in order to select proper corners. When speaking of "solutions" of the form (20), the position of the layer will be identified with the position of its center (i.e. with the surface on which $\psi = 0$).

Let us consider now a transition layer at a stationary position. Suppose that at some moment the layer was displaced along its normal to another position. For small displacement these two positions are parallel. In the new position according to Eq.(8) the layer cannot, in general, be steady and its position will evolve in time.

DEFINITION. *If after small parallel displacements along the normal directions the layer returns to its original position then it is said to be locally stable, otherwise it is said to be locally unstable.*

Let us compute the variation in the velocity of the layer which results from the variation of the layer position. Let $\delta x = \bar{n}\cdot\delta s$, then $\delta\bar{n} = 0$, $\delta\bar{v} = (\bar{n}\cdot\nabla)\bar{v}$ and as it follows from Eq.(8)

$$\delta c = \{(\bar{n}\cdot\nabla)\bar{v}\cdot\bar{n} + \frac{1}{H_{,q}}\,\bar{n}\cdot\nabla H\}\delta s, \tag{23}$$

where as before $q = \bar{v}\cdot\bar{n} - \bar{c}$. Therefore to assure the local stability it is required that at the steady position the following inequality

$$\bar{n}\cdot\nabla\bar{v} + \frac{1}{H_{,q}}\,\bar{n}\cdot\nabla H < 0 \qquad (24)$$

is satisfied.

Suppose now, that we are facing the situation when a corner appears at some point x_o. Suppose also, that this angle is formed of two pieces of a locally stable layer having $\bar{n}_1$ and $\bar{n}_2$ as unit normals in the vicinity of x_o (Fig.2). The problem can be reduced to two dimensions by taking the crossection of the angle with the two dimensional plane, call it π, which is spanned by $\bar{n}_1, \bar{n}_2$ and which passes through x_o. This is justified by the fact that the component of $\bar{v}$ which is perpendicular to $\bar{n}_1, \bar{n}_2$ will not appear in Eq.(8) or Eq.(9). It follows also from Eq.(9) which can be written in our case as

$$\bar{v}_\pi\cdot\bar{n} = F(x_o),$$

that the arms A and B on Fig. 2 are symmetric with respect to the direction $\bar{v}_\pi$ at x_o.

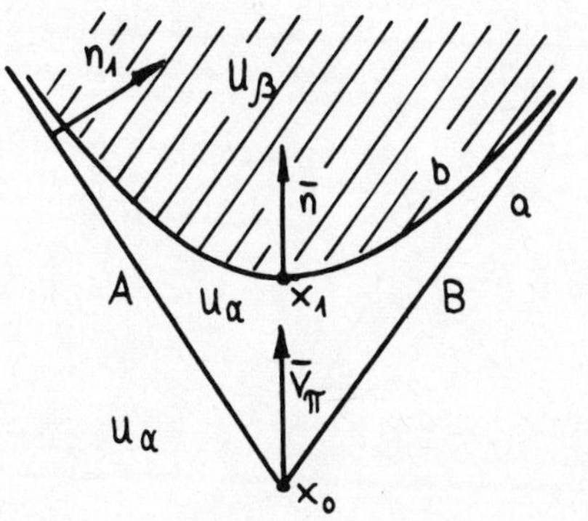

Fig. 2 *a)* The original corner position with a corner; *b)* The new, shifted position of the layer.

Let us consider a new position of the layer as it is shown on Fig. 2 and such that the distance $d(x_o,x)$ between x_o and x_1 (at x_1 the shifted layer is perpendicular to $\bar{v}$) is small enough. According to Corollary in section 3, $d(x_o,x)$ can be as small as $A\varepsilon^p$, with any $0 < p < 1$, still not violating the notion of the layer.

According to the convention made in section 2 for $u_\beta > u_\alpha$ the unit normal $\bar{n}_1, \bar{n}_2$ and $\bar{n}$ are directed as it is shown on Fig. 2. Take $d(x_o,x_1) \sim O(\varepsilon^p)$, $0 < p < 1$. Estimating from Eq.(9) the velocity c of the layer at the point x_1 we have

$$c = \bar{v}_\pi(x_o)\cdot\bar{n} - F(x_o) + O(\varepsilon^p). \tag{25}$$

On the other hand for any point $\tilde{x}$ of the original position which is also laying in $O(\varepsilon^p)$ neighbourhood of x_o we have

$$\bar{n}_1\cdot\bar{v}_\pi(x_o) - F(x_o) = O(\varepsilon^p). \tag{26}$$

(The original layer is supposed to be stationary.) From these equations it follows that the velocity of the shifted layer at the point x_1 is a quantity of order $O(1)$

$$c = \bar{v}_\pi(x_o)\cdot(\bar{n} - \bar{n}_1) + O(\varepsilon).$$

Since the velocity vector is $c\bar{n}$ we conclude, that the shifted layer can not return to its original position. *Thus for* $u_\beta > u_\alpha$ *the type of corners as one shown on Fig.2 cannot be stable.* Let now $u_\alpha > u_\beta$. Although in this case the directions of $\bar{n}_1$, $\bar{n}_2$ and $\bar{n}$ are reversed (as pointing always to larger u) yet, $c = \bar{v}\cdot(\bar{n} - \bar{n}_1)$ changes sign too; therefore $c\bar{n}$ remains unchanged. The stable cases arise if the orientation of the angle with respect to $\bar{v}_\pi$ is reversed. The analysis of different cases is summarized on Fig.3.

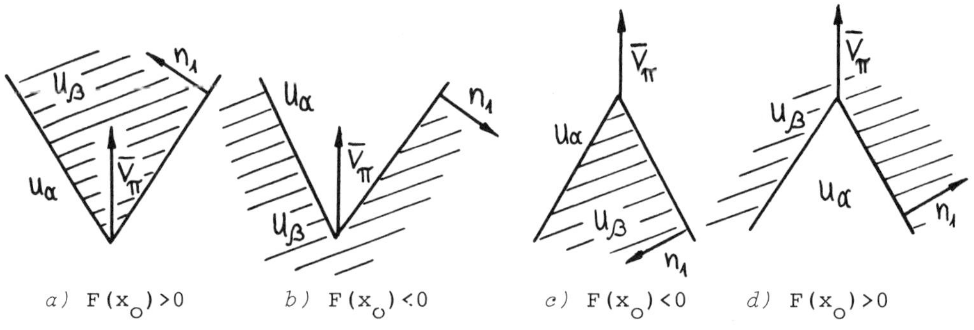

Fig.3 Assuming that in the shadoved region u is larger we have four possibilities of orientation of the corner. Two of them, i.e. *a)* and *b) are unstable* and the other two i.e. *c)* and *d) are stable.*

Other four cases, where the angle (that one which is less than π) is located on one side of the direction $\bar{v}_\pi$, must be rejected. As it was noticed the layer should be symmetrical with respect to $\bar{v}_\pi$.

One can tray to apply a similar reasoning in the case when the transition layer approaches the boundary of Ω and merges with the boundary layer (Fig.4). For example, in case *a)* of Fig.4 the transi-

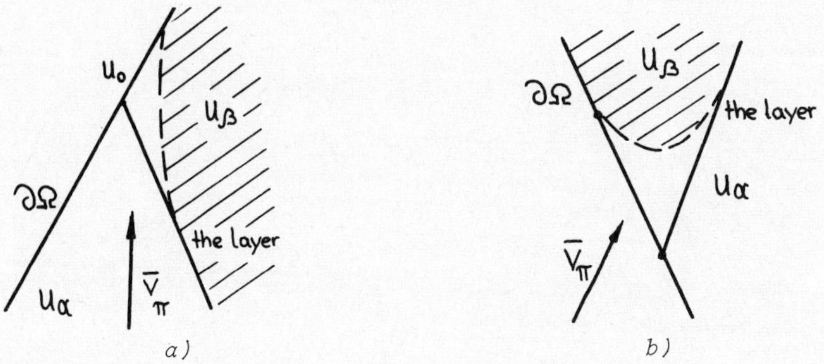

Fig. 4

tion layer exhibits some stability, whereas in case *b)* it seems to be unstable. However, the reasoning is not so clear as before. At the perturbed situations (dashed lines) we still have corners formed by two pieces of different type of boundary layer and the transition layer. Therefore the corner cannot be replaced by a smooth configuration.

6. Construction of the asymptotics

Let us come back to Eq.(8) or its equivalent form (9). According to these equations in the stationary case, the domain Ω can be divided into three regions Ω_-, Ω_o, Ω_+:

1° In Ω_- we have $\bar{v}(x)\cdot\bar{n} - F(x) < 0$ for any unit vector $\bar{n}$;

2° In Ω_o we have unit vectors $\bar{n}$ satisfying $\bar{v}(x)\cdot\bar{n} - F(x) = 0$;

3° In Ω_+ we have $\bar{v}(x)\cdot\bar{n} - F(x) > 0$ for any unit vector $\bar{n}$.

As there are no solutions of the Hamiltonian system (14) in Ω_+ and Ω_-, the regions Ω_+ and Ω_- are said to be impenetrable.

An important role in the construction of the asymptotics is played by the critical points, i.e. this points where $\bar{v}$ enters the impenetrable regions perpendicularly to their boundaries. In two dimensions, at those points, the two possible solutions of Eq.(9) for $\bar{n}$ merge together. In higher dimensions the cone of directions $\bar{p}$ satisfying $H(\frac{\bar{p}}{|p|},x) = 0$ degenerates to a plane which is tangent at x_o to $\partial\Omega_+$ or $\partial\Omega_-$ respectively. In order to illustrate this, let us consider a case as it is shown on Fig. 5. In the case *a)* the transition layer

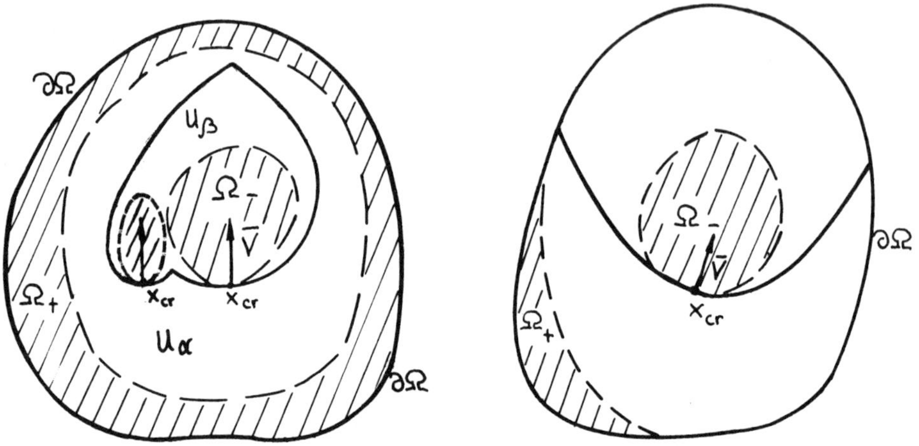

a) *b)*

Fig.5 The construction of the layer surface (impenetrable regions are shadowed)

cannot come to $\partial\Omega$ and therefore it must form a closed surface. But the closed surface without forbidden corners contains, as it can be easily noticed, at least one critical point. Thus in order to avoid forbidden corners (as *a)* and *b)* on Fig.3) the construction of the layer surface should be started from critical points.

In order to conclude our considerations we make a few remarks on the question of validity of the formal asymptotics. Having formal asymptotic solution u_A one may tray to construct so called super and subsolution, i.e. two functions $\tilde{u}_\varepsilon$ and $\underset{\sim}{u}_\varepsilon$ for $0 < \varepsilon < \varepsilon_o$ such that

a) $\bar{v}\cdot\nabla\tilde{u} - L\tilde{u} - f < 0$ (supersolution),

b) $\bar{v}\cdot\nabla\underset{\sim}{u} - L\underset{\sim}{u} - f > 0$ (subsolution).

If in addition $\tilde{u} \geqq \underset{\sim}{u}$ in $\bar{\Omega}$ then for any boundary condition such that $\tilde{u}_{|\partial\Omega} \geqq u_o \geqq \underset{\sim}{u}_{|\partial\Omega}$ there exists a solution, say u, of the equation

$$\bar{v}\cdot\nabla u = Lu + f. \tag{27}$$

Moreover, $\tilde{u} \geqq u \geqq \underset{\sim}{u}$ is satisfied [4,10].

In order to construct such super and subsolutions one may replace f by $f_\sigma = f + \sigma$, where σ is a parameter. The stable states are increasing with σ (for σ small enough). We need also a certain kind of expansion property, i.e. when σ increases the layers should expand in the direction, opposite to its normal $\bar{n}$. This is assured if the stability condition (24) is satisfied. Let S_σ be the new position of the layer (obtained by replacing f by f + σ). For σ positive all concave corners as d) on Fig.3 appearing in S_σ should be made smooth with the curvature of the order $O(\varepsilon^{\frac{1}{2}})$. Along the surface S_σ the transition layer can be placed according to the prescription (18). This produces a smooth function everywhere with the exception of some vicinities of the convex corners (as *c)* on Fig.3). Again, by making there the level surfaces smooth but of the curvature of $O(\varepsilon)$, one obtains a function, call it $u_A(\varepsilon,\sigma,t,x)$, which appears to be a supersolution of Eq.(27). For σ negative this construction must be repeated replacing the word convex by concave and vice versa. The function u_A resulting from this construction turns out to be now (for ε small, $\varepsilon<\varepsilon_o$) a subsolution of Eq.(27). Moreover, we have

$$u_A(-\sigma,\varepsilon,0,x) \leqq u^o(x) \leqq u_A(\sigma,\varepsilon,0,x)$$

for $0 < \sigma < \sigma_o$ and $0 < \varepsilon < \varepsilon_o(\sigma)$, where $u^o(x)$ is our zero-order approximation. This proves the existence of a genuine solutions approaching $u^o(x)$ when $\varepsilon\downarrow 0$.

The above construction is working in these cases when the transition layer does not comes to the boundary of Ω. Using the Serrin's sweeping principle the nonexistence of solutions with corners of types *a*) and *b)* on Fig.3 can be proved.

References

[1]. Eckhaus, W., van Harten, A., Peradzyński, Z. "A singularly perturbed free boundary problem describing a laser sustained plasma", SIAM J.Appl.Math., vol. 45, No. 1, 1985.

[2]. Eckhaus, W., van Harten, A., Peradzyński, Z. "Plasma produced by a laser in a medium with convection and free surface satisfying a Hamilton-Jacobi equation", (to be published).

[3]. Peradzyński, Z. "Singular perturbation with a free boundary", [in] New Problems in Mechanics of Continua, eds O. Brulin, R.K.T. Hsieh, University of Waterloo Press, 1983.

[4]. Sattinger, D.H. "Topics in Stability and Bifurcation Theory", Lecture Notes in Mathematics 309, Springer Verlag 1973.

[5]. Serrin, J. "Nonlinear Elliptic Equations of Second Order", AMS Symposium in Partial Differential Equations, Berkeley, 1971.

[6]. Trudinger, N.S. "Local estimates for subsolutions and supersolutions of general second order elliptic quasilinear equations", Invent. Math., vol. 61, (67-79), 1980.

[7]. Amann, H. "Invariant sets and existence theorems for semilinear parabolic and elliptic systems", J.Math.Anal.Appl., vol. 65, (432-467), 1978.

[8]. Schmitt, K. "Boundary value problems for quasilinear second order elliptic equations", Nonlinear Analysis Theory, Methods and Applications, vol. 2, (263-309), 1978.

[9]. Kaźmierczak, B., Peradzyński, Z. "Transition layers in solutions to elliptic systems", (in preparation).

[10]. Gilbarg, D., Trudinger, N.S. "Elliptic Partial Differential Equations of Second Order", Springer Verlag, Berlin-Heidelberg-New York, 1977.

Session V:

NON-NEWTONIAN FLUIDS

CONSTITUTIVE MODELS OF POLYMER FLUIDS: TOWARDS A UNIFIED APPROACH

H. Giesekus
Department of Chemical Engineering
University of Dortmund
P.O.Box 500500, D-4600 Dortmund 50 (F.R.G.)

1. Introduction

In rheology, the science of deformation and flow of materials, there exist two fundamentally different approaches for obtaining theoretical models: a phenomenological and a structural approach. In analogy to thermodynamics, the phenomenological theories start from some more or less general theorems which either have the character of principles (although not always strictly so) or define classes of ideal materials. In contrast, the structural theories are based on microscopic pictures of the respective materials. However, as a rule these are far too complicated to be applied directly and so they must be simplified with only the most significant features being retained. Moreover, the incomplete knowledge of structural details of these materials necessitates inclusion of more or less plausible suppositions which can only be tested by comparison of the resulting predictions with experiments.

The combination of simple structural features with empirical insertions results in so-called hybrid theories. Such constitutive models of polymer fluids are considered here, in particular for concentrated solutions and melts of flexible polymers. In spite of quite different ways of looking at the matter, remarkable similarity is observed between the predictions of these models. This stimulates an attempt to search for a description that unifies the various models of polymer fluids. However, before discussing these ideas more fully, some background information is provided in the form of a short outline of the phenomenological theory of so-called simple fluids.

2. Simple Fluids

A *simple material* is defined by the restriction that every local motion which maps material points (denoted by vectors $\underline{X}$ of the so-called reference configuration) into positions in space (denoted by one-parametric continuous manifolds $\underline{x}(t)$) is completely described by the history of the material deformation gradient $\underline{\underline{F}}(\underline{X},t)$ with components $\partial x_i(\underline{X},t)/\partial X_K$, defined by

$$d\underline{x}(t) = \underline{\underline{F}}(\underline{X},t)\cdot d\underline{X}, \quad dx_i(t) = \frac{\partial x_i(X_L,t)}{\partial X_K}\, dX_K \, , \tag{1}$$

so that the stress $\underline{\underline{T}}(\underline{x},t)$ is given as a functional

$$\underline{\underline{T}}(\underline{x},t) = \mathop{\mathcal{F}}_{s=0}^{\infty} [\underline{\underline{F}}(\underline{X},t-s)] \, . \tag{2}$$

If the principle of *material-frame indifference* (or *objectivity*) is introduced, this functional can be specialized as

$$\underline{\underline{T}}(\underline{x},t) = \underline{\underline{F}}(\underline{X},t)\cdot \mathop{\mathcal{G}}_{s=0}^{\infty} [\underline{\underline{C}}(\underline{X},t-s)]\cdot\underline{\underline{F}}^{\dagger}(\underline{X},t) \tag{3}$$

where

$$\underline{\underline{C}}(\underline{X},t) = \underline{\underline{F}}^{\dagger}(\underline{X},t)\cdot\underline{\underline{F}}(\underline{X},t), \quad C_{JK}(X_L,t) = \frac{\partial x_i(X_L,t)}{\partial X_J}\,\frac{\partial x_i(X_L,t)}{\partial X_K} \tag{4}$$

is the so-called (material) Cauchy-Green deformation tensor.

A *simple fluid* may be defined by the additional restriction that $\mathcal{G}$ should be commutable with every element $\underline{\underline{P}}(t)$ of the unimodular group (i.e. det $\underline{\underline{P}}(t) = 1$):

$$\underline{\underline{P}}(t)\cdot \mathop{\mathcal{G}}_{s=0}^{\infty} [\underline{\underline{C}}(t-s)]\cdot\underline{\underline{P}}^{\dagger}(t) = \mathop{\mathcal{G}}_{s=0}^{\infty} [\underline{\underline{P}}(t)\cdot \underline{\underline{C}}(t-s)\cdot\underline{\underline{P}}^{\dagger}(t)] \, . \tag{5}$$

This implies that the present configuration $\underline{x}(t)$ can be chosen as the reference configuration so that the constitutive equation simplifies to

$$\underline{\underline{T}}(\underline{x},t) = \mathop{\mathcal{H}}_{s=0}^{\infty} [\underline{\underline{C}}_{(t)}(t-s), \rho(t)] \tag{6}$$

where $\rho(t)$ denotes the density and

$$\underline{\underline{C}}_{(t)}(t') = \underline{\underline{F}}^{\dagger}_{(t)}(t')\cdot\underline{\underline{F}}_{(t)}(t') \tag{7}$$

the *relative* Cauchy-Green tensor derived from

$$d\underline{x}(t') = \underline{\underline{F}}_{(t)}(t')\cdot d\underline{x}(t), \quad dx_i(t') = \frac{\partial x_i(t')}{\partial x_k(t)}\, dx_k(t) \ . \tag{8}$$

If the incompressibility constraint is added, the density is no longer included, and eq. (6) reduces to

$$\underline{\underline{T}}(\underline{x},t) = -p\,\underline{\underline{1}} + \mathop{\mathcal{Q}}_{s=0}^{\infty}[\underline{\underline{C}}_{(t)}(t-s)] \tag{9}$$

where p is an indeterminate pressure parameter. For more details see e.g. Truesdell and Noll [1].

Using a theorem of Stone-Weierstrass the above functional (9) can be expanded into a series of multiple integrals of which the first members are given by

$$\begin{aligned}\underline{\underline{T}}(\underline{x},t) = -p\,\underline{\underline{1}} &+ \int_0^\infty f_1(s)\,\underline{\underline{C}}_{(t)}(t-s)ds \\ &+ \int_0^\infty\!\!\int_0^\infty f_{11}(s,s')\,[\mathrm{tr}\,\underline{\underline{C}}_{(t)}(t-s)]\,\underline{\underline{C}}_{(t)}(t-s')ds\,ds' \\ &+ \int_0^\infty\!\!\int_0^\infty f_2(s,s')\underline{\underline{C}}_{(t)}(t-s)\cdot\underline{\underline{C}}_{(t)}(t-s')ds\,ds' + \ldots\end{aligned} \tag{10}$$

This approximation is, however, very unwieldly under more general conditions though it contains some well-known approximations. For example, if only the first integral is retained a material-frame indifferent generalization of the linear theory is obtained, although not the only possible one.

Another approximation is obtained if one assumes that the multiple integral kernels contribute to the stresses only if $s \approx s'$ etc. and this is idealized by the substitution

$$f_{11}(s,s') = \delta(s-s')\bar{f}_{11}(s), \quad f_2(s,s') = \delta(s-s')\bar{f}_2(s), \quad \text{etc.} \tag{11}$$

Application of the Cayley-Hamilton theorem leads to the most general single-integral approximation

$$\underline{\underline{T}}(\underline{x},t) = -p\,\underline{\underline{1}} + \int_0^\infty[\varphi_{-1}(s;\,I_C,II_C)\underline{\underline{C}}^{-1}_{(t)}(t-s) + \varphi_1(s;\,I_C,II_C)\underline{\underline{C}}_{(t)}(t-s)]ds, \tag{12}$$

where I_C and II_C are the first and second invariants of $\underline{\underline{C}}_{(t)}$ (or $\underline{\underline{C}}^{-1}_{(t)}$ respectively) whereas the third invariant is unity because of the incompressibility constraint. This approximation is called the K-BKZ model [2,3].

There exist other approximations of the functional (9) which in some cases are more suitable for practical purposes. This is true, in particular, for the asymptotic expansion (in analogy to a Taylor series) in kinematic tensors (e.g. the so-called Rivlin-Ericksen tensors), but these are applicable only under relatively restricted kinematic conditions. Only for very simple classes of motions, in particular *simple shear* and *simple extensional* flows, does the constitutive functional simplify in such a way that a small number of functions (shear viscosity, first and second normal-stress coefficients, and extensional viscosity) provide a complete description of the rheological behaviour.

3. Statistical Models of Polymer Fluids

In the above phenomenological theory only very general material properties were introduced, such as fluid-like behaviour and incompressibility. To describe specialized classes of fluids, such as viscoelastic fluids, some more specific structural details have to be taken into consideration. Subsequently we specialize to polymer fluids, in particular concentrated polymer fluid systems including polymer melts as well as concentrated polymer solutions.

There exist two essentially different approaches. The first, somewhat non-specifically called the *molecular* approach, idealizes the polymer molecules by chain-like structures consisting of spherical "beads" which are connected by either "springs" (i.e. elastic bonds) or "rods" (i.e. rigid bonds). This idea was originally only used for dilute solutions. In contrast, the second so-called *network* approach was derived from the theory of solid rubbers by substituting the permanent junctions of a cross-linked network structure by transient ones, usually called "entanglements". Here only the main features of both approaches can be outlined with attention restricted to the simplest examples of each class. For a more detailed description and comparison see e.g. Giesekus [4] and the references cited there, in particular the monograph of Bird et al. [5].

3.1 Molecular Models

The simplest non-trivial model consisting of beads and springs is a dilute solution of so-called Hookean dumbbell molecules in a Newtonian solvent. These dumbbells are composed of two beads connected by a linear or Hookean spring, cf. Figure 1. The number of dumbbells per unit volume is assumed to be n, the distance vector between the centres of the beads $\underline{R}$ and the drag coefficient of each bead for the motion relative to the surrounding fluid ζ. A distribution function $\Psi(\underline{R},t)$ then exists which describes the number density of dumbbells expected to be found in an infinitesimal configuration volume $d\underline{R} = dR_1 dR_2 dR_3$ at time t, so that

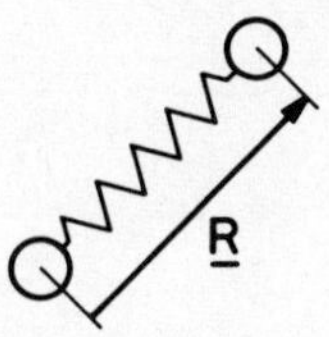

Figure 1. Elastic dumbbell molecule

$$\int_{\underline{R}} \Psi(\underline{R},t)\,d\underline{R} = n \;. \tag{13}$$

Because dumbbells are neither generated nor annihilated this distribution function obeys a continuity equation

$$\frac{\partial\Psi}{\partial t} + \frac{\partial}{\partial\underline{R}} \cdot (\dot{\underline{R}}\Psi) = 0 \tag{14}$$

called the *diffusion equation.*

To solve this equation $\dot{\underline{R}}$ has to be given as a function of the macroscopic flow and microscopic thermal (i.e. Brownian) motion. The velocity gradient of the macroscopic flow field at a point $\underline{x}$ is

$$\partial\underline{v}(\underline{x},t)/\partial\underline{x} = D\underline{\underline{F}}_{(t)}(t')/Dt'\big|_{t'=t} = \underline{\underline{L}} = \underline{\underline{D}} + \underline{\underline{W}} \tag{15}$$

where $\underline{\underline{D}}$ denotes the symmetric rate of deformation and $\underline{\underline{W}}$ the antisymmetric rate of rotation tensor. If $\dot{\underline{R}}$ is understood to be an infinitesimal quantity in comparison with the macroscopic length scale, then only $\underline{\underline{L}}$ is required for the description of the so-called affine part of this motion

$$\dot{\underline{R}} = \underline{\underline{L}}\cdot\underline{R} + \frac{2}{\zeta}\,\underline{F} \tag{16}$$

whereas the non-affine part originates from the forces exerted on the respective beads.

These forces can be described as the negative gradient of a Helmholtz free energy

$$W = \Phi(R) + kT \ln \Psi \tag{17}$$

with $\Phi(R)$ being the spring potential depending only on the magnitude $R = |\underline{R}|$, and $kT \ln \Psi$ the Brownian motion term calculated from momentum transfer in the near-equilibrium state. In the special case of Hookean dumbbells the spring potential is given by

$$\Phi(R) = \frac{H}{2} \underline{R}\cdot\underline{R} \ . \tag{18}$$

By introducing these expressions into eq. (14) the basic differential equation for the distribution function $\Psi(\underline{R},t)$ is obtained. Although it looks quite complicated, this equation can be solved exactly, at least in the case of steady flow conditions. However, it does not have to be solved to calculate the constitutive equation. Instead, multiplication with the tensorial product $\underline{\underline{RR}}$ and integration over the full space of $\underline{R}$ produces a very much simpler equation:

$$\frac{\mathcal{D}}{\mathcal{D}t} \langle RR \rangle = - \frac{4}{\zeta} [H \langle RR \rangle - kT \underline{\underline{1}}] \ . \tag{19}$$

The arrow-shaped brackets designate ensemble averages and $\mathcal{D}/\mathcal{D}t$ denotes the so-called *upper Oldroyd derivative* defined by

$$\frac{\mathcal{D}}{\mathcal{D}t} \langle \underline{\underline{RR}} \rangle = \frac{D}{Dt} \langle \underline{\underline{RR}} \rangle - [\underline{\underline{L}}\cdot \langle \underline{\underline{RR}} \rangle + \langle \underline{\underline{RR}} \rangle \cdot \underline{\underline{L}}^{\dagger}] \ . \tag{20}$$

For the equilibrium state ($\underline{\underline{L}} = 0$) the tensor $\langle \underline{\underline{RR}} \rangle$, which signifies the average distribution of the set of vectors $\underline{R}$, becomes isotropic:

$$\langle \underline{\underline{\overset{o}{R}\overset{o}{R}}} \rangle = \frac{1}{3} \langle \underline{\overset{o}{R}}\cdot\underline{\overset{o}{R}} \rangle \underline{\underline{1}} = \frac{kT}{H} \underline{\underline{1}} \ . \tag{21}$$

Because of this it seems appropriate to describe the transition from $\langle \underline{\underline{\overset{o}{R}\overset{o}{R}}} \rangle$ to $\langle \underline{\underline{RR}} \rangle$ by a mapping with a tensor $\underline{\underline{b}}$, called the *configuration tensor*, as follows

$$\langle \underline{\underline{RR}} \rangle = \underline{\underline{b}}\cdot \langle \underline{\underline{\overset{o}{R}\overset{o}{R}}} \rangle = \frac{1}{3} \langle \underline{\overset{o}{R}}\cdot\underline{\overset{o}{R}} \rangle \underline{\underline{b}} = \frac{kT}{H} \underline{\underline{b}} \ . \tag{22}$$

Therewith eq. (20) can be transformed to the *evolution equation*

$$\frac{\mathcal{D}\underline{\underline{b}}}{\mathcal{D}t} + \frac{1}{\lambda} (\underline{\underline{b}} - \underline{\underline{1}}) = 0, \quad \lambda = \zeta/4H \ . \tag{23}$$

The complete stress tensor $\underline{\underline{T}}$ consists of three terms:

$$\underline{\underline{T}} = -p\,\underline{\underline{1}} + 2\eta_o\,\underline{\underline{D}} + \underline{\underline{S}}\,, \tag{24}$$

where η_o is a Newtonian viscosity corresponding to the solvent including the influence of the beads themselves and $\underline{\underline{S}}$ is the excess stress tensor originating from the elastic connections between the beads. A straight-forward calculation enables this tensor to be given by

$$\underline{\underline{S}} = -n\,\langle\underline{R}\,\underline{F}\rangle = G(\underline{\underline{b}} - \underline{\underline{1}}),\quad G = nkT\,. \tag{25}$$

If this is introduced in eq. (23) one obtains at once the constitutive equation

$$\underline{\underline{S}} + \lambda\frac{\mathcal{D}\underline{\underline{S}}}{\mathcal{D}t} = 2\eta\,\underline{\underline{D}}\,,\quad \eta = G\lambda \tag{26}$$

which is well known as the constitutive equation of an *upper convected Maxwell fluid*.

As a result of a simple analysis, this model turns out to be too crude to deliver the finer details of the rheological properties of polymer fluids. However, it predicts at least the most striking deviations from Newtonian behaviour, namely a first normal-stress difference in steady simple shear flow and an extensional viscosity that increases with increasing rate of extension in steady simple extensional flow.

There exist many generalizations of this model to obtain a closer approximation to the microscopic structure of polymer molecules and, consequently, a better predictive quality. First, the Hookean dumbbell can be substituted by a so-called *Rouse chain molecule*, namely a chainlike structure of, say, N + 1 beads connected by N Hookean springs, see Figure 2. It is found that this leads to a set of N equations of the same type as eqs. (23) and (25), more precisely:

$$\underline{\underline{S}} = \sum_{j=1}^{N}\underline{\underline{S}}_j\,,\quad \underline{\underline{S}}_j = G(\underline{\underline{b}}_j - \underline{\underline{1}}),\quad \frac{\mathcal{D}\underline{\underline{b}}_j}{\mathcal{D}t} + \frac{1}{\lambda_j}(\underline{\underline{b}}_j - \underline{\underline{1}}) = 0 \tag{27}$$

with $\eta_j = G\lambda_j$, $G = nkT$ and a special distribution of relaxation times λ_j, see Bird et al. [5]. The configuration tensors $\underline{\underline{b}}_j$ are, however, not directly associated with the vectors $\underline{R}_j$ connecting neighbouring beads but to so-called normal co-ordinates $\underline{\underline{Q}}_j$ derived from these by a principal axes transformation.

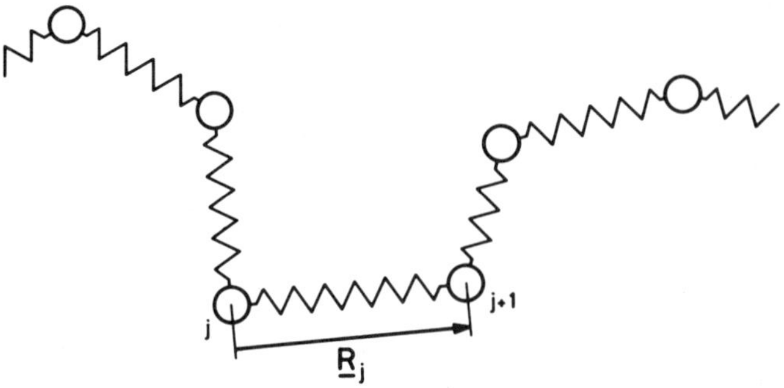

Figure 2. Elastic chain molecule

As a result, the constitutive equation of a solution of Rouse chains is equivalent to that of a multitude of dumbbell molecules with a special distribution of relaxation times. Constitutive equations of this type can easily be transformed into a single equation of integral type, which is included in eq. (12) as a special case with $\varphi_{-1} = \varphi_{-1}(s)$, $\varphi_1 = 0$, and called the *Walters fluid B'*.

A further generalization considers the so-called hydrodynamic interaction, i.e. the influence of the motion of a bead on the local fluid velocity at the positions occupied by the other beads. In principle this can be carried out by means of the *Oseen tensor*, but this method is only practicable for the most simple models. In particular, generalizations to more concentrated systems have so far only been successful using semi-empirical approaches. Two different paths have been followed. In the first, the upper Oldroyd derivatives are substituted by combined Oldroyd (or Gordon-Schowalter) derivatives

$$\frac{\mathbb{D}\underline{\underline{b}}}{\mathbb{D}t} = \frac{\mathcal{D}\underline{\underline{b}}}{\mathcal{D}t} + \xi(\underline{\underline{D}}\cdot\underline{\underline{b}} + \underline{\underline{b}}\cdot\underline{\underline{D}}), \quad 0 \leq \xi \leq 1. \tag{28}$$

In this way more realistic predictions are obtained, for example of shear-rate dependent viscosity as well as first and even second normal-stress differences, but in more complicated flows some peculiarities are found which do not occur with the upper convected Maxwell and Walters B' fluids, cf. Larson [6].

Following the second path, first taken by Giesekus [7], the constant scalar drag coefficient ζ is replaced by a configuration-dependent tensorial drag coefficient $\underline{\underline{\zeta}}$ or mobility $\underline{\underline{\zeta}}^{-1} = \overset{\circ}{\zeta}\underline{\underline{\beta}}$, where $\underline{\underline{\beta}}$ is called the relative mobility tensor. It can be understood to depend on either the actual configuration, as in the so-called reptation theories of Doi and Edwards [8] and - in a more explicit way - of Curtiss and

Bird [9], or on the mean configuration [7,10,11] represented by the configuration tensor $\underline{\underline{b}}$ or an equivalent quantity. The first interpretation leads, if certain restrictions are introduced, to an integral model of the K-BKZ type, eq. (12), but this shall not be further discussed here. The mean-configuration model, on the other hand, leads to a slight generalization of the evolution equation (23):

$$\frac{\mathcal{D}\underline{\underline{b}}}{\mathcal{D}t} + \frac{1}{2\lambda}\,[\underline{\underline{\beta}}\cdot(b - 1)+(b - 1)\cdot\underline{\underline{\beta}}^{\dagger}] = 0 \;. \tag{29}$$

Using the simplest non-trivial relation

$$\underline{\underline{\beta}} - \underline{\underline{1}} = \alpha(\underline{\underline{b}} - \underline{\underline{1}}) \;, \quad 0 \le \alpha \le 1 \tag{30}$$

eq. (29) can be transformed into the constitutive equation

$$(\underline{\underline{1}} + \frac{\alpha}{G}\,\underline{\underline{S}})\cdot\underline{\underline{S}} + \lambda\,\frac{\mathcal{D}\underline{\underline{S}}}{\mathcal{D}t} = 2\eta\underline{\underline{D}} \tag{31}$$

which contains the upper convected Maxwell model as a special case for $\alpha = 0$, but is otherwise non-linear in the stresses.

This model is very successful in predicting most characteristic features observed in simple test flows of moderately concentrated polymer solutions and melts of polymers with a more complicated steric structure. Only a few features are not described, such as stress overshoot in extensional start-up flow and extensional viscosities with a maximum at a certain rate of extension $\dot{\varepsilon}_{max} > 0$. However, these may also be included if a relaxation-type differential equation is substituted instead of the algebraic equation (30), namely

$$(\underline{\underline{\beta}} - \underline{\underline{1}}) + \varkappa\,\frac{\mathcal{D}}{\mathcal{D}t}\,(\underline{\underline{\beta}} - \underline{\underline{1}}) = \alpha(\underline{\underline{b}} - \underline{\underline{1}}) \;. \tag{32}$$

A detailed discussion of the predictions of this equation in connection with eqs. (25) and (29) is given in [12].

3.2 Network Models

The simplest transient network model, as proposed by Green and Tobolsky [13], consists of junctions connected by elastic *segments* (or *strands*), see Figure 3. In analogy to the Hookean dumbbell model, these elastic segments are also idealized by linear springs with

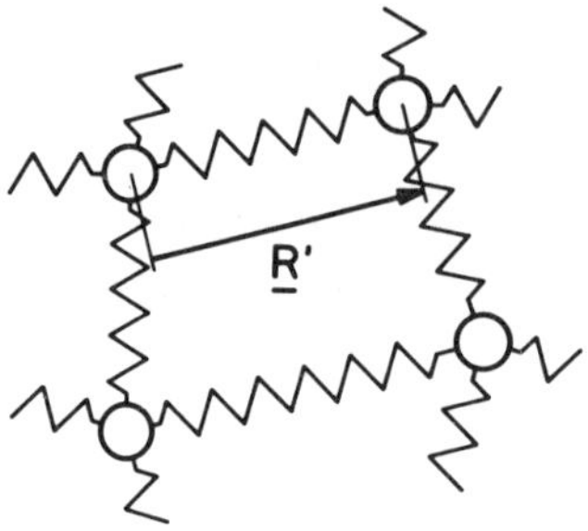

Figure 3: Elastic network

length and orientation designated by the vector $\underline{R}'$ and distributed independently according to the distribution function $\Psi'(\underline{R}',t)$. The number of segments per unit volume is assumed to be $n'(t)$ which, in contrast to the number of dumbbells, is generally considered to be time-dependent, and in analogy to eq. (13) the condition

$$\int_{\underline{R}'} \Psi'(\underline{R}',t)\,d\underline{R}' = n'(t) \qquad (33)$$

holds.

The basic assumption is that junctions are produced and destroyed, which leads to generation and loss rates for the segments of k and ℓ respectively. As a result $\Psi'(\underline{R}',t)$ does not only change because of a flux $\dot{\underline{R}}'\Psi'$ but also because of the existence of "sources" and "sinks". As a result the diffusion equation no longer retains the form of a continuity equation but rather is of the following type:

$$\frac{\partial \Psi'}{\partial t} + \frac{\partial}{\partial \underline{R}'} \cdot (\dot{\underline{R}}'\Psi') = k - \ell \, . \qquad (34)$$

The simplest assumption is that the junctions move affinely, i.e.

$$\dot{\underline{R}}' = \underline{\underline{L}} \cdot \underline{R}' \quad , \qquad (35)$$

whereas ℓ and k are proportional to the present distribution Ψ' and the isotropic equilibrium configuration $\overset{o}{\Psi}'$ respectively:

$$\ell = \frac{1}{\lambda'}\,\Psi' \quad , \qquad k = \frac{1}{\lambda'}\,\overset{o}{\Psi}' \quad . \qquad (36)$$

Since at equilibrium the rates of loss and generation must be equal, it can be concluded that the constants of proportionality for ℓ and k must be equal, with λ' having the character of an average lifetime of the junctions or a relaxation time.

Substituting eqs. (35) and (36) into eq. (34) provides the basic differential equation for the segment distribution function $\Psi'(\underline{R}',t)$. This equation looks quite different from the respective distribution equation for Hookean dumbbells and indeed their solutions differ markedly from one another as was shown recently by Larson [14]. It is therefore all the more surprising that multiplication of this equation with the tensorial product $\underline{R}'\underline{R}'$, integration over the $\underline{R}'$-space and introduction of a segment-configuration tensor $\underline{\underline{b}}'$ (defined in the same way as $\underline{\underline{b}}$ according to eq. (22)) leads to the evolution equation

$$\frac{\mathcal{D}\underline{\underline{b}}'}{\mathcal{D}t} + \frac{1}{\lambda'}(\underline{\underline{b}}' - \underline{\underline{1}}) = 0 \qquad (37)$$

which is of exactly the same form as eq. (23).

If the segments are assumed to be stressed in the same way as Hookean dumbbells under the combined influence of spring and Brownian forces, relations analogous to the excess stresses of dumbbell solutions, eq. (25), are obtained for the stress tensor:

$$\underline{\underline{S}} = \underline{\underline{G}}'(\underline{\underline{b}}' - \underline{\underline{1}}), \quad G' = n'kT \quad , \qquad (38)$$

and correspondingly the constitutive equation of an upper convected Maxwell fluid is obtained once again.

This result, though well known for many years, is very elucidating as it demonstrates that quite different structural concepts may lead to the same constitutive equation. This is the case not only for the simplest examples of the respective models but the equivalence is also maintained when the corresponding improvements are introduced into the network concept.

First, allowing for networks to consist of segments of different "complexity", i.e. with different relaxation times λ'_j, leads to an equation identical with eq. (27) (but, of course, without any restriction on the distribution of λ'_j). This is once again a Walters fluid B', although when derived in this way it is usually called the Lodge rubberlike liquid [15].

Next a "non-affine motion" of the junctions may be allowed for by substituting a combined derivative, according to eq. (28), for the upper Oldroyd derivative. This has been done independently by Johnson and Segalman [16] and Phan-Thien and Tanner [17]. If it is permitted that junctions not only be destroyed and regenerated but also that they move in a continuous manner, as is provoked by the original meaning of "entanglement", then the other generalizations introduced in the context of molecular models can also be transferred to network models, including that of configuration-dependent mobility, see e.g. Larson [18].

There is indeed a hitherto last generalization in the scope of network models, namely the introduction of a non-constant junction density, i.e. substitution of the constant λ' in the expressions for ℓ and k in eq. (36) by functions of the mean properties of the network represented by either the segment-configuration tensor $\underline{\underline{b}}'$ or the stress tensor $\underline{\underline{S}}$. If this concept is simplified by introduction of a "pseudo-steady state assumption", as is usually done tacitly, models are obtained which look quite similar to those mentioned above with the only difference being that λ' and η' are no longer constant coefficients but functions of the state of stress, i.e. $\lambda'(\underline{\underline{S}})$ and $\eta'(\underline{\underline{S}})$. If, however, the time dependence of the junction density $n'(t)$ (implying a time-dependent modulus, see eq. (38), and relaxation time) is taken seriously, this results in an additional differential equation for $x(t) = n'(t)/\overset{o}{n}'$, see Marrucci et al. [19].

This equation also bears some resemblance to the relaxation-type equation (32) connecting the relative mobility tensor $\underline{\underline{\beta}}$ with the configuration tensor $\underline{\underline{b}}$, although there are differences as $x(t)$ is a scalar quantity and therefore only invariants of $\underline{\underline{b}}'$ enter in the respective differential equation whereas in the analogous molecular model the differential equation is allowed to relate tensorial quantities. In principle such a tensorial relation could also be introduced in the network theories if the concept of entanglements is interpreted in the above mentioned wider sense. Conversely, in the molecular approach the constant modulus G could be generalized to depend on the configuration but there is no obvious physical reason for doing so.

4. Steps towards a Unified Approach

The main result of the foregoing section is that quite different structural approaches lead to equivalent or at least very similar constitutive equations. This stimulates an attempt to abstract the common characteristics from the individual features of each approach to arrive at a unified theory.

Such an attempt has already been undertaken by several authors, partly stimulated by a concept of Eckart [20] based on irreversible thermodynamics, with the network model as a background. In this case the starting point is either explicitly or implicitly the decomposition of the deformation gradient, introduced in eq. (1), into an elastic (or recoverable) part $\underline{\underline{F}}_{(e)}$ and an irreversible (also called irrecoverable, anelastic or plastic) part $\underline{\underline{F}}_{(p)}$, i.e.

$$\underline{\underline{F}}(t) = \underline{\underline{F}}_{(e)}(t) \cdot \underline{\underline{F}}_{(p)}(t) \; . \tag{39}$$

Here $\underline{\underline{F}}_{(e)}$ describes the actual deformation with respect to an associated "unloaded" state and $\underline{\underline{F}}_{(p)}$ the deformation of this unloaded state to the reference state, see e.g. Stickforth [21] or Leonov [22]. From this *multiplicative* decomposition there results an *additive* decomposition of the velocity gradient (cf. eq. (15))

$$\underline{\underline{L}}(t) = \underline{\underline{L}}_{(e)}(t) + \underline{\underline{L}}_{(p)}(t) \tag{40}$$

with

$$\underline{\underline{L}} = \frac{D\underline{\underline{F}}}{Dt} \cdot \underline{\underline{F}}^{-1} \; , \quad \underline{\underline{L}}_{(e)} = \frac{D\underline{\underline{F}}_{(e)}}{Dt} \cdot \underline{\underline{F}}_{(e)}^{-1} , \quad \underline{\underline{L}}_{(p)} = \underline{\underline{F}}_{(e)} \cdot \frac{D\underline{\underline{F}}_{(p)}}{Dt} \cdot \underline{\underline{F}}_{(p)}^{-1} \cdot \underline{\underline{F}}_{(e)}^{-1} \; , \tag{41}$$

and after a straight-forward calculation and introduction of the Finger tensor of recoverable strain

$$\underline{\underline{B}}_{(e)} = \underline{\underline{F}}_{(e)} \cdot \underline{\underline{F}}_{(e)}^{\dagger} \tag{42}$$

one obtains

$$\frac{D\underline{\underline{B}}_{(e)}}{Dt} = \underline{\underline{L}}_{(e)} \cdot \underline{\underline{B}}_{(e)} + \underline{\underline{B}}_{(e)} \cdot \underline{\underline{L}}_{(e)}^{\dagger} = (\underline{\underline{L}} - \underline{\underline{L}}_{(p)}) \cdot \underline{\underline{B}}_{(e)} + \underline{\underline{B}}_{(e)} \cdot (\underline{\underline{L}}^{\dagger} - \underline{\underline{L}}_{(p)}^{\dagger}) \tag{43}$$

which is equivalent to

$$\frac{\mathcal{D}\underline{\underline{B}}_{(e)}}{\mathcal{D}t} = \frac{D\underline{\underline{B}}_{(e)}}{Dt} - [\underline{\underline{L}} \cdot \underline{\underline{B}}_{(e)} + \underline{\underline{B}}_{(e)} \cdot \underline{\underline{L}}^{\dagger}] = - [\underline{\underline{L}}_{(p)} \cdot \underline{\underline{B}}_{(e)} + \underline{\underline{B}}_{(e)} \cdot \underline{\underline{L}}_{(p)}^{\dagger}] \; . \tag{44}$$

The tensor $\underline{\underline{B}}_{(e)}$ represents a mapping of the unloaded configuration onto the elastically deformed configuration and may therefore be identified with the configuration tensor $\underline{\underline{b}}$ defined in the foregoing section notwithstanding that in general it is not possible to associate a deformation gradient with this quantity, i.e. there is no general relation analogous to eq. (42). Making this identification and assuming at the same time that $\underline{\underline{L}}_{(p)}$ only depends on the symmetric tensor $\underline{\underline{B}}_{(e)} = \underline{\underline{b}}$,

$$\underline{\underline{L}}_{(p)} = \underline{\underline{L}}_{(p)}^{(b)} = \underline{\underline{D}}_{(p)}(\underline{\underline{b}}), \quad \underline{\underline{W}}_{(p)} = 0, \tag{45}$$

we obtain Leonov's basic formula [23]:

$$\frac{\mathcal{D}\underline{\underline{b}}}{\mathcal{D}t} = \underline{\underline{\varphi}}(\underline{\underline{b}}), \quad \underline{\underline{\varphi}}(\underline{\underline{b}}) = -2\underline{\underline{b}} \cdot \underline{\underline{D}}_{(p)}(\underline{\underline{b}}) . \tag{46}$$

This was also obtained using slightly different arguments by Dashner and VanArsdale [24], although without the specific interpretation of $\underline{\underline{\varphi}}(b)$.

To relate the configuration tensor $\underline{\underline{b}}$ to the excess stress tensor $\underline{\underline{S}}$ a Helmholtz free energy density (or elastic potential) w(b) is postulated to exist in analogy to an incompressible hyperelastic material, so that

$$\underline{\underline{S}} = 2[\frac{\partial w}{\partial I_b} (\underline{\underline{b}} - \frac{1}{3} I_b \underline{\underline{1}}) - \frac{\partial w}{\partial II_b} (\underline{\underline{b}}^{-1} - \frac{1}{3} II_b \underline{\underline{1}})] \tag{47}$$

where I_b and II_b are the first and second invariants of $\underline{\underline{b}}$. Some additional conditions must also be introduced to guarantee that the entropy production is always non-negative.

Recently Kwon and Shen [25] presented a "unified constitutive theory" which includes the above theories as special cases and additionally attempts to cover the effect of a varying density of network junctions. To this end an additional deformation gradient and associated quantities are introduced. These connect a fictitious current isotropic state (of reduced junction density) with the true equilibrium state. However, it is conjectured that such a transition from one isotropic state to another could more appropriately be described by variation of a scalar quantity analogous to a chemical potential (e.g. a dissociation potential) instead of a deformation process.

A different approach to a unified theory based on molecular as well as network models is now briefly outlined. In contrast to the foregoing approaches which start from a decomposition of the deformation and velocity gradient into a recoverable and an irrecoverable part (eqs. (39) and (40)), a clear distinction is made here from the start between the gross deformation of a volume element and the deformation of the individual molecules or network segments themselves. The former represents the relative motion of the centres of the polymer molecules and, if present, the surrounding solvent. It can be written $d\underline{X} \rightarrow d\underline{x}(t)$

and is described by the deformation gradient $\underline{\underline{F}}(t)$ according to eq. (1) or the velocity gradient $\underline{\underline{L}}(t)$ derived therefrom, cf. eq. (15). On the other hand, the deformation of the molecular structures, say $\langle \overset{o}{R}_i \overset{o}{R}_i \rangle \to \langle \underline{R}_i \underline{R}_i \rangle$, is signified by a set of configuration tensors $\underline{\underline{b}}_i$. They represent symmetric tensors like spatial measures of deformation but in contrast to these generally cannot be composed from deformation gradients. Because they are not associated with the material but define a "relative configuration volume", which is not subjected to a continuity equation, there is no a priori condition that their determinants should be restricted to unity in the case of incompressible fluid flow.

Additionally, a set of scalar functions $x_i(t)$, describing something like the relative junction density, and/or a set of tensorial functions $\underline{\underline{\beta}}_i$, describing something like the relative mobility, may also be introduced. These, however, depend in a way to be specified later on the set of configuration tensors $\underline{\underline{b}}_i$. If the respective sets each consist of more than one member, then these define a multi-mode model, whereas in the simplest case for which $i = 1$ only (so allowing the index to be omitted) a one-mode model is defined.

For the sake of simplicity, we first restrict attention to one-mode models. These are characterized by the existence of a Helmholtz free energy density $w(\underline{\underline{b}})$ which may under certain circumstances also depend on x but, it is believed, not additionally on $\underline{\underline{\beta}}$. The excess stresses are derived from this in the usual way. It is, however, necessary to take into account that these are not proportional to $III_b^{-1/2} = (\det \underline{\underline{b}})^{-1/2}$ because in this respect the analogy with a compressible hyperelastic material does not hold.

As far as the governing evolution equation is concerned two basically different cases must be distinguished. If x and $\underline{\underline{\beta}}$ are functions of the present value of $\underline{\underline{b}}$, there exists only one such equation

$$\frac{\mathcal{D}\underline{\underline{b}}}{\mathcal{D}t} = \underline{\underline{\varphi}}(\underline{\underline{b}}) \ , \qquad (48)$$

in which the upper Oldroyd derivative coupling the molecular deformation to the gross deformation may be replaced by another frame-indifferent time derivative. All models discussed in the foregoing section other than the generalized mean-configuration molecular model described by eq. (32) and the network models of the Marrucci type are covered by such an equation.

If, however, x and $\underline{\underline{\beta}}$ depend on the history of $\underline{\underline{b}}$ by means of a relaxation-type differential equation, then a pair of evolution functions are required. These are either of the form

$$\frac{\mathcal{D}\underline{\underline{b}}}{\mathcal{D}t} = \underline{\underline{\varphi}}(\underline{\underline{b}}, x), \quad \frac{Dx}{Dt} = \psi(\underline{\underline{b}}, x) \tag{49}$$

or

$$\frac{\mathcal{D}\underline{\underline{b}}}{\mathcal{D}t} = \underline{\underline{\varphi}}(\underline{\underline{b}}, \underline{\underline{\beta}}), \quad \frac{\mathcal{D}}{\mathcal{D}t}(\underline{\underline{\beta}} - \underline{\underline{1}}) = \underline{\underline{\chi}}(\underline{\underline{b}}, \underline{\underline{\beta}}). \tag{50}$$

The first of these can be understood to constitute a generalization of the Marrucci type of model and the second a generalization of the mean-configuration type of equation.

Turning briefly to multi-mode models we see at once the superiority of the concept of a priori independent introduction of gross deformation and molecular deformations over the concept of decomposition of the gross deformation in two factors according to eq. (39), since this latter concept can be generalized to multi-mode description only in a very artificial way, if at all. For the sake of simplicity we restrict attention here to the case in which all the above structural parameters depend on the present configuration only. Then a Helmholtz free energy density $w(\underline{\underline{b}}_1, \ldots, \underline{\underline{b}}_n)$ exists, from which partial stresses $\underline{\underline{S}}_i$ can be derived in the usual way with

$$\underline{\underline{S}} = \sum_{i=1}^{n} \underline{\underline{S}}_i . \tag{51}$$

Once again two cases must be distinguished for the evolution function, cf. also [10]. In the first, given as

$$\frac{\mathcal{D}\underline{\underline{b}}_i}{\mathcal{D}t} = \underline{\underline{\varphi}}_i(\underline{\underline{b}}_i), \tag{52}$$

all modes are uncoupled. Most of the multi-mode models so far discussed in the literature belong to this group; the one-mode specializations of some of these [15-17, 19, 23, 25] were mentioned in the foregoing and this section. If, however, the concept of configuration-dependent mobility is taken seriously, the respective tensors $\underline{\underline{\beta}}_i$ should depend on the full set of configuration tensors, and then all modes are coupled:

$$\frac{\mathcal{D}\underline{\underline{b}}_i}{\mathcal{D}t} = \underline{\underline{\varphi}}_i(\underline{\underline{b}}_1, \ldots, \underline{\underline{b}}_i, \ldots, \underline{\underline{b}}_n) . \tag{53}$$

Such a coupling also occurs when networks with segments of different complexities are described in a more realistic way, for every loss of a junction by which two polymer chains are linked together not only annihilates four segments but at the same time generates two new segments of higher complexity (i.e. greater length), and the reverse happens when a junction is generated, see e.g. Wiegel [26].

5. Conclusions

In spite of the result that a structural theory organised in the way outlined above could become very complicated if all possibilities were exhausted, this approach does enable the importance of various features to be estimated. In particular, it is possible to decide which terms should be included or omitted in the respective model of a polymer fluid to make it manageable without loosing any essential features. In any event a much narrower framework is supplied within which constitutive equations should be constructed than that provided by the general phenomenological theory of simple fluids.

Acknowledgement

The author is indebted to Dr. M. Hibberd for his help in improving the English.

References

1. C. Truesdell, W. Noll, in S. Flügge (ed.), Encyclopedia of Physics, Vol. III/2, Springer (Berlin 1965).
2. A. Kaye, College of Aeronautics, Cranfield, Note No. 134 (1962).
3. B. Bernstein, E.A. Kearsly, L.J. Zapas, Trans. Soc. Rheology 7, 391-410 (1963).
4. H. Giesekus, *A Comparison of Molecular and Network-Constitutive Theories for Polymer fluids*, in: J.A. Nohel, A.S. Lodge, M. Renardy (eds.): *Viscoelasticity and Rheology*, pp. 157-180, Academic Press (New York 1985).
5. R.B. Bird, O. Hassager, R.C. Armstrong, C.F. Curtiss, *Dynamics of Polymeric Liquids, Vol.* II. *Kinetic Theory*, John Wiley & Sons (New York 1977).

6. R.G. Larson, J. Non-Newtonian Fluid Mech. *13*, 279-308 (1983).

7. H. Giesekus, Rheol. Acta *5*, 29-35 (1966).

8. M. Doi, S.F. Edwards, J. Chem. Soc. Faraday Trans. II, *74*, 1789-1801, 1802-1817, 1819-1832; *75*, 38-54 (1979).

9. C.F. Curtiss, R.B. Bird, J. Chem. Phys. *74*, 2016-2025, 2026-2030 (1981).

10. H. Giesekus, Rheol. Acta *21*, 366-375 (1982).

11. H. Giesekus, J. Non-Newtonian Fluid Mech. *11*, 69-109 (1982), *12*, 367-374 (1983).

12. H. Giesekus, J. Non-Newtonian Fluid Mech. *17*, 349-372 (1985).

13. M.S. Green, A.V. Tobolsky, J. Chem. Phys. *14*, 80 (1946).

14. R.G. Larson, *Configuration distribution functions of polymer molecules.* Proc. Symp. Recent Developments in Structured Continua, University of Windsor, Ontario (Canada) May 29-31, 1985.

15. A.S. Lodge, Trans. Faraday Soc. *50*, 120 (1956).

16. M.W. Johnson, Jr., D. Segalman, J. Non-Newtonian Fluid Mech. *2*, 255-270 (1977).

17. N. Phan-Thien, R.I. Tanner, J. Non-Newtonian Fluid Mech. *2*, 353-365 (1977).

18. R.G. Larson, J. Non-Newtonian Fluid Mech. *13*, 279-308 (1983).

19. G. Marrucci, G. Titomanlio, G.C. Sarti, Rheol. Acta *12*, 269-275 (1973); D. Acierno, F.P. La Mantia, G. Marrucci, G. Titomanlio, J. Non-Newtonian Fluid Mech. *1*, 125-146 (1976).

20. C. Eckart, Phys. Rev. *73*, 373-382 (1948).

21. J. Stickforth, Int. J. Engng. Sci. *19*, 1775-1788 (1981).

22. A.I. Leonov, Rheol. Acta *21*, 683-691 (1982).

23. A.I. Leonov, Rheol. Acta *15*, 85-98 (1976).

24. P.A. Dashner, W.E. VanArsdale, J. Non-Newtonian Fluid Mech. *8*, 59-67 (1981).

25. T.H. Kwon, S.F. Shen, Rheol. Acta *23*, 217-230 (1984).

26. F.W. Wiegel, Physica *42*, 156-164 (1969).

APPLICATION OF HOMOGENIZATION TO THE STUDY

OF A SUSPENSION OF FORCE-FREE PARTICLES

T. Lévy
Université de Rouen
76130 Mont-Saint-Aignan
et Laboratoire de Mécanique Théorique (U. A. 229)
75230 Paris Cédex 05 - FRANCE

1. Introduction

Homogenization [1], [2] is an asymptotic two-scale method for studying physical and mechanical processes in media with microstructure on a scale much smaller than the macroscopic scale of interest. Under the assumption of a locally periodic structure of the medium, it furnishes a deductive procedure for obtaining the macroscopic equations of the limit phenomena as the ratio ε of the microstructure to the macrostructure tends to zero.

In the study of a non dilute suspension of solid particles S immersed in a viscous incompressible fluid, it is proved [3] that, if the structure of the mixture is locally periodic at the initial time, it evolves in time by keeping the locally periodic character ; consequently at any instant in the evolution of the system the required conditions of the homogenization techniques are satisfied with a period which is that of the structure at the same instant.

2. Formulation of the problem

The velocity $\vec{V}^\varepsilon(t, \vec{x})$ in the medium and the pressure $P^\varepsilon(t, \vec{x})$ in the fluid satisfy,

in the fluid : div $\vec{V}^\varepsilon = 0$

$$\rho_o \left(\frac{\partial V_i^\varepsilon}{\partial t} + V_j^\varepsilon \frac{\partial V_i^\varepsilon}{\partial x_j}\right) = \frac{\partial \sigma_{ij}^\varepsilon}{\partial x_j} + F_i^\varepsilon$$

$$\sigma_{ij}^\varepsilon = - P^\varepsilon \delta_{ij} + 2\mu D_{ij}(\vec{V}^\varepsilon) \quad \text{with} \quad D_{ij}(\vec{V}) = \frac{1}{2}\left(\frac{\partial V_i}{\partial x_j} + \frac{\partial V_j}{\partial x_i}\right)$$

in each particle S : $D_{ij}(\vec{V}^\varepsilon) = 0$

on the boundaries ∂S : $\vec{V}^\varepsilon$ is continuous, and for each particle S :

$$\int_S \rho_S^\varepsilon \frac{d\vec{V}^\varepsilon}{dt} dv = \int_S \vec{F}^\varepsilon dv - \int_{\partial S} \sigma_{ij}^\varepsilon n_j \vec{e}_i d\sigma$$

$$\int_S \rho_\varepsilon^S (\vec{x} - \vec{x}_G) \wedge \frac{d\vec{V}^\varepsilon}{dt} dv = \int_S (\vec{x}-\vec{x}_G) \wedge \vec{F}^\varepsilon dv - \int_{\partial S} (\vec{x}-\vec{x}_G) \wedge \sigma_{ij}^\varepsilon n_j \vec{e}_i d\sigma$$

$\vec{x}_G$ denotes the coordinate of the mass center G of the considered particle S and $\vec{F}^\varepsilon$ is the volumic density of applied forces. According to the general features of homogenization method, we introduce a new microscopic variable $\vec{y} = (\vec{x} - \vec{x}_G)/\varepsilon$ and we consider the $\vec{y}$ variable defined in a basic period Y (which depends on t and $\vec{x}$), homothetic to an actual period of the medium with the ratio ε^{-1}.

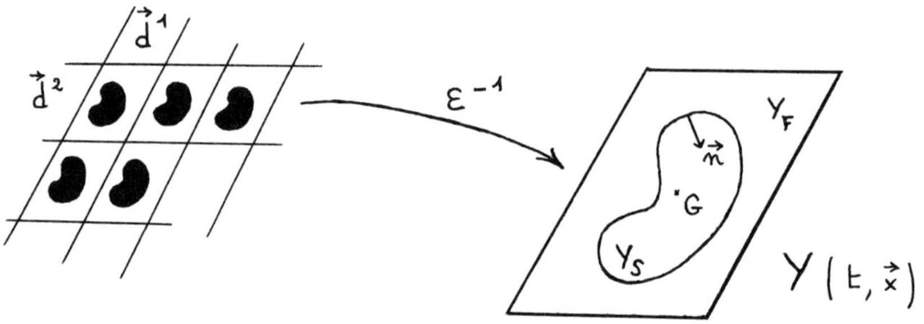

We suppose that the given quantities ρ_S^ε and $\vec{F}^\varepsilon$ are of the form $\rho_S(\vec{x}, \vec{y})$ and $\vec{F}(\vec{x}, \vec{y})$, Y-periodic in $\vec{y}$; and then taken into account the periodicity of the medium, we search $\vec{V}^\varepsilon(t, \vec{x})$ and $P^\varepsilon(t, \vec{x})$ in the form of double-scale asymptotic expansions when ε is very small

$$\vec{V}^\varepsilon(t, \vec{x}) = \vec{V}^0(t, \vec{x}, \vec{y}) + \varepsilon \vec{V}^1(t, \vec{x}, \vec{y}) + \ldots$$

$$P^\varepsilon(t, \vec{x}) = P^0(t, \vec{x}, \vec{y}) + \varepsilon P^1(t, \vec{x}, \vec{y}) + \ldots$$

with the functions $\vec{V}^i$, P^i Y-periodic with respect to the $\vec{y}$ variable. We suppose in this study that there is no action on the fluid, but on each particle S is applied a given couple proportional to its volume, precisely we take $\vec{F}$ of the form :

$$\vec{F}(\vec{x},\vec{y}) = \begin{cases} 0 & \text{if } \vec{y} \in Y_F \\ \dfrac{\phi(\vec{x}_G)\ \vec{f}(\vec{y})}{|S|^{1/3}} & \text{if } \vec{y} \in Y_S \quad \text{with } \int_{Y_S} \vec{f}(\vec{y})\,dy = 0 \end{cases}$$

3. Asymptotic expansion of the solution

We obtain for $\vec{V}^\varepsilon$:

$$\vec{V}^\varepsilon = \vec{V}^o(t,\vec{x}) + \varepsilon\left[- D_{ij}(\vec{V}^o)\ \vec{\chi}^{ij}(\vec{y}) + \phi(\vec{x})\ \vec{w}(\vec{y}) + \vec{A}(t,\vec{x})\right] + 0(\varepsilon)$$

where $\vec{\chi}^{ij}$ and $\vec{w}$ are the only solutions of the variational problems [4], [5] :

$$\begin{cases} \vec{\chi}^{ij} \in U_{ad}\ (\vec{P}^{ij}) = \left\{ \begin{array}{l} \vec{\psi} \in [H^1(Y)]^3,\ Y\text{-period},\ \int_Y \vec{\psi} dy = 0,\ \operatorname{div} \vec{\psi} = 0 \text{ in } Y_F, \\ D_{kl}(\vec{\psi}) = D_{kl}(\vec{P}^{ij}),\ \forall\ k,\ l,\ \text{in } Y_S \end{array} \right\} \\ \qquad\qquad \text{with } \vec{P}^{ij} \text{ such that } P_k^{ij} = y_j\ \delta_{ik} \\ \text{and } \int_{Y_F} 2\mu\, D_{kl}(\vec{\chi}^{ij})\ D_{kl}(\vec{\psi} - \vec{\chi}^{ij})\ dy = 0 \quad \forall\ \vec{\psi} \in U_{ad}(\vec{P}^{ij}) \end{cases}$$

$$\begin{cases} \vec{w} \in V_Y = \left\{ \begin{array}{l} \vec{u} \in [H^1(Y)]^3,\ Y\text{-period},\ \int_Y \vec{u}\ dy = 0,\ \operatorname{div} \vec{u} = 0 \text{ in } Y, \\ D_{ij}(\vec{u}) = 0,\ \forall\ i,\ j,\ \text{in } Y_S \end{array} \right\} \\ \text{and } \int_Y 2\mu\ D_{ij}(\vec{w})\ D_{ij}(\vec{u})\ dy = \dfrac{1}{|Y_S|^{1/3}} \int_Y \vec{f}(\vec{y})\ .\ \vec{u}\ dy \quad \forall\ \vec{u} \in V_Y. \end{cases}$$

Let us remark that $\vec{\chi}^{ij} = \vec{\chi}^{ji}$ and that in the solid part Y_S

$$\vec{\chi}^{ij} = \frac{1}{2}\,(\vec{P}^{ij} + \vec{P}^{ji}) + \vec{\alpha}^{ij} + \vec{\beta}^{ij} \wedge \vec{y},\ \vec{w} = \vec{a} + \vec{\omega} \wedge \vec{y}$$

with $\vec{\alpha}^{ij}$, $\vec{\beta}^{ij}$, $\vec{a}$, $\vec{\omega}$ independent of $\vec{y}$.

4. Macroscopic laws

The macroscopic equations are obtained by expressing the conservation laws in a macroscopic domain D made of whole periods but otherwise arbitrary [5], they are :

$$\operatorname{div} \vec{V}^o(t,\vec{x}) = 0$$

$$\tilde{\rho}\,\frac{\partial V_k^o}{\partial t} + \tilde{\rho}\, V_l^o\,\frac{\partial V_k^o}{\partial x_l} - \beta_{kijl}\, V_l^o\, D_{ij}(\vec{V}^o) + \phi\gamma_{kl}\, V_l^o = \frac{\partial <\sigma_{kl}^o>}{\partial x_l}$$

where $<\boldsymbol{\sigma}^o>$ is a macroscopic stress tensor defined by a mean surfacic value on the basic period Y of the first approximation of $\boldsymbol{\sigma}^\varepsilon(t, \vec{x})$. The homogenized coefficients are given by :

$$\tilde{\rho} = \frac{1}{|Y|}\,[|Y_F|\,\rho_o + \int_{Y_s} \rho_s\, dy]$$

$$\beta_{kijl} = \frac{1}{|Y|}\int_{Y_s} (\rho_s - \rho_o)\,\frac{\partial\, \chi_k^{ij}}{\partial y_l}\, dy, \quad \gamma_{kl} = \frac{1}{|Y|}\int_{Y_s} (\rho_s - \rho_o)\,\frac{\partial w_k}{\partial y_l}\, dy .$$

The calculation of $<\sigma_{kl}^o>$ [5] leads to the following macroscopic constitutive equation, where $\pi^o(t, \vec{x})$ is the mean volumic value of P^o in Y_F :

$$<\sigma_{kl}^o> = -\,\pi^o\,\delta_{kl} + a_{ijkl}\, D_{ij}\,(\vec{V}^o) + \phi b_{kl}$$

with $a_{ijkl} = \frac{2\mu}{|Y|}\int_{Y_F} D_{pq}\,(\vec{P}^{ij} - \vec{\chi}^{ij})\, D_{pq}\,(\vec{P}^{kl} - \vec{\chi}^{kl})\, dy$

$$b_{kl} = \frac{2\mu}{|Y|}\int_{Y_F} D_{pq}(\vec{w})\, D_{pq}\,(\vec{P}^{kl} - \vec{\chi}^{kl})\, dy + \varepsilon_{ipq}\,\beta_i^{kl}\,\Gamma_{qp} + \frac{\Gamma_{lk} - \Gamma_{kl}}{2}$$

and $\Gamma_{kl} = \frac{1}{|Y|\,|Y_s|^{1\,3}}\int_{Y_s} y_l\, f_k\, dy.$

The bulk stress tensor $<\boldsymbol{\sigma}^o>$ is not symmetric and the vector associated with its antisymmetrical part $\vec{P} = -\,\boldsymbol{\varepsilon} : <\boldsymbol{\sigma}^o>$ is the opposite of the density of couples by unit volume.

5. Deformation of the structure and conclusion

The deformation of the microstructure is given by the variation of the 3 vectors $\vec{d}^k$ defining the period and by the rotation rate $\vec{\Omega}(S)$ of each particle. Using the asymptotic expansion of $\vec{V}^\varepsilon$ we obtain [5]:

$$\frac{d\vec{d}^k}{dt} = \nabla\vec{V}^o\,.\,\vec{d}^k, \quad \vec{\Omega}(S) = \frac{1}{2}\,\mathrm{rot}\,\vec{V}^o - D_{ij}\,(\vec{V}^o)\,\vec{\beta}^{ij} + \phi\vec{\omega} .$$

So, in the evolution, the microstructure remains locally periodic. Furthermore the homogenized coefficients β_{kijl}, γ_{kl}, a_{ijkl}, b_{kl} which depend on the microstructure, are in fact depending on $\nabla\vec{V}^o$ and the applied couples. Particularly the dependence of the a_{ijkl}

on $\nabla\vec{V}^{o}$ points out the non-newtonian behavior of the bulk medium. We have elements to compute (at least in theory) the flow.

The limiting case of dilute suspensions ($c = |Y_s|/|Y| << 1$) is obtained at the first order with respect to the concentration c, from the preceeding results by an asymptotic analysis [6] [7]. Particularly when particles are identical spheres the Einstein viscosity and the vortex viscosity formula are found.

References

[1] BENSOUSSAN A., LIONS J. L., PAPANICOLAOU G., Asymptotic analysis for periodic structures, North-Holland, Amsterdam, 1978

[2] SANCHEZ-PALENCIA E., Non-homogeneous media and vibration theory, Lecture Notes in Physics, Vol.127, Springer-Verlag, Berlin, 1980

[3] LEVY T., SANCHEZ-PALENCIA E., Suspension of solid particles in newtonian fluid, J. Non-Newt. Fl. Mech., 13, p. 63-78, 1983

[4] LEVY T., Application de l'homogénéisation à l'étude d'une suspension de particules soumises à des couples, C. R. Acad. Sc., Paris, 299 II, p. 597-600, 1984

[5] LEVY T., Suspension de particules soumises à des couples, J. de Méca. to appear

[6] LEVY T., SANCHEZ-PALENCIA E., Suspension diluée dans un fluide visqueux de particules solides ou de gouttes visqueuses. C. R. Acad. Sc. Paris, 297 II, p. 193-196, 1983

[7] LEVY T., SANCHEZ-PALENCIA E., Einstein-like approximation for homogenization with small concentration, II Navier - Stokes equations. J. Non Lin. Anal., to appear.

ON THE ERICKSEN'S CONJECTURE

G.Mayné
Département de Mathématique - Université Libre de Bruxelles
Campus Plaine C.P.218/1 - Boulevard du Triomphe - 1050 Bruxelles

In 1956 [1], Ericksen formulated the following conjecture :
rectilinear, stationary flow in a tube (Poiseuille flow) of an incompressible non newtonian viscous fluid is only possible if the curves of constant speed have constant curvature.
Taking Cartesian axises with x_3 parallel to the flow direction, the velocity field is given by

$$v_1 = v_2 = 0 \qquad v_3 = 2\phi(x_1,x_2) \tag{1}$$

For such a flow the viscosity stress t_D of an incompressible simple fluid can be written [2] :

$$t_D = -\frac{1}{3}(k_3 + 2K^2k_2)1 + k_1V + k_2V^2 + k_3W \tag{2}$$

where k_i $(i = 1,2,3)$ are functions of $K = |\text{grad}\ \phi|$ and

$$V = \begin{pmatrix} 0 & 0 & \phi_{,1} \\ 0 & 0 & \phi_{,2} \\ \phi_{,1} & \phi_{,2} & 0 \end{pmatrix} \qquad W = \begin{pmatrix} 0 & 0 & 0 \\ 0 & 0 & 0 \\ 0 & 0 & 1 \end{pmatrix}$$

It can be shown [1] that the three equations of motion reduce to the two relations

$$\begin{cases} R' = (k_2\phi_{,j})_{,j} \\ -c = (k_1\phi_{,j})_{,j} \end{cases} \tag{3}$$

where R is an unknown function of ϕ related to the hydrostatic pressure p and the potential function of the volumic forces U :

$$p + U = -cx_3 + R(\phi) + \frac{1}{2}\int k_2 dK^2 - \frac{1}{3}(k_3 + 2K^2k_2) \tag{4}$$

Obviously, the two equations (3) are generally incompatible; in order to get compatibility conditions about the functions k_1, k_2 and R, we consider [2] the orthogonal curvilinear coordinate system (y_1,y_2) whose curves y_1 = constant are the curves of constant speed of the flow : $y_1 = \phi(x_1,x_2)$ (the curve $y_1 = 0$ is the boundary of the cross-section of the tube).
The corresponding metric tensor has two non vanishing components g_{11} and g_{22}.

$$K^2 \overset{\text{def}}{=} |\nabla\phi|^2 = \frac{1}{g_{11}} \tag{5}$$

R is a function of y_1; k_1, k_2 only depend on g_{11}.

Putting $g_1 = \sqrt{g_{11}}$, $g_2 = \sqrt{g_{22}}$, $T_1 = \frac{k_1}{k_2}$, $T_2 = \frac{g_1}{k_2}$, equations (3) become

$$\begin{aligned} g_{1,1} &= -\lambda - \mu R' \\ g_{2,1} &= g_2(\eta + \nu R') \end{aligned} \tag{6}$$

$$\begin{aligned} \text{where}\quad \lambda &= cg_1 \frac{T_2}{T_1'} \qquad \mu = g_1 \frac{T_1 T_2}{T_1'} \\ \eta &= -cg_1 \frac{T_2'}{T_1'} \qquad \nu = g_1\left(T_2 - \frac{T_2'}{T_1'} T_1\right) \end{aligned} \tag{7}$$

If $T_1' = 0$, $k_2 = \ell k_1$ (ℓ = constant), equations (3) are trivially compatible ($R' = -\ell c$) and Ericksen's conjecture fails.

Since (y_1, y_2) are plane curvilinear coordinates, g_{11} and g_{22} have to verify a zero total curvature condition :

$$\left(\frac{g_{2,1}}{g_1}\right)_{,1} + \left(\frac{g_{1,2}}{g_2}\right)_{,2} = 0 \tag{8}$$

$_{,1}$ and $_{,2}$ denote partial derivatives with respect to y_1 and y_2. So that we get a system of three equations (6) (8) for two unknowns g_1, g_2.

Multiplication of (8) by $\frac{1}{g_2} g_{1,2}$ and integration with respect to y_2 yields

$$\frac{1}{2}\left(\frac{g_{1,2}}{g_2}\right)^2 + \gamma + \delta R' + \varepsilon R'^2 + \varphi R'' = S(y_1) \tag{9}$$

where

$$\begin{aligned} \gamma' &\overset{\text{def}}{=} \frac{d\gamma}{dg_1} = c^2 \frac{g_1}{T_1'}\left(T_2 \frac{T_2'}{T_1'}\right)' \\ \delta' &= 2cg_1 \frac{T_1^2}{T_1'}\left(\frac{T_2}{T_1}\frac{T_2'}{T_1'}\right)' \\ \varepsilon' &= \frac{g_1 T_1^3}{2T_1'}\left[\frac{1}{T_1'}\left(\frac{T_2^2}{T_1}\right)'\right]' \\ \varphi' &= -\frac{T_1^2}{T_1'}\left(\frac{T_2}{T_1}\right)' \end{aligned} \tag{10}$$

S is an arbitrary function of y_1.

In order to investigate the compatibility of equations (6) and (9), we calculate $g_{1,12}$ by differentiation of (9) and (6.1) with respect to y_1 and y_2. We get a compatibility condition of the form

$$(\alpha + \beta R')(S - \gamma - \delta R' - \varepsilon R'^2 - \varphi R'') + (\lambda + \mu R')(\gamma' + \delta' R' + \varepsilon' R'^2 + \varphi' R'')$$

$$+ S' - \delta R'' - 2\varepsilon R' R'' - \varphi R''' = 0 \tag{11}$$

where $\alpha = 2(\eta + \lambda')$ $\beta = 2(\mu' + \nu)$

Relation (11) only depends on g_1 through functions represented by Greek letters and on y_1 through the functions R and S.

<u>Theorem</u> : if $T_1' \neq 0$ and if (11) depends explicitly on g_1 then Ericksen's conjecture is valid.

Proof : the curvature c_1 of the curves y_1 = constant is given by

$$c_1 = \frac{1}{g_1 g_2} g_{2,1} = \frac{1}{g_1}(\eta + \nu R')$$

If (11) depends explicitly on g_1, $g_{1,2} = 0$ and c_1 only depends on y_1 so that the curves y_1 = constant hare constant curvature. Ericksen's conjecture will fail when k_1, k_2 and R are such that (11) does not depend explicitly on g_1.
It can be shown that this will happen in the following cases :

<u>A : $c \neq 0$</u>

$$R' = k \qquad \left(\frac{g_1}{T_1'}\right)^2 (c + kT_1)^4 = p(\gamma + k\delta + k^2\varepsilon - \tfrac{1}{\ell}) \qquad S = \ell^{-1} \tag{12}$$

k, p, ℓ are constants.

<u>B : $c = 0$</u>

$$1. \quad wR''' + uR'^3 + vR'R'' + r\left(\frac{R'''}{R'}\right)' = 0 \tag{13}$$

T_1 and T_2 must satisfy the three equations :

$$\begin{aligned} \beta' f' - \beta'' f &= 0 \\ \beta' g' - \beta'' g &= 0 \\ \beta' \varphi'' - \beta'' \varphi' &= 0 \end{aligned} \tag{14}$$

where $f = (\varepsilon\beta - \mu\varepsilon')'$ $g = (\beta\varphi - \mu\varphi' + 2\varepsilon)'$

$$S = r\,\frac{R'''}{R'}$$

If the three constants u, v, w vanish, equations (14) will be compatible if

$$\left(\frac{T_2}{T_1}\right)^2 = 2ag_1^2 + 2b \tag{15}$$

$$\frac{1}{T_1} = \frac{1}{4hd^{n-2}}\left[\frac{e^{(2-n)x}}{n-2} - \frac{e^{-(2+n)x}}{n+2}\right] + \ell \qquad (16)$$

a, b, h, n, ℓ are constants; $a > 0$, $b > 0$, $d^2 = \frac{b}{a}$ and $g_1 = d \operatorname{sh} x$.

If $d = 0$

$$\frac{1}{T_1} = \frac{1}{2^n h(n-2)} g_1^{2-n} + \ell \qquad n \neq 2 \qquad (17)$$

$$\frac{1}{T_1} = -\frac{1}{4h} \ln g_1 + \ell \qquad n = 2 \qquad (18)$$

2. $R''' = pR'R'' + qR'^3$ (19)

T_1 and T_2 have to verify two relations :

$$(m - \varepsilon)\beta = -\mu\varepsilon' + q\varphi - nq$$
$$(20)$$
$$(n - \varphi)\beta = -\mu\varphi' + 2\varepsilon + p\varphi - 2m - pn$$

where p, q, m, n are constants.

3. $R'' = rR' + sR'^2$ (21)

T_1 and T_2 must satisfy only one relation :

$$\beta = t(\varepsilon\beta - \mu\varepsilon' + s\beta\varphi - s\mu\varphi' + 2s\varepsilon + 2s^2\varphi) + u \qquad (22)$$

where r, s, t, u are constants.

R.L. Fosdick and J.Serrin [3] proposed an analytic proof of Ericksen's conjecture based on the following theorem :

Let $\phi(x_1,x_2)$ be a solution of (3.2) in a bounded connected open set $\Omega \subset R^2$, where c is a non zero constant and k_1, k_2 satisfy the following conditions :

(i) k_1 and k_2 are analytic functions of K^2 near $K = 0$
k_1 is of class c^2 in K for all $K \geq 0$

(ii)
$$\frac{d(k_1 K)}{dK} > 0 \qquad \forall K \geq 0$$

If there is a point in Ω where $\nabla\phi = 0$ and if $k_2 \neq c^t k_1$ near $K = 0$, then ϕ must be either radially symmetric or plane symmetric that is :

$$\phi = \phi(r) \quad \text{or} \quad \phi = \phi(x_1) \qquad r^2 = x_1^2 + x_2^2$$

For such solutions, the curves of constant speed, of course, have constant curvature but the converse is not true; for instance, if $c = R = 0$, we have the solution $\phi = b\theta$.

In order to show that the hypotheses of the theorem are necessary, authors proposed some counter-examples for which theorem's assumptions being not fulfilled, the corresponding solutions are neither radially symmetric nor plane symmetric.

It is easy to check that for the five counter-examples considered (two for which $c \neq 0$ and three for which $c = 0$) k_1, k_2 and R are such that our relation (11) does not depend explicitly on g_1 and that these functions verify one of our set of conditions (12) → (22).

In conclusion, Ericksen's conjecture can be reformulated in a preciser statement as follows : Poiseuille flow of an incompressible non newtonian viscous fluid is only possible if the curves of constant speed have constant curvature except when $k_2 = \ell k_1$ with $R' = -\ell c$ or when k_1, k_2 and R are such that (11) does not depend explicitly on g_1.

[1] J.L.Ericksen. Quat.Appl.Math. 14 (1956), 318-321

[2] G.Mayné. Bull.Acad.R.Belg.Cl.Sc. 54 (1968), 90-104.

[3] R.L.Fosdick and J.Serrin. Proc.R.Soc.Lond. A 332 (1973), 311-333

LINEAR THERMODYNAMICS AND NON-LINEAR PHENOMENA IN FLUIDS

J. Verhás
Institute of Physics,
Technical University
Budapest H-1521 /Hungary/

I base my arguments on two simple, nevertheless, considerable facts. Firstly, the mechanical processes in a fluid are irreversible, secondly, the mechanical behaviour of fluids is intricate and diverse. Both of them invite us to investigate these phenomena with the aid of non-equilibrium thermodynamics. To make ahead during investigations, we turn to modelling. In non-equilibrium thermodynamics, modelling is done in three steps. The first step is the decision which state variables and which entropy function are our choice. In the case of fluids, the state variables are often the specific internal energy, the specific volume and some dynamic /internal/ variables.

$$s=s/u, v, \xi_1, \xi_2, \dots/ \qquad /1/$$

In a lot of applications, the physical meaning of the dynamic variables is of little importance, they merely define the topology of the thermodynamic state space. Hence, a transformation, the existence of which is guaranteed by Morse's lemma, leads to a standardized entropy function.

$$s = s^e/u, v/ - \frac{1}{2}\sum_i \xi_i^2 \qquad /2/$$

From now we suppose that the dynamic variables are tensors of second order. The next step of modelling relates to the balance equation for internal energy, which, most commonly, reads

$$\varrho \frac{dU}{dt} + \mathrm{div}\, \bar{J}_q = \underline{\underline{t}} : \overset{\circ}{\underline{\underline{d}}} \, . \qquad /3/$$

Here ϱ is the density, $\bar{J}_q$ the heat flow, $\underline{\underline{t}}$ is Cauchy's stress and $\overset{\circ}{\underline{\underline{d}}}$ stands for the symmetric part of the velocity gradient.

The entropy balance is obtained as

$$\varrho \frac{ds}{dt} + \mathrm{div}\left(\frac{1}{T}\bar{J}_q\right) = \sigma_s \geq 0 \qquad /4/$$

with

$$\sigma_s = \bar{J}_q \ \mathrm{grad}\frac{1}{T} + \frac{1}{T}\underline{\underline{t}} : \overset{\circ}{\underline{\underline{d}}} - \varrho \sum_i \underline{\underline{\xi}}_i : \overset{\circ}{\underline{\underline{\xi}}}_i \qquad /5/$$

Now we apply Onsager's law. For the sake of simplicity, isotherm and isochoric flow is treated with. The thermodynamic equations governing the processes are

$$\begin{aligned} \underline{\underline{t}}_D &= L_{oo}\frac{1}{T}\overset{\circ}{\underline{\underline{d}}} - \sum_i L_{oi}\varrho\,\underline{\underline{\xi}}_i \\ \overset{\circ}{\underline{\underline{\xi}}}_i &= L_{io}\frac{1}{T}\overset{\circ}{\underline{\underline{d}}} - \sum_i L_{ik}\varrho\,\underline{\underline{\xi}}_k \end{aligned} \qquad /6/$$

with the Onsager-Casimir reciprocal relations

$$L_{oi} = \pm L_{io}\ , \qquad L_{ik} = \pm L_{ki} \qquad /7/$$

Here $\underline{\underline{t}}_D$ is the stress deviator. The coefficients L are tensors of 4-th order in general and can depend on a lot of variables but particularly they can be constant scalars as well. In the latter case, we speak of the linear theory of thermodynamics, which gives good approximations, at least in a neighbourhood of an equilibrium. As the last step of modelling, we choose the coefficients. When doing it, the methods and rules of approximation theory are very practical. It is worth to mention that even the linear approximation gives account on a number of phenomena, such as viscoelastic behaviour, non-linear relation between stress and shear rate, normal stress effect, plastic flow, creep as well as elastic deformations before and after a plastic flow, etc. It seems that the sources of non-linearities are rather in the rotations occuring with a flow than in the non-linearity of the thermodynamic equations of motion.

To illustrate this, I refer to the most simple model with a single second order tensor as a dynamic variable and with constant Onsager-coefficients. Having eliminated the dynamic variable from the equations we get

$$\underline{\underline{t}}_D + \tau_t \overset{\circ}{\underline{\underline{t}}}_D = 2\eta(\overset{\circ}{\underline{\underline{d}}} + \tau_d \overset{\circ\circ}{\underline{\underline{d}}}) \qquad /8/$$

where the small circles refer to Jaumann-derivates, τ_t, τ_d and η

are material coefficients. The most simple solutions of this system of partial differential equations refer to linear viscoelasticity. The solutions for simple shear flow are of more interest. The viscometric functions are of the form

$$\tau = \eta \varkappa \frac{1 + \tau_d \tau_t \varkappa^2}{1 + \tau_t^2 \varkappa^2} \qquad /9/$$

$$-\sigma_y = \sigma_x = \frac{\eta(\tau_t - \tau_d)\varkappa^2}{1 + \tau_t^2 \varkappa^2} \qquad /1o/$$

The shear stress function and the normal stress functions are plotted in Figs 1. and 2. respectively.

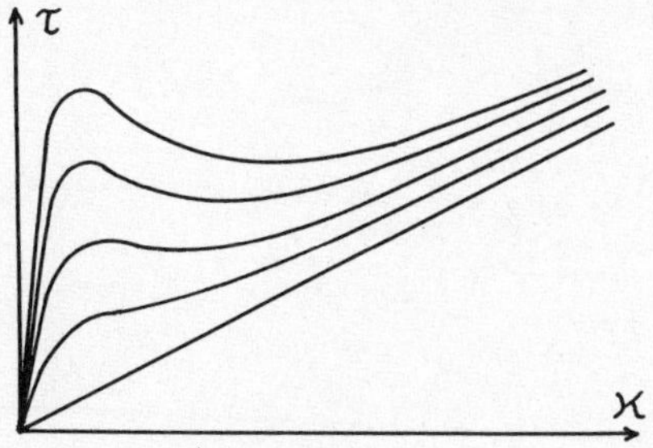

Fig.1.
Shear stress functions

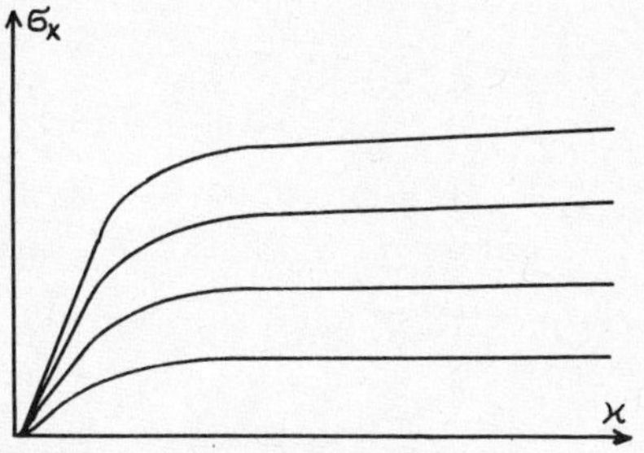

Fig.2.
Normal stress functions

A computational method based on complex numbers makes the calculations easier. It displays the Cox-Merz rule immediately and connects the viscometric functions with each other by Hilbert-transformation. It turns out, too, that the Cox-Merz rule and the linear Onsager-equations have a common field of validity. The complex number representation is suitable for treating with transient processes.

Analysing the stability of the stationary solutions, we find that the decreasing section of the shear stress function is unstable, /such one exists if $\tau_t/\tau_d > 9$ / hysteresis is performed. If the ratio of τ_t to τ_d is large enough the phenomenon turns into plasticity.

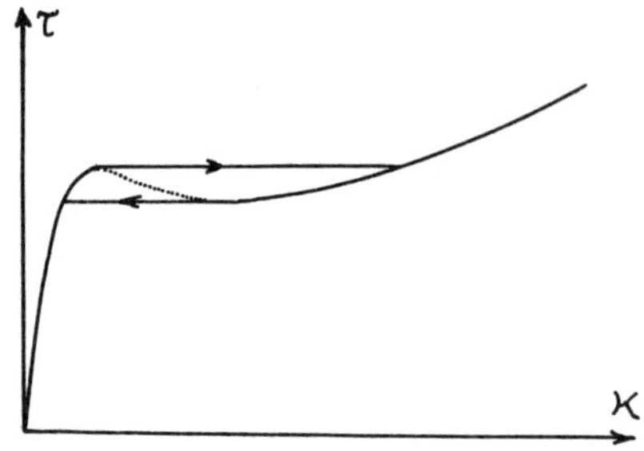

Fig.3
Hysteresis in shear flow

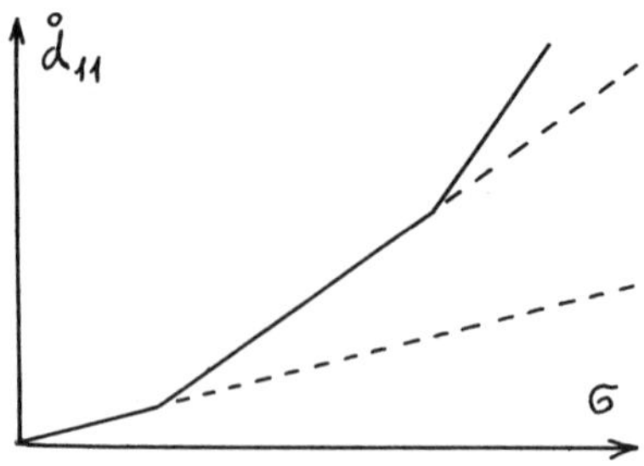

Fig.4.
Plastic flow at uniaxial tension

Very interesting solutions belong to uniaxial tension and compression. A slow motion, say creep, is stable up to a limit of load,

$$\overset{\circ}{\underline{\underline{d}}} = \frac{1}{2\eta}\underline{\underline{t}}_D \;; \qquad \sigma \le \frac{2\eta}{\tau_t - \tau_d} \qquad /11/$$

and no angular velocity occurs. At higher loads, an angular velocity perpendicular to the traction enters and the rate of deformation speeds up. At loads

$$\sigma > \frac{6\eta}{\tau_t - \tau_d} \qquad /12/$$

the angular velocity starts to rotate around the traction and a further acceleration takes place.

Sometimes the creep term is negligible and eq.8. results in

$$\overset{\circ}{\underline{\underline{t}}}_D = \frac{2\eta}{\tau_t - \tau_d}\left(\overset{\circ}{\underline{\underline{d}}} + \tau_d \overset{\circ\circ}{\underline{\underline{d}}} \right) \qquad /13/$$

which can be integrated easily in a corotating frame

$$\underline{\underline{t}}_D = \frac{2\eta}{\tau_t - \tau_d}\left(\underline{\underline{d}} + \tau_d \overset{\circ}{\underline{\underline{d}}} - \underline{\underline{d}}_o \right) \qquad /14/$$

Here $\underline{\underline{d}}_o$ is a constant of integration. This equation describes elastic /or more precisely viscoelastic/ deformations before and after a plastic flow.

The other models mentioned before give way to deal with more complicated materials as well. I hope that the thermodynamic method will be favoured better in the future.

Acknowledgement

The author is indebted to the University Stuttgart for the kind invitation and the support that makes possible to take part in the conference.

References

De Groot,S.R.-Mazur,P.: Non-Equilibrium Thermodynamics. North-Holland Publ.Co. Amsterdam 1962.

Gyarmati,I.: Non-Equilibrium Thermodynamics. Springer-Verlag, Heidelberg, Berlin, New-York 197o.

Kluitenberg,G.A.: Plasticity and Non-Equilibrium Thermodynamics. in: The Constitutive Law in Thermoplasticity. ed.by Th.Lehman CISM Courses and Lectures No 281 Springer Verlag Wien New-York 1984.

Garrod,C.-Hurley,J.: Symmetry Relations for the Conductivity Tensor. Phys.Rev. A 27 1487...149o /1983/

Hurley,J.-Garrod,C.: Generalization of the Onsager Reciprocity Theorem. Phys.Rev.Lett. 48 1575...1577 /1982/

Verhás,J.: On the Thermodynamic Theory of Deformation and Flow. Periodica Polytechnica Ser.Chem.Engng./Budapest/ 21 319...332 /1977/

Verhás,J.: On the Entropy Current. J.Non-Equilib.Thermodyn. 8 2o1...2o6 /1983/

Verhás,J.: An Extension of the Governing Principle of Dissipative Processes to Non-Linear Constitutive Equations. Ann.d.Phys. 4o 189...193 /1983/

Verhás,J.: A Thermodynamic Approach to Viscoelasticity and Plasticity Acta Mechanica /Wien/ 53 125...139 /1981/

Verhás,J.: Thermodynamics and Rheology /in Hungarian/ Müszaki Könyvkiadó Budapest 1985.

SOME REMARKS ON THE LIMIT OF VISCOELASTIC FLUIDS AS THE RELAXATION TIME TENDS TO ZERO

Jean-Claude Saut
Laboratoire d'Analyse Numérique, Université Paris-Sud
Bâtiment 425, 91405 Orsay (France)

1. Introduction. Incompressible viscoelastic fluids of Oldroyd type with zero retardation time are governed by the differential constitutive law :

$$(1.1) \qquad \lambda \frac{\mathcal{D}\underline{T}}{\mathcal{D}t} + \underline{T} = 2\,\eta\,\underline{D}\ , \ \operatorname{Tr}\underline{D} = 0\ ,$$

where $\underline{T}$ is the extra-stress, $\underline{D} = \frac{1}{2}(\nabla\underline{u} + \nabla\underline{u}^T)$ the symmetric part of the velocity gradient, λ is the relaxation time and η an elastic viscosity. The Oldroyd derivative $\frac{\mathcal{D}}{\mathcal{D}t}$ is given by

$$(1.2) \qquad \frac{\mathcal{D}\underline{T}}{\mathcal{D}t} = \frac{\partial \underline{T}}{\partial t} + (\underline{u}.\nabla)\underline{T} + \underline{T}\,\underline{W} - \underline{W}\,\underline{T} - a(\underline{D}\,\underline{T} + \underline{T}\,\underline{D})\ ,$$

where $\underline{W} = \frac{1}{2}(\nabla\underline{u} - \nabla\underline{u}^T)$ is the vorticity tensor and $a \in [-1,1]$. When the relaxation time λ is zero, (1) reduces to the constitutive law of Newtonian fluids :

$$(1.3) \qquad \underline{T} = 2\,\eta\,\underline{D}\ , \ \operatorname{Tr}\underline{D} = 0\ ,$$

which, together with the law of motion

$$(1.4) \qquad \rho\left(\frac{\partial \underline{u}}{\partial t} + (\underline{u}.\nabla)\underline{u}\right) + \nabla p = \nabla.\underline{T} + \rho\,\underline{b}$$

(where $\rho > 0$ is the density and $\underline{b}$ are body forces), lead to the classical system of Navier-Stokes equations. One is lead naturally to the following question : does a solution of an initial-boundary value problem associated to (1.1)(1.4) converge as $\lambda \to 0$ to a solution of the corresponding problem for the Navier-Stokes equations. Apart from its own theoretical interest, this question could explain why numerical computations of viscoelastic fluids flows work well when the Weissenberg number (here $\frac{\lambda U}{L}$, where U and L are a typical velocity and length in the flow) is small [2]. This singular perturbation problem possesses a difficulty which is not present in other singular limit problems of fluid mechanics (see [7]) : the system (1)(4), although of "hyperbolic" type, can loose evolutionarity (see [4],[5],[6]), i.e. Hadamard instabilities can appear and the Cauchy problem can become ill-posed. Moreover, even in situation where (1)(4) is of evolution type, no general results concerning initial-boundary value problems (involving physical boundary conditions) are known. In this Note we address two simple examples where the aforementioned question can be positively answered. More general situations will be considered elsewhere.

2. Linearized equations for flows perturbing uniform motion.

We shall consider the linearized equations for flows perturbing a uniform motion with constant velocity $U\underline{e}_1$ in the direction x. One obtains from (1.1), (1.2), (1.4) the system (where subscrpts denote differentiation) :

$$(2.1)\qquad \begin{aligned} &\lambda(\underline{T}_t + U\,\underline{T}_x) + \underline{T} = 2\,\eta\,\underline{D} \\ &\rho(\underline{u}_t + U\,\underline{u}_x) + \nabla p = \nabla\cdot\underline{T} \\ &\nabla\cdot\underline{u} = 0\,. \end{aligned}$$

We differentiate $(2.1)_2$ with respect to t and take the divergence of $(2.1)_1$ to get a system involving $\underline{u}$ and a pressure q as unknown :

$$(2.2)\qquad \begin{aligned} &\lambda\rho(\underline{u}_{tt} + 2\,U\,\underline{u}_{xt} + U^2\,\underline{u}_{xx}) - \eta\,\Delta\,\underline{u} + \rho\,\underline{u}_t \\ &\qquad + \rho\,U\,\underline{u}_x + \nabla q = 0\,. \end{aligned}$$

The system (2.2) is of evolution type although the steady system is elliptic iff

$$(2.3)\qquad \eta - \lambda\,U^2\,\rho > 0$$

(or equivalently if and only if $M^2 < 1$, where the viscoelastic Mach number M is defined by $M^2 = \dfrac{U^2}{C^2}$, $C^2 \underset{\text{def}}{=} \eta/\lambda\rho$).

When $\lambda = 0$ we have the limit problem :

$$(2.4)\qquad \begin{aligned} &\rho\,\underline{u}_t - \eta\,\Delta\,\underline{u} + \rho\,U\,\underline{u}_x + \nabla\,\tilde{q} = 0 \\ &\nabla\cdot\underline{u} = 0. \end{aligned}$$

In order to state a precise convergence result, we introduce a functional setting. We shall consider (2.1) (or (2.2), (2.4)) in the cylinder $\Omega\times(0,T)$, where Ω is a smooth open set in $\mathbb{R}^3$ which is bounded in one direction, and use the classical spaces
$H = \{\underline{u} \in L^2(\Omega)^3,\ \operatorname{div}\underline{u} = 0,\ \underline{u}.\underline{n} = 0 \text{ on } \partial\Omega\}$, $\underline{n}$ unit outward normal on $\partial\Omega$, $V = \{\underline{u} \in H_0^1(\Omega)^3\ ;\ \operatorname{div}\underline{u} = 0\}$. Moreover, we denote by P the orthogonal projector in $L^2(\Omega)^3$ onto H . We complete (2.2), (2.4) by the boundary condition :

(2.5) $\underline{u}_{|\partial\Omega} = 0$

and the initial conditions :

(2.6) $\underline{u}(x,0) = \underline{u}_0 \; , \; \underline{u}_t(x,0) = \underline{u}_1$ for (2.2)

(2.7) $\underline{u}(x,0) = \underline{u}_0 \; ,$ for (2.4) .

We summarize our results in the

Theorem 2.1. (i) For $\underline{u}_0 \in H^3(\Omega)^3 \cap V$ satisfying

(2.8) $$\eta\, P\, \Delta\, \underline{u}_0 + \rho\, U \frac{\partial}{\partial x} \underline{u}_0 = 0 \quad \text{on} \quad \partial\Omega \; ,$$

there exists a unique $\underline{u} \in C([0,T];H^3(\Omega)^3 \cap V)$ solution of (2.4), (2.5), (2.7).

(ii) If λ is chosen such that (2.3) holds, let $\underline{u}_0 \in H^2(\Omega)^3 \cap V$, $\underline{u}_1 \in V$; then there exists a unique solution $\underline{u}^\lambda$ of (2.2), (2.5), (2.6) satisfying $\underline{u}^\lambda \in C([0,T];H^2)^3 \cap V)$, $\underline{u}^\lambda_t \in C([0,T;V)$.

(iii) We assume moreover that $\underline{u}_0 \in H^3(\Omega)^3 \cap V$, satisfies (2.8) and that $\underline{u}_1 \in H^2(\Omega)^3 \cap V$. Then there exists $C > 0$, independent of λ , such that

(2.9) $$\|\underline{u}^\lambda(t) - \underline{u}(t)\|_V \leq C\lambda \quad , \quad \forall\, t \in [0,T],$$

(2.10) $$\|\underline{u}^\lambda_t(t) - (\underline{u}_t(t) + \underline{\theta}^\lambda_t(t))\|_H \leq C\lambda^{1/2} \quad , \forall\, t \in [0,T] \; ,$$

where $\underline{\theta}^\lambda(t) = \lambda(\underline{u}_1 - \underline{u}_t(0))\,(1 - \exp(-t/\lambda))$.

(iv) Let $\underline{T}^\lambda$ be the stress corresponding to $\underline{u}^\lambda$ in (2.1), with the initial data $\underline{T}^\lambda(0) = \underline{T}_0 \in H^1(\Omega)^6$. Under the hypothesis (ii), (iii), there exists $C > 0$ independent of λ such that

(2.11) $$\|\underline{T}^\lambda(t) - (2\eta\, \underline{D}[\underline{u}(t)] + \underline{\Theta}^\lambda(t))\|_{H^1} \leq C\lambda \; ,$$

where $\underline{\Theta}^\lambda$ is the solution of

(2.12) $$\begin{aligned} &\lambda(\Theta^\lambda_t + U\, \Theta^\lambda_x) + \Theta^\lambda = 2\eta\, \lambda\, \underline{D}[\underline{g}] \\ &\Theta^\lambda(0) = \underline{T}_0 - 2\eta\, \underline{D}[\underline{u}_0] \; . \end{aligned}$$

Here $\underline{g}$ is defined by $\underline{u}^\lambda = \underline{u} + \lambda \underline{g}$.

□

Theorem 2.1 is proven by using classical techniques in singular perturbation theory (see [8]). The condition (2.8) is a compatibility condition (cf [10]) which is used to obtain the smoothness of u required to get estimates (2.9)(2.10)(2.11) ; the correctors $\underline{\theta}^\lambda$ and $\underline{\Theta}^\lambda$ take into account the loss of an initial condition at $t = 0$.

Remark 2.1. The Giesekus model [3] corresponds to $a = 1$ in (1.2) and the right hand side of (1.1) replaced by $2\eta \underline{D} + \frac{\lambda\alpha}{\eta} \underline{T}^2$ where $0<\alpha\leqq 1$. For the flow under consideration here, one gets exactly the system (2.2) and Theorem 2.1 is still valid.

Remark 2.2. In linearization at uniform motion, the extra-stress in fluids with instantaneous elasticity of the Coleman-Noll type [1] is given by (see [5]) :

$$\underline{T}(t) = \int_{-\infty}^{t} G(t-s)\, \underline{A}\, [\underline{u}(\underline{\chi},s)]ds \,, \tag{2.13}$$

where $\underline{A} = 2\,\underline{D}$, G (the relaxation kernel) is smooth, positive, monotone decreasing, and $\underline{\chi} = [x - U(t-s), y, z]^T$.

The system corresponding to (2.1) is now

$$\begin{aligned} &\underline{T}_t + U\,\underline{T}_x - G(0)\,\underline{A}\,[\underline{u}(\underline{x},t)] - \int_{-\infty}^{t} G'(t-s)\,\underline{A}\,[\underline{u}(\underline{\chi},s)]ds \\ &\rho(\underline{u}_t + U\,\underline{u}_x) + \nabla p = \nabla.\underline{T} \,, \quad \nabla.\underline{u} = 0 \,, \end{aligned} \tag{2.14}$$

which is exactly (2.1) when $G(s) = \frac{\eta}{\lambda}\exp(-s/\lambda)$. One obtains immediately a system involving still $\underline{u}$ and a pressure q as unknown :

$$\begin{aligned} &\rho(\underline{u}_{tt} + 2\,U\,\underline{u}_{xt} + U^2\,\underline{u}_{xx}) - G(0)\,\Delta\underline{u} \\ &\qquad - \int_{-\infty}^{t} G'(t-s)\,\Delta\underline{u}(\chi,s)ds + \nabla q = 0 \\ &\nabla.\underline{u} = 0 \end{aligned} \tag{2.15}$$

which is of evolution type if and only if $\rho U^2 < G(0)$, and reduces to (2.2) when $G(s) = \frac{\eta}{\lambda}\exp(-\frac{s}{\lambda})$. The Newtonian case is obtained formally when $G(s) = \delta(s)$, the delta function. It would be interesting to study the singular perturbation problem arising when one takes a general smooth approximation of δ.

3. Shear flows.

The shear flows of Oldroyd models (1.1) are governed by the first order system

$$\begin{aligned} &\sigma_t - (a+1)\tau v_x + \frac{\sigma}{\lambda} = 0 \\ &\tau_t + \left[\frac{(1-a)\sigma}{2} - \frac{(1+a)\gamma}{2} - \frac{\eta}{\lambda}\right] v_x + \frac{\tau}{\lambda} = 0 \\ &\gamma_t + (1-a)\tau v_x + \frac{\gamma}{\lambda} = 0 \\ &\rho v_t - \tau_x = 0 \,, \end{aligned} \tag{3.1}$$

where $v(x,t)$ is the rectilinear velocity in the direction y and

$(\sigma,\gamma,\tau) = (T^{<yy>}, T^{<x,x>}, T^{<x,y>})$. (The pressure appears only in the remaining equation of motion which is not coupled to these). We notice furthermore that $(3.1)_1$ and $(3.1)_2$ lead to the equation

$$[(1-a)\sigma + (1+a)\gamma]_t + \frac{1}{\lambda}[(1-a)\sigma + (1+a)\gamma] = 0 ,$$

and choosing $\sigma_0 = \sigma(x,0)$ and $\gamma_0(x) = \gamma(x,0)$ such that $(1-a)\sigma_0 + (1+a)\gamma_0 \equiv 0$, we can reduce (3.1) to

(3.2)
$$\sigma_t - (a+1)\tau v_x + \frac{\sigma}{\lambda} = 0$$
$$\tau_t + [(1-a)\sigma - \frac{\eta}{\lambda}]\, v_x + \frac{\tau}{\lambda} = 0$$
$$\rho v_t - \tau_x = 0$$

which we supplement with the boundary conditions

(3.3) $$v(0,t) \equiv 0 \quad , \quad v(1,t) \equiv U$$

and the initial conditions

(3.4) $$v(x,0) = v_0 \ , \ v_t(x,0) = v_1 \ , \ \sigma(x,0) = \sigma_0 \ , \ \tau(x,0) = \tau_0 \ .$$

This system is strictly hyperbolic if and only if

(3.5) $$\frac{\eta}{\lambda} - (1-a)\sigma > 0 \ .$$

It is instructive to consider first the case $a = 1$ (upper convected Maxwell model). Then (3.2) can be decoupled (this occurs also for $a = -1$) and reduces to

(3.6)
$$\lambda \rho w_{tt} - \eta w_{xx} + \rho w_t = 0$$
$$\lambda \tau_t - \eta w_x + \tau - \eta U = 0$$
$$\lambda \sigma_t - 2\lambda\tau w_x + \sigma - 2\lambda\tau U = 0$$

where we have set $v = w + Ux$ (then $w(0,t) = w(1,t) = 0$) . This elementary singular perturbation problem is easily solved yielding results similar to those in Theorem 2.1.

The general case $a \neq \pm 1$ is more delicate and is treated in [9].

<u>References</u>.

[1] B. Coleman, W. Noll, Fondation of viscoelasticity, Rev. Mod. Phys., 33 (1961), 239-249.

[2] M.J. Crochet, A.R. Davies, K. Walters, Numerical simulation of non-newtonian flow, Elsevier, Amsterdam, 1984.

[3] H. Giesekus, A unified approach to a variety of constitutive models for polymer fluids based on the concept of configuration dependent molecular mobility, Rheol. Acta, 21 (1982), 366-375.

[4] D.D. Joseph, Hyperbolic phenomena in the flow of viscoelastic fluids, Proceedings of the Conference on Viscoelasticity and Rheology, Madison (1984), J. Nohel ed., Academic Press, to appear.

[5] D.D. Joseph, M. Renardy, J.C. Saut, Hyperbolicity and change of type in the flow of viscoelastic fluids, Arch. Rational Mech. Anal., 87 (1985), 213-251.

[6] D.D. Joseph, J.C. Saut, Change of type and loss of evolution in the flow of viscoelastic fluids, J. Non Newtonian Fluid Mech., to appear.

[7] S. Klainerman, A. Majda, Singular limits of quasilinear hyperbolic systems with large parameters and the incompressible limit of compressible fluids, Comm. Pure Appl. Math., XXXIV (1981), 481-524.

[8] J.L. Lions, Perturbations singulières dans les problèmes aux limites et en contrôle optimal, Lecture Notes in Mathematics n° 323, Springer-Verlag, 1973.

[9] J.C. Saut, A nonlinear singular perturbation problem in viscoelastic fluids, in preparation.

[10] R. Temam, Behaviour at time t = 0 of the solutions of semi-linear evolution equations, J. Diff. Eq., 43 (1982), 73-92.

HYDRODYNAMICS OF RIGID MAGNETIC SUSPENSIONS

R.K.T. Hsieh

Department of Mechanics
Royal Institute of Technology
S-100 44 Stockholm, Sweden

ABSTRACT: Using a model of dilute suspension of rigid spherical magnetic particles, the hydrodynamic behaviour of a magnetic fluid is given. Emphasis is being laid on the anisotropic properties of a magnetic fluid under the exertion of an applied magnetic field.

1. INTRODUCTION AND SUMMARY

In a previous paper, see Hsieh [1], an engineering model of a magnetic fluid which extended the validity of the model developed by Rosensweig & Neuringer [2] has been given. This model, given within the frame of hydrostatics, is the continuum approximation of a dilute suspension of rigid spherical magnetic particles and it has led to industrial applications. As in the case for the Rosensweig & Neuringer model, however, it could not describe some experimentally observed phenomena like the dependence of the viscosity on the applied magnetic field or the dependence of the susceptibility on the applied magnetic field [3]. It is the aim of this paper to show that such phenomena can only be explained by accounting for the dynamical behaviour of the fluid. Using the same model of a dilute suspension of rigid spherical magnetic particles, a set of equations for the description of the system is derived. The emphasis is being laid on the contribution due to the magneto-mechanical coupling. It is shown that the simplest engineering model would be the consideration of a Newtonian fluid in which the viscosity is augmented with a term depending on the applied magnetic field.

2. EQUATIONS OF MOTION

If the density ρ of the magnetic fluid is defined by the relation

$$\rho = n\mu + n^{(Nm)}\mu^{(Nm)} = n(\mu + \frac{n^{(Nm)}}{n} \mu^{(Nm)}) \tag{1}$$

where μ and $\mu^{(Nm)}$ are respectively the mass of magnetic particle (monodonain size) and nonmagnetic particle, n an $n^{(Nm)}$ are respectively the number of magnetic and nonmagnetic particles in a unit volume V of a magnetic fluid. This volume is larger than the volume of the magnetic particle V^m. For $n^{(Nm)}/n$ being constant, there is no diffusion particle flow.

The macroscopic continuum balance equations write [1]

$$\frac{d}{dt}\int_{V(t)} \rho dV = 0 \tag{2}$$

$$\frac{d}{dt}\int_{V(t)} \rho \underset{\sim}{v} dV = \int \underset{\sim}{t} dS + \int \rho \underset{\sim}{f} dV \tag{3}$$

$$\frac{d}{dt}\int_{V(t)} (I\underset{\sim}{\Omega} + \frac{\underset{\sim}{M}}{\gamma}) dV = \int \underset{\sim}{N} dS + \int \rho \underset{\sim}{\ell} dV + \int \underset{\sim}{x} \times \underset{\sim}{t} dS \tag{4}$$

$$\frac{d}{dt}\int \frac{1}{2}(\rho v_i v_i + I\Omega_i\Omega_i)dV + \frac{d}{dt}\int \rho U dV = \int (\rho f_i v_i + (\underset{\sim}{H}^L \times \underset{\sim}{M})_i\Omega_i + (\underset{\sim}{M} \times \underset{\sim}{H}^L)_i\Omega_i^o + \rho\ell_i\Omega_i + \rho r)dV + \int (t_i v_i + N_i\Omega_i + q_i n_i)dS \tag{5}$$

where $I\underset{\sim}{\Omega}$ is the angular moment per unit volume of the magnetic fluid. $\underset{\sim}{I} = ni$ is the moment of inertia of magnetic fluid per unit volume, i is the moment of inertia of a spherical particle, $\underset{\sim}{\Omega}$ is the angular velocity of the magnetic fluid particle, $\underset{\sim}{M}/\gamma$ is the spin momentum of the electrons involved into the magnetization $\underset{\sim}{M}$, and $\underset{\sim}{M} = n\underset{\sim}{m}$ with $\underset{\sim}{m}$ being the magnetic moment of a monodomain magnetic particle, γ is the gyromagnetic ratio.

Eqs. (2) - (4) are respectively the balance of the continuity of mass, of linear momentum and of angular momentum. Eq. (5) is the balance of energy, [4]. For isotropic homogeneous magnetic fluids, the stress $t_i = t_{ij}n_j$ and the couple stress $N_i = N_{ij}n_j$, where n_i is the unit vector, write

$$t_{ij} = -[p^* + (\lambda - \frac{2}{3}\mu)v_{k,k}]\,\delta_{ij} + \mu(v_{i,j} + v_{j,i}) + 2\xi\varepsilon_{ijk}[(\nabla\times\underset{\sim}{v})_k - \Omega_k] \tag{6}$$

$$N_{ij} = \gamma^1\Omega_{k,k}\,\delta_{ij} + \gamma^2(\Omega_{i,j} + \Omega_{j,i}) + \gamma^3(\Omega_{i,j} - \Omega_{j,i}) \tag{7}$$

and the heat flux writes

$$q_i = -\lambda T_{,i} \tag{8}$$

In quasi-steady approximation and in the absence of electrical displacement and electric current, the electromagnetic field equations write

$$\int \underset{\sim}{B}\cdot d\underset{\sim}{S} = 0 \tag{9}$$

$$\int \underset{\sim}{E}\cdot d\underset{\sim}{\ell} = -\frac{d}{dt}\int \underset{\sim}{B}\cdot d\underset{\sim}{S} \tag{10}$$

$$\int \underset{\sim}{H}\cdot d\underset{\sim}{\ell} = 0 \tag{11}$$

and

$$\underset{\sim}{B} = \mu_o(\underset{\sim}{M} + \underset{\sim}{H}) \tag{12}$$

In general to complete the description of the ferrohydrodynamic interaction, it is necessary to consider the equations governing the magnetic field $\underset{\sim}{H}$. The magnetic intensity $\underset{\sim}{H}$ is the sum of the applied field $\underset{\sim}{H}^o$ assumed to be known and the induced field $\underset{\sim}{H}_{ind}$ produced by the distribution of induced polarization, i.e.,

$$\underset{\sim}{H} = \underset{\sim}{H}^o + \underset{\sim}{H}_{ind}$$

We therefore have $\nabla\cdot\underset{\sim}{B}^o = 0$ and $\nabla\times\underset{\sim}{H}^o = 0$ in vacuum. With the use of these relations, in matter we shall have $\nabla\cdot\underset{\sim}{H}_{ind} = -\nabla\cdot\underset{\sim}{M}$ and $\nabla\times\underset{\sim}{H}_{ind} = 0$ or $\underset{\sim}{H}_{ind} = -\nabla^2\phi = \nabla\cdot\underset{\sim}{M}$, i.e. ϕ satisfies the Poisson equation.

In local form, accounting for the incompressibility condition $\nabla\cdot\underset{\sim}{v} = 0$ and with the neglecting of the spin and mechanical angular momentum, often encountered for in physics, we shall get

$$\rho\frac{d\mathbf{v}}{dt} = -\nabla p^* + \mu\,\nabla^2\mathbf{v} + 2\xi\nabla \times (\nabla \times \mathbf{v} - \boldsymbol{\Omega}) + \mu_o(\mathbf{M}\cdot\nabla)\,\mathbf{H} \quad (13)$$

$$4\xi(\nabla \times \mathbf{v} - \boldsymbol{\Omega}) = \mu_o(\mathbf{M} \times \mathbf{H}) \quad (14)$$

$$\rho c\left(\frac{dT}{dt}\right) = -\lambda\nabla^2 T + \rho r \quad (15)$$

$$\nabla\cdot\mathbf{B} = 0 \quad (16)$$

$$\nabla\times\mathbf{H} = 0 \quad (17)$$

where the friction of the rotation coefficient ξ can be estimated from the model of suspension with rigid spherical particles

$$\xi = \frac{3}{2}\,\Phi\,\xi^o$$

Here $\Phi = nV$ is the hydrodynamic concentration of the particles and ξ^o is the dynamic viscosity of the carrier liquid. In eq. (15) neglection has been made of magnetocalorific effects. We see that equations (12)-(17) form a coupled magneto-thermo-mechanical system. In hydrostatics it was shown that the problem can be solved "mechanically" by defining an effective pressure depending of both the density and the magnetic field as well as of an effective gravitational field. In dynamic case, the matter is more complicated, this will be the topic of the next section. Stability problems are not being discussed. In such case the vorticity equation has to be used, see e.g. [5].

3. DYNAMIC ANISOTROPIES OF MAGNETIC FLUIDS

In the absence of an applied field, magnetic fluid behaves macroscopically like an isotropic fluid. In the presence of a magnetic field, it has been however experimentally observed that the fluid has anisotropic behaviour. In statics, this anisotropy is manifested through the dipolar interaction energy and is accounted for through the factor $\lambda' = m^2/r^3kT$. However, within this assumption both, e.g., the experimentally observed anisotropic properties of the magnetic fluid susceptibility or its viscosity could not be described. This is due to the condition of $\mathbf{M}$ being permanently maintained parallel with the applied field $\mathbf{H}$ (the value of $\mathbf{M}$ is determined by local instant values of temperature and field). In fact the establishment of equilibrium magnetization requires finite time $\tau(\sim\tau_B)$ and hence the complete set of ferrohydrodynamical equation must contain a dynamical equation for $\mathbf{M}(t)$. The Neuringer-Rosensweig approximation is therefore based on an instantaneous relaxation of fluid magnetization.

Before proposing a modelling including this dynamical behaviour, let us repeat that the aim is the extension of the validity of the engineering model with emphasis on the magneto-mechanical coupling (see [1]). As said before, the point is now on the mechanism of rotation of the fluid magnetization in the presence of an applied field.

As done in [1], let us first look to the picture at the magnetic domain level. If we suppose that we have uniaxial ferromagnets, the crystalline anisotropy energy is given by

$$E^a = -(KV^m/m^2)(\mathbf{m}.\mathbf{e})^2 \quad (18)$$

Here K is the anisotropy constant, e_i is the unit vector of the easiest axis frozen into the particle. For K<0 the axis is preferred, for $K>0$ it is more favourable to lie in the plane perpendicularly to the axis. In the wall separating the two antiparallel domains, the magnetization must rotate through a region increasing the anisotropy energy. Thus the magneto-crystalline anisotropy is minimized as the wall thickness is reduced. However as the wall gets thinner, the exchange energy increases. For $T<T_c$, the magnetization is large and in bulk samples, is directed along an axis of the crystal which minimizes the anisotropy energy. For finite sample of volume V^m, the total anisotropy is KV^m. If $KV^m < kT < kT_c$, the magnetization is developed but relatively decoupled from the crystal axes, here k is the Boltzmann constant and T_c is the Curie temperature. This is the case of superparamagnetism. Thus the monodomain ferromagnetic grains here considered depending on the ratio $\sigma = KV/kT$ may be superparamagnetic. While the anisotropy of a ferrofluid in presence of a static magnetic field has been related to the dipolar character of the material, for a rotating field, the dynamic aspect of the anisotropy comes from the behaviour of the individual grains themselves. Two different mechanisms exist, rotation of the magnetic moment inside the grains by overcoming energy barrier KV^m between different directions of easy magnetization or rotation of the grains with its magnetic moment fixed with respect to the crystallographic axis. These two are related to two relaxations of the magnetization during the rotation. The first relaxational mechanism, originally pointed out by Neel is specifically inherent to subdomain particles. The probability of such a transition depends exponentially on the particle volume V^m. The characteristic time of the Neel's relaxation process can be written [6]

$$\tau_N = \tau_o \sigma^{-1/2} e^{\sigma} \text{ (for } \sigma>2) \text{ and } \tau_N = \tau_o \sigma \text{ (for } \sigma \ll 1) \qquad (19)$$

Here $\tau_o = M^d/2\beta\gamma K_1$, M^d is the bulk magnetization of the magnetic domain and $\beta \sim 10^{-1}$ is a dimensionless precession damping parameter. If the duration t of the magnetization measuring test is $t>\tau_N$, each particle is superparamagnetic. For $t<\tau_N$, each of them will be ferromagnetic.
The second relaxational mechanism is related to the equilibrium orientation of the magnetic moments in suspensions by rotation of the particles themselves with respect to the liquid matrix. This relaxational mechanism is characterized by the Brownian time of rotational diffusion

$$\tau_B = \frac{3V\xi^o}{kT} \qquad (20)$$

where ξ^o is the dynamic viscosity of the carrier fluid.

Thus, ferrofluid magnetization is based on two fluctuation mechanisms. One is determined by the properties of ferromagnets and the other by the viscosity of the fluid.

Using irreversible thermodynamics, a two-parameter relaxation model can be written [7]

$$\dot{\underline{M}} = \underline{\Omega} \times \underline{M} - \frac{1}{\tau_m}(\underline{M} - \kappa(\underline{H} + \underline{A})) - \underline{M}\nabla\cdot\underline{v} \qquad (21)$$

$$\dot{\underline{A}} = \underline{\Omega} \times \underline{A} - \frac{1}{\tau_a}(\underline{A} - \beta\underline{M}) \qquad (22)$$

where $\underset{\sim}{A}$ describes the orientation of the particle anisotropy, the vector $\underset{\sim}{A}$, obtained from microstructural consideration, is related to the local magnetic field $\underset{\sim}{H}^L$ introduced in a phenomenological way, see eq. (5) and [1]. τ_m and τ_a are two relaxation times, the parameters are related by $\kappa = \chi/(1-\beta\chi)$ and β is between 0 and 1, χ is the equilibrium magnetic fluid susceptibility.

Expanding the equations (21)-(22) with respect to the small parameters $\tau_m t$ and $\tau_a t \ll 1$ where t is the characteristic time of the problems and taking only linear terms of the small parameter expansion series, we get (quasi-equilibrium approximation)

$$\underset{\sim}{M} = \underset{\sim}{M}_o - \tau_{//} H \underset{\sim}{e} - \frac{1}{2}\tau_{\perp} H(\dot{\underset{\sim}{e}} - (\nabla \times \underset{\sim}{v}) \times \underset{\sim}{e}) \tag{23}$$

$$\tau_{//} = \chi \frac{\partial(\beta M_o)}{\partial H} \tau_a + (1+\beta\chi) \frac{\partial M_o}{\partial H} \tau_m \tag{24}$$

$$\tau_{\perp} = \frac{4\xi \;\chi[\chi\tau_a\beta + (1+\beta\chi)\tau_m]}{4\xi + \mu_o[\beta\chi\tau_a + (1+\beta\chi)\tau_m]HM_o} \tag{25}$$

M_o is the equilibrium magnetization, [1], and the magnetization $\underset{\sim}{M}$ has been decomposed into a component parallel to the applied field and one perpendicular to it, $\underset{\sim}{M} = M_{//}\underset{\sim}{e} + \underset{\sim}{M}_{\perp}$. This relaxation equation preserves its form for various physical models of dynamic magnetization, only the material constants $\tau_{//}$, $\tau_{\perp}$ are changing. It can therefore be considered as the simplest phenomenological hypothesis.

Substituting eq. (14) into eq. (13), the relaxation contribution to the magnetic ponderomotive force can then be written as

$$\underset{\sim}{F} = \frac{1}{2}\nabla \times \mu_o\tau_{\perp}H^2(\underset{\sim}{e} \times \dot{\underset{\sim}{e}} - \frac{1}{2}(\nabla\times\underset{\sim}{v}) + (\underset{\sim}{e}\cdot(\frac{1}{2}\nabla\times\underset{\sim}{v})\underset{\sim}{e}) - \mu_o\tau_{\perp}H(\underset{\sim}{e} - ((\frac{1}{2}\nabla\times\underset{\sim}{v})\times\underset{\sim}{e})\cdot\nabla)\underset{\sim}{H} - \tau_{//}\dot{H}\nabla H \tag{26}$$

For isothermal flows in a uniform steady-state magnetic field, this force further reduces to

$$\underset{\sim}{F} = \frac{1}{2}\nabla \times \mu_o\tau_{\perp} H^2(\underset{\sim}{e}(\frac{1}{2}(\nabla\times\underset{\sim}{v})\cdot\underset{\sim}{e}) - \frac{1}{2}(\nabla\times\underset{\sim}{v})) \tag{27}$$

Together with the incompressibility condition $\nabla\cdot\underset{\sim}{v} = 0$, the equations of motion rewrite

$$\rho\dot{\underset{\sim}{v}} = -\nabla p^* + (\mu+\xi^r)\nabla^2\underset{\sim}{v} + \xi^r\nabla\times(\underset{\sim}{e}\cdot\nabla\times\underset{\sim}{v})\underset{\sim}{e} + \rho\underset{\sim}{g} \tag{28}$$

where the rotational viscosity ξ^r is given by $\xi^r = \tau_{\perp}\frac{H^2}{4}$.

Substituting the value of $\tau_{\perp}$ from eq. (25) into the rotational viscosity, we obtain

$$\xi^r = \frac{3}{2}\Phi\;\xi^o G(\alpha) \tag{29}$$

where $G(\alpha) = \dfrac{\beta^* F(\alpha)L^2(\alpha)}{2F(\alpha) + \beta^* L^2(\alpha)}$, $F(\alpha) = \dfrac{\alpha\,L(\alpha)}{2+\alpha\,L(\alpha)}$

and $\beta^* = (\frac{\mu_o nm^2}{kT})\,\beta$ is a parameter characterizing the freezing-in of the magnetic moment into a particle, $L(\alpha)$ is the Langevin function and $\alpha = \mu_o mH/kT$.

Equations (13)-(17), together with eq. (23) form an engineering model for the description of the hydrodynamics of magnetic fluids within the quasi-equilibrium approximation.

For $|\frac{1}{2} \nabla \times \underline{v}|\tau \ll 1$, eq. (23) reduces to the first term on its RHS. It is easily seen that for constant applied field normal to the hydrodynamic vortex, the relaxation effect would be to augment the viscosity of the Newtonian fluid with an apparent viscosity ξ^r. For nonuniform magnetic field, such conclusion cannot be drawn and heat and mass transfer considerations have to be accounted for, see e.g. [9].

REFERENCES

[1] Hsieh, R.K.T., Hydrostatics of rigid magnetic suspensions in Continuum Models of Discrete Systems 5, ed. by A.J.M. Spencer (Balkema, Rotterdam)

[2] Neuringer, J.L. and R.E. Rosensweig, Phys.Fluids 7 (1964) 1927

[3] Shliomis, M.I. and Yu.L. Raikher, IEEE Trans.Magn. MAG 16 (1980) 237

[4] Hsieh, R.K.T., IEEE Trans. MAG 16 (1980) 207

[5] Curtis, R.A., Phys.Fluids 14 (1971) 2096

[6] Neel, L., Ann.Geophys. 5 (1949) 99

[7] Raikher, Yu.L. and M.I. Shliomis, JETP 67 (1978) 14

[8] Kashevsky, B.E., Magnit.Girodinam. 4 (1978) 14

Non-Newtonian Fluids of Second Grade - Rheology, Thermodynamics and Extended Thermodynamics

K. Wilmanski
Polish Academy of Sciences
Warsaw

1. Introduction

Thermodynamic theories of non-newtonian fluids, developed during the last decade, have brought to attention the problem of thermodynamic stability of Nth grade fluids. Dunn & Fosdick [1] have shown that standard arguments of thermodynamics lead to the conclusion that second grade fluids have an unstable thermodynamic equilibrium state.

In the recent paper [2] Ingo Müller and myself have shown that the so-called extended thermodynamics does not have this fallacy but, at the same time, it favours a rate-type model rather than an Nth order fluid.

Also an analysis of one-dimensional models (see R.S. Rivlin, K. Wilmanski [3]) reveals serious drawbacks of Nth order fluids in non-steady flows. Indpendently of the order N, such models always predict

- infinite speed of shear pulses,
- cut-off in frequency spectrum at $\nu = \frac{1}{2\pi\lambda}$, where λ denotes the characteristic time of a non-newtonian fluid ($\lambda = 0.1 - 1$ sec for typical non-newtonian fluids),
- critical damping of all frequencies for N = 2 and 3 (i.e. $\nu \equiv 0$) yielding unrealistic restrictions for wave length.

In this article I present a compact review of those results and show the consequences of two thermodynamic requirements - second law in the form of entropy inequality, and stability of thermodynamic equilibrium - in two different cases: 2nd order fluid as an approximation of the maxwellian fluid and 2nd grade fluid as an exact model.

2. Rheology

The one-dimensional maxwellian model of non-newtonian fluids can be introduced in, at least, two different ways

- as a rate-type material, described by a differential relation between stresses and their derivatives and Rivlin-Ericksen tensors, representing the shear rate and its derivatives,
- as a material with memory whose stress is given by a functional on the space of shear rate histories.

For the purpose of this paper, we choose the first method and define the <u>one-dimensional maxwellian model</u> by the following equation

$$\left(1 + \lambda \frac{\partial}{\partial t}\right) \tau = \mu \varkappa , \tag{2.1}$$

where τ is the shear stress, $\varkappa$ - shear rate

$$\varkappa = \frac{\partial v}{\partial x} , \tag{2.2}$$

v being a non-zero component of velocity, λ - a characteristic time of a non-newtonian fluid, μ - viscosity coefficient. The former may be easily related to the parameter α_1 , measured in steady-state normal stress experiments (see, e.g. Ginn & Metzner [4])

$$\lambda = - \frac{\alpha_1}{\mu} . \tag{2.3}$$

Experiments as well as some theoretical considerations within rheology indicate

$$\mu \geqslant 0 \, , \, \lambda > 0 \quad \Rightarrow \quad \alpha_1 < 0 . \tag{2.4}$$

Assuming incompressibility (ϱ = const) the momentum balance equation together with (2.1) and (2.2) yields the following equation for the velocity field

$$\frac{\partial^2 v}{\partial x^2} - \frac{\lambda \varrho}{\mu} \frac{\partial^2 v}{\partial t^2} = \frac{\varrho}{\mu} \frac{\partial v}{\partial t} . \tag{2.5}$$

This equation predicts the finite speed of shear pulse

$$V = \sqrt{\frac{\mu}{\lambda \varrho}} . \tag{2.6}$$

It is common in rheology to replace the relation (2.1) by its approximate solution

$$\tau \cong \mu \sum_{k=0}^{N-1} (-\lambda)^k \varkappa^{(k)} , \qquad \varkappa^{(k)} := \frac{\partial^k \varkappa}{\partial t^k} \tag{2.7}$$

defining an <u>Nth order fluid.</u> This relation, together with the momentum balance leads to the following equation for the velocity field

$$\frac{\partial^2 v}{\partial x^2} + \frac{\partial^2}{\partial x^2} \sum_{k=1}^{N-1} (-\lambda)^k \frac{\partial^k v}{\partial t^k} = \frac{\varrho}{\mu} \frac{\partial v}{\partial t} . \tag{2.8}$$

In contrast to the equation (2.5), this equation predicts an infinite speed of propagation of shear pulses - independently of the order N.

It is obvious that the relation (2.7) approximates the maxwellian fluid if the series $\sum_{k=0}^{\infty} (-\lambda)^k \varkappa^{(k)}$ converges. This imposes, however, certain restrictions on solutions of (2.8) which we call convergence conditions. To demonstrate those conditions, let us consider a wave motion, described by the following relation

$$v(x,t) = V_0 e^{i(\chi x - \omega t)}, \tag{2.9}$$

where V_0 is an amplitude, χ - wave vector, ω - frequency and all of them may be complex. The substitution of that relation in the field equations yields the following dispersion relations

- for the maxwellian fluid

$$\frac{\varrho}{\mu} i\omega(1 - i\lambda\omega) = \chi^2 \tag{2.10}$$

- for the Nth order fluid

$$\frac{\varrho}{\mu} i\omega(1 - i\lambda\omega) = \chi^2\left[1 - (i\lambda\omega)^N\right]. \tag{2.11}$$

The convergence of the Nth order fluids to the maxwellian fluid for $N \to \infty$ yields

$$\lim_{N\to\infty} (i\lambda\omega)^N = 0 \quad \Rightarrow \quad (\mathrm{Re}\,\omega)^2 + (\mathrm{Im}\,\omega)^2 < \frac{1}{\lambda^2}. \tag{2.12}$$

The condition (2.12) indicates the existence of the upper bound of the frequency spectrum and this bound is independent of the order N.

The relation (2.7) may also be considered as a definition of an exact model, in which case we speak of an Nth grade fluid.

In order to show the basic defferences of the maxwellian model and the Nth order fluid let us consider the flow in a layer of thickness h, whose initial velocity profile is given by the relation

$$v(x,0) = V_0 \sin\left(k\sqrt{\frac{\varrho}{\mu\lambda}}\,x\right), \quad k := \frac{\pi}{h}\sqrt{\frac{\mu\lambda}{\varrho}}, \quad x \in \langle 0,h\rangle, \tag{2.13}$$

and the boundary conditions have the form

$$v(0,t) = v(h,t) = 0. \tag{2.14}$$

It is easy to check that the solution - a particular case of (2.8) - is as follows

$$v(x,t) = V_0 e^{-\frac{\beta}{\lambda}t} \sin\left(k\sqrt{\frac{\varrho}{\mu\lambda}}\,x\right)\cos\left(\frac{\nu}{\lambda}t\right), \tag{2.15}$$

where ν must be zero for the 2nd order fluid and it may either be zero, or

$$k^2 = \nu^2 + \frac{1}{4} \tag{2.16}$$

in the case of maxwellian fluid. The relation between β and k for those two models is shown in Fig. 1.

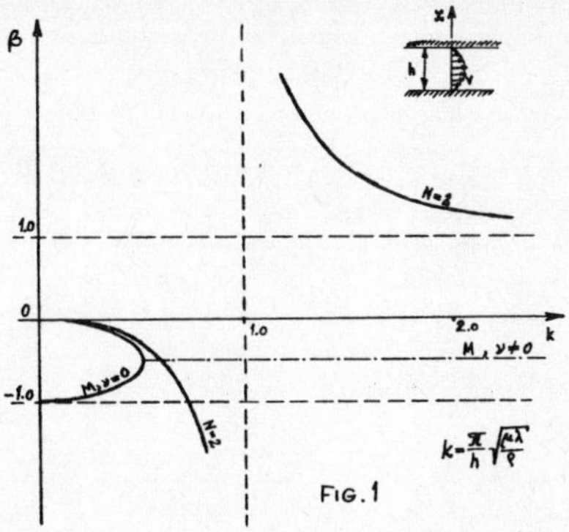

FIG. 1

As we see both models predict the critical damping ($\nu \equiv 0$) for very thick layers ($k < 0.5$). Otherwise the maxwellian model predicts damped vibrations, while 2nd order model continues to yield the critical damping. Moreover, for thin layers ($k > 1$), 2nd grade fluid leads to an instability ($\beta > 1$), discovered by Coleman and Mizel [5]. However, in the case of 2nd order fluid approximating the maxwellian model, the convergence condition (2.12) yields

$$|\beta| < 1 \tag{2.17}$$

and this means that there is no solution of the problem (2.13-14) for such thin layers.

The above example shows that 2nd order fluids describe very badly non-steady flows of non-newtonian fluids. Moreover, as we show further, the 2nd grade fluid, considered as an exact model, violates thermodynamic requirements.

Similar conclusions can be drawn for arbitrary finite N - as we show in paper [3].

3. Thermodynamics of isothermal processes

Let us start our thermodynamic considerations with the maxwellian model, always assuming incompressibility. In this case, the basic fields are velocity v, temperature T and shear stress τ. They are supposed to satisfy the following field equations

$$\begin{aligned} &\varrho \frac{\partial v}{\partial t} = \frac{\partial \tau}{\partial x}, \\ &\varrho \frac{\partial \varepsilon}{\partial t} = \tau \varkappa, \\ &\left(1 + \lambda \frac{\partial}{\partial t}\right)\tau = \mu \varkappa, \end{aligned} \tag{3.1}$$

where

$$\varepsilon = \varepsilon_0(T) + \varepsilon_1(T)\,\tau^2 \tag{3.2}$$

is the internal energy, ε_0 being its value in thermodynamic equilibrium ($\tau = 0$). The quadratic expression (3.2) can be justified by the structure of equation $(3.1)_3$. The material coefficients $\lambda, \mu, \varepsilon_0, \varepsilon_1$ are supposed to be chosen in such a way that the second law of thermodynamics is satisfied. The latter is assumed to have the form of the entropy inequality

$$\frac{\partial \eta}{\partial t} \geqslant 0, \quad \eta = \eta_0(T) + \eta_1(T)\,\tau^2 \tag{3.3}$$

where η is entropy density, η_0 - its equilibrium value.

Making use of the procedure, in which the field equations (3.1) are considered to be constraints imposed on solutions of the inequality (3.3) (see: I. Müller[6]), we easily arrive at the following relations

$$\mu \geqslant 0, \quad \frac{\partial \varepsilon_0}{\partial T} - T\frac{\partial \eta_0}{\partial T} = 0, \quad \varepsilon_1 - T\eta_1 = \frac{\lambda}{2\varrho\mu}, \tag{3.4}$$

thus exhausting the consequences of the second law in the present case.

Apart from the entropy inequality, thermodynamics requires that, in isolated systems, entropy reaches its maximum value in the equilibrium state. This condition, called a stability condition of thermodynamic equilibrium, is equivalent, for isothermal processes, to the requirement that the Helmholtz free energy

$$\psi := \varepsilon - T\eta = \psi_0(T) + (\varepsilon_1 - T\eta_1)\,\tau^2 \tag{3.5}$$

has a minimum in this state. According to the formula $(3.4)_3$ we have then

$$\lambda > 0. \tag{3.6}$$

The inequalities $(3.4)_1$ and (3.6) show that the maxwellian fluid fulfils automatically the thermodynamic conditions (compare the inequalities (2.4)).

Let us turn now our attention to a 2nd order fluid. We investigate first the case, when it is considered to be an exact model. In such a case the basic fields are velocity v and temperature T. They are supposed to satisfy the field equations

$$\varrho\frac{\partial v}{\partial t} = \frac{\partial \tau}{\partial x}, \quad \varrho\frac{\partial \varepsilon}{\partial t} = \tau\varkappa, \tag{3.7}$$

where

$$\varepsilon = \varepsilon_0 + \varepsilon_1\mu^2\varkappa^2, \quad \tau = \mu\left(\varkappa - \lambda\frac{\partial \varkappa}{\partial t}\right) \tag{3.8}$$

and, for convenience, we denote the coefficient of $\varkappa^2$ by ($\varepsilon_1\mu^2$). Otherwise they are quite arbitrary.

Let us now impose on those fields the requirement that they satisfy the entropy inequality

$$\frac{\partial \eta}{\partial t} \geqslant 0 , \qquad \eta = \eta_0 + \eta_1 \mu^2 \varkappa^2 \tag{3.9}$$

where, again, the coefficient of $\varkappa^2$ is denoted by ($\eta_1 \mu^2$). It is easy to prove the following consequences of this condition

$$\mu \geqslant 0 , \qquad \frac{\partial \varepsilon_0}{\partial T} - T \frac{\partial \eta_0}{\partial T} = 0 , \qquad \varepsilon_1 - T\eta_1 = -\frac{\lambda}{2\varrho\mu} . \tag{3.10}$$

Hence the stability of equilibrium would yield the condition

$$\lambda < 0 \tag{3.11}$$

contrary to all observations. This fallacy of 2nd grade fluid has been pointed out by Dunn & Fosdick [1].

One can also show that the same condition (3.11) follows for arbitrary order N, if Nth grade fluid is considered to be an exact model.

Although the governing set of equations remains unchanged thermodynamic requirements change dramatically, if we assume that the 2nd order fluid is supposed to approximate the maxwellian fluid. Assuming for simplicity the monotonicity of that approximation, we see from (2.7) that the following convergence condition should hold

$$|\varkappa| > |\lambda \varkappa^{(1)}| > |\lambda^2 \varkappa^{(2)}| > \dots \tag{3.12}$$

This means, however, that the field v must satisfy not only the field equation $(3.7)_1$ but also the above requirement for its derivatives. This fact provides for a considerable change in the procedure of exploitation of the entropy inequality. Instead of relations (3.10), we then obtain

$$\mu \geqslant 0 , \qquad \frac{\partial \varepsilon_0}{\partial T} - T \frac{\partial \eta_0}{\partial T} = 0 , \qquad \frac{\lambda}{\varkappa} \frac{\partial \varkappa}{\partial t} \leqslant \frac{1}{2} \tag{3.13}$$

where the relation $(3.4)_3$ has been used.

In t he case of the solution (2.15), the inequality $(3.13)_3$ indicates, for instance

$$\beta < \frac{1}{2} . \tag{3.14}$$

This condition is, obviously, satisfied (see: Fig. 1).

Hence, we conclude that, if the 2nd order fluid is an approximation of the maxwellian model of a non-newtonian fluid, it satisfies the second law. At the same time, however, the entropy inequality does not predict any specific structure of the Helmholtz free energy and - consequently - does not say anything about thermodynamic stability of such a model. The sign of λ and the stability are both the consequences following from the requirements imposed on the maxwellian model rather than on the 2nd order fluid.

4. Extended thermodynamics of isothermal processes

One of the fundamental differences between thermodynamics of the maxwellian fluid and thermodynamics of the Nth order fluid is the choice of independent basic fields:

maxwellian fluid	Nth order fluid
v, T, τ	v, T

This difference provides the need for the construction of an additional field equation in the case of the maxwellian model. In our presentation, we have taken that additional equation $(3.1)_3$ from rheology.

There is, however, a systematic method of constructing field equations by use of balance laws - the method is suggested by the kinetic theory. If the stress deviator or - additionally, for non-isothermal processes - the heat flux vector become basic fields, we have to use more balance equations than the standard ones which are given by the conservations laws of mass, momentum and energy. These additional equations appear in a natural way in extended thermodynamics [6].

In the simple one-dimensional isothermal process, the additional balance equation referred to an inertial frame of reference has the form [2]:

$$2\varrho \frac{\partial}{\partial t}(e_2 \tau) - p\varkappa = P, \qquad P = \beta_1 \tau + \beta_2 \frac{\partial \tau}{\partial t} + o(\tau^2) \tag{4.1}$$

where p is the pressure, P-production term, e_2, β_1, β_2 - material coefficients. We have used here already simple constitutive relations turning the original balance equation into the field equation for τ. The important difference between balance equation (4.1) and the classical conservation laws of mass, momentum and energy is the presence of the production term P.

We can easily identify the material coefficients in (4.1) and we arrive at the relation

$$\left(1 + \lambda \frac{\partial}{\partial t}\right)\tau = \mu\varkappa + \frac{2\mu\varrho e_2}{p}\frac{\partial \tau}{\partial t} + o(\tau^2) \tag{4.2}$$

with

$$\beta_1 = -\frac{p}{\mu}, \qquad \beta_2 = -\frac{p\lambda}{\mu}. \tag{4.3}$$

If we limit our attention to linear terms in (4.2), the difference between this equation and equation (2.1) appears to lie only in the second term on the right-hand side. The following modification of characteristic time λ

$$\tilde{\lambda} = \lambda - \frac{2\mu\varrho e_2}{p} \tag{4.4}$$

transforms equation (4.2) into (2.1).

We have argued elsewhere (see: [2]) that, in normal circumstances, the above correction of characteristic time is negligibly small. This means that extended thermodynamics leads, essentially, to the same results as our considerations of section 3. By keeping the small term, however, we gain a considerable extension of physical insight - particularly connected with kinetic theory of non-newtonian fluids.

References

[1] Dunn, E., Fosdick, R.L. Thermodynamics, Stability and Boundedness of Fluids of Complexity 2 and Fluids of Second Grade. A.R.M.A., 56, (1974).

[2] Müller, I., Wilmanski, K. Extended Thermodynamics of a Non-Newtonian Fluid. Rheologica Acta (to be published)

[3] Rivlin, R.S., Wilmanski, K. On One-Dimensional Nth Order Models of Non-Newtonian Fluids (forthcoming)

[4] Ginn, R.F., Metzner, A.B. Measurements of Stresses Developed in Steady Laminar Shearing Flows of Viscoelastic Media. Trans. Soc. Rheol., 13:14 (1969)

[5] Coleman, B.D., Mizel, V.J. Breakdown of Laminar Shearing Flows for Second-Order Fluids in Channels of Critical Width. ZAMM, 46 7 (1966)

[6] Müller, I. Thermodynamics. Pitman Publ.(1985)

MOLECULAR MECHANISMS OF NON-LINEAR RUBBER ELASTICITY

Andrzej Ziabicki

Polish Academy of Sciences, Institute of Fundamental Technological Research, Warsaw, Poland

INTRODUCTION

Polymer networks - vulcanised rubbers, gels, swollen cross-linked polymers - exhibit a wide range of non-linear elastic properties. These properties can be described using a number of phenomenological relations, but the molecular mechanisms involved are not quite clear. Molecular models based on the statistical properties of a single polymer chain lead to a narrow class of constitutive equations, incapable of describing real polymer systems.

In the search of molecular mechanisms which could produce more flexible constitutive relations, we will analyse the effects of entropy and internal energy of a *single polymer chain, intra-*, and *inter-chain* interactions, as well as *topological constraints* manifesting themselves as "chain entanglements". We will also discuss briefly topological effects related to the way in which polymer chains are connected in junction points ("crosslinks") to produce a coherent network. References to original papers containing detailed solutions will be given. Some of the problems discussed in this review have been presented at the *Network Group Meetings* in *Jablonna* [1] and *Manchester* [2].

PHENOMENOLOGICAL EQUATIONS OF RUBBER ELASTICITY

We will confine our discussion to homogeneous, isotropic, solid materials, subjected to uniform strain $\underset{\sim}{\Gamma}$. In equilibrium, *the elastic potential*, W, reduces to a scalar function of three principal invariants of the *strain tensor* (left Cauchy Green tensor), $\underset{\sim}{\Gamma}$. For a homogeneous material which does not undergo any phase transition, the potential W is continuous and can be presented in the form of a series expansion

$$W(\underset{\sim}{\Gamma}) = \sum_i \sum_j \sum_k A_{ijk} (I_1 - 3)^i (I_2 - 3)^j (I_3 - 1)^k \tag{1}$$

I_1, I_2, I_3 denote invariants of the strain tensor $\underset{\sim}{\Gamma}$, viz.

$$\begin{aligned} I_1 &= \mathrm{tr}\underset{\sim}{\Gamma} \\ I_2 &= (1/2)[(\mathrm{tr}\,\underset{\sim}{\Gamma})^2 - \mathrm{tr}\,\underset{\sim}{\Gamma}^2] \\ I_3 &= \det \underset{\sim}{\Gamma} \end{aligned} \tag{2}$$

The expansion (1) is chosen so, that the potential disappears in the unstrained state ($\underset{\sim}{\Gamma}=\underset{\sim}{1}$). Eq. (1) *does not include assumption of an incompressible material, or isochoric deformation.* For isochoric deformation ($\det\underset{\sim}{\Gamma}=1$) the terms with k>0 disappear, leaving the relation unchanged. This leads to contitutive equations significantly different to those based on the theory of *Rivlin* [3], including the assumption of isochoric deformation.

For two most popular strain geometries one obtains

$$\underset{\sim}{\Gamma} = \begin{bmatrix} \lambda^2 & 0 & 0 \\ 0 & \lambda^{-2\mu} & 0 \\ 0 & 0 & \lambda^{-2\mu} \end{bmatrix} \tag{3}$$

$$I_1 = \lambda^2 + 2\lambda^{-2\mu}; \quad I_2 = 2\lambda^{2-2\mu} + \lambda^{-4\mu}; \quad I_3 = \lambda^{2-4\mu} \tag{4}$$

for *uniaxial extension* ($\lambda>1$) or *compression* ($\lambda<1$), and the Poisson coefficient μ, and

$$\underset{\sim}{\Gamma} = \begin{bmatrix} 1+\gamma^2 & \gamma & 0 \\ \gamma & 1 & 0 \\ 0 & 0 & 1 \end{bmatrix} \tag{5}$$

$$I_1 = I_2 = 1 + \gamma^2; \quad I_3 = 1 \tag{6}$$

for *simple shear*.

When the magnitude of deformation (i.e. deviation of the strain tensor $\underset{\sim}{\Gamma}$ from the unit tensor $\underset{\sim}{1}$) is considered

$$||\underset{\sim}{E}|| = (\mathrm{tr}\ \underset{\sim}{E}\underset{\sim}{E}^{T})^{1/2} \tag{7}$$

$$\underset{\sim}{E} = \underset{\sim}{\Gamma} - \underset{\sim}{1} \tag{8}$$

I_1 introduces terms of the first order of $||\underset{\sim}{E}||$, I_2 - terms of the second order, and I_3 - of the third order. Consequently, eq.(1) can be rewritten as a sum of terms with increasing order, p, viz.

$$W(\underset{\sim}{\Gamma}) = \sum_{p=1}^{\infty} \sum_{j=0}^{p/2} \sum_{k=0}^{(p-2j)/3} A^{p}_{jk} (I_1-3)^{p-2j-3k} (I_2-3)^{j} (I_2-1)^{k} \tag{9}$$

The first summation (over p) classifies terms with increasing magnitude of deformation. The first-order equation (p=1) applicable for infinitely small deformations ($||E||<<1$) includes a single linear term, known as *Neo-Hookean*

$$W(\underset{\sim}{\Gamma}) = A^{1}_{00}\ (I_1-3) \tag{10}$$

The second-order equation includes three terms

$$W(\underset{\sim}{\Gamma}) = A^{1}_{00}\ (I_1-3) + A^{2}_{00}\ (I_1-3)^2 + A^{2}_{10}\ (I_2-3) \tag{11}$$

The third invariant, I_3, appears in the third-order equation, together with two other p=3 terms

$$A^3_{00}\,(I_1-3)^3,\ A^3_{01}\,(I_3-1),\ \text{and}\ A^3_{10}\,(I_1-3)(I_2-3) \qquad (12)$$

It may be noted that our second-order equation (eq. 11) is different from one derived by *Rivlin* [3], linear both in I_1 and I_2

$$W(\underset{\sim}{\Gamma}) = C_1\,(I_1-3) + C_2\,(I_2-3) \qquad (11a)$$

Eq. (11a) was also postulated empirically by *Mooney* [4] to account for the observed properties of some vulcanised rubbers: linear behaviour in shear, non-linear in uniaxial extension. Eq. (11a) known as *Mooney-Rivlin Equation* is widely used (and often misused) for the interpretation of experimental data. The material constant C_2 is usually determined from extrapolation of stress to infinite (!) extension ($\lambda\to\infty$), in spite of the fact that deformation of any crosslinked system is limited, and the equation itself is applicable to small strains only (second order of smallness of $||\underset{\sim}{E}||$). A more correct procedure for the determination of material constants in the *Mooney-Rivlin equation* was suggested in ref. [2].

In the non-isochoric model, the second-order equation includes both a I_2 term, and a quadratic I_1^2 term, allowing for non-linear behaviour both in extension and shear. Linear shear behaviour combined with non-linear extension, actually observed in many (though not all) rubbers, should be ascribed to disappearance of the material constant A^2_{00}, rather than identified with a special form of the constitutive equation. Also incompressibility of real materials, providing justification for the assumption of isochoric deformation, is not universally observed. The "complete" (non-isochoric) second-order equation of rubber elasticity (eq.11) was derived from the molecular considerations by *Ishihara* [5] and used for the analysis of experimental data by *Zahorski* [6]. The experimental data analysed did not indicate a *Mooney-Rivlin behaviour*, i.e. the A^2_{00} constant was far from zero.

A good constitutive equation of rubber elasticity should satisfy two conditions. First, single equation should adequately describe behaviour of the same material in *various geometries of deformation* (uniaxial extension and compression, shear, bending, torsion, etc.). Second, one equation should be capable of describing many *different materials*. Passing from one material to another, one should find different material constants without any change of the constitutive relation. E.g. the same equation should account for the *Mooney-Rivlin behaviour* ($A^2_{00}=0$) in some rubbers, and for a different behaviour in others. We will show that most molecular mechanisms assumed responsible for rubber elasticity do not offer enough flexibility in the determination of the constitutive relations. The *material constants*, A^p_{jk} are either *linearly dependent*, or *combined with a small parameter*, which reduces importance of higher-order terms to a narrow range of structures. We will show that the desired decoupling leading to flexible constitutive relations can be realised by consideration of a network containing both permanent, *localised crosslinks*, and *topological constraints* - entanglements.

MOLECULAR MECHANISMS

Elastic behaviour of a single polymer chain

A linear polymer chain composed of n chemical bonds of length ℓ, valence angles α, and rotation angles ϕ (Figure 1a) represents a statistical system with many degrees of freedom. Extension of the chain is realised, first of all, by rotation of chemical bonds and, to a lesser extent, by deformation of valence angles and stretching of valence bonds.

Linear chain molecule is usually modelled by a system of N rigid, *freely-jointed statistical segments* of fixed length L, ($N<n$, $L>\ell$), each exhibiting two rotational degrees of freedom (Figure 1b). The parameters of such a model (N and L) are calculated from the molecular characteristics of the real chain - number, and length of chemical bonds, (n,ℓ), valence angles, α, and the potential of rotation $U(\phi)$ [7,8]. The "macroscopic" state of such a microsystem is characterised by the *end-to-end distance*, h. Displacement of chain ends leading to the change of h affects entropy and internal energy of the chain, thus contributing to its elasticity. The main contribution to deformation-dependent free energy is *conformational entropy;* the number of different microstates available at given end-to-end distance decreases with increasing h. Internal energy changes are related to *rotational isomers*. Both effects yield free energy of an isolated chain, f_{ch}, as a function of even powers of the end-to-end distance, h

$$f_{ch}/kT = \sum_{p=1}^{\infty} a_p \, (h^2/\langle h_0^2\rangle)^p \tag{13}$$

$\langle h_0^2\rangle$ is average square end-to-end distance in the unperturbed state. The theories of conformational entropy of a freely-jointed chain [9], as well as those related to internal energy of rotational isomers [10] indicate that each coefficient a_p is combined with a (p-1)-st power of a small parameter, viz. *reciprocal number of statistical chain segments*, $\varepsilon=1/N$

$$a_p = a_{po}\,\varepsilon^{p-1} \tag{14}$$

which automatically truncates the series (13) to one, or two first terms, unless very short, or very rigid chains are considered. For a freely-jointed chain without deformation-dependent internal energy, the

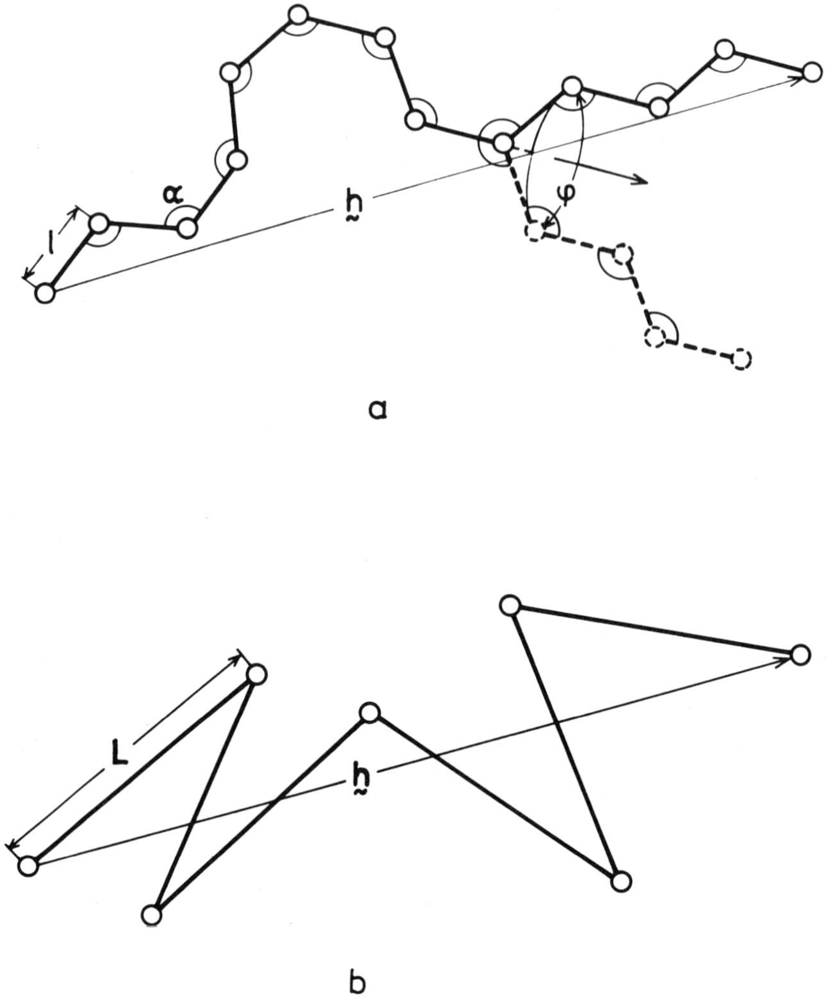

Figure 1.

Linear polymer chain

a. Real chain with fixed valence angles and hindered rotation in the main chain. b. Model chain with freely-jointed statistical segments

coefficients in eq.(13) are simply (p-1)-st powers of the parameter ε multiplied by constants

$a_1 = 3/2$ (*Gaussian statistics of conformations*),

$a_2 = (9/20)\ \varepsilon,$

$a_3 = (99/350)\ \varepsilon^2.$

In the case of non-zero rotational energy, the length and number of statistical segments depend on the end-to-end distance h, and the coefficients a_p acquire additional substructure

$$a_p = \mathrm{const}\ \varepsilon_0^{p-1}\ (1 + b_1\ \varepsilon_0 + b_2\ \varepsilon_0^2 + \ldots) \tag{15}$$

without changing their basic properties [10]. The same can be said about the effect of intramolecular interactions (with a r^{-2m} potential) between individual atoms or atomic groups in the polymer chain [11]. In eq.(15) $\varepsilon_0 = \varepsilon(h=0)$ denotes parameter ε related to the number of statistical chain segments at zero end-to-end distance.

When thermodynamic properties of the network are described as a sum of contributions from N_{ch} individual network chains with affinely displaced ends and uniformly deformed vectors $\underset{\sim}{h}$,

$$\underset{\sim}{h}_i = \underset{\sim}{\Lambda}\underset{\sim}{h}_{,o} \qquad \text{for all } i \tag{16}$$

($\underset{\sim}{\Lambda}$ is deformation gradient, and $\underset{\sim}{\Gamma} = \underset{\sim}{\Lambda}\underset{\sim}{\Lambda}^T$), the resulting elastic potential is calculated as

$$W(\underset{\sim}{\Gamma}) = \sum_{i=1}^{N_{ch}} [f_{ch}(\underset{\sim}{h}_i;\underset{\sim}{\Gamma}) - f_{ch}(\underset{\sim}{h}_i;\underset{\sim}{1})]$$

and assumes a special form

$$W(\underset{\sim}{\Gamma}) = A_{10}\,(I_1-3) + A_{20}\,\varepsilon\,(I_1^2-4/3I_2-5) + A_{30}\,\varepsilon^2\,(I_1^3- $$

$$- 12/5\, I_1 I_2 + 8/5\, I_3 - 7) + \dots \qquad (17)$$

in which powers of various invariants of the same order p are *grouped together*, and *combined with* (p-1)-st powers of the *small parameter* ε.

Combination of invariants into groups results from averaging of p-powers of the ratio $h^2/\langle h^2\rangle$

$$\langle (h^2/\langle h_0^2\rangle)\rangle = I_1/3$$
$$\langle (h^2/\langle h_0^2\rangle)^2\rangle = (I_1^2 - 4/3\, I_2)/3$$
$$\langle (h^2/\langle h_0^2\rangle)^3\rangle = (I_1^3 - 12/5\, I_1 I_2 + 8/5\, I_3)/3$$

... etc.

Eq.(17) confines the predicted non-linear behaviour to very short and/or very rigid chains (small N, Large ε), and excludes description of different kinds of non-linear behaviour by a single equation. The existing experimental data indicate that eq.(17) cannot adequately describe the elastic behaviour of real rubbers.

Phantom Networks

The theory of *phantom networks* - systems in which polymer chains are connected in some points (*junctions*, or *crosslinks*) without other interactions), was formulated by *James and Guth* [12].

The first difference between a *phantom network* and a collection of *independent chains* concerns the reference (underformed) state. Reduction of the translational degrees of freedom of chain ends leads to *contraction of chains* as they join the network. The *contraction factor*, $A_{con} = \langle h^2_{net}\rangle/\langle h_0^2\rangle$ is related to the way in which network chains are connected to network junctions (crosslinks). For a system with *Gaussian statistics of conformations* (applicable to long, flexible chains) the contraction factor A_{con} was derived by *Walasek* [13]

$$A_{con} = (N_{jct}-1)/N_{ch} \tag{18}$$

For large systems, $N_{jct} >> 1$

$$A_{con} \simeq N_{jct}/N_{ch} \tag{18a}$$

N_{jct} denotes the number of junctions (crosslinks) in the system.

For an *ideal, s-functional network* (i.e. one with *s* chains issuing from each junction)

$$A_{con} = N_{jct}/N_{ch} = 2/s \tag{18b}$$

which for the most popular tetrafunctional system (s=4), reduces to 1/2. This value was used in the early theory of phantom networks by *James and Guth* [12]. For non-ideal networks with more complex connectivity pattern, A_{con} assumes values intermediate between 2/s and unity, and depends on the topological structure of the system.

The second consequence of combining N_{ch} chains into a network with N_{jct} junctions is, that not all network chains (or junctions) are elastically effective, i.e. not all junctions transmit forces applied to the boundary of the system. A junction is *elastically effective*, if at least three paths starting from the junction, and following network chains, lead to the boundary. Consequently, the number of elastically effective chains and junctions depends on the connectivity pattern of the network. We will discuss this problem in one of the following sections.

Free energy of a phantom network can be obtained from the statistical integral

$$Z = \int\int\int \exp[-\sum_i \sum_{p=1}^{\infty} a_p (h^2/\langle h^2_{net}\rangle)^p \; d\underset{\sim}{u} \ldots d\underset{\sim}{u}_{N_{jct}} \tag{19}$$

where summation is performed over all elastically effective chains ($i=1,2,\ldots N_{ch}$) and over various orders of the molecular deformation, p. Integration is performed over the positions of elastically effective junctions, $\underset{\sim}{u}_j$, $(j=1,2,\ldots,N_{jct})$. Powers of the end-to-end distance, h^{2p}, are reduced by the average end-to-end distance in the network, $\langle h^2_{net}\rangle$, rather than $\langle h^2_0\rangle$, characteristic of an isolated chain. The coefficients a_p are taken from an isolated chain (eqs. 13-15) and include effects of conformational entropy, internal energy and entropy of rotational isomers, and intramolecular potential interactions. The macroscopic strain tensor $\underset{\sim}{\Gamma}$ appears through the boundary conditions: terminal junctions (those located on the boundary of the system) are displaced uniformly, following the macroscopic deformation gradient $\underset{\sim}{\Lambda}$.

Consideration of a phantom network instead of isolated chains, does not introduce anything new to the constitutive relations. The factor determining non-linear elasticity is the same as before - elastic response of a single chain (eq.13) controlled solely by even powers of the end-to-end distance h. The constitutive equation for a phantom network has the same form as eq.(17), with invariant expressions for each order p coupled into groups and combined with (p-1)-st powers of the small parameter ε. The only difference lies in the numerical values of the material constants and their substructure (detailed dependence on the parameter ε).

Rod-rod interactions of chain segments

In the search of molecular mechanisms responsible for elastic behaviour different to that characteristic of isolated chains or phantom networks, two special models have been analysed.

The first model is based on the *rod-like shape* of the molecular segments forming network chains. Deformation of the network affects *orientation distribution of chain segments* and, consequently, their interactions. Deformation-controlled *molecular field* is created, and *particle-field* interactions contribute to the elastic potential, W.

The concept of rod-rod interactions with a cosine-square potential, widely used in the theory of liquid crystals, was suggested as an additional mechanism of rubber elasticity by *Guth* [14]. More recently, *Walasek* [15] reconsidered the problem, starting from the classical liquid-crystal theory with a rod-rod interaction potential $U(\theta)$ dependent on the single angle θ

$$U(\theta) = - B\langle P_2\rangle P_2(\theta) \tag{20}$$

B is interaction constant, and P_2 is the second Legendre function

$$P_2(\theta) = (3\cos^2\theta - 1)/2 \tag{21}$$

The average P_2 in eq.(20) characterises intensity of the induced molecular field. Combination of the interaction potential with orientation distribution controlled by conformational entropy of a freely-jointed polymer chain leads to the additional free energy term in the constitutive equation of rubber elasticity

$$\Delta W(\underset{\sim}{\Gamma}) = \text{const.}\ B\varepsilon(I_1^2 - 3I_2)/(1 - B/5kT) + 0(\varepsilon^2) \tag{22}$$

The interactions introduce a second order effect, both in terms of the parameter ε, and of the magnitude of deformation (I_1^2, I_2), but the groups of invariants which appear, are different to those in the theory of isolated chains, or phantom networks. Consequently, in a network with rod-rod interactions the *invariant terms are decoupled*, though still combined with (p-1)-st powers of the small parameter ε

$$W(\underset{\sim}{\Gamma}) = A^1_{00}(I_1-3) + A^2_{00}\varepsilon(I_1-3)^2 + A^2_{10}\varepsilon(I_2-3) + O(\varepsilon^2) \qquad (23)$$

The effect of rodlike interactions, consistent with the early result of *Guth*, modifies the shape of the constitutive equation, making it more flexible, and admitting more diversified non-linear behaviour. The appearance of (p-1)-st powers of the small parameter ε at all non-linear terms, still limits importance of this effect to networks composed of short and/or rigid chains.

Topological constraints - chain entanglement

Impenetrability of real polymer chains, neglected in the theory of phantom networks, leads to *topological constraints* (entanglements) and related deformation-dependent free energy. In the early treatments of this problem, both theoretical [16-18] and experimental [19], permanent, or temporary *entanglements* were considered as *additional crosslinks*, equivalent in their elastic response to localised junctions. The theory of chain entanglement is still far from being complete, but several simple models suggest that the *elastic response of an entanglement is different from that of a localised junction.*

To illustrate this point we will discuss a simple model consisting of four polymer chains connected in one point (localised junction) (Figure 2a). The model is identical with one introduced by *Flory and Rehner* [20], but given a different interpretation. *Flory and Rehner* considered their tetrahedron a representative cell of a large, uniform network; we are using it as a self-consistent, one-junction system.

Free ends of the four chains are subjected to uniform displacement $\underset{\sim}{\Lambda}$.

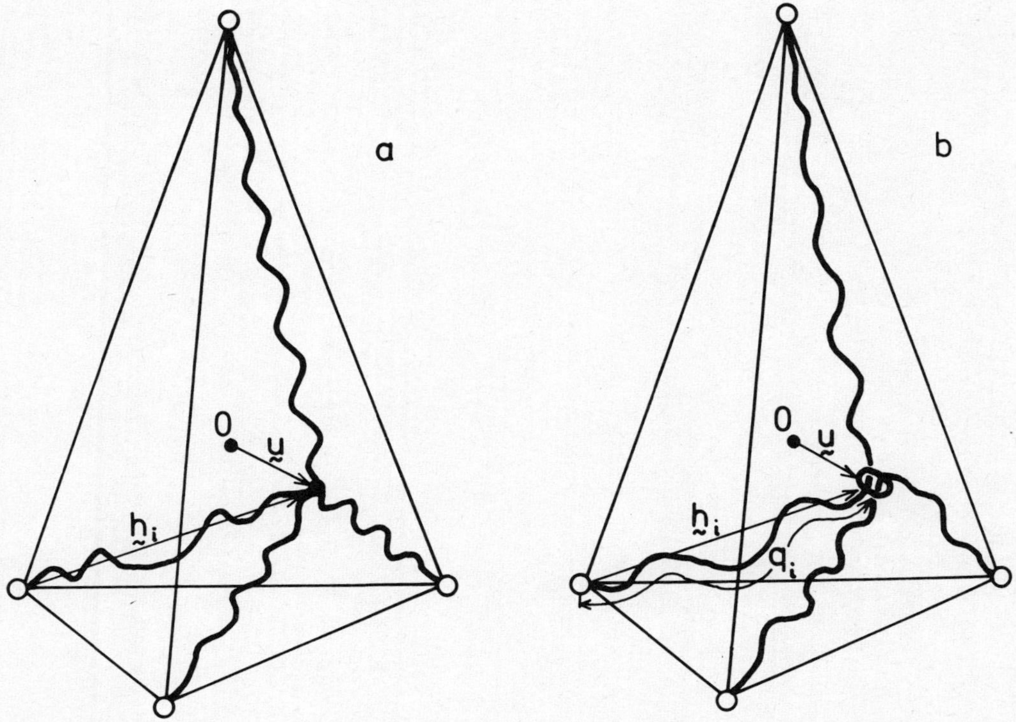

Figure 2.

One-junction models of crosslinked systems

a. Fixed (localised) junction: permanent crosslink   b. Non-localised (sliding) contact: entanglement

The central junction has three degrees of freedom, and performs fluctuations in space. Statistical integral for such a four-chain system with chain free energy f_{ch} from eq.(13)

$$Z = \int \exp[-\sum_i f_{ch}(h_i)/kT]\, d\underset{\sim}{u} \qquad (24)$$

can easily be found. Integration is performed over the position of the central junction, $\underset{\sim}{u}$. The model describes behaviour of a typical *phantom network* and leads to the constitutive equation (17).

When fixed, *localised junction* is replaced by a *sliding ring* (Figure 2b), the system represents *an entanglement*. Actually, the model overestimates the effect of topological constraints : real entanglements possess more degrees of freedom because entangled chains need not be in a permanent contact, as implied in the ring model. Therefore the elastic effects predicted provide upper bound to the real elastic behaviour.

The entanglement junction modelled by a sliding ring has five degrees of freedom. In addition to three translational degrees of freedom exhibited by a fixed junction (position of the contact in the Euclidean space, $\underset{\sim}{u}$), there appear also *positions of the ring in the space of the entangled chains*, q_k (cf. Figure 2b). The total contour lengths of the entangled chains, $Q_1 = q_1 + q_2$ and $Q_2 = q_3 + q_4$ being constants, position of the entanglement can be characterised by two dimensionless parameters $x,y \in (-1,1)$

$$x = (q_1-q_2)/(q_1+q_2)$$
$$y = (q_3-q_4)/(q_3+q_4)$$

The minimum free energy of deformation of the entangled chains (Fig.2b) is generally lower than that for chains combined in a fixed junction (Fig.2a), because of the increased number of the degrees of freedom (sliding along the chain contours). Entangled system can better adjust to deformation applied to its boundary, than network with fixed crosslinks. The difference between $F_{min,ent}$ and $F_{min,fix}$ is the larger, the larger is strain [21]. Free energy and the elastic potential W for an entangled network can be obtained from the statistical integral

$$Z = \iiint \exp[-\Sigma f_{ch}(h_i)/kT] \, d\underset{\sim}{u} \, dx \, dy \tag{25}$$

taken over all five degrees of freedom (three components of the position vector $\underset{\sim}{u}$, two positions along the contour, x,y). The elastic potential

$$W = -kT[\ln Z(\underset{\sim}{\Gamma}) - \ln Z(\underset{\sim}{1})] \tag{26}$$

can be obtained by expansion of the function $\ell n Z(I_1, I_2, I_3)$ in power series of the small parameter ε (reciprocal number of segments in the entangled chain) and/or powers of invariants of the strain tensor $\underset{\sim}{\Gamma}$ [22,23]. It can be shown that the resulting constitutive equation significantly differs from the relations found for isolated chains or phantom networks (eqs. 17 and 23). The *invariant expressions* for any order of deformation are *not grouped;* there is *no relation* between the *magnitude of deformation* (order p), and the order of the *molecular parameter,* ε. Material constants A^{p}_{jk} in the original equation (1) are independent of each other, and not combined with any power of the parameter ε

$$W(\underset{\sim}{\Gamma}) = A^1_{00}(1 + a_1\varepsilon + a_2\varepsilon^2 + \ldots)(I_1-3) + A^2_{00}(1 + b_1\varepsilon + b_2\varepsilon^2 + \\ + \ldots)(I_1-3)^2 + A^2_{10}(1 + c_1\varepsilon + c_2\varepsilon^2 + \ldots)(I_2-3) + \ldots \tag{27}$$

It can be observed that terms with arbitrarily high order of deformation appear without the molecular parameter ε, i.e. various kinds of non-linear behaviour are admitted in systems composed of very long and flexible chains (small ε). Even *polymer chains with Gaussian statistics* (first term in eq.13, $\varepsilon \to 0$) which yielded linear (Neo-Hookean) elasticity in phantom networks, produce *non-linear elastic potential* (of arbitrarily high order) *in an entangled system.* Chain entanglement, leading to completely decoupled material constants in the constitutive relations can explain variety of non-linear properties observed in polymer networks. This strongly supports the opinion that the role of topological constraints in rubber elasticity is important, if not determining

TOPOLOGICAL STRUCTURE AND ELASTICITY OF POLYMER NETWORKS

It has been noted that the *connectivity pattern* in the polymer network, i.e. the way in which chain molecules are attached to network junctions (crosslinks), affects conformation in the undeformed (reference) state, the number of elastically effective junctions and chains, and the magnitude of elastic constants.

The fact that part of the crosslinked material can be elastically ineffective because of incomplete, or defective attachment to crosslinks (dangling chains, loops, etc.) has been observed by many authors [16,24-29] who have tried to estimate the magnitude of this effect. These early attempts did not include any systematic and complete treatment of topological structures. More recent works [1,13] discussed all possible configurations of an s-functional crosslink, and proposed a method of their evaluation.

Distribution of Crosslink Types

Different *structural elements* can issue from a network junction (crosslink). Seven such elements are possible in a tetrafunctional network: a *free-end chain*, not attached to any other crosslink;a *singlet* i.e. a single chain spanning two different crosslinks; a *loop*, - chain issuing from, and returning to the same crosslink; a *doublet*, a *triplet*, and a *quadruplet*,i.e. a pair, a triad, and a tetrad of chains all attached with one end to a common crosslink "i", and with the other end to another (common) crosslink "j". Finally, one, or more functionalities of the crosslink can be *void*, not saturated by any polymer chain. These seven structural elements in a tetrafunctional network (s+3 different elements appear in an s-functional system) can be combined in various ways to produce *34 topologically different crosslinks* (Figure 3). The distribution of such crosslinks determines contraction factor, number of elastically effective chains and crosslinks, material constants in the constitutive equation of elasticity, swelling behaviour, etc. [1, 13].

Actual structure of a crosslinked polymer system is controlled by chemical structure and crosslinking conditions. There is no good way of theoretical prediction, or experimental determination of the distribution of all 34 types of junctions; we have proposed, instead, a simple model, based on the assumption of *random distribution of structural elements* among crosslinks. Simple combinatorial analysis, taking into account the existing constraints,leads to the fractions of various types of crosslinks

$$n(s^{\alpha}f^{\beta}v^{\gamma}d^{\delta}\ell^{\lambda}t^{\mu}q^{\nu}) =$$

$$= [2(4/3)^{\mu}(\alpha+\beta+\gamma)!(\Sigma-t/3)^{2(1-\mu-\nu)-\delta-\lambda} /(2-\delta-$$

$$-\lambda)!\ \alpha!\beta!\gamma!\delta!\lambda!(\Sigma+\Delta-t/3)^{1-\mu-\nu}\ \Sigma^{\alpha+\beta+\gamma}]\ s^{\alpha}f^{\beta}v^{\gamma}d^{\delta}\ell^{\lambda}t^{\mu}q^{\nu} \qquad (28)$$

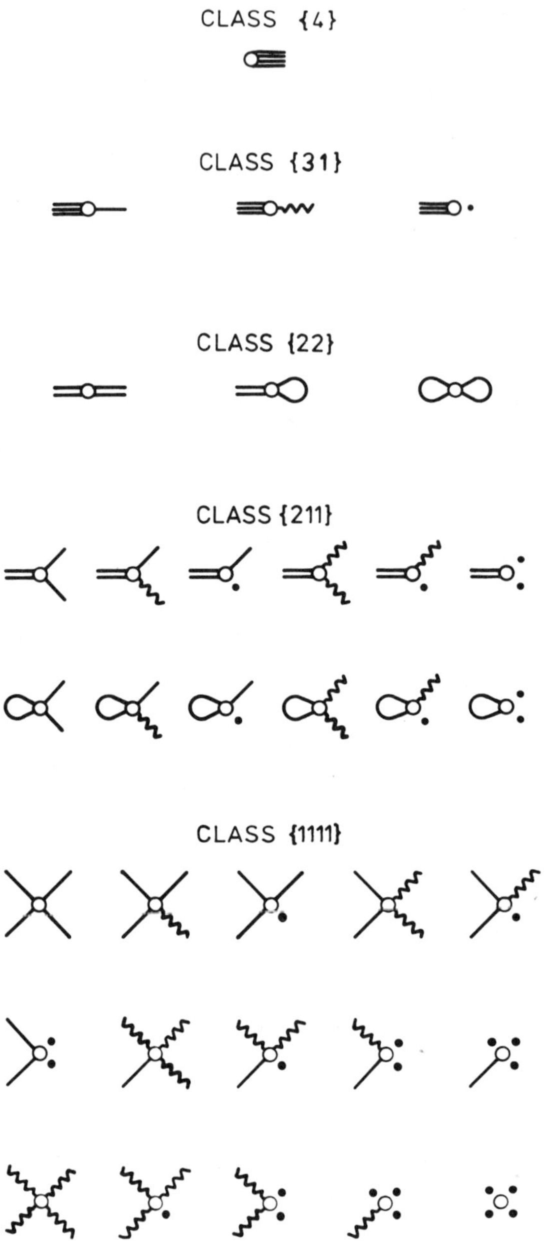

Figure 3.

Different crosslink types in a tetrafunctional network

where the exponents $\alpha,\beta,\gamma,\delta,\lambda,\mu,\nu$ satisfy the condition

$$\alpha + \beta + \gamma + 2(\delta+\lambda) + 3\mu + 4\nu = 4 \qquad (29)$$

and assume integer values from the range

$$\alpha,\beta,\gamma \in \{0,1,2,3,4\}$$
$$\delta,\lambda \in \{0,1,2\}$$
$$\mu,\nu \in \{0,1\}$$

$n(s^{\alpha}f^{\beta}v^{\gamma}\ldots)$ is a fraction of junctions with "α" singlets, "β" free-end chains, "γ" void functionalities, etc. s,f,v,d,ℓ,t,q denote fractions of crosslink functionalities in the system saturated, respectively, with singlets, free-end chains, voids, doublets, loops, triplets, and quadruplets. $\Sigma = s + f + v$, is the sum of functionalities occupied by monofunctional elements, $\Delta = d + \ell$, the sum of functionalities consumed by bifunctional elements (doublets and loops)

The fractions of junctions, $n(\ldots)$, and fractions of functionalities saturated by individual structural elements are both normalised to unity

$$n(s^4) + n(s^3f) + \ldots + n(q) = 1 \qquad (30)$$
$$s + f + v + d + \ell + t + q = 1 \qquad (31)$$

It may be noted that, in spite of the assumed randomness, junction types do not form a multinomial distribution; this results from arrangement of structural elements with different functionalities (1,2,3,4). Junctions are first classified into several topological groups ("1111" - four monofunctional elements, "211" - one bifunctional, and two monofunctional elements, etc.); multinomial distribution holds within each group. The resulting global distribution (eq.28) includes additional normalisation factors. Junction type distribution, $n(\ldots)$, together with the total number of crosslinks, N_{jct}, determines elastic, and swelling properties of the network. Using the random model (eq.28) and the normalisation condition (31), one can *reduce the number of parameters from 35* (34 fractions of junction types plus total number of junctions) *to seven* (six independent fractions of structural elements, s,f,v,...,q, and N_{jct}). This is still more than can be determined from chemical and physical data. Further reduction of the number of independent parameters is based on the postulate of *equilibrium conformation in the conditions of crosslinking*. This assumption yields four, out of six fractions of structural elements: quadruplets, triplets, doublets, and loops [1,13]. This leaves us with a tractable model characterised completely by *three parameters* (s,f,N_{jct}) which determine topological structure, elasticity, and other properties of the network.

Distribution of Crosslinks in the Space of Network Chains

An important structural characteristic of crosslinked systems is the *average distance between neighbour crosslinks*. Two different distances should be distinguished. The average distance between *spatial neighbours* (i.e. crosslinks closest in the Euclidean space), $\langle r^2 \rangle$, is controlled solely by the number of crosslinks, N_{jct} in the sample volume V

$$\langle r^2 \rangle = \text{const.}\ (V/N_{jct})^{2/3} \qquad (32)$$

where const. is a geometrical factor related to the shape of a representative network cell. Obviously, $\langle r^2 \rangle$ decreases with increasing concentration of junctions, (N_{jct}/V). It has been observed by the present author and *Klonowski* that average distance between *topologically neighbour crosslinks*, i.e. crosslinks closest in the space of the chain, $\langle h^2 \rangle$, is generally different to $\langle r^2 \rangle$. More than that, identification of both distances leads to unphysical prediction about the internal stress produced in the process of crosslinking [30]. The distance between topologically neighbour junctions, $\langle h^2 \rangle$, is identical with the average square end-to-end distance of network chains. The most probable, equilibrium value of $\langle h^2 \rangle$ which minimises elastic free energy in the conditions of crosslinking, represents unperturbed dimensions of the network chain, $\langle h^2_0 \rangle = NL^2$, multiplied by the contraction factor

$$\langle h^2 \rangle_{eq} = A_{con} NL^2 \tag{33}$$

The average number of statistical chain segements in a network chain, N, depends on the volume fraction of polymer in the system, v_p, and on the number of network chains, N_{ch}. Realising that $A_{con} = N_{jct}/N_{ch}$, one arrives at

$$\langle h^2 \rangle_{eq} = A_{con}(v_p/v_o)(V/N_{jct})L^2 \tag{34}$$

where v_o denotes molecular volume of a single chain segemnt. The ratio of the distances of topological and spatial neighbours increases with polymer concentration, v_p, and decreases with increasing crosslink density, N_{jct}/V; it depends also on the topological structure of the system

$$\langle h^2 \rangle_{eq}/\langle r^2 \rangle = \text{const.}A_{con} L^2 (v_p/v_o)(N_{jct}/V)^{-1/3} \tag{37}$$

The necessary condition of formation of a continuous network is the average distance between topologically neighbour crosslinks larger than, or equal to $\langle r^2 \rangle$. The condition $\langle h^2 \rangle/\langle r^2 \rangle = 1$ determines critical situation with minimum polymer concentration, or maximum crosslink density.

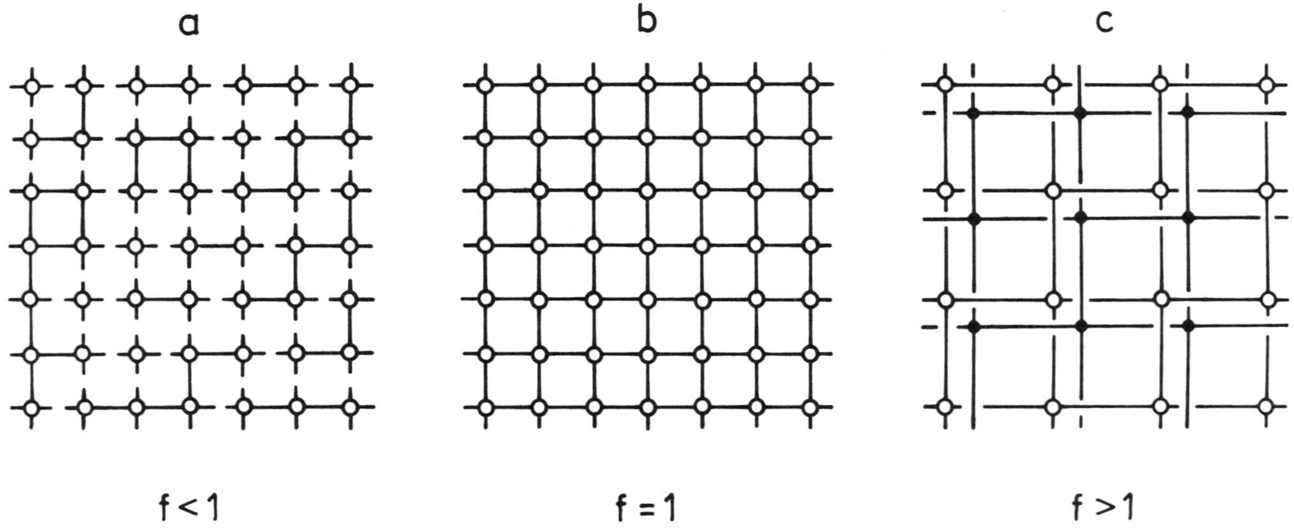

Figure 4.

Scheme of crosslinked systems with different ratio f= $\langle h^2 \rangle / \langle r^2 \rangle$

a. f<1: Discontinuous system, b. f=1: Simple network, c. f>1 : Interpenetrating network

Figure 4. presents three crosslinked systems with different values of the ratio $f = \langle h^2\rangle/\langle r^2\rangle$. $f<1$ characterises a *collection of junctions* (crosslinks) *without continuous connection* into a network. $f=1$ corresponds to critical crosslinking conditions, and yields a *simple (single) network*. $f>1$ - charateristic of real crosslinked systems - describes a system of *multiple*, or *interpenetrating networks*. $f^{3/2}$ - ratio of the volume swept out by ends of the network chain, and volume of the system shared by a single chain, is a measure of *multiplicity*, or *complexity* of an interpenetrating network. E.g. a system with $(\langle h^2\rangle/\langle r^2\rangle)^{3/2} = 2$ is equivalent to two simple networks mutually interpenetrated. Interpenetration seems to be a basic property of all polymer networks. It has no direct effect on elasticity: modulus of a "double" interpenetrating network is, in first approximation, equal to that of a single network with double number of crosslinks, provided that the reference conformation of network chains, $\langle h^2_{net}\rangle$ is the same. It should be realised, however that $\langle h^2_{net}\rangle$ is controlled by the distance between topological, rather than spatial, neighbours.

REFERENCES

1. A. Ziabicki, *Polymer*, 20, 1973, (1979)
2. A. Ziabicki and J. Walasek, *Brit.Polymer J.*, 17, 116, (1985)
3. R.S. Rivlin, *Philos.Trans.Royal Soc.(London)*, A241, 379, (1948); ibid. A242, 173, (1949)
4. M. Mooney, *J.Appl.Phys.*, 11, 582, (1940); *J.Colloid Sci.*, 6, 96, (1951)
5. A. Ishihara, N. Hashitsume and M. Tatibana, *J.Chem Phys.*, 19, 1508, (1951)
6. S. Zahorski, *Bull.Acad.Polon.Sci.*, *Ser.Sci.Tech.*, 10, 421, (1962)
7. M.V. Volkenstein, *Configurational Statistics of Polymeric Chains*, Interscience, New York, 1963
8. P.J. Flory, *Statistical Mechanics of Chain Molecules*, Interscience, New York, 1969

9. W. Kuhn and F. Grün, *Kolloid Z.*, 101, 248, (1942)

10. L. Jarecki, *Colloid & Polymer Sci.*, 257, 711, (1979)

11. A. Isihara, *Adv. Polymer Sci.*, 7, 450, (1971)

12. H.M. James and E. Guth, *J.Chem.Phys.*, 21, 1048, (1953)

13. J. Walasek, *Ph.D. Thesis*, Warsaw 1978; A.Ziabicki and J.Walasek, *Macromolecules*, 11, 471, (1978)

14. E. Guth, *Proc. III Rubber Conference*, London, 1954, p. 364; *J.Polymer Sci.*, C12, 89, (1966)

15. J. Walasek, *IFTR Reports*, #30, (1985)

16. N.R. Langley, *Macromolecules*, 1,348, (1968)

17. S.F. Edwards, *Proc. Phys. Soc. (London)*, 91, 513, (1967); ibid.92, 9, (1967)

18. W.W.Graessley and D.S. Pearson, *J.Chem.Phys.*, 66, 3363, (1977)

19. O. Kramer, R.L. Carpenter, V. Ty, and J.D. Ferry, *Macromolecules*, 7, 79, (1974)

20. P.J. Flory and J. Rehner, *J.Chem.Phys.*, 11, 512, (1943)

21. A. Ziabicki, *Colloid & Polymer Sci.*, 254, 1, (1976)

22. J. Walasek, *IFTR Reports*, #21, (1985)

23. A. Ziabicki, *Colloid & Polymer Sci.*, 252, 767, (1974)

24. P.J. Flory, *Chem.Revs,*, 35, 51, (1944)

25. A.V. Tobolsky, D.J. Metz and R.B., Mesrobian, *J.Am.Chem.Soc.*,72, 1946, (1950)

26. L. Mullins and A.G. Thomas, *J.Polymer Sci.*, 43, 13, (1960)

27. J. Scanlan, *J.Polymer Sci.*, 43, 501, (1960)

28. L.C. Case, *J.Polymer Sci.*, 43, 397, (1960)

29. A. Ziabicki and W.Klonowski, *XIV Prague Microsymposium on Macromolecules*,August 1974; A. Ziabicki and W.Klonowski, *Rheol. Acta*, 14, 113, (1975)

30. A. Ziabicki and W.Klonowski, *Rheol. Acta*, 14, 105, (1975).

SOME MATHEMATICAL PROBLEMS ARISING IN MODERN DEVELOPMENTS IN NON-NEWTONIAN FLUID MECHANICS

M. Brennan, R. S. Jones and K. WALTERS
Department of Applied Mathematics
University College of Wales
Aberystwyth
U.K.

SUMMARY

We consider a number of mathematical problems which have arisen in modern developments in non-Newtonian fluid mechanics, especially those which emanate from recent work on numerical simulation. Particular attention is given to the problem of specifying extra boundary conditions when the order of the governing equations for the visco-elastic problem is higher than that in the corresponding Newtonian problem. It is argued that in such problems, the so-called hierarchy expansion of the Simple Fluid is inappropriate.

1. INTRODUCTION

In the development of non-Newtonian fluid mechanics, several distinctive thrusts of activity are evident, originating either from major research contributions by individuals or such external factors as the speed and capacity of computers. One such major contribution was the pioneering work of Oldroyd [1] on the formulation of rheological equations of state (constitutive equations). It contained general formulation principles, which have often been rediscovered or restated, but not superseded. Oldroyd illustrated his general theory by appeal to simple models (like the so called Oldroyd B model). It is interesting to note that while the use of these models was for many years deprecated by those seeking some form of generality, they have recently become widely used within the context of the numerical simulation of non-Newtonian flow.

A complementary approach to formulating rheological equations of state was developed in the late fifties and early sixties, with major contributions from Coleman, Noll, Truesdell, Green, Rivlin and Ericksen [2-5]. Most of this work was noteworthy for its search for generality in constitutive modelling, the most notable being that of Coleman and Noll with their attempts to find general equations of state for restricted classes of flow. Their 'hierarchy expansion', valid for 'slow flow', was especially influential, the so-called second-order fluid being very popular amongst those interested in the solution of flow problems. The resulting analyses invariably involved perturbation expansions, the base solution being provided by the (first-order) Newtonian solution. The perturbation parameter was either the speed of flow, a characteristic relaxation time, or a product of these.

Practical and industrial rheologists, after a brief flirtation with the hierarchy expansion and similar developments, became increasingly disillusioned on account of the limited range of applicability of the resulting solutions, and the period between the mid-sixties and mid-seventies saw little significant progress in non-Newtonian fluid mechanics. The advent of the high speed computers of the mid-seventies opened up the prospect of solving problems involving complex geometries and highly elastic liquids, well beyond the range of the perturbation expansions of earlier days (see, for example, [6,7]). Admittedly, the rheological models employed had to be relatively simple for reasons of tractability and there was (and still is) a renewed interest in the so-called Oldroyd/Maxwell models originating from Oldroyd's 1950 paper.

The numerical simulation of non-Newtonian flow in complex geometries has given a significant boost to research in the field [7] and has also generated its fair share of interesting mathematical problems, some of which are still unresolved. We address some of these in the present paper.

2. EQUATIONS OF STATE

It is convenient to begin by writing down the constitutive equations for a general incompressible fluid model:

$$p^{ik} = -p\delta^{ik} + T^{ik}, \dagger \tag{1}$$

$$T^{ik}(t) = \mathop{F}_{-\infty}^{t}\left[C^{j\ell}(t')\right], \tag{2}$$

where p^{ik} is the stress tensor, p an arbitrary isotropic pressure, δ^{ik} is the kronecker delta and $C^{j\ell}$ is a suitable strain tensor which in the present context we take as the Finger tensor; t is the present time, with $-\infty < t' \leq t$, and F is a tensor-valued functional.

Except for certain very simple flows, (2) is too general to be useful in solving flow problems. Approximations are therefore unavoidable if progress is to be made and these have followed two distinct lines.

(I) One can consider special choices of the response functional F, which clearly lead to *particular* equations, that are nevertheless valid under *all* conditions of motion and stress. Examples of approximations of this kind are the so-called Oldroyd/

† We use standard tensor notation throughout. Covariant sufficies are written below, contravariant sufficies above, and the usual convention for repeated sufficies is assumed.

Maxwell models, and by way of illustration, we write down the equations for the Oldroyd B fluid:

$$T^{ik} + \lambda_1 \frac{\delta}{\delta t} T^{ik} = 2\eta_0 \left[1 + \lambda_2 \frac{\delta}{\delta t}\right] e^{(1)ik}, \quad (3)$$

where $\delta/\delta t$ is the (upper convected) derivative introduced by Oldroyd [1] and $e^{(1)ik}$ is the usual rate-of-strain tensor. η_0 is a constant with the dimensions of viscosity and λ_1 and λ_2 are constants with the dimensions of time. λ_1 is usually referred to as the relaxation time and λ_2 is often referred to as the retardation time. Models like (3) for non-zero λ_2 are outside the scope of the early development of Simple Fluid theory carried out by Coleman and Noll, but they are within the orbit of the recent work of Saut and Joseph [8]. The Oldroyd B model will play a prominent part in the discussion which follows in §§ 2-4.

(II) There are approximations arising from simplifications in the flow, so that $C^{j\ell}$ has a relatively simple form. These approximations lead to *general* equations of state for *restricted* classes of flow. The most well known example of this type of approximation is provided by the hierarchy equations of Coleman and Noll [3]. The relevant expansion is based on speed-of-flow, the first order approximation being the Newtonian fluid and the so-called second-order fluid has equations of state which are essentially given by

$$T^{ik} = 2\eta_0 e^{(1)ik} - 2k e^{(2)ik} + 4\zeta e^{(1)i}{}_j e^{(1)jk}, \quad (4)$$

where η_0, k and ζ are constants and

$$e^{(2)ik} = \frac{\delta}{\delta t} e^{(1)ik} = \frac{\partial e^{(1)ik}}{\partial t} + v^m \frac{\partial e^{(1)ik}}{\partial x^m} - \frac{\partial v^i}{\partial x^m} e^{(1)mk} - \frac{\partial v^k}{\partial x^m} e^{(1)im}. \quad (5)$$

In the present development, the higher rates of strain $e^{(n)ik}$ are related to the so-called White-Metzner tensors rather than the more widely used (covariant) Rivlin-Ericksen tensors, but there is no essential issue of principle involved in this choice of kinematic variable.

We note that because of the $v^m \frac{\partial}{\partial x^m}$ term in the derivative $\frac{\delta}{\delta t}$, the use of the second (and higher) order fluids in flow problems is likely to lead to governing differential equations which are of higher order than those arising from the use of the Navier-Stokes equations. Extra boundary conditions are therefore likely to be necessary. The perturbation methods of solution used by the vast majority of workers in the field masked this interesting possibility, and, apart from a few isolated papers including one of relevance by Giesekus [9], the problem has not been given the prominence it deserves.

Many of the arguments which follow will be based on a consideration of the Oldroyd B model given by (3) or, what is equivalent

$$T^{ik} = 2\eta_0 \frac{\lambda_2}{\lambda_1} e^{(1)ik} + 2\eta_0 \frac{(\lambda_1-\lambda_2)}{\lambda_1^{\,2}} \int_{-\infty}^{t} e^{-\frac{(t-t')}{\lambda_1}} \frac{\partial x^i}{\partial x'^m} \frac{\partial x^k}{\partial x'^r} e^{(1)mr}(\underline{x}',t')dt', \tag{6}$$

where x'^i is the position at time t' of the element which is instantaneously at x^i at time t.

An alternative form to (6), which so far as we can determine has not been considered before, is given by

$$T^{ik} = 2\eta_0 e^{(1)ik} - 2\eta_0 \frac{(\lambda_1-\lambda_2)}{\lambda_1} \int_{-\infty}^{t} e^{-\frac{(t-t')}{\lambda_1}} \frac{\partial x^i}{\partial x'^m} \frac{\partial x^k}{\partial x'^r} e^{(2)mr}(\underline{x}',t')dt'. \tag{7}$$

The advantage of (7) is that the first term on the right hand side can be identified as a Newtonian component and the second is a viscoelastic component.

Integrating (7) successively by parts, yields

$$T^{ik} = 2\eta_0 \Big[e^{(1)ik} - (\lambda_1-\lambda_2)e^{(2)ik} + \lambda_1(\lambda_1-\lambda_2)e^{(3)ik} - \dots (-1)^{n-1}\lambda_1^{\,n-2}(\lambda_1-\lambda_2)e^{(n)ik}$$

$$+ (-1)^{n}\lambda_1^{\,n-2}(\lambda_1-\lambda_2)\int_{-\infty}^{t} e^{-\frac{(t-t')}{\lambda_1}} \frac{\partial x^i}{\partial x'^m} \frac{\partial x^k}{\partial x'^r} e^{(n+1)mr}(\underline{x}',t')dt' \Big]. \tag{8}$$

When $n = 2$, we have

$$T^{ik} = 2\eta_0 \left[e^{(1)ik} - (\lambda_1-\lambda_2)e^{(2)ik} + R \right],$$

where

$$R = (\lambda_1-\lambda_2)\int_{-\infty}^{t} e^{-\frac{(t-t')}{\lambda_1}} \frac{\partial x^i}{\partial x'^m} \frac{\partial x^k}{\partial x'^r} e^{(3)mr}(\underline{x}',t')dt'. \tag{10}$$

Neglecting R we call (9) the second-order equivalent (SOE) of the Oldroyd B model (cf. equation (4)). Note that the remainder term R is given in useful closed form.

For future reference, we note that the SOE of the upper-convected Maxwell model ($\lambda_2 = 0$) is given by

$$T^{ik} = 2\eta_0 \left[e^{(1)ik} - \lambda_1 e^{(2)ik} + R \right], \tag{11}$$

where

$$R = \lambda_1 \int_{-\infty}^{t} e^{-\frac{(t-t')}{\lambda_1}} \frac{\partial x^i}{\partial x'^m} \frac{\partial x^k}{\partial x'^r} e^{(3)mr}(\underline{x}',t')dt'. \quad (12)$$

In the next section we shall outline some of the current problems in non-Newtonian fluid mechanics by appealing to the Oldroyd B model. We realize that the consideration of simple fluid models will provide at best only a partial resolution of the problem and that ideally we would wish to obtain general results which would be independent of the fluid model. We do not see these two approaches as mutually exclusive and we firmly believe that the final resolution of the problems will involve both a consideration of simple fluid models (to elucidate the main features) and general analysis to complete the picture. The search for conclusions of general validity may be very difficult. Indeed, we feel that even our own work in the present paper on simple fluid models is in some ways incomplete.

3. OUTSTANDING MATHEMATICAL PROBLEMS ARISING IN THE NUMERICAL SIMULATION OF NON-NEWTONIAN FLUIDS

The subject of the Numerical Simulation of Non-Newtonian Flow has developed rapidly over the last ten years [6,7] and there is no sign of any diminution of effort. Within reason, there is now no limitation on the complexity of the fluid model which can be employed although most of the published work in the field has been concerned with relatively simple fluid models, the upper convected Maxwell model and the Oldroyd B model being the most popular.

Potentially, current work is of significant practical application, but before the fruits of this can be fully realized, certain mathematical problems have to be resolved. The most important and well known is the so-called 'high Weissenberg number' problem, which is covered in detail in the text by Crochet, Davies and Walters [7]. Quite simply, all reputable algorithms presently available fail to converge at some relatively small value of the relevant elasticity number. This breakdown is broadly independent of the fluid model and the precise numerical technique employed. Amongst the reasons for the breakdown which have already been suggested are (i) change of type of the governing equations (from elliptic to hyperbolic) as the elasticity number increases, (ii) bifurcation and the appearance of limit points, (ii) the magnification of discretization errors by non-linear coupling.

Significant progress in resolving the problem has already been made and the scope of existing algorithms (even without a final resolution of the high Weissenberg number problem) is now such that some meaningful practical problems can be tackled.

Another problem in modern developments in the Numerical Simulation of Non-Newtonian Flow concerns the precise form of the singularity near sharp re-entrant corners.

(Recall that even in the Newtonian case the velocity gradients and the stress components can be singular in the region of a re-entrant corner [10,11]). Physical variables change very rapidly in the neighbourhood of a re-entrant corner, and it is anticipated that where they occur, terms involving the convected derivative $\delta/\delta t$ will dominate the behaviour of the model. This means that, for the Oldroyd B model, we would anticipate a nearly Newtonian fluid-like behaviour given by

$$T^{ik} = 2\eta_0 \frac{\lambda_2}{\lambda_1} e^{(1)ik}, \tag{13}$$

whereas for the UCM Model (with $\lambda_2 = 0$), we expect a completely different *solid*-like response:

$$T^{ik} = \frac{2\eta_0}{\lambda_1} e^{ik}, \tag{14}$$

where e^{ik} is a suitable *strain* variable. It is clear therefore that the retardation time λ_2 plays a significant role in determining the asymptotic response of models near abrupt changes in geometry. However, it is nevertheless true that the search for a consistent asymptotic response, especially for the UCM model, is far from complete. There is enough experimental evidence (see for example [12]) to demonstrate the dramatic influence the precise shape of re-entrant corners can have on flow characteristics and it is certainly true that numerical algorithms for highly elastic liquids will not be considered reliable until the corner problem is adequately resolved.

We have already alluded to another problem which we shall address in the present communication. It concerns the possibility of the governing equations for elastic liquids being of higher order than the corresponding equations arising from the Navier-Stokes equations, with the associated problem of specifying further boundary conditions. In general terms, the basic problem is self evident; elastic liquids have a built-in memory of past deformation and conditions within a given domain will often require information about conditions *outside* that domain. Having said that, there is still a need to devise a consistent approach to the extra-boundary conditions problem and we attempt to make some headway in this direction in the next section.

4. BOUNDARY CONDITIONS

To illustrate how the necessity for extra boundary conditions enters into non-Newtonian flow problems, we concentrate on a problem first discussed in detail by Giesekus [9] (see also Rajagopal [13] and Walters [14]). We shall approach the problem from a slightly different standpoint and, in the main, limit attention to the case where the density can be set equal to zero in the governing equations.

The two-dimensional flow is shown schematically in Fig 1.

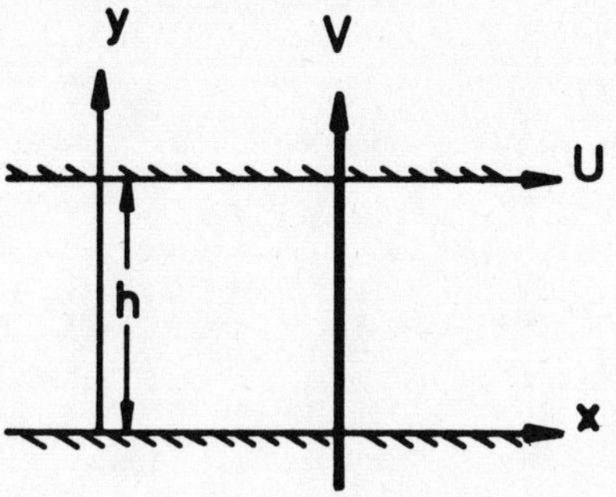

Figure 1: The flow geometry.

The bottom plate at y = 0, of infinite extent, is at rest. The top plate at y = h, also of infinite extent, moves with a velocity U in the x-direction. There is injection with velocity V over y = 0 and suction at the same velocity over y = h. The boundary conditions are therefore

$$\left.\begin{aligned} u(0) &= 0, \quad u(h) = U \\ v(0) &= V, \quad v(h) = V. \end{aligned}\right\} \tag{15}$$

In the related problem discussed by Rajagopal [13], the boundary conditions are

$$u(0) = 0, \quad u(\infty) = U, \quad v(0) = V \ (V < 0). \tag{16}$$

The boundary conditions (15) suggest that we take as a plausible assumption:

$$u = u(y), \quad v = V(\text{a constant}), \tag{17}$$

which automatically satisfies the conservation of mass equation. The corresponding displacement functions (x',y') are given by

$$\left.\begin{aligned} x' &= x - \int_0^s u[y-Vs']ds', \\ y' &= y - Vs, \end{aligned}\right\} \tag{18}$$

where s is the time lapse (t-t'). We note that if we insert the relevant non-dimensional solution to the Newtonian problem which we shall see is u = y, v = V, the displacement functions are given as a finite power series in the time lapse:

$$\begin{aligned} x' &= x - ys + \frac{Vs^2}{2}, \\ y' &= y - Vs. \end{aligned} \tag{19}$$

At this point it is convenient to work out the shear stress component T^{xy} for the general integral model sometimes called liquid B', with equations of state

$$T^{ik} = 2\int_{-\infty}^{t} \Psi(t-t') \frac{\partial x^i}{\partial x'^m} \frac{\partial x^k}{\partial x'^r} e^{(1)mr}(\underline{x}',t')dt'. \tag{20}$$

Substituting (18) into (20) we obtain for T^{xy}:

$$T^{xy} = \frac{1}{V}\int_{-\infty}^{y} \Psi\left(\frac{y-y'}{V}\right)\frac{du}{dy'}\, dy'. \tag{21}$$

The corresponding stress component for the Oldroyd B model can be written in the alternative forms:

$$T^{xy} = \eta_0 \frac{\lambda_2}{\lambda_1}\frac{du}{dy} + \eta_0 \frac{(\lambda_1-\lambda_2)}{\lambda_1^2 V}\int_{-\infty}^{y} e^{-\frac{(y-y')}{\lambda_1 V}}\frac{du}{dy'}\, dy', \tag{22}$$

or

$$\left[1 + \lambda_1 V\frac{d}{dy}\right] T^{xy} = \eta_0\left[1 + \lambda_2 V\frac{d}{dy}\right]\frac{du}{dy} . \tag{23}$$

The relevant stress equation of motion has the form

$$\frac{dT^{xy}}{dy} = \rho V\frac{du}{dy} , \tag{24}$$

where ρ is the density.

We now introduce the non-dimensional variables

$$\left.\begin{aligned} &\lambda_1{}^* = \frac{\lambda_1 V}{h}, \quad \lambda_2{}^* = \frac{\lambda_2 V}{h}, \quad u^* = \frac{u}{U}, \quad V^* = \frac{V}{U}, \\ &y^* = \frac{y}{h}, \quad T^* = \frac{hT^{xy}}{\eta_0 U}, \quad R_e = \frac{Uh\rho}{\eta_0}, \end{aligned}\right\} \tag{25}$$

and the governing equations for the Oldroyd B fluid become (on immediately dropping the star notation for convenience of presentation)

$$\left[1 + \lambda_1 \frac{d}{dy}\right] T = \left[1 + \lambda_2 \frac{d}{dy}\right]\frac{du}{dy} , \tag{26}$$

$$\frac{dT}{dy} = R_e V\frac{du}{dy} , \tag{27}$$

subject to

$$u(0) = 0, \quad u(1) = 1. \tag{28}$$

At this point it is convenient to introduce one possible boundary condition by specifying the stress at entry to the domain, namely

$$T = T_0 \quad \text{when } y = 0. \tag{29}$$

We confine attention to flows for which $R_e V = 0$ and (26) and (27) can then be combined to yield.

$$u' + \lambda_2 u'' = T_0, \tag{30}$$

where the dash refers to differentiation with respect to y. For arbitrary T_0 the boundary value problem represented by equations (28), (29) and (30) is a singular perturbation problem since the associated Newtonian problem ($\lambda_2 = 0$) has no solution unless $T_0 = 1$ when the solution is $u = y$. It is clear then that the solution cannot have a regular perturbation expansion of the form

$$u = \sum_{n=0}^{\infty} u_n(y)\lambda_2^n, \tag{31}$$

because u_0 does not exist. In the limit $\lambda_2 \to 0$ we have a boundary layer at $y = 0$ with inner solution

$$u = (1 - T_0)(1 - e^{-\xi}), \text{ where } \xi = \frac{y}{\lambda_2},$$

and outer solution

$$u = T_0 y + (1 - T_0).$$

It is not difficult to show that the solution to (30) satisfying (28) and (29) is

$$u = y + (1 - T_0)\left[\frac{1-e^{-\frac{y}{\lambda_2}}}{1-e^{-\frac{1}{\lambda_2}}} - y\right]. \tag{32}$$

As yet there is no experimental evidence to suggest the existence of the sort of boundary layer that would occur if T_0 could be chosen arbitrarily for all λ_2. Suppose we stipulate that $T_0 \to 1$ as $\lambda_2 \to 0$, then the right-hand side of (32) tends to y as $\lambda_2 \to 0$. However it still does not follow that u(y) has an expansion of the form (31), and it is clear from solution (32) that, in general, $d^n u/dy^n$ will be singular as $\lambda_2 \to 0$ for some $n > 1$. Of course if $T_0 = 1$ the expansion consists solely of the zero order solution $u_0 = y$ and $u_n = 0$ $(n > 0)$, (this is not the case if one includes inertia). But a counter example to the argument that we must choose $T_0 = 1$ in order for the solution (32) to tend uniformly to the Newtonian solution (in the sense that $u \to y$, $u' \to 1$ and $u^{(n)} \to 0$ $(n > 1)$) as $\lambda_2 \to 0$ is the solution (32) with $T_0 = 1 + \overline{T}e^{-1/\lambda_2^2}$ where $\overline{T}$ is arbitrary. Solutions (32) for which $T_0 \to 1$ as $\lambda_2 \to 0$ are important since they reduce to the Newtonian solution as $\lambda_2 \to 0$ and require knowledge of the stress at $y = 0$ in their determination. Without that knowledge (or some consistent alternative) the problem is under specified. Analysis similar to that above can be carried out when inertia is included. However in this case for a particular choice of $T_0(R_e, \lambda_2)$ (corresponding to $T_0 = 1$) there is a non-terminating perturbation expansion in λ_2.

The case of the Upper Convected Maxwell model is also interesting, since the solution for the velocity field has the same form as that for a Newtonian liquid with no need to invoke an extra boundary condition. However, this *is* necessary to determine the extra stress component T^{xx} which has the form

$$T^{xx} = 2\lambda_1 + ce^{-\frac{y}{\lambda_1}}, \tag{33}$$

where c is a constant.

In normal circumstances, it might be thought permissible to expand the Maxwell model in SOE form to yield (11) with R = 0, this equation providing a meaningful approximation to the Maxwell model for small λ_1. Equation (11) taken together with $dT/dy = 0$ gives

$$u' - \lambda_1 u'' = T_0 , \tag{34}$$

which has as a solution

$$u = y + (1 - T_0)\left[\frac{1-e^{\frac{y}{\lambda_1}}}{1-e^{\frac{1}{\lambda_1}}} - y\right] . \tag{35}$$

This is again a singular perturbation problem with a boundary layer forming on the plate y = 1 as $\lambda_1 \to 0$ (unless $T_0 = 1$). If, as before, we impose the condition $T_0 \to 1$ as $\lambda_1 \to 0$ then the solution (35) is such that $u \to y$ as $\lambda_1 \to 0$. In contrast to the Maxwell model we *are* able to employ the extra boundary condition and the solution is clearly different from the known exact solution, except of course when $T_0 = 1$. The reason for this discrepancy is immediately apparent and we can state it in two ways. First, there is implicit in the expansion to obtain the SOE the assumption that the variables can be expanded as a power series in λ_1. The resulting solution (35) is clearly not of this form. Indeed, if we did assume a power series form for the variables

$$\left.\begin{aligned} u &= u_0 + \lambda_1 u + \lambda_1^2 u_2 + \ldots, \\ T_0 &= 1 + \lambda_1 T_{01} + \lambda_1^2 T_{02} + \ldots, \end{aligned}\right\} \tag{36}$$

and systematically solve for the coefficients in the usual way, we would recover

$$u = y, \quad T_0 = 1$$

as required.

Secondly, the SOE equivalent to the Maxwell model is obtained by neglecting the remainder term in (11) and if we now use the solution (35) to obtain the various terms in the full explicit form of the UCM model, we obtain

$$T = \left[\frac{du}{dy} - \lambda_1 \frac{d^2u}{dy^2} + R\right], \tag{37}$$

where

$$\left.\begin{aligned}
\frac{du}{dy} &= T_0 - \frac{(1-T_0)}{\lambda_1\left(1-e^{\frac{1}{\lambda_1}}\right)} e^{\frac{y}{\lambda_1}}, \\
\frac{d^2u}{dy^2} &= -\frac{(1-T_0)}{\lambda_1^2\left(1-e^{\frac{1}{\lambda_1}}\right)} e^{\frac{y}{\lambda_1}}, \\
R &= -\frac{(1-T_0)}{2\lambda_1\left(1-e^{\frac{1}{\lambda_1}}\right)}\left[e^{\frac{y}{\lambda_1}} - e^{-\frac{y}{\lambda_1}}\right],
\end{aligned}\right\} \tag{38}$$

Clearly, for $T_0 \neq 1$, the remainder term cannot be neglected since $R = \frac{1}{\lambda_1} O(u' - \lambda_1 u'')$. It is not surprising therefore that the SOE method is giving a meaningless solution to the general problem in this case. This is important since it is an illustration of a general problem whenever the flow problem is such as to require extra boundary conditions. In general, the solution will have exponential terms which do not have a power series expansion in the small parameter (like e^{y/λ_1} in (35)). This might be used as an argument to choose the stress at entry T_0 in such a way as to ensure that there is a power series solution, but we have already seen that a consistent solution with $T_0 \neq 1$ can be obtained for the Oldroyd B model and, clearly, limiting the choice to $T_0 = 1$ is an unnecessary and unwarranted restriction.

We are led to offer the following conjecture:

In any steady flow problem, the hierarchy equations cannot be used as approximations to the general functional equation for the Simple Fluid if the resulting equations governing the flow are higher order than those obtained from the Navier-Stokes equations. Exceptions to this general rule may be obtained for special choices of the extra boundary conditions required for the viscoelastic flow problem.

There is clearly scope here for a detailed and thorough study of this conjecture in situations which are not limited to particular fluid models in special flows.

We now turn to another matter of concern within the general area of the boundary condition problem. It is a trivial matter to choose a kernel function Ψ in (21) for which the equation for the stress has the form

$$\left[1 + \lambda_1^{(1)} \frac{d}{dy} + \lambda_1^{(2)} \frac{d^2}{dy^2} + \ldots\right] T = \left[1 + \lambda_2^{(1)} \frac{d}{dy} + \lambda_2^{(2)} \frac{d^2}{dy^2} + \ldots\right] \frac{du}{dy} \quad . \tag{39}$$

This taken together with the stress equation of motion yields

$$u'' + \lambda_2^{(1)} u''' + \lambda_2^{(2)} u^{iv} + \ldots = 0. \tag{40}$$

We now need more than three boundary conditions and it is *not* sufficient to simply supply the 'Newtonian' velocity boundary conditions and the stress at entry, as suggested, for example, by Crochet, Davies and Walters [7]. Depending on the number of derivatives in the governing equation, we need more and more information about the variable u and its derivatives at y = 0, and this is equivalent to knowing its form outside the domain of interest, which is in keeping with the whole idea of 'fluid memory'. Certainly, in the general case, we will need complete information about the function u in a small neighbourhood of y = 0.

Turning to the implication of the statement of Crochet, Davies and Walters that a knowledge of the flow pre history is equivalent to a knowledge of stress at entry in non-Newtonian flow problems, we remark that in the present context this is indeed the case for the Maxwell model, where we may write

$$T = T_0 e^{-\frac{y}{\lambda_1}} + \frac{1}{\lambda_1}\int_0^y e^{-\frac{(y-y')}{\lambda_1}} \frac{du}{dy'} dy' . \tag{41}$$

However it may be shown that for more complicated models, 'pre history' means more than a knowledge of stress at entry.

ACKNOWLEDGMENT

We are happy to acknowledge helpful discussions with Dr A R Davies and Dr H Holstein.

REFERENCES

1. OLDROYD, J G: On the formulation of rheological equations of state. Proc.Roy. Soc. A200, (1950), 523-541.

2. TRUESDELL, C, NOLL, W: *The Non-Linear Field Theories of Mechanics*. Springer. (1965).

3. COLEMAN, B D, NOLL, W: An approximation theorem for functionals, with applications in continuum mechanics. Arch.Rat.Mech.Anal. 6,(1960), 355-370.

4. GREEN, A E, RIVLIN, R S: The mechanics of non-linear materials with memory. Part 3. Arch.Rat.Mech.Anal. 4,(1960), 387-404.

5. RIVLIN, R S, ERICKSEN, J L: Stress deformation relations for isotropic materials. J.Rat.Mech.Anal. 4,(1955), 323-425.

6. CROCHET, M J, WALTERS, K: Numerical methods in non-Newtonian fluid mechanics. Annual Reviews of Fluid Mechanics. 15,(1983), 241-260.

7. CROCHET, M J, DAVIES, A R, WALTERS, K: *Numerical Simulation of Non-Newtonian Flow*. Amsterdam. Elsevier. (1984).

8. SAUT, J C, JOSEPH, D D: Fading Memory. Arch.Rat.Mech.Anal. 81,(1983), 53-95.

9. GIESEKUS, H: Zur Formuliering der Randbedingungen in Strömungen viskoelastischer Flüssigkeiten mit Injektion und Absauging an den Wänden. Rheol.Acta. 9,(1970), 474-487.

10. MOFFAT, H K: Viscous and resistive eddies near a sharp corner. J.Fluid Mech. 18, (1964), 1-18.

11. HOLSTEIN, H, PADDON, D J: A singular finite difference treatment of re-entrant corner flow. Part 1. Newtonian fluids. J. non-Newtonian Fluid Mech. 8,(1981), 81-93.

12. WALTERS, K, WEBSTER, M F: On dominating elastico-viscous response in some complex flows. Philos.Trans.Roy.Soc. London Ser. A308,(1982), 199-218.

13. RAJAGOPAL, K R: On the creeping flow of the second-order fluid. J. non-Newtonian Fluid Mech. 15,(1984), 239-246.

14. WALTERS, K: "Overview of Macroscopic Viscoelastic Flow" to appear in *'Viscoelasticity and Rheology'*. Academic Press. (1985).

SHEAR FLOWS OF NON-LINEAR VISCO-ELASTIC FLUIDS

M. Slemrod
Department of Mathematical Sciences
Rensselaer Polytechnic Institute
Troy, New York 12180

Acknowledgement

*This research was supported in part by the Air Force office of Scientific Research, Air Force Systems Command, USAF, under Contract/ Grant No. AFORS-81-0172. The United States Government is authorized to reproduce and distribute reprints for government purposes not withstanding any copyright herein.

0. Introduction

The purpose of this note is to review earlier works [1]-[3] which which showed how simple shearing flow of a visco-elastic fluid can have radically different behavior via change in the non-linear constitutive relation. Specifically we show how the choice for the extra shearing stress given by

$$T_E^{xy}(t) = \sigma\left(\int_0^\infty e^{-\alpha s} v_x(x,t-s)ds\right) \tag{0.1}$$

$\sigma'>0$, $\sigma''\neq 0$ can lead to formation of a vortex sheet where as

$$T_E^{xy}(t) = \int_0^\infty e^{-\alpha s} \sigma(v_x(x,t-s)) \tag{0.2}$$

$\sigma'>0$, $\sigma''\neq 0$ can lead to formation of a vortex shock. If σ' can take on both positive and negative values we show the flow can exhibit hysteritic phase changes.

1. Equations of motion

First let us set up our basic assumptions on the flow. If in a fixed Cartesian coordinate system x,y,z the velocity fields of a flowing fluid body has the form

$$v = (v^x, v^y, v^z)$$

where

$$v^x = 0, \quad v^y = v(x,t), \quad v^z = 0, \tag{1.1}$$

we say the motion is a rectilinear shearing flow. For such a flow the condition of incompressibility

$$\operatorname{div} \underline{v} = 0$$

is automatically satisfied.

Coleman & Noll [4] have shown that if the fluid is a simple fluid then the components of the Cauchy stress obey the relations

$$T = -pI + T_E$$

where the extra stress T_E satisfies

$$\begin{aligned}
T_E^{xy} &= T_E^{yx} = \mathop{s_0}_{s=0}^{\infty} [\Lambda_t(s)] , \\
T_E^{xx} &= T_E^{yy} = \mathop{s_1}_{s=0}^{\infty} [\Lambda_t(s)] , \\
T_E^{yy} &= T_E^{zz} = \mathop{s_2}_{s=0}^{\infty} [\Lambda_t(s)] , \\
T_E^{zy} &= T_E^{yz} = T_E^{xz} = T_E^{zx} = 0 , \\
T_E^{xx} &= T_E^{yy} = T_E^{zz} = 0 .
\end{aligned} \tag{1.2}$$

Here p is an indeterminate hydrostatic pressure and s_0, s_1, s_2 are real-valued, generally non-linear functions of the relative shearing history $\Lambda_t(s)$, where

$$\Lambda_t(s) = - \int_{t-s}^{t} v_x(x,\tau)d\tau \qquad (0 \leq s < \infty) .$$

We assume the fluid is isotropic. Isotropy implies the functionals s_0, s_1, s_2 satisfy the relations

$$\begin{aligned}
\mathop{s_0}_{s=0}^{\infty} [-\Lambda_t(s)] &= -\mathop{s_0}_{s=0}^{\infty} [\Lambda_t(s)] , \\
\mathop{s_i}_{s=0}^{\infty} [-\Lambda_t(s)] &= \mathop{s_i}_{s=0}^{\infty} [\Lambda_t(s)] , \quad i = 1,2 .
\end{aligned} \tag{1.3}$$

If we substitute conditions (2) into the equations of balance of linear momentum we find

$$0 = \frac{\partial T_E^{xx}}{\partial x} - \frac{\partial p}{\partial x} + b_1(x,t) \tag{1.4a}$$

$$\rho\, v_t(x,t) = \frac{\partial T_E^{yx}}{\partial x} - \frac{\partial p}{\partial y} \tag{1.4b}$$

$$0 = \frac{\partial p}{\partial z}\ . \tag{1.4c}$$

We assume the fluid density ρ is a positive constant. Here b_1 is a component of the body force $b = (b_1,0,0)$ and T_E^{xx}, T_E^{yx} are functions of x and t only. From (4a) we see $\frac{\partial^2 p}{\partial y \partial x} = 0$ are hence $\frac{\partial p}{\partial y}$ depends at most on y and t. However, (4b) further restricts $\frac{\partial p}{\partial y}$ to depend only on t. We set $\frac{\partial p}{\partial y} = \gamma(t)$ (the driving force per unit volume on the fluid). From now on $b_1(x,t)$ will be the component of body force necessary to preserve (4a) as an identity. Hence, the only relevant dynamic equation is (4b) which we rewrite as

$$\rho\, v_t(x,t) = \frac{\partial T_E^{yx}}{\partial x} - \gamma(t)\ . \tag{1.5}$$

We will assume our viscoelastic fluid is confined between two parallel walls of infinite extent at $x = -1$ and $x = +1$. The top wall at $x = 1$ moves with velocity V and the bottom wall is stationary. If we assume non-slip boundary conditions we have

$$v(-1,t) = 0, \qquad v(1,t) = V\ . \tag{1.6}$$

2. Constitutive assumptions

We shall make some assumptions on the nature of the functional s_0. For this analysis we assume s_0 has two particularly simple forms, i.e.

$$\underset{s=0}{\overset{\infty}{s_0}}\ (\Lambda_t(s)) = \sigma\left[\int_0^\infty e^{-\alpha s} v_x(x,t-s)\,ds\right]\ , \tag{2.1}$$

and

$$\underset{s=0}{\overset{\infty}{s_0}}\ (\Lambda_t(s)) = \int_0^\infty e^{-\alpha s}\, \sigma(v_x(x,t-s))\,ds\ . \tag{2.2}$$

Here σ is a real valued odd smooth function defined on the real line and α is a positive constant. In (2.1) s_0 is a (generally) non-linear function of linear functional

$$\int_0^\infty e^{-\alpha s}\, v_x(x,t-s)ds \;.$$

In (2.2) we see s_0 is a linear functional of non-linear function $\sigma(v_x(x,t-s))$.

3. Shearing perturbations of a steady shearing flow

Assume either (2.1) or (2.2) holds and the driving force $\gamma(t) \equiv 0$. Then for constitutive equations (2.1) and (2.2) the balance of linear momentum equation (5) implies respectively

$$\rho\, v_t(x,t) = \sigma\Big(\int_0^\infty e^{-\alpha s}\, v_x(x,t-s)ds\Big)_x \quad (\text{for } (2.1)) \tag{3.1}$$

$$\rho\, v_t(x,t) = \int_0^\infty e^{-\alpha s}\, \sigma(v_x(x,t-s))_x ds \quad (\text{for } (2.2)) \tag{3.2}$$

Systems (3.1),(1.6) and (3.2),(1.6) admit steady rectilinear flow

$$v(x) = \frac{V}{2}\,(x+1) \tag{3.3}$$

To study the stability of the flow against shearing perturbations we set

$$\hat{v}(x,t) = v(x,t) - \frac{V}{2}\,(x+1) \;.$$

Observe that for constitutive relation (2.1) this implies

$$\rho\, \hat{v}_t(x,t) = \sigma\Big(\int_0^\infty e^{-\alpha s}\, \hat{v}_x(x,t-s)ds + \frac{V}{2\alpha}\Big)_x \tag{3.4}$$

with boundary conditions

$$\hat{v} = 0 \quad \text{at} \quad x = -1,1 \;. \tag{3.5}$$

For constitutive equation (2.2) we see $\hat{v}$ satisfies

$$\rho\, \hat{v}_t(x,t) = \int_0^\infty e^{-\alpha s}\, \sigma\Big(\hat{v}_x(x,t-s) + \frac{V}{2}\Big)ds \tag{3.6}$$

along with boundary condition (3.5).

In each case we assume a smooth velocity history

$$\hat{v}(x,\tau) = \hat{v}_0(x,\tau), \qquad -\infty<\tau\leq 0$$

consistent with the governing dynamical equations and boundary conditions.

4. Analysis of the flow equations for relation (2.1)

Consider equation (3.4). Define

$$\hat{\sigma}(\xi) = \frac{1}{\rho}\left[\sigma(\xi + \frac{V}{2\alpha}) - \sigma(\frac{V}{2\alpha})\right] ,$$

$$u(x,t) = \int_0^\infty e^{-\alpha s}\, \hat{v}_t(x,t-s)ds ,$$

$$w(x,t) = \int_0^\infty e^{-\alpha s}\, \hat{v}_t(x,t-s)ds .$$

Integration by parts implies

$$u(x,t) = \hat{v}(x,t) - \alpha \int_0^\infty e^{-\alpha s}\, \hat{v}(x,t-s)ds$$

so that

$$u_t = \hat{v}_t - \alpha u . \tag{4.1}$$

Combining (4.1) and (3.4) and using the definitions of u,w we see that

$$u_t = \hat{\sigma}(w)_x - \alpha u , \tag{4.2}$$

$$w_t = u_x , \tag{4.3}$$

with boundary conditions

$$u(-1,t) = u(1,t) = 0 , \tag{4.4}$$

and initial conditions

$$u(x,0) = u_0(x), \qquad w(x,0) = w_0(x) \tag{4.5}$$

where u_0,w_0 are obtained from $\hat{v}_0(x,\tau)$ via the definitions of u,w and $\hat{\sigma}(0) = 0$.

5. Analysis of the flow equations for (2.2)

Consider equation (3.6) set

$$\tilde{\sigma}(\xi) = \frac{1}{\rho}\left[\sigma(\xi + \frac{V}{2}) - \sigma(\frac{V}{2})\right]$$

$$\tilde{u} = v_t ,$$

$$\tilde{w} = v_x .$$

Now note that (3.6) can be expressed as

$$\hat{v}_t = \int_{-\infty}^{t} e^{-\alpha(t-\tau)} \tilde{\sigma}(\hat{v}_x(x,\tau))_x d\tau \tag{5.1}$$

which upon differentiating with respect to t yields

$$\hat{v}_{tt} = -\alpha \int_{-\infty}^{t} e^{-\alpha(t-\tau)} \tilde{\sigma}(\hat{v}_x(x,\tau))_x d\tau + \tilde{\sigma}\,(\hat{v}_x(x,t))_x , \tag{5.2}$$

From (5.1) we see that (5.2) can be written as

$$\hat{v}_{tt} + \alpha\,\hat{v}_t = \tilde{\sigma}(v_x(x,t))_x . \tag{5.3}$$

If we use the above definitions of u, w we see

$$\tilde{u}_t = \sigma(\tilde{w})_x - \alpha\tilde{u} , \tag{5.4}$$

$$\tilde{w}_t = \tilde{u}_x , \tag{5.5}$$

$$\tilde{u}(-1,t) = \tilde{u}(+1,t) = 0 , \tag{5.6}$$

$$\tilde{u}(x,0) = \tilde{u}_0(x), \qquad \tilde{w}(x,0) = \tilde{w}_0(x) \tag{5.7}$$

where again $\tilde{u}_0$, $\tilde{w}_0$ are obtained from $\hat{v}_0(x,\tau)$, and $\tilde{\sigma}(0) = 0$.

Notice (4.2)-(4.5) and (5.4)-(5.7) are identical except for the slight differences in the sigmas.

6. Breakdown of smooth solutions

Let us impose the condition

$$\sigma' > 0 \tag{6.1}$$

(which of course forces $\hat{\sigma}' > 0$ and $\tilde{\sigma}' > 0$) .

Then both (4.2)-(4.5) and (5.4)-(5.7) are of the form

$$U_t = K(w)_x - \alpha U \;, \tag{6.2}$$

$$W_t = U_x \;, \tag{6.3}$$

$$U(-1,t) = U(+1,t) = 0 \tag{6.4}$$

$$U(x,0) = U_0(x), \; W(x,0) = W_0(x) \;, \tag{6.5}$$

with $K(0) = 0$, $K' > 0$. The condition $K' > 0$ makes the system (6.2) (6.3) strictly hperbolic (i.e. the matrix

$$\begin{bmatrix} 0 & K' \\ 1 & 0 \end{bmatrix}$$

has two real distinct eigenvalues.)

If we impose the condition of genuine non-linearity at zero $K''(0) \neq 0$ breakdown of smooth solutions will occur. This result is made precise below.

Define Riemann invariants

$$\left\{ {r \atop s} = U \pm \Phi(W), \qquad {r_0 \atop s_0} = U_0 \pm \Phi(W_0) \right.$$

where

$$\Phi(W) = \int_0^W \sqrt{K'(s)}\, ds \;.$$

If $|r_0|, |s_0|$ are sufficiently small and $K''(0) > 0$ and $r_{0,x}$ or $s_{0,x}$ is positive and sufficiently large at some point x, then (6.2),(6.3) has a solution (U,W) in $C^1(-1,1) \times C^1(-1,1)$ for only a finite time. A similar result holds if $K''(0) < 0$ and $r_{0,x}$ and $s_{0,x}$ is sufficiently negative at some point x.

Since

$${r_x \atop s_x} = U_x \pm \Phi'(W) W_x$$

and $\Phi'' > 0$ we have

$$\begin{aligned}
&\text{(i)} \quad r_{0x} \text{ large if } U_{0x} \text{ and/or } W_{0x} \text{ is large },\\
&\text{(ii)} \quad s_{0x} \text{ large if } U_{0x} \text{ and/or } -W_{0x} \text{ is large },\\
&\text{(iii)} \; -r_{0x} \text{ large if } -U_{0x} \text{ and/or } -W_{0x} \text{ is large },\\
&\text{(iv)} \; -s_{0x} \text{ large if } -U_{0x} \text{ and/or } W_{0x} \text{ is large },
\end{aligned} \tag{6.6}$$

7. What does break down mean physically?

First consider the case of constitutive relation (2.1). In this case

$$\hat{\sigma}''(0) = \frac{1}{\rho}\sigma''\left(\frac{V}{2\alpha}\right).$$

So if $\sigma''(\frac{V}{2\alpha}) \neq 0$ i.e., σ is quadratically non-linear at $\frac{V}{2\alpha}$ and u_{0x} and w_{0x} are sufficiently (positively or negatively) large as required by Section 6.1 we know $|u_x|+|w_x| \to \infty$ in finite time. However, since $\hat{v}_t = \sigma(w)_x$ and $\hat{v}_x = u_x + \alpha w$ we see $|u_x|+|w_x|\to\infty$ implies $|v_t|+|v_x|\to\infty$. Hence breakdown of smooth solutions for (6.2)-(6.5) implies that (3.4)-(3.5) will have

$$|\hat{v}_t|+|\hat{v}_x|\to\infty \quad \text{in finite time.}$$

This suggest (but doesn't prove) that v and hence v form a jump discontinuity in finite time. The surface across which discontinuity exists is called a vortex sheet

For the constitutive equation (2.2) the story is different. In this case assume $\tilde{\sigma}''(0) = \sigma''(\frac{V}{2}) \neq 0$. Then since $\tilde{u} = \hat{v}_t$, $\tilde{w} = \hat{v}_x$ we see that if $\tilde{u}_{0x} = \hat{v}_{tx}(x,0)$, $\tilde{w}_{0x} = \hat{v}_{tx}(x,0)$ is appropriately suffi_ ciently large (positively or negatively). Section 6 implies $|v_{tx}|+|v_{xx}|\to\infty$ in finite time. Again this suggests the formation of a jump discontinuity in v_t or v_x in finite time. Actually the Rankine-Hugoniot jump condition for the singular surface $x = s(t)$ across is

$$-\left(\frac{ds}{dt}\right)[v_t] = \frac{1}{\rho}[\sigma(v_x)]$$

Hence, if $[v_x] \neq 0$ then $[\sigma(v_x)] \neq 0$ and hence $[v_t] \neq 0$ and vice-versa. Thus a jump in v_t occurs if and only if there is a jump in v_x.

A propagating singular surface which supports a jump in the acceleration v_t is called an acceleration wave. We call a propagating singular surface across which the vorticity curl (v^x,v^y,v^z) experiences a jump discontinuity a vortex shock. In our problem the vorticity is equal to $v_x(x,t)e_z$. We have shown if $\sigma''(\frac{V}{2}) \neq 0$ and for appropriate initial data we can expect the formation of a vortex shock or equivalently an acceleration wave.

8. A third constitutive relation: non-monotone σ

In this section we consider the case where σ has the shape shown in Figure 1. Here σ has the shape shown in Figure 1. Here σ is such that

$$\sigma' > 0 \quad \text{or} \quad [0,\alpha),\ (\beta,\infty)\ ,$$

$$\sigma' < 0 \quad \text{or} \quad (\alpha,\beta)\ ,$$

$$\sigma(a) = \sigma(\beta) = \gamma_a\ ,$$

$$\sigma(\alpha) = \sigma(b) = \gamma_b\ .$$

We will assume constitutive relation (2.2). From equation (1.5) we know that if there is a constant applied driving force γ $v(x,t)$ satisfies

$$\rho\, v_t(x,t) = \int_0^\infty e^{-\alpha s}\, \sigma(v_x(x,t-s))ds - \gamma\ . \tag{8.1}$$

We assume

$$v = 0 \quad \text{at} \quad x = -1,1\ . \tag{8.2}$$

Equivalently (8.1) can be written as

$$\rho\, v_t(x,t) = \int_{-\infty}^{t} e^{-\alpha(t-\tau)}\, \sigma(v_x(x,\tau))_x d\tau - \gamma\ . \tag{8.3}$$

If we differentiate (8.3) with respect to t we find

$$\rho\, v_{tt}(x,t) = \sigma(v_x(x,t))_x - \alpha \int_{-\infty}^{t} e^{-\alpha(t-\tau)}\, \sigma(v_x(x,\tau))_x d\tau\ . \tag{8.4}$$

We shall assume the walls are stationary so that

$$v = 0 \quad \text{at} \quad x = -1,1\ . \tag{8.6}$$

We follow the standard nomenclature of phase transitions. Specifically we say that when v_x takes on values on $[0,\alpha)$ the fluid is in α phase and when v_x takes on values on (β,∞) the fluid is in the β-phase. Of course for different values of x the fluid may be in the α and β phases simultaneously.

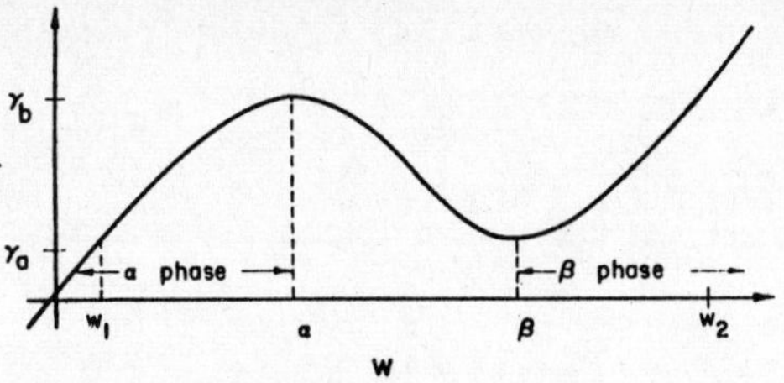

Fig.1

Since the function σ is not globally invertible we denote by $\sigma_\alpha{-1}$ and $\sigma_\beta{}^{-1}$ the respective inverses of σ in the α and β phases i.e.

$\sigma_\alpha{}^{-1}(\xi)$ $[0,\alpha]$, for $0 \leq \xi \leq \gamma b$,

$\sigma_\beta{}^{-1}(\xi)$ $[\beta,\infty]$, for $\gamma a \leq \xi \leq \infty$.

We shall admit solutions with values in the α and β phases only. Solutions with values in the region (α,β) are unstable and not admissible.

9. Steady motion

For steady flow (8.5) implies

$$\sigma(v_x)_x = \alpha\gamma$$

which upon integration yields

$$\sigma(v_x) = \alpha\gamma x + \text{const} . \qquad (9.1)$$

By symmetry of the flow about x=0 v(x) should be an even function of x and hence v_x should be odd. Also σ is an odd function by isotropy and hence $\sigma(v_x) - \alpha\gamma x$ is odd. The only constant which is odd in x is zero so (9.1) reduces to

$$\sigma(v_x) = \alpha\gamma x . \qquad (9.2)$$

If $0 \leq \gamma \leq \frac{\gamma a}{\alpha}$ then $0 \leq {}^\alpha\gamma x \leq \gamma a$ when x $[0,1]$. In this case (9.2) has a unique solution

$$v(x) = \int_1^x \sigma_\alpha{-1}(\alpha\gamma s)ds , \qquad (9.3)$$

$$v_x(x) = \sigma_\alpha{-1}(\alpha\gamma x) . \qquad (9.4)$$

For $\frac{\gamma_a}{\alpha} \leq \gamma$ the lack of a unique inverse allows (9.2) to process non-unique solutions. For example we may have

$$v(x) \int_{x^*}^{x} \sigma_\alpha^{-1}(\alpha\gamma s)ds + \int_{1}^{x} \sigma_\beta^{-1}(\alpha\gamma s)ds \ ,$$

$$v_x(x) = \sigma_\alpha^{-1}(\alpha\gamma x) \ , \qquad 0 \leq x \leq x^* \ ;$$

$$v(x) = \int_{1}^{x} \sigma_\beta^{-1}(\alpha\gamma s)ds \ , \tag{9.5}$$

$$v_x(x) = \sigma_\beta^{-1}(\alpha\gamma x) \ , \quad x^* \leq x \leq 1 \ .$$

The function given by (9.5) is continuously differentiable on $(0,x^*),(x^*,1)$, continuous on $(0,1)$ and satisfies the boundary condition $v(1) = 0$. The values for $x \quad [-1,0)$ are obtained by symmetry. Here x^* is an arbitrary number between $\frac{\gamma_a}{\alpha\gamma}$ and 1. Hence (9.5) represents an infinite number of solutions of (9.2) each possessing a jump discontinuity in v_x at x^*. Note that $(0,x^*)$ the fluid is in α-phase and $(x^*,1)$ the fluid is in the β-phase. The line $x=x^*$ is a singular surface where the fluid possess an interphase jump in v_x. However, we have even more lack of uniquess. For if $\frac{\gamma_a}{\alpha} \leq \gamma \leq \frac{\gamma_b}{\alpha}$ (9.3),(9.4) provides a continuously differentiable solution of (9.2). Finally for $\gamma > \frac{\gamma_a}{\alpha}$ (9.2) can possess only solutions existing in both α and β phases. In this case a jump discontinuity on v at some interphase surface is inevitable.

Now consider the following "thought" experiments. Let γ be raised from zero. For $\gamma \quad [0,\gamma_a/\alpha)$, (9.2) possesses only C^1 solutions given by (9.3), (9.4). When γ is raised above γ_a/α we face the possibility of a spatial bifurcation occuring with the appearance of a solution of the form (9.5). Continuing to raise γ beyond γ_b guarantees the formation of a solution possessing a jump discontinuity in v_x. If we then decrease γ we should eventually see this α-phase solution (9.3),(9.4) when γ is less than γ_a/α. Hence we expect to see both bifurcation and coalescense of fluid phases.

Local dynamical analysis can be used to predict precisely how the above scenario should take place. The idea is simple. We imbed the original non-Newtonian fluid flow in one which contains a small Newtonian viscous term μv_x in the shearing stress:

$$T_E^{xy}(t) = \int_0^\infty e^{-\alpha s}\, \sigma(v_x(x,t-s))_x ds + \mu v_x \tag{9.6}$$

The equation for balance of linear momentum

$$\rho\, (v_{tt} + \alpha v_b) = \sigma\mu(v_x(x,t))_x + \mu v_{xxt} - \alpha\gamma \tag{9.7}$$

where $\sigma^\mu(v_x) = \sigma(v_x) + \alpha^\mu v_x$. We then use the usual theory of struc-

shocks to decide which of the singular surfaces described above is admissible. The details are rather lengthy and may be found in [3].

REFERENCES

1. M. Slemrod, "Instability of steady shearing motions in a non-linear viscoelastic fluid", Archive for Rational Mechanics and Analysis, Vol. 68, August 1978, p. 211-225.

2. M. Slemrod, Appendix A: Breakdown of smooth shearing flow for two constitutive relations: the vortex sheet or the vortex shock, in "Hyperbolic phenomena in the flow of viscoelastic liquids" by D. D. Joseph, to appear Proc. Symposium on Visco-elasticity and Rheology, Oct. 16-18, 1984, Mathematics Research Center, Univ. of Wisconsin, Madison.

3. J. Hunter and M. Slemrod, "Unstable visco-elastic fluid flow exhibiting hysteretic phase changes", Physics of Fluids, Vol. 26 Sept. 1983, p. 2345-2351.

HYPERBOLIC PHENOMENA IN THE FLOW OF VISCOELASTIC LIQUIDS

D. D. Joseph
Department of Aerospace Engineering and Mechanics
The University of Minnesota
Minneapolis, Minnesota 55455

This paper is dedicated to the memory of

Professor Ernst Becker"

Summary

In this paper I discuss concepts of viscosity, elasticity, hyperbolicity, Hadamard instability and change of type in the flow of viscoelastic fluids

1. Constitutive Equations

Constitutive equations relate stress and deformation. Too many constitutive equations have been proposed by people to get one model which will describe all the possible motions of a fluid. Since the variety of responses which are available to viscoelastic fluids is very great, a single equation which accounts for everything may be too abstract to be of much practical use. Eqs. (1.10) and (1.12) are examples of too abstract equations. More specific models are useful only when the domain of deformations in which they live is specified. Therefore, in an ideal world we could have a model valid within a prescribed class of deformations.

There is a great simplification in the problem of constitutive modeling when the deformations are a small perturbation of states of rest. These deformations depend on a Newtonian viscosity μ and a smooth relaxation function $G(s)$, where $G(s) > 0$, $G'(s) < 0$ for $0 \leq s=t-\tau \leq \infty$, and τ is the past time. The stress $\boldsymbol{\tau}$ is given by

$$\boldsymbol{\tau} = 2\mu\mathbf{D}[\mathbf{u}] + 2\int_0^\infty G(s)\mathbf{D}[\mathbf{u}(\mathbf{x},t-s)]\, ds \tag{1.1}$$

where μ is the Newtonian viscosity, $\mathbf{u}(\mathbf{x},t)$ is the velocity and $\mathbf{D}[\mathbf{u}]$ the symmetric part of the velocity gradient. Equation (1.1) is a Jeffreys' type of generalization of Boltzmann's equation of linear viscoelasticity in which the presence of a Newtonian contribution is acknowledged. Eq. (1.1) also holds in the class of small perturbations of rigid motions. We might think that (1.1) models a polymeric solution in which the solvent is Newtonian and the polymers add elasticity.

A constitutive equation of the rate type may be obtained as the time derivative of (1.1)

$$\frac{\partial \tau}{\partial t} = 2\mu \frac{\partial \mathbf{D}}{\partial t} + 2G(0)\mathbf{D} + 2\int_0^\infty G'(s)\mathbf{D}[\mathbf{u}(\mathbf{x},t-s)]ds. \tag{1.2}$$

Jeffreys' model is a special case of (1.2) in which

$$G(s) = \frac{\eta}{\lambda} e^{-s/\lambda} \tag{1.3}$$

where λ is the relaxation time and η is the elastic viscosity. Combining (1.2) and (1.3) we get

$$\lambda \frac{\partial \tau}{\partial t} = 2\mu\lambda \frac{\partial \mathbf{D}}{\partial t} + 2\tilde{\mu}\mathbf{D} - \tau. \tag{1.4}$$

A retardation time

$$\Lambda = \mu\lambda/\eta \tag{1.5}$$

is usually defined for (1.4). When $\mu = 0$, (1.4) gives rise to a Maxwell model

$$\lambda \frac{\partial \tau}{\partial t} = 2\eta\mathbf{D} - \tau. \tag{1.6}$$

Fluids with $\mu = 0$ are like relaxing elastic solid. They propagate shock waves. Fluids with $\mu \neq 0$ are diffusive; they smooth shocks.

Equations (1.1) and (1.2) are perturbation equations and are naturally not invariant under changes of frame which do not satisfy the same conditions of linearization. Various invariant theories which are said to be linear have been proposed. For example Coleman and Noll [1961] linearized a functional depending on the history of the right relative Cauchy-Green strain tensor. Naturally they arrive at a linear expansion, linear in this non-linear tensor. They call this "the finite linear theory of viscoelasticity." When applied to incompressible fluids they get (1.1) with $\mu = 0$ and $\mathbf{D}$ replaced with the s derivative of $\mathbf{C}_t(\mathbf{x},t-s) - \mathbf{1} = \mathbf{G}(s)$, $\mathbf{G}(0) = 0$. The linearization of $\dot{G}(s)$ around $\mathbf{0}$ is $\mathbf{D}$. If the kernel $G(s) = \frac{\eta}{\lambda} e^{-s/\lambda}$ is of Maxwell's type, then Coleman and Noll's equation is a lower convected Maxwell model. If we suppose that the stress functional depends on the finger tensor, rather than the Cauchy tensor, we arrive at Lodges theory, which is the same as an upper convected Maxwell model when the kernel is of Maxwell's type. Saut and Joseph [1985] under different hypotheses than Coleman and Noll arrived at (1.1) with $\dot{G}(s)$ in the place of **D** under the integral and $\mu \neq 0$. If Saut and Joseph had used $\mathbf{H}(s) = \mathbf{C}_t^{-1}(\mathbf{x},t-s) - \mathbf{1}$ instead of $\mathbf{G}(s)$ they would have $\dot{H}(s)$ replacing **D** under the integral. The rate equations for an equation of the Saut-Joseph type, with a kernel of Maxwell type, is an Oldroyd B. None of these so called linear equations are completely linearized. When they are completely linearized they reduce universally to (1.1) and (1.2). These two equations are model independent. They apply to all viscoelastic fluids in motions which perturb rest. This shows that the Newtonian viscosity μ and the relaxation function G(s) are genuine material parameters which are also model independent.

To our knowledge, the first person to introduce a rate equation with a Newtonian viscosity and relaxing elasticity was H. Jeffreys (1929, p. 265). Most of the models arising from molecular modeling of polymeric solutions, like those of Rouse and Zimm, have a Newtonian contribution from the solvent and are of the Jeffreys' type. An invariant formulation of rate equations containing relaxation and retardation (Newtonian viscosity) effects evidently first appears in the celebrated 1950 paper of Oldroyd. Green and Rivlin (1960) appear to have been the first to introduce Newtonian viscosity to integral models. They get rate terms from integrals by allowing delta funcitons and their derivatives in the kernels. Saut and Joseph (1983) derived integral expressions of the type introduced by Green and Rivlin from a theory of fading memory in which the ensemble of all possible linearized stresses coincides with certain topological dual of a domain space (say, a Sobolev space) for allowed deformations. Maxwell models and the generalization of these embodied in the theory of fading memory of Coleman and Noll (see SJ for references) cannot contain a Newtonian viscosity. These models are all instantaneously elastic. Various kinds of hyperbolic phenomena, waves, shock waves, loss of evolution, Hadamard instabilities, change of type arise in fluids with instantaneous elasticity (see Joseph, Renardy and Saut (1985), Joseph (1985A), Joseph and Saut (1985) and this review). Many distinguished scientists of the 19th and early 20th century, Poisson, Maxwell, Poynting, Boltzmann believed that liquids were closer to solids than to gases, with instantaneous and relaxing elasticity, and there is also a line of interesting experiments of this same period which explore this idea (see Joseph, 1985B for a recent historical perspective). The results given in this paper are an entry in this history. The notion of instantaneous and relaxing elasticity can be reconciled with polymers in Newtonian solvents by supposing that the solvents are elastic and not Newtonian.

We have argued that the response to motions perturbing rest (or rigid motions) is completely determined when the Newtonian viscosity μ and the relaxation function G(s) are known. G(s) gives the relaxation for $s \geq 0$ of stresses after a sudden step in displacement. The name "shear stress modulus" or "shear modulus" or "elastic modulus" will be reserved for the largest value G(0) of G(s).

It is helpful that we understand viscosity in the following way. Suppose that we are in the case of steady shearing with one component of velocity $u(x)$ depending on one variable x. The shear stress $\tau(\kappa) = \tau_{12}$ of $\boldsymbol{\tau}$ depend then on the rate of shear $\kappa(x) = D_{12}$ of $\mathbf{D}$ and (1.1) reduces to $\tau = (\mu + \eta)\kappa$ where

$$\tilde{\mu} = \mu + \eta \tag{1.7}$$

is the zero shear viscosity and

$$\eta = \int_0^\infty G(s)\, ds \tag{1.8}$$

is the elastic viscosity. Newtonian fluids have $\eta = 0$, $\tilde{\mu} = \mu$. Elastic fluids have $\mu = 0$, $\tilde{\mu} = \eta$. In general

$$\tilde{\mu} \geq \eta \tag{1.9}$$

with equality for elastic fluids. It is easy to measure the zero shear viscosity $\tilde{\mu}$, but the measurement leave μ and η undetermined. Elastic fluids ($\mu = 0$) with short memories can appear to be Newtonian in standard rheometrical tests.

We turn now to the problem of constitutive modeling in the general case. A framework for such modeling could start form Noll's theory of a simple incompressible fluid. In this, the stress at a particle $\mathbf{x}$ is given by a functional on the history of right relative Cauchy tensor $\mathbf{C}_t(\tau)$ or the history of $\mathbf{G}(s) = \mathbf{C}_t(\tau) - \mathbf{1}$; $\boldsymbol{\tau} = \mathbf{F}[\mathbf{G}]$. To assign meaning to $\mathbf{F}[\ ]$ it is necessary to specify its domain. The Coleman-Noll theory of fading memory is a consequence of the assumption $\mathbf{G}(s)$ lies in a weighted $L_h^2(0,\infty)$ Hilbert space with weight h which makes the large s values of $\mathbf{G}(s)$ irrelevant. However, this large domain excludes some well established material models and phenomena. Saut and Joseph (1983) showed that by restricting the allowed domain of $\mathbf{F}$, the topological dual may be enlarged leading to distributions in the dual and rate terms in the constitutive model. Another method, presented here, is to keep the large domain with $\mathbf{G}(s)$ $L_h^2(0,\infty)$ but to add $\mathbf{f}(\mathbf{D})$ where $\mathbf{f}$ is an ordinary isotropic symmetric tensor valued function of $\mathbf{D}(\mathbf{x},t)$. Thus

$$\boldsymbol{\tau} = \mathbf{f}(\mathbf{D}) + \mathop{\mathbf{F}}_{s=0}^{\infty}[\mathbf{G}(s)]\,. \tag{1.10}$$

In this decomposition $\mathbf{f}(\mathbf{D})$ is the viscous part, and $\mathbf{F}$ the elastic. It may be assumed without losing much generality that $\mathbf{f}(\mathbf{D})$ is a quadratic polynomial in $\mathbf{D}$ with coefficient which depend on the invariant scalars of **D**. Equation (1.1) follows easily form linearizing (1.10) on states of rest, by representing linear functionals with scalar products (integrals) in the weighted $L^2(0,\infty)$ the domain space of **G**(s).

Joseph, Renardy and Saut (1985), hereafter called JRS, derived the general form of the constitutive equation of rate type for any elastic liquid in any motion. This equation arises from calculations following the time differentiation of the stress functional and it may be written as

$$\frac{d\tau_{ij}}{dt} = S_{ijkp}D_{kp}[\mathbf{u}] + A_{ijkp}\Omega_{kp}[\mathbf{u}] + N_{ij} \tag{1.11}$$

where all tensors are symmetric in (i,j), D_{kp} is the symmetric part of the velocity gradient $\partial u_k/\partial x_p$ and Ω_{kp} is the skew symmetric part. The fourth order tensors **S** and **A** are expressible by integrals and N_{ij} is of "lower order" in the sense of hyperbolice analysis. More discussion of this equation can be found in the paper of Joseph (1985).

When the liquid also has a viscous response we may replace (1.11) by

$$\frac{d\tau_{ij}}{dt} = R_{ijkp}\frac{dD_{ij}}{dt} + S_{ijkp}D_{kp} + A_{ijkp}\Omega_{kp} + N_{ij} \tag{1.12}$$

where **R** is a fourth order tensor valued function $R_{ijk\ell} = \partial f_{ij}/\partial D_{k\ell}$. In the fully linearized case, $R_{ijk\ell} = 2\mu\delta_{ik}\delta_{j\ell}$. JRS identified a class of models which are more general than (1.11) but contain all the special models of rate equations which appear in the rheological literature. In their model, the symmetric fourth order tensor **S** is expressed as the most general form involving any second order tensor **P**. When this is applied to (1.12) we find that

$$\frac{d\boldsymbol{\tau}}{dt} = \mathbf{R}\frac{d\mathbf{D}}{dt} + \mathbf{P}\cdot\mathbf{D} + (\mathbf{P}\cdot\mathbf{D})^T + \frac{1}{2}[\mathbf{A}\cdot\boldsymbol{\Omega} + (\mathbf{A}\cdot\boldsymbol{\Omega})^T] + \mathbf{N} \tag{1.13}$$

No assumption is made about the fourth order tensor **A**. The special models which are studied in the rheological literature are such that the tensors **P**, **A** and **N** are expressible in terms of the extra stress $\boldsymbol{\tau}$. These models include those of Oldroyd, Maxwell, Giesekus, Leonev, Phan Thien and Tanner and many others (see Joseph, (1985))

When R = 0 the dynamical system associated with (1.13) supports hyperbolic waves of vorticity provided that the stresses do not enter a forbidden region in which the Cauchy problem is no longer well posed. These and some other phenomena are related to the type of a partial differential equation.

(1) The unsteady quasilinear problem is called evolutionary if, roughly speaking, the Cauchy problem for it is well posed (this is a notion strictly weaker than hyperbolicity). The loss of evolution is an instability of the Hadamard type in

which short waves will sharply increase in amplitude. For many models, those treated here, the problem of evolution may be conveniently framed in terms of vorticity.

(2) The steady quasilinear system may be analyzed for type. It is neither elliptic nor hyperbolic. On the other hand, the vorticity is either hyperbolic or elliptic, and it may change type, hyperbolic in some regions of flow and elliptic in others, as in transonic flow. We shall show that the full unsteady quasilinear system will undergo a loss of stability in the sense of Hadamard when the steady vorticity equation for inertialess flow is hyperbolic.

We consider a number of examples. Some models are always evolutionary and do not change type in unsteady flow. The vorticity equation for steady flow of such models can and does change type. Other models can become non-evolutionary and therefore undergo Hadamard instability. Some flows of these models are evolutionary, e.g.; shearing flows, while others are not evolutionary. In either case, the steady problem can undergo a change of type. Loss of evolution is impossible in flow perturbing uniform motion. To lose evolution it is necessary that certain stresses should exceed critical values. In this sense the loss of evolution can be identified with the problem of failure of numerical simulations at high Weissenberg numbers.

2. Loss of Evolution

The loss of evolution is a concept associated with the well posedness of the Cauchy problem. Let us consider a quasilinear system of the form

$$\mathbf{A}\,\frac{\partial \mathbf{u}}{\partial t} + \sum_{j=1}^{n} B_j\,\frac{\partial \mathbf{u}}{\partial x_j} + \mathbf{b} = 0\,, \tag{2.1}$$

where $\mathbf{A}$, $\mathbf{B}$, ..., B_n are m x m matrix valued functions and $\mathbf{b}$ is an m vector depending on $\mathbf{u},\mathbf{x},t$. The system (2.1) is evolutionary in some domain D of $\mathbb{R}^m \times \mathbb{R}^n \times \mathbb{R}$ in the t direction if for every fixed $\mathbf{u},\mathbf{x},t$ in D and any unit n-vector $\boldsymbol{\nu}$, the eigenvalue problem

$$\left[-\lambda \mathbf{A} + \sum_{j=1}^{n} \nu_j\,B_j\right]\mathbf{v} = 0 \tag{2.2}$$

has only real eigenvalues. The system (2.1) is hyperbolic in the t direction if (2.2) has m real eigenvalues, not necessarily distinct, and a set of m linearly independent eigenvectors.

To justify these definitions, let us consider the simple case where $\mathbf{b} = 0$ and the matrices $\mathbf{A}$, $\mathbf{B}$, ..., $\mathbf{B}_m$ are independent of $\mathbf{u},\mathbf{x},t$. Let $\boldsymbol{\nu}$ be a unit vector in $\mathbb{R}^n$. If (2.1) is evolutionary, then any plane wave solution of (2.1) propagating in the $\boldsymbol{\nu}$-direction $\mathbf{v}(\mathbf{x},t) = \mathbf{v} \exp(ik(\boldsymbol{\nu}\cdot\mathbf{x} - \lambda t))$, k real, has necessarily λ real. This prevents the so called Hadamard instability, i.e., the fact that at any time t the amplitude of $\mathbf{u}$ could become arbitrarily large, even if $\mathbf{u}$ is bounded (but highly oscillatory) at time $t = 0$. In this context hyperbolicity means that in every direction in space, m independent plane waves can propagate.

The quasilinear systems for the velocity, stresses and pressure of fluids with instantaneous elasticity are not hyperbolic in the usual sense. For these systems it is proper to think of the loss of evolution. On the other hand, the unsteady equation for the vorticity is hyperbolic when it is evolutionary and is evolutionary when hyperbolic.

The use of plane waves to study the well posedness of the Cauchy problem is justified, in general, for waves so short that $\mathbf{A}$, $\mathbf{B}_j$ and $\mathbf{b}$ have constant components in a small period cell defined by the wave.

Hadamard instabilities are much stronger than those studied in bifurcation theory. One cannot expect a secondary flow whose dynamics would be governed by the evolution of one or several modes. Rather, if the initial value problem becomes ill-posed, there are flow fields which would not occur even as transient states, since "random" disturbances containing all modes would blow up instantly. The importance of loss of evolution in the flow of viscoelastic fluids was first recognized by Rutkevitch [1969, 1970, 1972]. Many other results were given by Joseph, Renardy and Saut [1985], called JRS. Independent results, following in part out of discussions leading to the paper by JRS, were given by Ahrens, Joseph, Renardy and Renardy [1984], and Renardy [1984, 1985]. Dupret and Marchal [1984, 1985] have also given some new results on the problem of loss of evolution in three dimensional problems.

3. Loss of Evolution and Change of Type for Models with Hyperbolic Vorticity.

We could consider all the models which give rise to a vorticity equation which can be hyperbolic in the steady case. These were identified by Joseph, Renardy and Saut [1985] as JRS models. They include all constitutive models whose principal parts are of Oldroyd type.

$$\lambda \frac{\boldsymbol{\tau}}{t} = 2\eta\, \mathbf{D}[\mathbf{u}] + \boldsymbol{\ell}\,, \tag{3.1}$$

where $\boldsymbol{\ell}$ is of lower order (see JRS) and where λ is the relaxation time, $\boldsymbol{\tau}$ is the extra stress, η is the elastic viscosity, $\mathbf{D}[\mathbf{u}]$ is the symmetric part of the velocity gradient, and

$$\frac{\tau}{t} = \frac{\partial\tau}{\partial t} + (\mathbf{u}\nabla)\tau + \tau\boldsymbol{\Omega} - \boldsymbol{\Omega}\tau - a(\mathbf{D}\tau + \tau\mathbf{D}) , \tag{3.2}$$

where $\boldsymbol{\Omega} = \frac{1}{2}(\nabla\mathbf{u} - \nabla\mathbf{u}^T)$ is the skew symmetric part of $\nabla\mathbf{u}$ and a $[-1,1]$.

The lower order terms $\boldsymbol{\ell}$ in (3.1) may depend on $\mathbf{u}$ and τ, but not on their derivatives. The upper convected, corotational and lower convected Maxwell models arise when $\boldsymbol{\ell} = -\tau$ and $a = [1,0,-1]$. Different models can be obtained by different theories of the lower order terms $\boldsymbol{\ell}$. One version of the model by Phan-Thien and Tanner [1977] may be expressed by (3.1) with $\boldsymbol{\ell} = -(1 + c \operatorname{tr} \tau)\tau$ where c is a constant. The Johnson-Segalman model [1977] with an exponential kernel is a special case of Phan-Thien and Tanner with $c = 0$ and in fact is one of the Maxwell type of Oldroyd models. A simple Giesekus model [1982] is given by (3.1) with $a = 1$, $\boldsymbol{\ell} = -(\tau + \frac{\alpha}{\mu}\tau^2)$ where $0 \leq \alpha \leq 1$ is a constant related to the relative mobility tensor.

We now turn to the analysis of evolutionarity for systems governed by (3.1). The method used is a kind of linearized stability analysis for short waves of the type already given in Section 2.

We shall study the problem of evolution of the ten field variables $[\mathbf{u},\tau,p]$ satisfying

$$\lambda \frac{\tau}{t} = \eta(\nabla\mathbf{u} + \nabla\mathbf{u}^T) + \boldsymbol{\ell} ,$$

$$\rho\left[\frac{\partial \mathbf{u}}{\partial t} + (\mathbf{u}\cdot\nabla)\mathbf{u}\right] + \nabla p - \operatorname{div} \tau = 0 , \tag{3.3}$$

$$\operatorname{div} \mathbf{u} = 0 ,$$

where $\frac{\tau}{t}$ is given by (3.2). We decompose the motion

$$[\mathbf{u},\tau,p] = [\hat{\mathbf{u}},\hat{\tau},\hat{p}] + [\mathbf{u}',\tau',p'] ,$$

where the roof functions satisfy (3.3) and the prime functions are small. We take the liberty of calling the roof flow basic; but in fact it is an arbitrary solution of (3.3). After linearizing, we find that

$$\lambda\left[\frac{\partial\tau'}{\partial t} + (\hat{\mathbf{u}}\cdot\nabla)\tau' + \hat{\tau}\Omega' - \Omega'\hat{\tau} - a(D'\hat{\tau} + \hat{\tau}D')\right] = 2\eta D' + \ell' ,$$

$$\rho\left[\frac{\partial\mathbf{u}'}{\partial t} + (\hat{\mathbf{u}}\cdot\nabla)\mathbf{u}'\right] + \nabla p' - \operatorname{div}\tau' = -\rho(\mathbf{u}'\cdot\nabla)\hat{\mathbf{u}} , \tag{3.4}$$

$$\operatorname{div}\mathbf{u}' = 0 ,$$

where ℓ' does not involve derivatives of $\mathbf{u}'$, τ'.

We next fix our attention at a point x_0 of the field and define $\chi = x - x_0$. Then we imagine that the basic flow and the derivatives of it in (3.3) are constant and equal to their value at x_0. We may then represent the cartesian components of the disturbance as

$$[\mathbf{u}',\tau',p'] = [\omega_i,\sigma_{ij},q]\exp i(k_\ell\chi_\ell - \omega t) ,$$

where the ten amplitudes $[\omega_i,\sigma_{ij},q]$ depend on the basic flow at x_0. The amplitudes are governed by

$$c\sigma_{ij} - \frac{1}{2}\hat{\tau}_{i\ell}[(1-a)\omega_\ell n_j - (1+a)\omega_j n_\ell] - \frac{1}{2}[\omega_i n_\ell(1+a) - \omega_\ell n_i(1-a)]\hat{\tau}_{\ell j}$$

$$+ \mu(n_j\omega_i + n_i\omega_j) = O(1/k) ,$$

$$-\rho c\omega_i = -qn_i + \sigma_{ij}n_j , \tag{3.5}$$

$$\omega_i n_i = 0 ,$$

where $\mathbf{n} = \mathbf{k}/k$, $k = \sqrt{k_1^2 + k_2^2 + k_3^2}$ and

$$c = \frac{\omega}{k} - \hat{\mathbf{u}}\cdot\mathbf{n} . \tag{3.6}$$

These equations were first derived by Rutkevitch [1970] for the three values a = [1,0,-1]. When $k \to \infty$, the problem (3.5) can be regarded as an eigenvalue problem for ω/k or c (see (2.2)).

We can find the ten amplitudes $[\omega_i,\sigma_{ij},q]$ if and only if the determinant Δ of the coefficients vanishes, where

$$\Delta = \left[-\rho c^2 + \mu - \frac{1}{2}\tau_{22}(1-a) + \frac{1}{2}\tau_{11}(1+a)\right]\left[-\rho c^2 + \mu + \frac{1}{2}\tau_{11}(1+a) - \frac{1}{2}\tau_{33}(1-a)\right]. \tag{3.7}$$

The derivation of (3.7) follows along lines set down by Rutkevitch; special coordinates are selected such that $n_1 = 1$, $n_2 = n_3 = 0$, $\hat{\tau}_{23} = 0$ and $\hat{\tau}_{22} > \hat{\tau}_{33}$.

The nontrivial values c^2 are then given by

$$c_+^2 = \frac{1}{\rho}\left[\mu + \frac{1}{2}\,\hat{\tau}_{11}(1 + a) - \frac{1}{2}\hat{\tau}_{33}(1 - a)\right] ,$$
$$c_-^2 = \frac{1}{\rho}\left[\mu + \frac{1}{2}\hat{\tau}_{11}(1 + a) - \frac{1}{2}\hat{\tau}_{22}(1 - a)\right] . \tag{3.8}$$

Departing slightly now from Rutkevitch, using the result proved in Section 4, we call c the velocity of propagation of wave fronts of short waves of vorticity.

In any case, one has $c^2 = f$, where f is real valued. If $f > 0$, we get propagation. If $f < 0$, then

$$c = \pm\, i\sqrt{f} = \pm\, i\mathrm{Im}\left[\frac{\omega}{k}\right] . \tag{3.9}$$

Equation (3.9) shows that if $f < 0$, then there exist short waves of rapidly growing amplitude, the flow undergoes a Hadamard instability.

If we now suppose that at x_0 the system is in principal coordinates of $\hat{\tau}$, the eigenvalues of $\hat{\tau}$ satisfying

$$\hat{\tau}_1 \geq \hat{\tau}_2 \geq \hat{\tau}_3 . \tag{3.10}$$

Then $f > 0$, (and the system is of evolution type, stable to short waves) if and only if

$$\mu + \frac{a}{2}(\hat{\tau}_1 + \hat{\tau}_3) - \left[\frac{\hat{\tau}_1 - \hat{\tau}_3}{2}\right] > 0 . \tag{3.11}$$

Among the Maxwell models ($\boldsymbol{\ell} = \boldsymbol{\tau}$) only the upper ($a = 1$) and lower ($a = -1$) convected models are always evolutionary. This follows from the integral form of these two models. The integrals are expressed by positive definite tensors restricting the range of $\boldsymbol{\tau}$ to evolutionary regions (see JRS, 1985). Dupret and Marchal [1985] have shown that if the criterion for evolution is satisfied initially it will not fail subsequently. We will show in Section 7 that Maxwell fluids $\boldsymbol{\ell} = -\boldsymbol{\tau}$ and with $a \neq \pm 1$ can lose evolution in certain flows. The models Phan-Thien and Tanner, [1977], Johnson and Segalman [1977], Leonov [1976] and Giesekus [1982] may also lose evolution in certain flows.

4. Evolution of the Vorticity

The vorticity equation for (3.3) in 3-D flows may be written as (see (6.4) in Joseph, 1985):

$$\rho\left[\frac{\partial^2\zeta_k}{\partial t^2} + 2(\mathbf{u}\cdot\nabla)\frac{\partial\zeta_k}{\partial t} + u_e u_j \frac{\partial^2\zeta_k}{\partial x_e \partial x_j}\right] + \frac{1}{2}(a - 1)e_{kej}\tau_{jq}\frac{\partial[\operatorname{curl}\boldsymbol{\zeta}]_q}{\partial x_e}$$

$$- \frac{1}{2}(a + 1)\tau_{mp}\frac{\partial^2\zeta_k}{\partial x_m \partial x_p} - \frac{\eta}{\lambda}\nabla^2\zeta_k + \ell_k = 0\,, \tag{4.1}$$

where e_{kej} is the alternating tensor and ℓ_k are all terms of order lower than two derivatives of the vorticity $\boldsymbol{\zeta} = \operatorname{curl}\mathbf{u}$.

The analysis for stability to short waves which was given in Section 3 may be applied to (4.1). We find exactly the same formula $\Delta = 0$ given by (3.7).

It follows that the quasilinear system (3.3) is of evolution type if and only if the vorticity equation (4.1) is of evolution type.

5. First Order Quasilinear Systems for Plane Flow

In plane flow, we have six equations in six unknowns.

$$\begin{aligned}
&\sigma_t + u\sigma_x + v\sigma_y + \tau(v_x - u_y) - a[2\sigma u_x + \tau(u_y + v_x)] - 2\mu u_x = \ell_1\,,\\
&\tau_t + u\tau_x + v\tau_y + \frac{1}{2}(\sigma - \gamma)(u_y - v_x) - \frac{a}{2}(\sigma + \gamma)(u_y + v_x) - \mu(u_y + v_x) = \ell_2\,,\\
&\gamma_t + u\gamma_x + v\gamma_y + \tau(u_y - v_x) - a[2\gamma v_y + \tau(u_y + v_x)] - 2\mu v_y = \ell_3\,,\\
&\rho(u_t + uu_x + vu_y) + p_x - \sigma_x - \tau_y = 0\,,\\
&\rho(v_t + uv_x + vv_y) + p_y - \tau_x - \gamma_y = 0\,,\\
&u_x + v_y = 0\,,
\end{aligned} \tag{5.1}$$

where $\boldsymbol{\tau} = \begin{bmatrix}\sigma & \tau\\ \tau & \gamma\end{bmatrix}$, $\mathbf{u} = (u,v)$, where ℓ_1, ℓ_2, ℓ_3 depend on $\boldsymbol{\tau}$ and possibly on $\mathbf{u}$, but not on their derivatives.

The analysis of evolution follows exactly along the lines laid out in Section 3 and Section 6. Since there are only two normal stresses, we replace (3.10) with $\hat{\tau}_1 \ge \hat{\tau}_2$ and (3.11) becomes

$$\mu + \frac{a(\hat{\tau}_1 + \hat{\tau}_2)}{2} - \frac{\hat{\tau}_1 - \hat{\tau}_2}{2} > 0 \; . \tag{5.2}$$

There is only one speed of propagation (3.8) in the plane. The condition (5.2) for evolution is necessary and sufficient for the stability of solutions of (5.1) to short waves. The same condition, but written relative to general coordinates:

$$\tau^2 - \left[\mu - \frac{\gamma}{2}(1 - a) + \frac{\sigma}{2}(1 + a)\right]\left[\mu - \frac{\sigma}{2}(1 - a) + \frac{\gamma}{2}(1 + a)\right] < 0 \; ,$$

$$\frac{1}{2}\gamma(1 - a) - \frac{1}{2}\sigma(1 + a) - \mu < 0 \tag{5.3}$$

was derived by Joseph, Renardy and Saut [1985] for the vorticity equation associated with (5.1) (see also Section 6).

$$\lambda\Big[\rho \frac{\partial^2\zeta}{\partial t^2} + 2\rho(\mathbf{u}\cdot\nabla)\frac{\partial\zeta}{\partial t} + \left[\rho u^2 - \mu - \frac{\sigma}{2}(1 + a) + \frac{\gamma}{2}(1 - a)\right]\frac{\partial^2\zeta}{\partial x^2}$$

$$+ 2(\rho uv - \tau)\frac{\partial^2\zeta}{\partial x \partial y} + \left[\rho v^2 - \mu + \frac{\sigma}{2}(1 - a) - \frac{\gamma}{2}(a + 1)\right]\frac{\partial^2\zeta}{\partial y^2}\Big]$$

$$= \tilde{\ell}_1 \text{ (of lower order)} \; . \tag{5.4}$$

Condition (5.2) (or(5.3)) implies that the unsteady equation for vorticity is hyperbolic.

The loss of evolution of system (5.1) should not be confused with the possible change of type of the steady problem. We shall examine the connection with these two phenomena now. The difference between the steady and unsteady problem is most easily explained in terms of the vorticity. The analysis of steady problems for the quasilinear system (5.1) may be written as

$$\mathbf{H}\mathbf{q}_x + \mathbf{J}\mathbf{q}_y + \boldsymbol{\ell} = 0 \; .$$

where $\mathbf{q}$ is a column vector with components $[u, v, \sigma, \gamma, \tau, p]$ and $\mathbf{H}$, $\mathbf{J}$, $\boldsymbol{\ell}$ depend on $\mathbf{q}$ but not on its derivatives. To analyze the type of this system we look for characteristics $\theta(x,y) = \text{const.}$, $\theta_x dx + \theta_y dy = 0$.

The analysis is straightforward. The characteristics are given by $\frac{dy}{dx} = \alpha$, where α is a solution of

$$\det[-\alpha\mathbf{H} + \mathbf{J}] = 0 \; . \tag{5.5}$$

This leads us to (11.2) of JRS:

$$(1 + \alpha^2)(-\alpha u + v)^2\{\rho(-\alpha u + v)^2 + \frac{(\gamma - \sigma)}{2}(\alpha^2 - 1)$$

$$+ 2\tau\alpha - (\alpha^2 + 1)(\mu + a\frac{(\gamma + \sigma)}{2})\} = 0 \,. \tag{5.6}$$

There are imaginary roots $\alpha = \pm i$, double real roots along streamlines $\alpha = \frac{v}{u}$ and two roots for the last factor:

$$\alpha = \frac{B}{A} \pm \frac{\sqrt{B^2 - AC}}{A} \,, \tag{5.7}$$

where

$$A = \mu - \rho u^2 + \frac{1}{2}\,\sigma(1 + a) - \frac{1}{2}\,\gamma(1 - a) \,,$$

$$B = \tau - \rho uv \,, \tag{5.8}$$

$$C = \mu - \rho v^2 + \frac{1}{2}\,\sigma(a - 1) + \frac{1}{2}\,\gamma(a + 1) \,.$$

Wherever

$$B^2 - AC = -\mu^2 + \rho[\mu + a\sigma + a\gamma](u^2 + v^2) + \frac{\rho}{2}(\gamma - \sigma)(u^2 - v^2)$$

$$+ \tau^2 + \frac{\sigma^2}{4}(1 - a^2) + \frac{\gamma^2}{4}(1 - a^2) - \mu a(\sigma + \gamma) - 2\rho\tau uv > 0 \,,$$

we have two more real characteristics.

Our system (5.5) is therefore of a mixed type: it has imaginary characteristics and therefore is not hyperbolic; it has real characteristics and therefore is not elliptic. This is not an unusual situation in fluid mechanics (the steady Euler equations of incompressible inviscid fluids are of mixed type) but gives rise to mathematical difficulties: the study of such systems is not well developed (see Saut [1985]). What is new is the fact that the characteristics associated with the last factor in (5.6) can be real or complex in the flow. JRS showed that the roots (5.7) are in fact associated with the steady vorticity equation. This equation is either hyperbolic ($B^2 - AC > 0$) or elliptic ($B^2 - AC < 0$). The roots can be elliptic in one region of flow and hyperbolic in other regions, as in the case of transonic flow. The other characteristics have a simple interpretation: the imaginary roots $\alpha = \pm i$

are associated with the equation $\Delta\psi = -\zeta$, where ζ is the vorticity and ψ the stream function.

6. Vorticity in Plane Flow

In plane flow, there is one nonzero component of vorticity satisfying

$$\rho \frac{\partial^2\zeta}{\partial t^2} + 2\rho(\mathbf{u}\cdot\nabla)\frac{\partial\zeta}{\partial t} - A\frac{\partial^2\zeta}{\partial x^2} - 2B\frac{\partial^2\zeta}{\partial x\partial y} - C\frac{\partial^2\zeta}{\partial y^2} + \ell = 0 \,, \tag{6.1}$$

where ℓ is of lower order and A, B, C are defined by (5.4), i.e.,

$$A = -\rho u^2 + \mu + \frac{\sigma}{2}(1 + a) - \frac{\gamma}{2}(1 - a) \,,$$

$$C = -\rho v^2 + \mu - \frac{\sigma}{2}(1 - a) + \frac{\gamma}{2}(a + 1) \,, \tag{6.2}$$

$$B = \tau - \rho uv \,.$$

The analysis of evolution is most easily framed relative to (6.1). Let us start with a general definition.

A linear partial differential operator of second order

$$L\zeta = P(\mathbf{x},t,\zeta_{tt},\zeta_{tx_1},\ldots,\zeta_{tx_n},\zeta_{x_1x_1},\zeta_{x_1x_2},\ldots,\zeta_{x_nx_n}) + \text{lower order terms}$$

is evolutionary with respect to t is some domain D of $\mathbb{R}^n$ if for every unit vector $\mathbf{k} = (k_1,\ldots,k_n)$ in $\mathbb{R}^n$, for any $t \quad \mathbb{R}$ and any $\mathbf{x} \quad D$, the quadratic polynomial in α

$$P(\mathbf{x},t,-\alpha^2,-\alpha k_1,\ldots,-\alpha k_n,-k_1^2,-k_1k_2,\ldots,-k_n^2) = 0$$

has only <u>real zeros</u>. In the case of constant coefficients, this definition implies that there are no plane wave solutions with arbitrarily large amplitude, i.e., there are no Hadamard instabilities.

The polynomial P = 0 evaluated for (6.1) becomes

$$\rho\alpha^2 + 2\alpha\lambda(uk_1 + vk_2) - Ak_1^2 - 2Bk_1k_2 - Ck_2^2 = 0 \,.$$

This must have real zeroes for every unit vector $\mathbf{k} = (k_1,k_2)$. This leads to

$$(A + \rho u^2)k_1^2 + 2(B + \rho uv)k_1k_2 + (C + \rho v^2)k_2^2 > 0, \ \forall\, k_1,k_2 \,,$$

which implies

$$A + \rho u^2 > 0 \text{ and } (B + \rho uv)^2 - (C + \rho v^2)(A + \rho u^2) < 0\ .$$

Using (6.2), this is equivalent to (5.3).

The same relationship arises from analysis of stability to short waves. We fix $\mathbf{u},\tau$ (hence A,B,C) at their values at x_0 and put $\ell = 0$ and introduce $\boldsymbol{\chi} = \mathbf{x} - \mathbf{x}_0$, writing

$$\zeta(x,y,t) = \hat{\zeta}(x_0,y_0)\exp i[k_1(x - x_0) + k_2(y - y_0) - \omega t]. \tag{6.3}$$

We find that, with $k^2 = k_1^2 + k_2^2$

$$c^2 = \left[\mu + (\sigma + \gamma)\frac{a}{2}\right]k^2 + \frac{1}{2}(\sigma - \gamma)(k_1^2 - k_2^2) + 2\tau k_1 k_2\ , \tag{6.4}$$

where

$$c^2 = \omega^2 + 2\omega(uk_1 + vk_2) + u^2k_1^2 + v^2k_2^2 + 2uvk_1k_2\ .$$

For evolution it is necessary that $c^2 > 0$ for all $k_1,k_2 \in \mathbb{R}$ when $k = (k_1^2 + k_2^2)^{1/2} \to \infty$. If $c^2 < 0$, then

$$\text{Im}\ \frac{\omega}{k} = \pm\ \text{positive constant}$$

and we have a nasty instability to short waves. The criterion $c^2 > 0$ is exactly (5.3). When expressed in principal coordinates $\sigma \geq \gamma$ and $\tau = 0$, we find that the vorticity is of evolution type provided that

$$\mu + \frac{a(\sigma + \gamma)}{2} + \frac{\gamma - \sigma}{2} > 0. \tag{6.5}$$

We may relate the criterion for a change of type in steady flow to the criterion for loss of evolution. Consider $\Delta \underset{\text{def}}{=} B^2 - AC$ when $\rho = 0$. Therefore,

$$\Delta = -f_1 f_2 + \tau^2 ,$$

$$f_1 = \mu + a\,\frac{\gamma + \sigma}{2} - \frac{\sigma - \gamma}{2} , \tag{6.6}$$

$$f_2 = \mu + a\,\frac{\gamma + \sigma}{2} - \frac{\sigma - \gamma}{2} .$$

If we suppose that the system (5.1) (or equivalently the unsteady vorticity equation (6.1)) is of evolution type, then we have (5.3) which clearly implies $B^2 - AC < 0$ for $\rho = 0$, i.e., the steady vorticity equation with $\rho = 0$ is elliptic. Thus hyperbolicity of the steady vorticity equation with $\rho = 0$ implies that the full quasilinear system (5.1) is not evolutionary.

The converse is not true. The equation for vorticity with $\rho = 0$ in the steady case is elliptic when

$$\tau^2 - \left[\mu + \frac{\sigma - \gamma}{2} + a\,\frac{(\sigma + \gamma)}{2}\right]\left[\mu + \frac{\gamma - \sigma}{2} + a\,\frac{(\sigma + \gamma)}{2}\right] < 0 .$$

This inequality with $a \neq 0$ does not imply the condition (5.3) for evolution when viewed in principal coordinates $\sigma \geq \gamma$, $\tau = 0$; for example, (5.3) is violated when $a < 0$; $\tau = 0$; $\gamma >> 1$; $\sigma >> 1$; $0 < \sigma - \gamma << 1$. To understand this, consider the equation

$$\frac{\partial^2 \phi}{\partial t^2} - A\,\frac{\partial^2 \phi}{\partial x^2} - C\,\frac{\partial^2 \phi}{\partial y^2} = 0 . \tag{6.7}$$

The steady equation is elliptic when $AC > 0$. But the unsteady equation is evolutionary (hyperbolic) with respect to t, if and only if $AC > 0$ <u>and</u> $A > 0$. If $AC > 0$ and $A < 0$, (6.7) is an elliptic equation!

We recall that the quasilinear system (5.1) is evolutionary if and only if (6.1) is evolutionary. We can study loss of evolution by using results from the study of change of type in steady inertialess flow. It is perhaps useful to remark that <u>we must have a loss of evolution, instability to short waves, whenever the vorticity of an inertialess steady flow becomes hyperbolic. Conversely, if the vorticity of an inertialess steady flow is elliptic and</u> $A > 0$, <u>where</u> $A = \mu + \frac{\sigma - \gamma}{2} + a\,\frac{\sigma - \gamma}{2}$, <u>then the system (5.1) is evolutionary.</u>

All of the models considered here, except the upper and lower convected Maxwell models may change type in an inertialess steady flow.

7. Examples Taken From Linear Theory

In fact, the theory of evolution is based on equations linearized on an arbitrary flow, called basic. We may evaluate the criteria for evolution and change of type on the basic flow. Many examples of this procedure were given in JRS [1985] and by Yoo, Ahrens and Joseph [1985] for the study of change of type in steady flow.

It is of interest to examine the relationship of change of type in steady flow to the study of short wave instability in unsteady flow. Section 11 of JRS gives analysis for change of type in motions for an upper convected Maxwell model perturbing shear flow, extensional flow, sink flow and circular Couette flow. All these problems are elliptic when $\rho = 0$ and all undergo a change of type for $\rho \neq 0$.

A similar type of analysis, using an upper convected maxwell model, was given by Yoo, Ahrens and Joseph [1985] of the three dimensional sink flow and by Yoo, Joseph and Ahrens [1985] for Poiseuille flow in a channel with wavy walls. These flows also change type when $\rho \neq 0$ and are always evolutionary.

It is of interest to study these problems in cases in which it is possible to lose evolution. We shall examine the examples treated in JRS for Oldroyd models ($a \neq \pm 1$, $\ell = -\tau$ in (3.1)) and some new examples. The corotational Maxwell model ($a = 0$) seems to lose evolution at the lowest levels of stress (the smallest Weissenberg numbers).

Joseph and Saut [1985] studied these problems in cases for which it is possibl to have the short wave instability which leads to a loss of well posedness. They found that flows perturbing plane Couette and Poiseuille flow of Oldroyd models [$-1 \leq a \leq 1$] can change type in steady flow but are always well posed as initial value problems. Flows perturbing extensional flows of Oldroyd models, those of Phan-Thien and Tanner and a popular model by Giesekus, change type in steady flow and under other conditions can become unstable and lose well posedness. Flows perturbing sink flows, flows of upper convected, lower convected and corotational Maxwell models always change type in steady flow but only the corotational model can become unstable.

8. ELASTICITY AND VISCOSITY

We have seen that fluids with instantaneous elasticity may undergo Hadamard instabilities to short waves at high levels of stress (high Weissenberg numbers). We already noted in Section 2 that these short wave instabilities may be avoided by introducing various regularizing terms. One effective method for regularization which is also natural for viscoelastic fluids is to add a viscosity term to the constitutive equation (for an example, see Dupret, Marchal and Crochet, 1985). Many popular models of fluids have a Newtonian viscosity. The models of Jeffreys, Oldroyd, Rouse and Zimm and molecular models of solutions with Newtonian solvents lead to Newtonian contributions to the stress. To make this method useful it is necessary that the viscosity used should be appropriate to the fluid under study.

To decide about elasticity and viscosity we could consider ever more dilute solutions of polymer chains of large molecules in solvents which might be thought to be Newtonian. What happens when we reduce the amount of polymer? There are two good ideas which are in collision. The first idea says that there is always a viscosity and some elasticity with an ever greater viscous contribution as the amount of polymer is reduced. On the other hand, we may suppose that liquid is elastic so that $\mu = 0$ and the viscosity η is the area under the graph of the relaxation function. Since η is finite in all liquids, we have $\eta = G(0)\bar{\lambda}$, where $\bar{\lambda}$ is a mean relaxation time. Maxwell's idea is that the limit of extreme dilution is such that the rigidity $G(0)$ tends to infinity and $\bar{\lambda}$ to zero in such a way that their product η is finite. Ultimately, when the polymer is gone, we are left with an elastic liquid with an enormously high rigidity. This idea apparently requires anomalous behavior because $G(0)$ appears to decrease with polymer concentration when the concentration is finite.

The contradiction between the two foregoing ideas and the apparent anomaly can be ameliorated by replacing the notion of a single mean relaxation time with a distribution of relaxation times. This notion is well grounded in structural theories of liquids in which different times of relaxation correspond to different modes of molecular relaxation. It is convenient again to think of polymers in a solvent, but now we can imagine that the solvent is elastic, but with an enormously high rigidity. In fact many of the so called Newtonian solvents have a rigidity of the order 10^9 Pascals, which is characteristic of glass, independent of variations of the chemical characteristics among the different liquids (for example, see Harrison, 1976). To find this glassy modulus it is necessary to use very ingenious high frequency devices operating in the range 10^9 Hertz and to supercool the liquids to temperatures near the glassy state. In these circumstances the liquid acts like a glassy solid, the molecular configurations cannot follow the rapid oscillations of stress, the liquid cannot flow. For slower processes it is possible for the liquid to flow and if the relaxation is sufficiently fast the liquid will appear to be Newtonian in more normal flows. For practical purposes there is no difference between Newtonian liquids and liquids with rigidities of order 10^9 and mean relaxation times of 10^{-10} seconds or so. In fact it is convenient to regard such liquids as Newtonian, even though $\mu = 0$ and $\tilde{\mu} = \eta$.

The presence of polymers would not allow the liquid to enter the region of viscous relaxation at such early times. Instead much slower relaxation processes associated with the polymers would be induced. The second epoch of relaxation occurs in a neighborhood of very early times $t = t_1$ (or at very high frequencies). An effective modulus $G(t_1)$ may be defined at $t = t_1$ or for any t in the neighborhood of t_1.

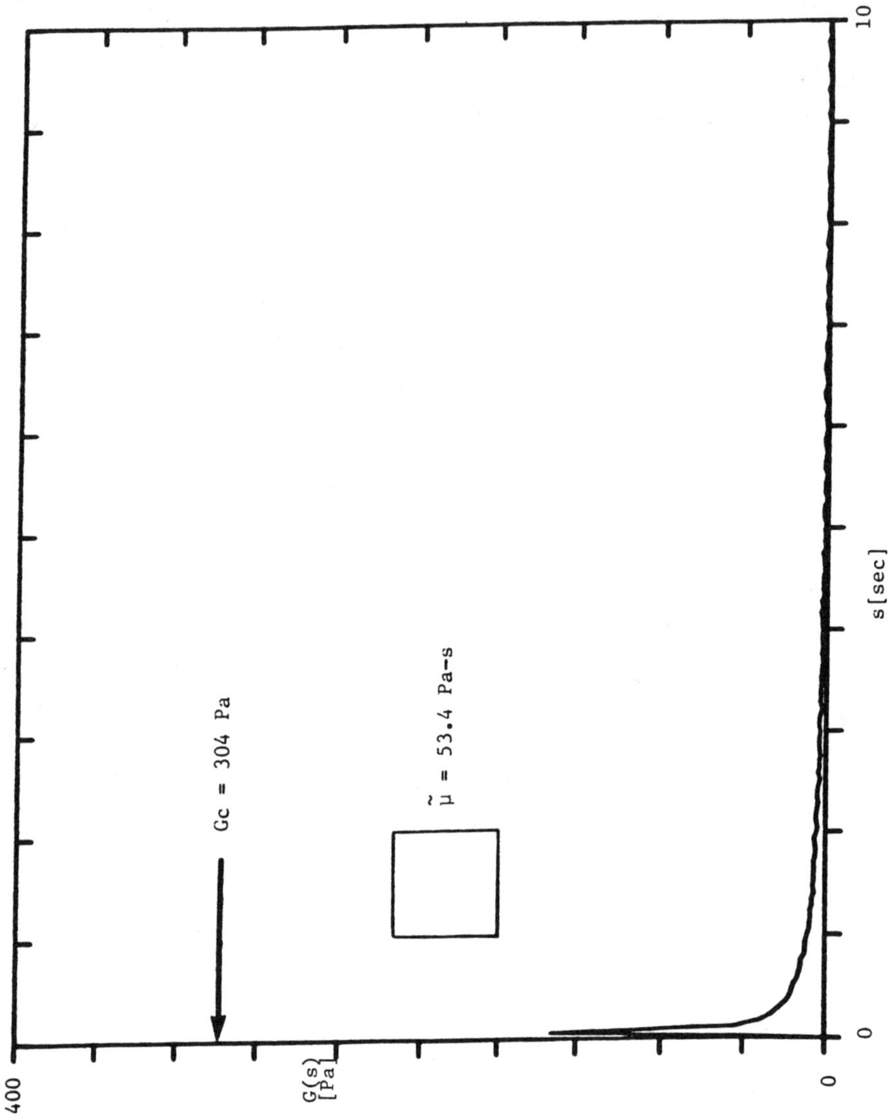

Figure 1: (Joseph, Riccius, Arney, 1985). Shear modulus G_c = 304 Pa and relaxation function G(s) for an observed wave speed c = 50.6 cm/sec in a 1 carboxymethyl cellulose (CMC) solution in 49 water and 50 glycerin. E. H. Lieb (1975) photographed shear waves in the same solution using tracers. He estimated c > 8 cm/sec.

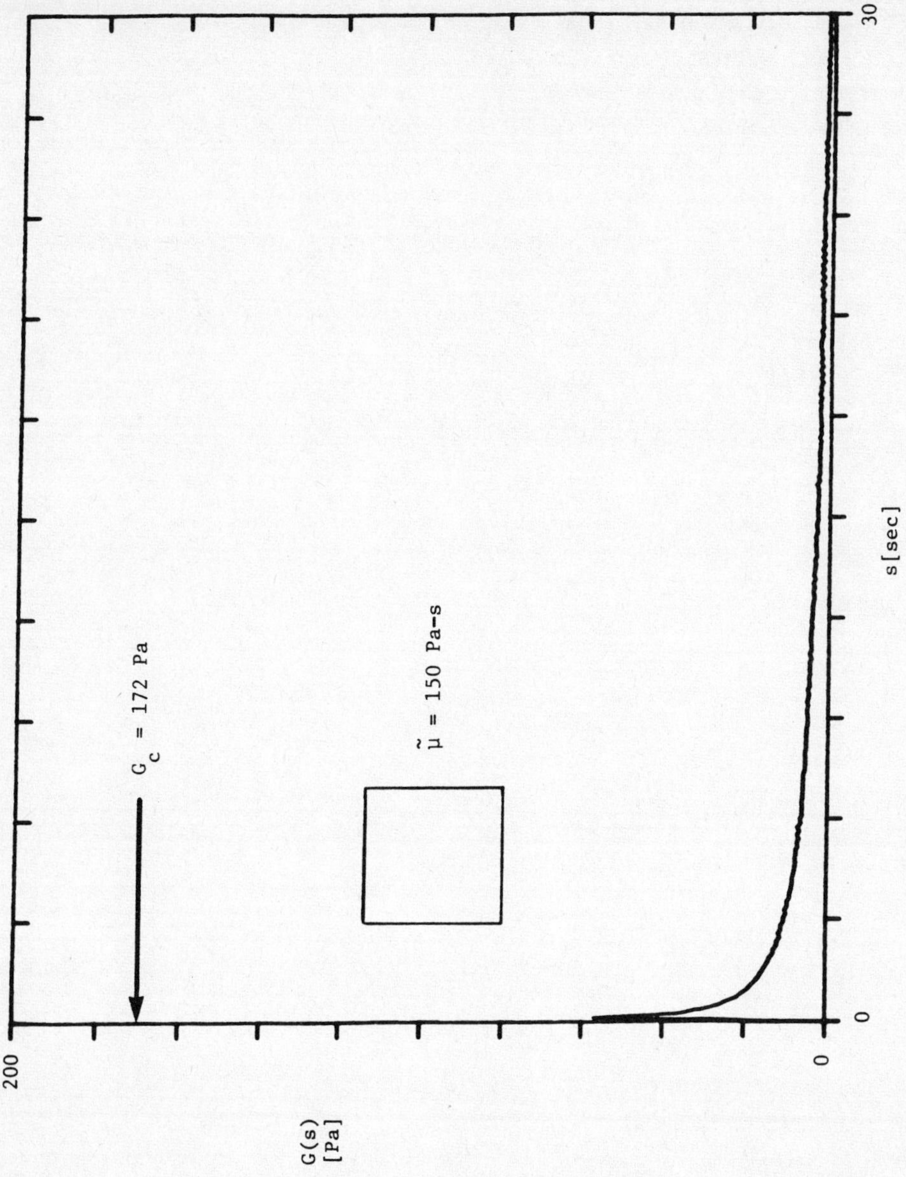

Figure 2: (Joseph, Riccius, Arney, 1985). Shear modulus G_c = 172 Pa and relaxation G(s) for an observed wave speed c = 38.4 in a 1.5 Poly(acrylamide) Separan AP30 in 50 glycerin and 48.5 water. The zero shear viscosity, $\tilde{\mu}$ = 160 Pa-s is the area of the square. Bird, Armstrong and Hassager (1977) exhibit frequency data ($\omega \leq 1000$ rad/sec) of Huppler, et al. for this solution. They say that the storage modulus $G'(\omega)$ is nearly at its limiting with $G'(\omega)$ = 140 Pa at frequencies of ω = 100.

The relaxation function may be measured on standard cone and plate rheometers, using, for example, stress relaxation after a sudden strain. Examples of such stress relaxation, taken on a Rheometric System 4 rheometer is shown in Figs. 1 and 2. The rise time of this instrument is roughly 0.01 sec and the more rapid part of the stress relaxation cannot be obtained with such devices. The modulus G_c was measured by Joseph, Riccius and Arney (forthcoming) using a wave speed meter. They measure transit times of impulsively generated shear waves into a viscoelastic liquid at rest. A Couette apparatus is used; the outer cylinder is moved impulsively; the time of transit of the shear wave from the outer to inner cylinder is measured. They set up criteria to distinguish between shear waves and diffusion. One criterion is that transit times δt should be reproducible without large standard deviations and such that $d = c\delta t$, where d is gap size, and c, the wave speed is a constant independent of d. In other words, transit times are independent of gap size. Then, using theoretical results for propagation of shear wave into rest $c = \sqrt{G_c/\rho}$. We could regard G_c as the effective modulus or rigidity.

This paper has been prepared for Amorphous Polymers Workshop, March 5-8, 1985, held at the Institute for Mathematics and its Applications. The work was supported by the U.S. Army Research Office, Math and by the National Science Foundation, Fluid Mechanics. Many of the results given here are taken from previous works with various collaborators, but most especially from a recent [1985] work with Jean Claude Saut.

References

R. B. Bird, B. Armstrong, O. Hassager, Dynamics of Polymeric Liquids. Wiley, 1977.

B. D. Coleman and W. Noll, Foundations of linear viscoelasticity, Rev. Mod. Physics **33**, 239-249 (1961). Erratum op. cit. **36**, 1103 (1964).

F. Dupret, J. M. Marchal, Sur le signe des valeurs propres du tenseur des extra-constraints dans un ecoulement de fluide de Maxwell, forthcoming.

F. Dupret, J. M. Marchal, Proceedings of the fourth workshop on numerical methods in non-Newtonian flows, Spa, Belgium, June 3-5, 1985 (to appear in JNNFM).

F. Dupret, J. M. Marchal, and M. J. Crochet, On the consequence of discretination errors in the numerical calculation of viscoelastic flow, J. Non Newtonian Fluid Mech. **18**, 173-186 (1985).

H. Giesekus, A simple constitutive equation for polymer fluids based on the concept of deformation-dependent tensorial mobility, J. Non Newtonian Fluid Mech. **11** (1982), 69-109.

A. E. Green and R. Rivlin, The mechanics of non-linear materials with memory, Part III, Arch. Rational Mech. Anal. **4**, (1960), 387.

G. Harrison, The Dynamic Properties of Supercooled Liquids. Academic Press, 1976.

H. Jeffreys, The Earth. Cambridge University Press, 1929.

M. W. Johnson, D. Segalman, A model for viscoelastic fluid behavior which allows non-affine deformation, J. Non Newtonian Fluid Mech. **2** (1977), 255-270.

D. D. Joseph, Hyperbolic phenomena in the flow of viscoelastic fluids. Proceedings of the Conference on Viscoelasticity and Rheology, U of WI (1984), edited by J. Nohel, A. Lodge, M. Renardy, Academic Press, (to appear, 1985 A). See also MRC Report 2782.

D. D. Joseph, Historical perspectives on the elasticity of liquids, (to appear in JNNFM, 1985 B).

D. D. Joseph, M. Renardy, J. C. Saut, Hyperbolicity and change of type in the flow of viscoelastic fluids, Arch. Rational Mech. Anal. **87** (1985), 213-251.

D. D. Joseph and J. C. Saut, Change of type and loss of evolution in the flow of viscoelastic fluids, (to appear in JNNFM, 1985).

A. I. Leonov, Nonequilibrium thermodynamics and rheology of viscoelastic polymer media, Rheol. Acta **15** (1976), 85-98.

N. Phan-Thien, R. I. Tanner, A new constitutive equation derived from network theory, J. Non Newtonian Fluid Mech. **2** (1977), 353-365.

M. Renardy, Singularly perturbed hyperbolic evolution problems with infinite delay and an application to polymer rheology, SIAM J. Math. Anal. **15** (1984), 333-349.

M. Renardy, "A local existence and uniqueness theorem for K-BKZ fluid, Arch. Rational Mech. Anal. **88** (1985), 83-94.

I. M. Rutkevitch, Some general properties of the equations of viscoelastic incompressible fluid dynamics, PMM **33**, No. 1 (1969), 42-51.

I. M. Rutkevitch, The propagation of small perturbations in a viscoelastic fluid, J. Appl. Math. Mech. **34** (1970), 35-50.

I. M. Rutkevitch, On the thermodynamic interpretation of the evolutionary conditions of the equations of mechanics of finitely deformable viscoelastic media of Maxwell type, J. Appl. Math. Mech. **36**, (1972), 283-295.

J. C. Saut, Mathematical problems associated with equations of mixed type for the flow of viscoelastic fluids. Proceedings of the fourth workshop on numerical methods in non-Newtonian flows, Spa, Belgium, June 3-5, 1985 (to appear in JNNFM).

J. C. Saut, D. D. Joseph, Fading memory, Arch. Rational Mech. Anal. **81**, 53-95 (1983).

J. Y. Yoo, M. Ahrens, D. D. Joseph, Hyperbolicity and change of type in sink flow, J. Fluid Mech. **153** (1985), 203-214.

J. Y. Yoo, D. D. Joseph, Hyperbolicity and change of type in the flow of viscoelastic fluids through channels, J. Non Newtonian Fluid Mech. **19**, (1985), 15-41.

RHEOLOGY OF SHAPE MEMORY ALLOYS

Ingo Müller

FB 9 - Hermann-Föttinger-Institut, TU Berlin

ABSTRACT

A model is described that is capable of simulating the load-deformation-temperature behaviour of materials with shape memory. The model considers the rheological properties of these materials as activated processes.

1. Introduction

Shape memory alloys are characterized by a strong dependence of the load-deformation diagrams on temperature. At low temperatures the behaviour of such an alloy is much like that of a plastic body with a virginal elastic curve, a yield limit, creep and residual deformation. At high temperature the behaviour is pseudoelastic with a hysteresis in the first and third quadrant of the load-deformation diagram. The complex behaviour is the consequence of a martensitic-austenitic phase transition with twin formation in the martensitic phase.

The paper describes the typical response of a shape memory alloy under different conditions of dynamic and thermal loading. It introduces a model which is supposed to simulate the observed behaviour of such an alloy and investigates its properties by subjecting the model to the same kind of dynamic and thermal loading as the body. There is good qualitative agreement between the predictions of the model and the observations of the alloy itself.

Yield and creep of the alloys are considered as activated processes for whose description simple rate laws are formulated that permit the simulation of the rheological properties of shape memory alloys.

2. PHENOMENOLOGY

Typical load-deformation diagrams of memory alloys are shown schematically in fig. 1 whose curves are abstracted from the articles in the books by Perkins (1975) and Delaey & Chandrasekharan (1982). These curves represent quasistatic isothermal experiments with shape memory alloys. At low temperatures there is an original elastic curve through the origin, which is the natural state, and a yield limit at which the body yields deformation without increase of load. The yield ends on a lateral elastic line that allows loading far beyond the yield limit. Unloading provides residual deformation.

At higher temperature this behaviour is qualitatively unchanged but the yield limit is decreased.

When the temperature is raised further, we observe a very different load-deformation diagram. There is still an elastic curve through the origin and a yield limit. Also we still have the lateral elastic line but unloading along this line does not lead to a residual deformation. Rather there is recovery of the yielded deformation when the load falls below the recovery load. Unloading will bring the body back to its natural state along the initial elastic curve. This behaviour is called pseudoelastic. It is elastic in that the body returns to its natural state. But it is only pseudoelastic, because there is a hysteresis in the loading-unloading cycle.

At a still higher temperature the pseudoelastic behaviour persists but there are quantitative changes: The yield limit and the recovery limit grow and both grow closer together so that the hysteresis loop becomes smaller.

It is clear in which sense the diagrams of fig. 1 imply "memory". Indeed, if at low temperature we gave the body a residual deformation after unloading, a simple rise

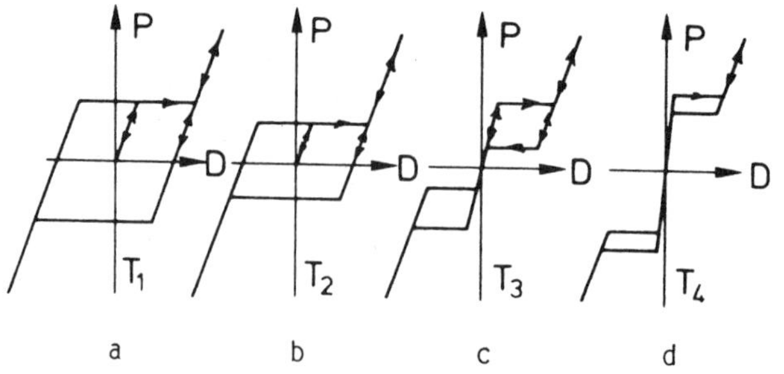

Fig. 1. Schematic load deformation curves at different temperatures. Arrows indicate possible directions of loading and unloading.

in temperature will bring it back to its natural state, because that is its only possible deformation under zero load. We say that the body "remembers" its natural state.

The range of temperatures covered by the diagrams of fig. 1 typically is 50 K around room temperature. A typical recoverable deformation is 6%.

Metallurgists have determined that the peculiar load-deformation-temperature behaviour of memory alloys is accompanied by an austenitic-martensitic phase change and martensitic twin formation. At low temperatures the body is martensitic and in the natural state it consists of equal proportions of the martensitic twins; on the lateral elastic lines one or the other twin prevails. At high temperature the body is austenitic at small loads; however, a big load can still force it into the martensitic phase with one twin prevailing.

Of particular interest in this paper are the rheological properties of shape memory alloys, i.e. the processes of creep in yielding and recovery. Instructive examples for these processes are shown in figs. 2 and 3. The input in those figures consists of an oscillating tensile load and of an external temperature which first increases and then decreases. The resulting deformation is recorded and we observe that it oscillates along with the load.

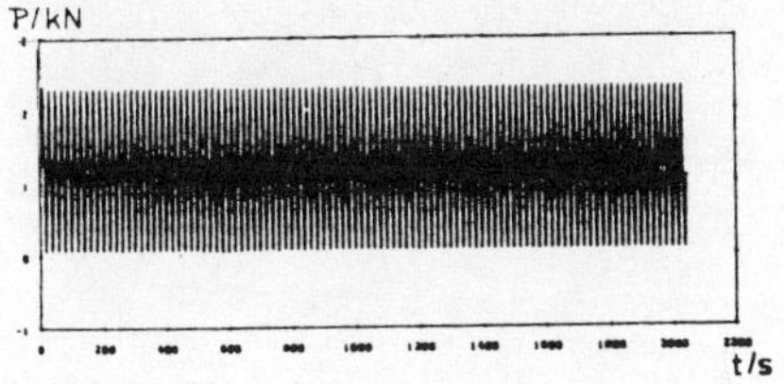

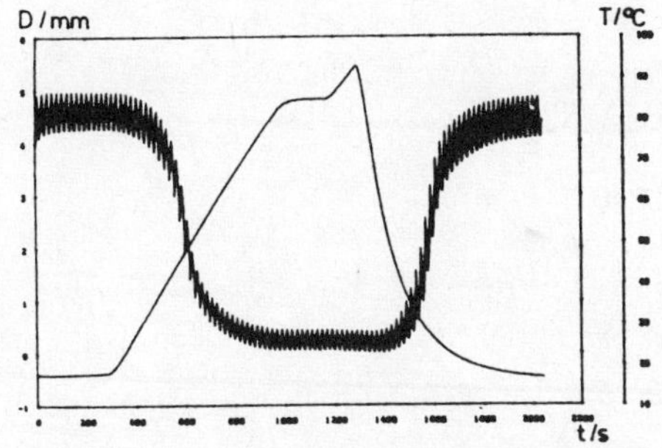

Fig. 2. Deformation as a result of an oscillating tensile force and a varying temperature. The initial phase is martensitic with one twin prevailing. The figure is taken from H. Ehrenstein (1985) who exploits such plots so as to determine the material characteristics of memory alloys.

In fig. 2 we see that at first, at a low temperature the mean value of the deformation is big suggesting that the body oscillates up and down along the right lateral elastic line of fig. 1a. So the body is martensitic and one twin prevails. As the temperature increases the mean value of the deformation decreases and we conclude that the body oscillates along the elastic austenitic line through the origin. A decrease of temperature brings the deformation back to its former large value.

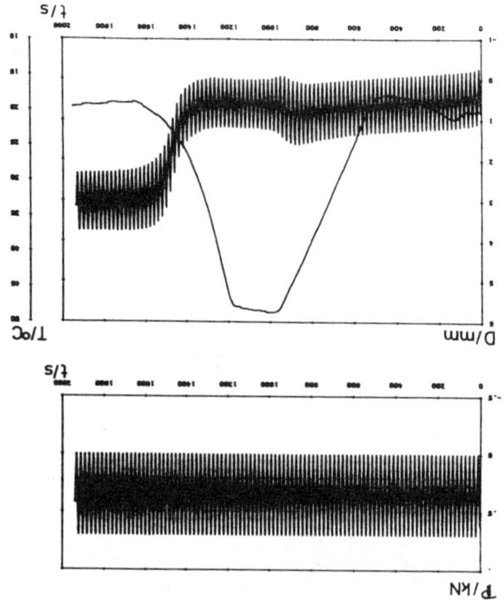

Fig. 3. Deformation as a result of an oscillating tensile force and a varying temperature. The initial phase is martensitic with both twins present.

In fig. 3 the deformation is small at first which indicates that the body is near its natural state with an equal proportion of the different types of martensitic twins. The mean value of the deformation increases slowly as the temperature rises which must be due to the fact that, as the yield limit decreases, there is creep which is due to the creation of one martensitic twin at the expense of the others. When the temperature rises high enough the deformation decreases and we conclude from that that the body has become austenitic. Upon a subsequent decrease of temperature we observe an increase of deformation just like in fig. 2.

The purpose of this paper is the presentation of a model that is capable of simulating the load-deformation-temperature behaviour described above. The model has been developed and perfected by Achenbach, Müller and Wilmanski in several papers (e.g. see Müller & Wilmanski (1981), Achenbach & Müller (1986) and Müller (1985)).

3. THE MODEL AND ITS QUALITATIVE BEHAVIOUR

3.1. Basic Element

The basic element of the model is a lattice particle, a small piece of the metallic lattice of the body, which is shown in fig. 4 in three different equilibrium configurations denoted by $M_{\pm}$ for the martensitic twins and by A for austenite. Clearly the martensitic twins may be considered as sheared versions of the austenitic particle. Intermediate shear lengths are also possible, of course, and the upper part of fig. 4 shows the postulated form of the potential energy for a given shear length Δ. The lateral minima correspond to the martensitic phase and the central metastable minimum corresponds to the austenitic phase. In between these minima there are energetic barriers.

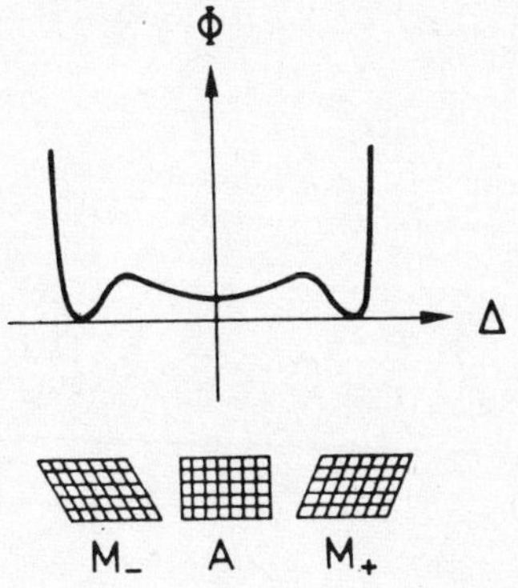

Fig. 4. Lattice particles and their potential energy.

3.2. The Body as a Whole and its Deformation

The lattice particles are arranged in layers and the layers are stacked in the manner shown in fig. 5a, which represents the body in the martensitic phase with alternating layers of M_+ and M_-. We consider that configuration as the natural configuration of the body at low temperature.

For a proper appreciation of the model we proceed to describe what happens when the stack of layers is lightly loaded in the vertical direction. The layers are then subject to shear stresses and the M_- layers become steeper, while the M_+ layers become flatter. Each layer contributes the vertical component of its shear length of the model which we take to be a measure for the deformation $D - D_0$. We have

$$D - D_0 = \frac{1}{\sqrt{2}} \sum_{i=1}^{N} \Delta_i , \tag{3.1}$$

where the summation extends over all layers. Removal of the load lets the layers fall back to their original orientations and the body contracts, i.e. the deformation was elastic under a small load.

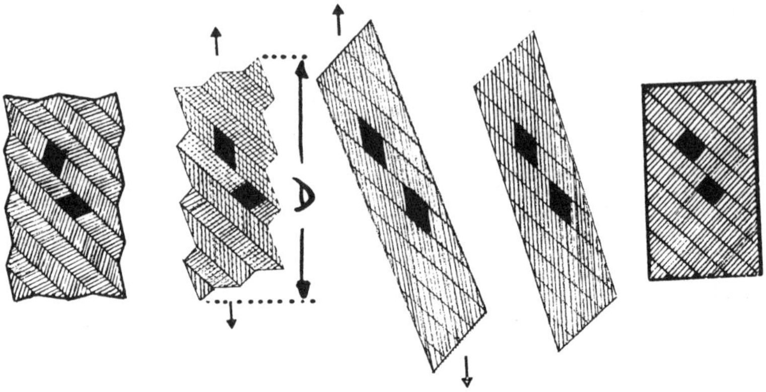

Fig. 5. Model of a body built from martensitic and austenitic layers

However, a critical bigger load will be able to flip the M_- layers and thus achieve a large deformation because the flipping goes along with a big increase of the shear lengths of the flipping layers. Once all M_- layers have flipped, the body can be loaded beyond the critical load. Removal of the load now will let all layers fall back to the initial orientation of the M_+ layers and thus leave the body with a considerable residual deformation.

In this manner we understand the initial elastic branch of the low temperature load-deformation diagrams, the yield, the lateral elastic curves, and the residual deformation.

In a manner to be described below the unloaded configurations of fig. 5a and 5d will turn into the austenitic configuration shown in fig. 5e upon heating. To the naked eye that configuration is identical to the one of fig. 5a, it is only the shape of the surface that differs. And indeed, a decrease of temperature will lead from the configuration of fig. 5e to the one of fig. 5a without change of external shape. Thus the sequence of graphs in fig. 5 gives a suggestive interpretation of the observations that describe the shape memory effect.

3.3. Energetic Considerations and the Role of Thermal Fluctuation

The considerations of the previous section can be repeated in terms of energetic considerations that are illustrated in fig. 6. The left hand side of that figure refers to a low temperature. Initially half of the layers lie in the left minimum and half in the right one, so that the body is in its natural state. If a load is applied, the potential energy of the load, which is a linear function of Δ , must be added to $\phi(\Delta)$ of fig. 4 and the new potential energy is thus deformed and assumes the form shown by the second diagram down in fig. 6. The barrier on the left hand side is decreased but it is still there and prevents the layers from flipping. Flipping becomes inevitable when the load is so big that the left minimum is eliminated. This is indicated by the lower diagram on the left hand side of fig. 6.

If the temperature is higher, the layers participate in the thermal fluctuation. On the right hand side of fig. 6 this situation is illustrated by the pools of points in the martensitic minima. The height of the pools indicates the mean kinetic energy of the layers. Without load the barriers are still high enough to prevent layers from flipping. But the intermediate load lowers the M_- barrier sufficiently to enable the particles to flip, even though there is still a barrier. This is indicated in the second diagram of the right hand side of fig. 6 and thus we understand that the yield limit is decreased by an increase of temperature.

3.4. On the Motivation of the Shape of the Potential Energy

The figs. 7 refer to the unloaded body and, in particular in fig. 7a shows the pool of fluctuating layers at a high temperature where the barriers are easily overcome. The average position of the layers is in the centre and we may say that the body is austenitic. If the temperature decreases, so does the height of the pool of the fluctuating layers and fig. 7b shows a situation where the layers can still easily overcome the barriers between the central minimum and the lateral ones. But once the layers have become martensitic, they find it difficult to return to the austenitic phase because the barriers for a jump from $M_\pm$ to A are higher than those for a jump from A to $M_\pm$. Thus it occurs that as the temperature continues to drop, the unloaded body will become martensitic and in all likelihood it will assume its natural state with half of the layers in either martensitic minimum.

In order for this to be so the potental $\phi(\Delta)$ has been postulated in the form of fig. 4 with the austenitic minimum metastable and the martensitic ones stable.

3.5. Interfacial Energies

The argument used so far in the description of the model have considered the layers as quite independent. It turns out, however, that this is not quite sufficient for a successful simulation of the observed phenomena, in particular, for the simulation

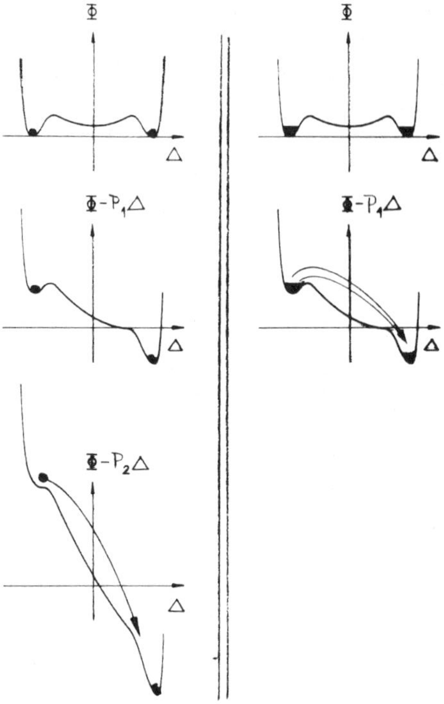

Fig. 6. Potential energy at different loads. Situation of lattice layers at small and elevated temperature.

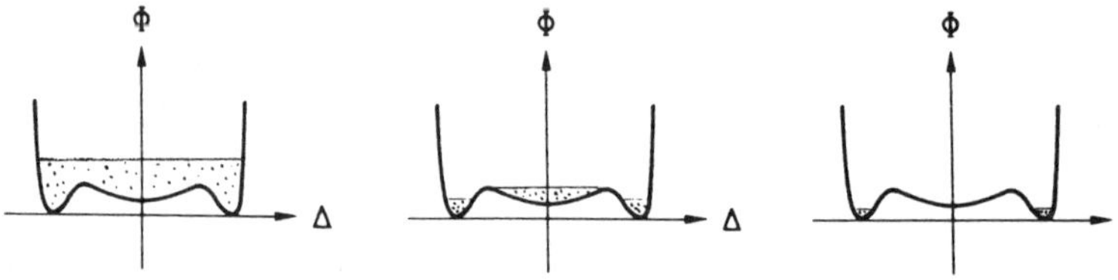

Fig. 7. The creation of austenite and martensite.

of the hystereses. One way of improving the model is to assume that there is an interfacial energy between the austenitic and martensitic layers. That energy is supposed to reflect the energy stored in the lattice distortions which occurs when a highly symmetric austenitic layer is in coherent contact with a martensitic layer*).

*)Note that there is a lattice distortion between martensitic twins as well but this is comparatively slight and we ignore it.

If e is the energy per interface and K is the number of austenitic-martensitic layers, the interfacial energy of the model is Ke and it remains to relate this value to the phase fractions x_A and $x_M = x_{M-} + x_{M+}$ of austenitic and martensitic layers.

Rather obviously there is no tight relation between K and x_A, x_M. Indeed, for given values of K, x_A and x_M there are

(3.2) $$W_k = \binom{Nx_A - 1}{K/2 - 1}\binom{Nx_M - 1}{K/2 - 1}$$

possibilities to realize that number of interfaces**). We shall assume that the number K is in fact observed for which W_K is maximal. This number can easily be calculated and we obtain

(3.3) $$K_{m.p} = 2N x_A (1 - x_A)$$

Note that of course there are no interfaces for the purely austenitic phase $x_A = 1$ and for the purely martensitic one $x_A = 0$. According to (3.3) the number of interfaces is maximal for $x_A = x_M = 1/2$. Thus we see that the consideration of interfacial energies will tend to stabilize the pure phases.

The argument that has led to the equation (3.3) is essentially probabilistic, because the number K for which W_K is maximal is the most probable one. For that argument to be valid, however, the temperature must be high. because the thermal fluctuations must be big enough to lead from one K to another one often enough, so that the most probable one can be established.

4. MATHEMATICAL EVALUATION OF THE MODEL AND ITS PREDICTIONS

4.1. Rate Laws for Phase Fractions and Temperature

The idea described above in Section 3.3. that thermal fluctuations permit the layers of the model to jump across barriers and thus contribute to the deformation is tantamount to saying that the observed processes are thermally activated processes. The theory of such processes was first developed in the context of chemical reactions but it can be adapted to other situations where thermal motion overcomes barriers.

Here we shall assume that the phase factors x_{M-}, x_A and x_{M+} satisfy the rate laws

**) The problem of determining this number is equivalent to the calculation of the number of possibilities for distributing Nx_A particles over $K/2$ cells and Nx_M over $K/2$ cells, so that no cell remains empty.

$$\dot{x}_{M_-} = -\overset{-0}{\mu}\, x_{M_-} + \overset{0-}{\mu}\, x_A$$

(4.1) $$\dot{x}_A = +\overset{-0}{\mu}\, x_{M_-} - \overset{0-}{\mu}\, x_A - \overset{0+}{\mu}\, x_A + \overset{+0}{\mu}\, x_{M_+}$$

$$\dot{x}_{M_+} = \qquad\qquad + \overset{0+}{\mu}\, x_A - \overset{+0}{\mu}\, x_{M_+} .$$

Thus the change of x_{M_-} has two causes, a gain due to particles that jump from the central minimum to the left, and a loss due to particles that jump from the left minimum to the middle. The gain is supposed to be proportional to x_A and the loss is proportional to x_{M_-}. The factors of proportionality are called transition probabilities and their form will presently be discussed. What holds for $\dot{x}_{M_-}$ will also hold, mutatis mutandis, for $\dot{x}_{M_+}$ and for x_A. Of course, the equation for $\dot{x}_A$ is a little more complex, because the central minimum can exchange particles with both sides.

The transition probability $\overset{-0}{\mu}$ (say) is assumed to be proportional to the probability of an M_--layer to be on the top of the left barrier, whose shear length Δ is equal to m_L

(4.2) $$\frac{e^{-\frac{\phi(m_L)-Pm_L}{kT}}}{\int_{-\infty}^{m_L} e^{-\frac{\phi(\Delta)-P\Delta}{kT}}\, d\Delta} .$$

This must be multiplied by the factor $\sqrt{\frac{kT}{2\pi m}}$ which represents the mean speed of the layers of mass m. But even with this factor the product is not quite equal yet to the transition probability. Indeed, so far the interfacial energy has not been taken into account. As was discussed in Section 3.5 this energy provides an additional barrier whose height depends on the actual value of the phase fraction. Thus, if

(4.3) $$E_i = 2eN x_A x_M$$

is the interfacial energy, a transition $M_- \rightarrow A$ will change its value by

(4.4) $$\overset{-0}{\Delta} E_i = \frac{1}{N}\frac{\partial E_i}{\partial x_{M_-}} = 2e(1-2x_M) .$$

Therefore the expression (4.2), which was already multiplied by $\sqrt{\frac{kT}{2\pi m}}$, must be multiplied by the additional Boltzmann factor exp $\left\{-\frac{\overset{-0}{\Delta} E_i}{kT}\right\}$, if we wish to obtain the transition probability. We thus have

(4.5)
$$\overset{-0}{p} = \sqrt{\frac{kT}{2\pi m}}\;\frac{e^{-\frac{\Phi(m_L)-Pm_L}{kT}}}{\int_{-\infty}^{m_L} e^{-\frac{\Phi(\Delta)-P\Delta}{kT}}\,d\Delta}\; e^{-\frac{2e}{kT}(1-2x_M)}$$

and, of course, the other transition probabilities are constructed in an analogous manner.

There is also a rate law for the temperature which is a simplified form of the balance of internal energy. It reads

(4.6)
$$C\dot{T} = -\alpha\,(T-T_E) - \left(\dot{x}_{M_-} H^-(P) + \dot{x}_{M_+} H^+(P)\right),$$

where C is the heat capacity and α is the coefficient of heat transfer between the body and its surroundings. By (4.6) there is a rate of change of temperature when the body temperature differs from the external temperature T_E and when the phase fractions vary. That latter contribution is due to the fact that a lattice particle when it jumps from one well into the other will generally convert potential energy into kinetic energy. i.e. heat. or vice-versa. Fig. 8 shows how $H^{\pm}$ in (4.6) are de-

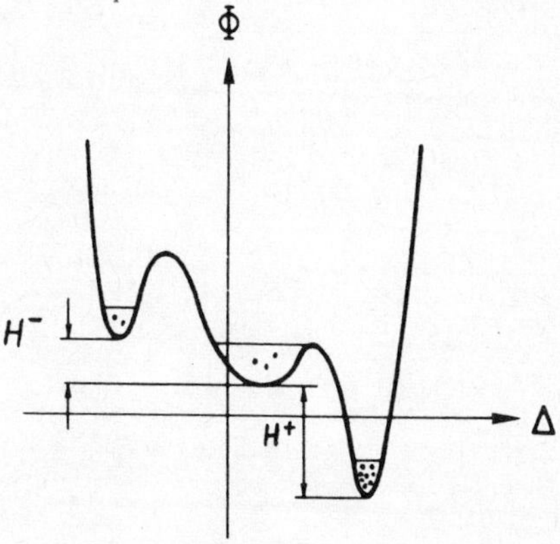

Fig. 8. On the definition of the dissipated energies

fined and, of course, these energy values will depend on the load P, since that load determines the depths of the potential wells.

Given the external temperature T_E and the load P as functions of time the rate laws (4.1) and (4.6) may serve for t he calculation of the phase fractions $x_{M_\pm}$, x_A and of the body temperature T as functions of time. Of course, this calculation cannot be done analytically. because of the strongly non-linear character of that set of differential equations. But numerically the integration is quite easy in a step-by step procedure.

4.2. Calculation of the Deformation

The main objective, however, of the dynamic theory is not the calculation of $x_{M_\pm}(t)$, $x_A(t)$ and $T(t)$ for given functions $T_E(t)$ and $P(t)$, but rather it is the calculation of the deformation D as a function of time.

The calculation of $D(t)$ proceeds from the formula (3.1). We assume that, if a layer belongs to the phases $M_\pm$ or A respectively, its shear length is given by the expectation values

$$(4.7)\quad \bar{\Delta}_{M_-} = \frac{\int_{-\infty}^{m_L} \Delta e^{-\frac{\phi-P\Delta}{kT}} d\Delta}{\int_{-\infty}^{m_L} e^{-\frac{\phi-P\Delta}{kT}} d\Delta}, \quad \bar{\Delta}_{M_+} = \frac{\int_{m_R}^{\infty} \Delta e^{-\frac{\phi-P\Delta}{kT}} d\Delta}{\int_{m_R}^{\infty} e^{-\frac{\phi-P\Delta}{kT}} d\Delta}, \quad \bar{\Delta}_A = \frac{\int_{m_L}^{m_R} \Delta e^{-\frac{\phi-P\Delta}{kT}} d\Delta}{\int_{m_L}^{m_R} e^{-\frac{\phi-P\Delta}{kT}} d\Delta}$$

where m_L and m_R are the shear length of the left and right maximum of the potential function $\phi-P\Delta$. The deformation is then given by the formula

$$(4.8)\quad D - D_0 = \frac{N}{\sqrt{2}}\left(x_{M_-}\bar{\Delta}_{M_-} + x_A\bar{\Delta}_A + x_{M_+}\bar{\Delta}_{M_+}\right),$$

because $x_{M_\pm}$ and x_A are the probabilities for finding a layer in the phases $M_\pm$ and A respectively.

Inspection of (4.7) and (4.8) shows that once $P(t)$ is given and $x_{M_\pm}(t)$, $x_A(t)$ and $T(t)$ have been calculated from (4.1) and (4.6) we may calculate $D(t)$. Here again the calculation requires a numerical integration of course.

4.3. Some Results

Fig. 10, which is taken from Achenbach & Müller (1986), gives a full account of all numerical calculations: $P(t)$ was prescribed as an alternating tensile and compressive load while the external temperature was fixed on different levels as indicated by the numbers Θ, which are proportional to the absolute temperature. The various curves on the left hand side of fig. 10 show the calculated resulting functions $T(t)$, $x_{M_\pm}(t)$, $x_A(t)$ and $D(t)$ as indicated.

On the right hand side of fig. 10 the time has been eliminated between $P(t)$ and $D(t)$ so that load-deformation diagrams have appeared. For the four different temperatures these curves must be compared to the schematic curves of fig. 1 and we con-

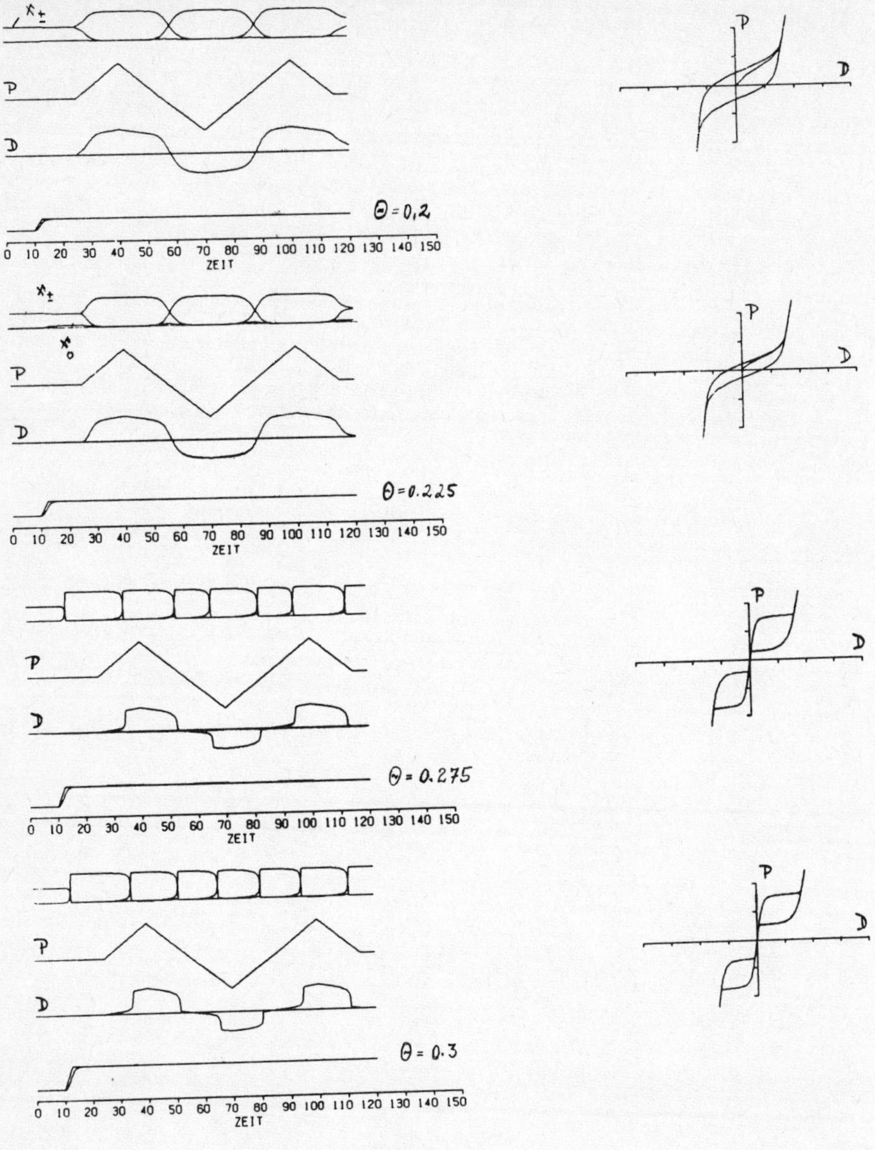

Fig. 9: Predictions of the model for an alternating tensle and compressive load $P(t)$ and $T_E(t)$ = const.

clude that there is good qualitative agreement. *)

Of course, once the system of equations (4.1), (4.6), (4.8) is there, we may evaluate them for arbitrary functions $P(t)$ and $T_E(t)$. In particular, we may simulate the reaction of the body to the input of Ehrenstein's standard test program, viz. an oscillating tensile force and a variable temperature as shown in fig. 2. The calculated deformation resulting from that input is shown in the lower curve of fig. 10 and we conclude by comparison with fig. 2 that there is good qualitative agreement.

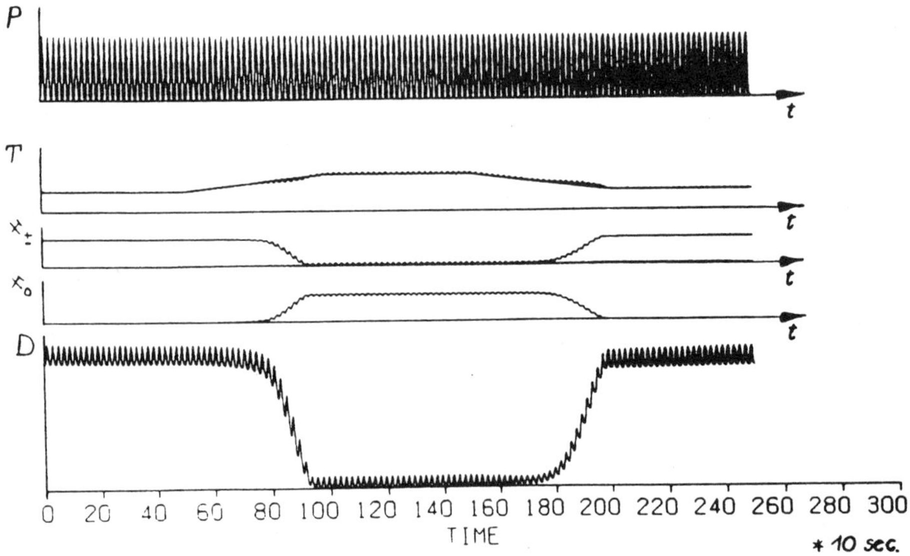

Fig. 10. Simulation of the standard test program (see fig. 2).

Of course, the numerical procedure can do more than the experiment. It can calculate the body temperature and the phase fractions and these are also listed in fig.10. The actual temperature differs from the external one by little spikes, that can be observed in the second curve of fig. 10. It turns out that x_{M_-} is practically always equal to zero, while x_{M_+} gives way to x_A when the temperature rises as is to be expected from the sharp decrease of $D(t)$. As the temperature decreases again, M_+ is reappearing and the body extends sharply.

In the simulation represented by fig. 10 the initial phase fractions were set at $x_{M_+} = 1$, $x_{M_-} = x_A = 0$ and this led to the initial large deformation. It is also

*) Note that the diagrams of fig. 10 are non-symmetric in tension and compression. This is due to the fact that the actual calculations have been done with a slightly more complex model that accounts for the rotation of layers in the deformation. That model has been described by Achenbach, Atanackovic and Müller (1985).

possible to start with the initial conditions $x_{M_+} = x_{M_-} = \frac{1}{2}$ and $x_A = 0$. Fig. 11 shows the resulting behaviour when the input is the same as in fig. 10 viz. an oscillating tensile load and an external temperature which first increases and then decreases. The curve $D(t)$ in fig. 11 must be compared to the corresponding curve of fig. 3 and we see that all qualitative features of those two curves are alike. In particular we see now that indeed the creep in the initial period is due to a slow conversion of M_- into M_+.

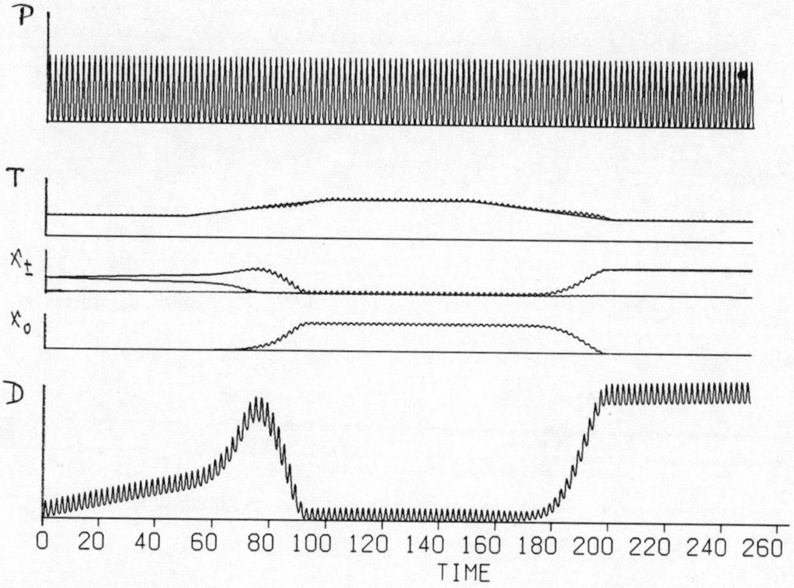

Fig. 11. Simulation of the standard test programm with both twins present initially (see fig. 3)

Fig. 12 shows a similar simulation as fig. 11 except that the frequency of the oscillating load is lower. An interesting feature occurring in the $T(t)$ -curve is clearly visible in that figure: As M_+ layers are converted into A layers upon heating the temperature sinks below the external temperature, because kinetic energy is converted into potential energy. Later when A-layers are converted back into M_+-layers T is bigger than T_E , because now potential energy is converted into kinetic energy, i.e. heat.

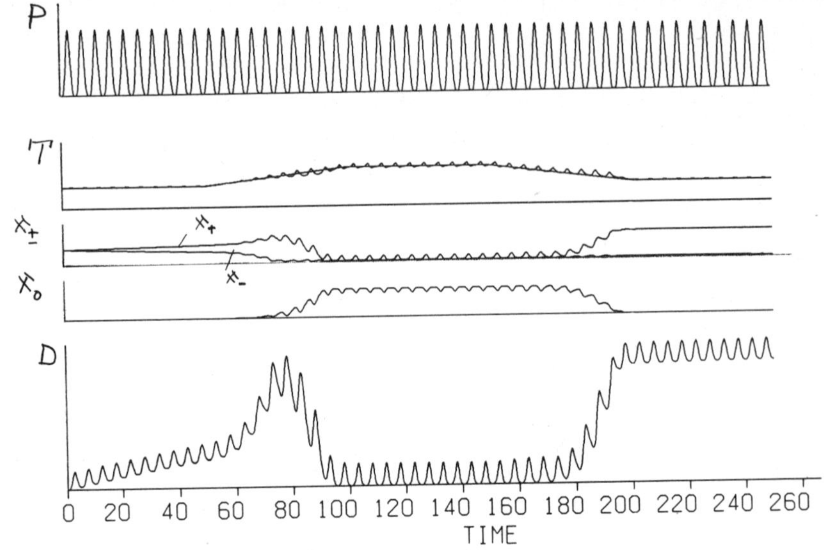

Fig. 12. Simulation of the standard test program at low frequency of load.

Another interesting aspect of the flow properties of the model is represented by fig. 13 which shows two cycles of tensile stress at a monotonically increasing external temperature. We see that, as long as the temperature is small, the load converts M_- into M_+ which persists even after unloading so that a residual deformation occurs. Then, as the temperature gets higher we still have M_+ as the load is big, but unloading will produce A which upon reloading is again converted into M_+.

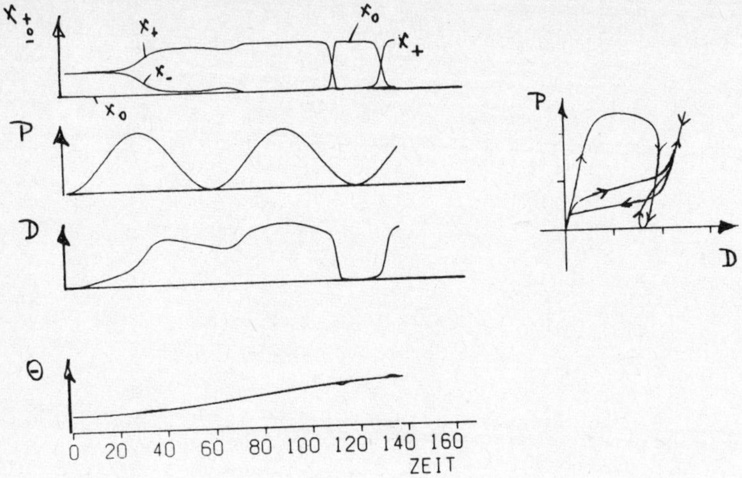

Fig. 13. Two tensile load cycles at increasing temperature

References

Achenbach, M., Atanackovic, T., Müller, I. (1985), "A Model for Memory Alloys in Phase Strains", Int. J. of Solids and Structures.

Achenbach, M., Müller, I. (1986), "Simulation of Material Behaviour of Alloys with Shape Memory", Archives of Mechanics, (in press)

Delaey, L., Chandrasekharan, L. (eds.) (1982) "Conf. on Martensitic Transformation, Leuven, J. de Physique 43

Müller, I., (1985), "Pseudoelasticity in Shape Memory Alloys - An Extreme Case of Thermoelasticity", IMA Preprint, Minneapolis, 1

Perkins, J., (1975), Shape Memory Effects in Alloys, Plenum Press, New York, London

Acknowledgement: The numerical calculations leading to the plots of the figures 9 through 13 have been performed by Mr. M. Achenbach, who has also been active in formulating and improving the model.

Session VI:

MISCELLANEOUS

A UNILATERAL MODEL TO THE EVALUATION OF THE COLLAPSE LOAD OF MASONRY SOLIDS.

M. COMO * and A. GRIMALDI **

* Istituto di Tecnica delle Costruzioni, University of Naples, Italy

** Department of Civil Engineering, II University of Rome, Italy.

ABSTRACT

In this paper the collapse of masonry solids is examined. For the masonry material a constitutive elastic model with zero tensile strength is assumed. The existence of solutions of the elastic equilibrium problem is analyzed and a definition of the collapse condition is provided. The corresponding kinematical and statical theorems, for the evaluation of the collapse load, are established.

1. INTRODUCTION

Very urgent is to day the request to give rational formulations to the problems of strength evaluation of masonry structures. Whole cities are masonry made and today is urgent the demand of knowledge about their preservation and safety. This request is producing scientific considerations involving research in Mechanics and Mathematics.

Aim of this paper is to establish a method for the collapse load evaluation of masonry solids. This problem is of worthy interest in the analysis of the behaviour of masonry buildings under earthquakes loadings.

To develop this analysis a crucial starting point is the choice of the constitutive equations of the masonry materials. It is well known, on the other hand, that they exhibit very low tensile strengths. Collapse of masonry walls under horizontal forces occurs with the formation of cracks due more to tensile failures than to crushings. It can be useful, therefore, to assume as constitutive model the elastic material with zero tensile strength. Actually this unilateral model has been examined in some theoretical and applied studies (1-8).

Having recalled the constitutive equations of the assumed model, this paper analyzes the problem of the elastic equilibrium of the masonry solid and provides a definition of the collapse condition. The corresponding kinematical and statical theorems are then established. They are very similar to the well known theorems of standard limit analysis of elasto-plastic structures.

This paper syntetizes the results given in (8) while a first development of the approach here presented was been given in (2) where the physical and technical aspects of the problem were mainly examined.

2. THE CONSTITUTIVE MODEL OF THE ELASTIC MASONRY MATERIAL WITH NO TENSILE STRENGTH

The assumed contitutive model is based on the following assumptions:

- tensile stresses are not allowed. Consequently, the principal values of the stress tensor σ are not positive. This condition will be referred to as

$$\sigma \leq 0 \tag{1}$$

- The deformation tensor is given by the sum of an elastic and an anelastic part, this last due to material cracking

$$\varepsilon = \varepsilon_e + \varepsilon_c \tag{2}$$

  where

$$\varepsilon_e = C^{-1}\sigma \tag{3}$$

  and C is the elastic matrix of the material.

- The cracking strains ε_c must produce always only dilatations of the material at any direction. This statement implies that the principal values of the cracking strain tensor ε_c will be always not negative

$$\varepsilon_c \geq 0 \tag{4}$$

- The normality condition

$$\sigma \cdot \varepsilon_c = 0 \tag{5}$$

is also assumed. The cracking strains will be zero along the compression principal directions of the stress tensor. This assumptions implies coaxiality between tensors σ and ε_c.
Hence we can write the following constitutive equations

$$\sigma = C\,(\varepsilon - \varepsilon_c) \tag{6_1}$$

$$\sigma \leq 0 \tag{6_2}$$

$$\varepsilon_c \geq 0 \tag{6_3}$$

$$\sigma \cdot \varepsilon_c = 0 \tag{6_4}$$

In the case of the uniaxial state, the assumed constitutive model yields the following stress-strain law

$$\sigma = \Theta \; E \; \varepsilon \tag{7}$$

where

$$\theta = \begin{cases} 0 & \varepsilon \geq 0 \\ 1 & \varepsilon < 0 \end{cases} \tag{8}$$

The assumed model therefore implies that the stress tensor σ is a single-valued function $\sigma\,(\varepsilon)$ of the strain tensor ε and defines a non linear elastic behaviour of the material (fig. 1).

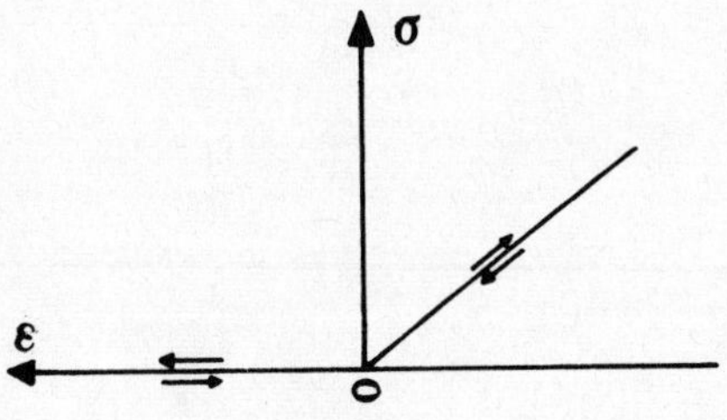

Fig. 1

This model is of course very different from the elasto-plastic scheme with zero tensile strength for which dilatations are not reversible.

3. ANALYSIS OF THE EXISTENCE PROBLEM OF ELASTIC EQUILIBRIUM SOLUTIONS

Let us consider a masonry solid loaded by vertical dead loads g and live loads λ q, increasing with a load parameter λ (fig. 2). For sake of simplicity we will assume that the loads act on the boundary of the region occupied by the masonry solid.

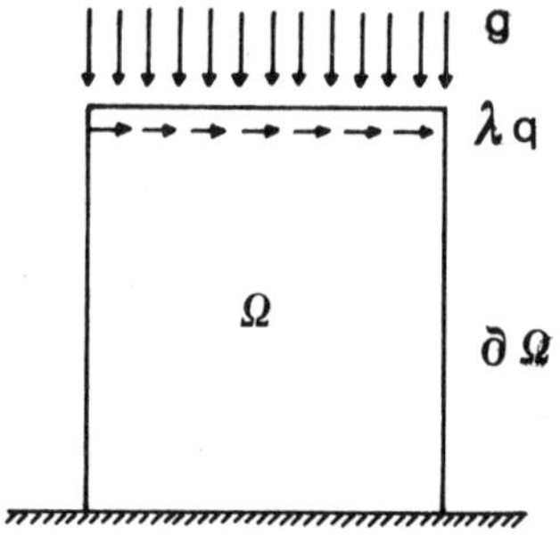

Fig. 2

Because of the assumed constitutive model, the existence of solutions of the elastic equilibrium of the masonry body is connected to the existence of stress fields σ, in equilibrium with the loads g + λ q and satisfying the admissibility condition (1). Consequently, a restriction condition for the external loads is required. For instance, in the case of fig. 2, solutions of the elastic equilibrium without the vertical compression loads g cannot exist.

- In absence of body forces the elastic solution has to satisfy the following equations:
- equilibrium:

$$\mathrm{Div}\ \sigma = 0 \tag{9}$$

- constitutive model:

$$\varepsilon = C^{-1}\sigma + \varepsilon_c \qquad \sigma \le 0$$

$$\varepsilon_c \ge 0 \qquad \sigma \cdot \varepsilon_c = 0 \tag{10}$$

- strain compatibility

$$\varepsilon = D\, u \tag{11}$$

where D is the differential operator that connects displacements and strains.

On the boundary of the masonry solid, displacements u must satisfy the constraint conditions while the stresses σ must be in equilibrium with the applied loads.

- The research of solution can be performed by using the variational approaches corresponding to the minimum principles of the total potential energy or of the complementary energy. The total potential energy functional of the loaded masonry solid is defined by the differentiable and convex functional

$$E\,(u, \varepsilon_c) = \tfrac{1}{2} \left\langle Du - \varepsilon_c, Du - \varepsilon_c \right\rangle_V - \tfrac{1}{2} \left\langle g + \lambda\, q, u \right\rangle_S \tag{12}$$

The first term represents the strain energy produced in the body by the elastic strains while the second one the potential energy of the surface external loads. The functional $E\,(u, \varepsilon_c)$ is defined, of course, in suitable functional spaces V and W of the variable u and ε_c. The solution of the elastic problem, i.e. the solution of eqs. (9, 10, 11), is equivalent to finding the minimum of the functional $E\,(u, \varepsilon_c)$ in the space of the displacements u and of the cracking strains $\varepsilon_c \geq 0$.

- The existence of the solution, as previously pointed out, is strictly connected to satisfying a compatibility condition for loads. Such a condition can be stated as

$$\left\langle g + \lambda\, q, u \right\rangle \leq 0 \qquad u \in M \tag{13}$$

where

$$M = \{u : Du \geq 0\} \tag{13'}$$

The set M defines strain fields which produce only dilatation of the material; i.e. displacement fields with zero strain energy. They will be defined as "<u>mechanisms</u>".

It is easy to prove that <u>condition</u> (13) <u>is necessary to the existen-</u>

ce of the solution.

This statement immediately comes out if we take into account that, for $u \in M$ and $\varepsilon_c = Du$, it is

$$E(u, \varepsilon_c) = -\langle g + \lambda q, u \rangle \tag{14}$$

Hence, when

$$\langle g + \lambda q, u \rangle > 0 \qquad u \in M \tag{15}$$

with

$$\bar{u} = \alpha u \qquad \alpha > 0 \tag{16}$$

we get

$$\lim_{\alpha \to \infty} E(\bar{u}, \varepsilon_c) = -\infty \tag{17}$$

which means that the functional $E(u, \varepsilon_c)$ cannot admit minimum.

No solution of the elastic equilibrium exists if, at least for one mechanism displacement u, condition (15) holds.

This result is also evident by observing that, in term of stresses, the solution σ has to be found in the set S of the admissible stress fields $\sigma \leq 0$ and in equilibrium with the applied loads $g + \lambda q$. The statically admissible set S is therefore defined as the set of the stress field σ which satisfies the relations

$$\langle \sigma, \delta \varepsilon \rangle = \langle g + \lambda q, \delta u \rangle \qquad \forall \delta u \in V \tag{18}$$

$$\sigma \leq 0 \tag{18'}$$

Hence, if a solution σ exists, of course in the set S, from condition (13') fur $\delta u \in M$ we have $\delta \varepsilon \geq 0$.

Thus, taking in account (18'), we get

$$\langle g + \lambda q, \delta u \rangle \leq 0 \qquad \forall \delta u \in M \tag{19}$$

A detailed analysis of the relation between the inequality (19) and the existence and uniqueness of the solution has been recently deveped by G. Romano and M. Romano (3).

The research of the elastic solution can be also worked out by using the minimum principle of the complementary energy

$$E^{*}(\sigma) = \tfrac{1}{2} \left\langle C^{-1} \sigma, \sigma \right\rangle_{V} \qquad (20)$$

Likewise to the linear elastic case it is possible to show that the research of the solution is equivalent to the evaluation of the statically admissible stress field σ which minimizes the functional $E^{*}(\sigma)$. Therefore another necessary condition to the existence yields.

The set S of the statically admissible stress fields cannot be empty.

The necessary condition of the existence of the solution of the elastic equilibrium of the loaded masonry solid is the existence of at least one admissible stress field σ in equilibrium with the applied loads. Because of the differentiability and strict convexity of $E^{*}(\sigma)$ this condition is also sufficient to the existence (3, 7). From this property we also get that the stress field si unique. However uniqueness of the stress does not imply, as a rule, uniqueness of the corresponding displacements and strains.

4. COLLAPSE CONDITIONS OF THE MASONRY SOLID UNDER DEAD AND LIVE LOADS

We will now examine the evolution of the existence problem along the loading process

$$\lambda q \qquad \lambda \geq 0 \qquad (21)$$

of the live forces q.

At $\lambda = 0$, i.e. for the masonry solid only loaded by dead forces g, we assume that the corresponding stress solution does exist and is unique. Consequently we have that

$$\langle g, u \rangle \leq 0 \qquad \forall u \in M \qquad (22)$$

Increasing the load multiplier λ, the existence condition (13) will be violated at a value λ_c. Then, beyond λ_c, the problem will not admit solution any more. In fact it is easy to recognize that if solution does'nt exist, for instance at $\lambda = \lambda_1$, solutions cannot exist for any $\lambda \geq \lambda_1$.

To prove this statement it is equivalent to show that if solution exist at $\lambda = \lambda'$, we have solutions for any λ such that $0 \leq \lambda \leq \lambda'$. In fact the stress field σ' solution at $\lambda = \lambda'$ satisfies the conditions:

$$\langle \sigma', \delta \varepsilon \rangle = \langle g, \delta u \rangle + \lambda' \langle q, \delta u \rangle \quad \forall \delta u \in V \tag{23}$$

$$\sigma' \leq 0$$

Similarly for the solution σ_g at $\lambda = 0$ we have

$$\langle \sigma_g, \delta \varepsilon \rangle = \langle g, \delta u \rangle \qquad \forall \delta u \in V \tag{23'}$$

$$\sigma_g \leq 0$$

Let us consider, for $0 \leq \lambda \leq \lambda'$, the stress field

$$\sigma(\lambda) = \sigma_g \left(1 - \frac{\lambda}{\lambda'}\right) + \frac{\lambda}{\lambda'} \sigma'$$

From eqs. (23) and (23') we get

$$\langle \sigma(\lambda), \delta \varepsilon \rangle = \langle g, \delta u \rangle + \lambda \langle q, \delta u \rangle \quad \forall \delta u \in V \tag{24}$$

and

$$\sigma(\lambda) \leq 0$$

Therefore, for any λ in the interval $0 \leq \lambda \leq \lambda'$, the set S of the statically admissible stress fields is not empty, and this is a sufficient condition for the existence of a solution.

Therefore it is of great importance the value λ_c of λ, defined as collapse multiplier of the distribution of live loads λq, that marks out the transition from existence to non existence conditions. Hence, the collapse multiplier λ_c defines the distribution of loading

$$g + \lambda q \tag{25}$$

where for $\lambda < \lambda_c$, equilibrium configurations exist, while for $\lambda > \lambda_c$ elastic equilibrium cannot be satisfied. The evaluation of λ_c is therefore centered in verifying existence conditions.

Finding the collapse multiplier λ_c for a masonry structure represents a problem of great importance because it provided its lateral ultimate strength. It can be performed according the kinematical and the statical approaches that reproduce the standard limit analysis of structures.

- The kinematical approach

We define as kinematical multiplier

$$\lambda^+ (u) \tag{26}$$

corresponding to the mechanism u, the value

$$\lambda^+ (u) = - \frac{\langle g, u \rangle}{\langle q, u \rangle}, \qquad u \in M \tag{27}$$

that marks the transition, from negative to positive values, of the work of external loads $g + \lambda q$ along the mechanism u.

For $\lambda > \lambda^+$ the necessary condition for the existence is violated along the mechanism u and

$$\lambda^+ (u) \geq \lambda_c \tag{28}$$

The solution of the elastic equilibrium does exist only when condition (13) holds for any mechanism. The collapse multiplier thus will be given by

$$\lambda_c = \inf_{u \in M} \lambda^+ (u) \tag{29}$$

- The statical approach

The stress solution σ exist iff the set S of the statically admissible stress fields is not empty. Thus we define statically admissible multiplier the value $\lambda^- (\sigma)$ of λ for which an admissible stress field σ, in equilibrium with the load $g + \lambda^- q$, does exist. As previously shown, if solution exist for $\lambda = \lambda^-$, solution will exist also for $\lambda < \lambda^{-}$. Hence

$$\lambda_c \geq \lambda^{-}(\sigma) \tag{30}$$

The collapse load multiplier λ_c can be therefore defined as

$$\lambda_c = \sup_{\sigma \in S} \lambda^{-}(\sigma) \tag{31}$$

The proposed approach has been successfully applied to the evaluation of the lateral strength of masonry panels and walls (2, 8).

REFERENCES

1. J. Heyman, "The stone skeleton" Int. Journ. of Solids and Structures, 2, 1966.

2. M. Como, A. Grimaldi, "Analisi limite di pareti murarie sotto spinta", Università di Napoli, Atti Ist. Tecnica delle Costruzioni n. 546, nov. 1983.

3. G. Romano, M. Romano, "Elastostatics of Structures with Unilateral Conditions on Strains and Displacements", Unilateral Problems in Structural Analysis, Ravello, sept. 22-24, 1983.

4. A. Baratta, R. Toscano, "Stati tensionali in pannelli di materiale non resistente a trazione", AIMETA, VI Congr. Naz.le Genova, 7-8-9 Ottobre 1982.

5. S. Di Pasquale, "Questioni di meccanica dei solidi non reagenti a trazione", AIMETA, VI Congr. Naz.le Genova, 7-8-9 Ottobre 1982.

6. P. Villaggio, "Stress Diffusion in Masonry Walls", Journ. Struct. Mech. 9 (4), 1981.

7. E. Giusti, M. Giaquinta, "Researches on the Statics of Masonry Structures", Atti Ist. Mat. U. Dini, Firenze, 1983/84-10.

8. M. Como, A. Grimaldi, "An unilateral model for the limit Analysis of Masonry Walls", Proc. Congr. "Unilateral Problems in Struct. Analysis" Ravello, Sept. 1983, Pubbl. CISM.

ON THE KORN TYPE INEQUALITY AND PROBLEM OF JUSTIFICATION OF REFINED THEORIES FOR ELASTIC PLATES

T.S. Vashakmadze
I.N. Vekua Institute of Applied Mathematics
Tbilisi University , Tbilisi, 380043, USSR

The paper deals to problems related with the justification of methods of reducing multidimensional problems to problems of less dimension on an example of boundary value problems of elasticity theory for bodies of small thickness for which the study of this problem, as is well known, is one of the major importance (see, e.g. [I - 3]). The material presented announces and generalizes a number of results, stated in [4-5] .
Let the initial three-dimensional problem of elasticity have the form (for convenience some notations 3-5 will be used):

$$\sigma_{ij,j} = f_i, \quad e_{ij} = \lambda \delta_{ij} \sigma_{kk} + 2\mu \sigma_{ij} = \frac{1}{2}(\partial_i u_j + \partial_j u_i), \quad x \in \Omega_h, \tag{1}$$

$$lu = g, \quad S = \partial \mathcal{D}(x_1, x_2) \times]-h, h[, \tag{2}$$

$$\sigma_3 - du = g^{\pm}, \quad S^{\pm} = \mathcal{D} \times \{\pm h\}. \tag{3}$$

At first we shall consider the two-dimensional problems, corresponding to the system (1)-(3) and constructed by the projective method of Vekua-Kantorovich. The first 2N+1 vector equations of this systems will be called Vekua's reduced model.
The following theorem holds:
Theorem. Let on the lateral boundary S the boundary conditions, corresponding to the problem (1)-(3), be homogeneous and such that the equalities:

$$(\overset{n}{u}_{\alpha,\beta}, \overset{m}{u}_3) = \int_{\mathcal{D}} \overset{n}{u}_{\alpha,\beta} \overset{m}{u}_3 \, dx_1 dx_2 = -(\overset{n}{u}_\alpha, \overset{m}{u}_{3,\beta}), \tag{4}$$

where $\overset{n}{u}_i$ are the unknown expansion coefficients of the function u_i , are fulfilled. Then the operator of the theory of plates, corresponding to Vekua's reduced model (V.r.m.) satisfies a Korn type inequality with the constant independent of N.
We use V.r.m. on the form (8) [6] . Multiplying this system by $\overset{n}{u}$, summing up from 0 to 2N+1 and carrying out some transformations, we obtain:

$$-(L_N U_N, U_N) \geq \left(-(\Delta U_N, U_N) + h^{-2}\|U_N\|_2^2\right). \qquad (5)$$

Here L_N is a linear operator, U_N is an approximate solution, corresponding to the reduced model,

$$\left(\overset{n}{u}, \overset{n}{v}\right)_1 = \frac{1}{2n+1}\left(\overset{n}{u}, \overset{n}{v}\right), \quad \left(\overset{n}{u}, \overset{n}{v}\right)_2 = (2n+1)\left(\sum_{i \geq n(2)} \overset{i+1}{u}, \sum_{i \geq n(2)} \overset{i+1}{v}\right),$$

$$\sum_{i \geq n(2)} \overset{i}{u} = \overset{n}{u} + \overset{n+2}{u} + \cdots .$$

Inequality (5) can be used for proving the existence and uniqueness of the genèralized solution of V.r.m. with the boundary conditions on ∂D, satisfying relation (4) and the positive definitness of the corresponding operator when the restrictions, known in the literature are imposed on the domain $D(x_1, x_2)$.
Moreover, by vitrue of the positive definiteness of V.r.m. operator and the approximating theorems [5] it is not difficult to estimate an error and to prove the convergence of the corresponding process in the Sobolev space $W_p^{\alpha+2}(\Omega_h)$ where $\alpha \geq 0$, $p \in (4/3, 4)$.
Though the above-stated holds for the case, when the body Ω_h is an anisotropic nonhomogenous plate or a shallow shell, but in applying it the serious difficulties arise even for (1) - (3), connected with the inversion of operator L_N.
Therefore we deduce a new r.m. which is based on the exact nonlinear representations (e.n.r.) [4-5]. The solution of such r.m. is reduced to the inversion of the simple structure operator N-times. N denotes the number of the pseudolayer and defines the accuracy of approximation of the problem (1)-(3) by the r.m.
Let $\Omega_H = D \times]0,H[$, $d=0$ $h_i = ih$ ($i=0,1,2,\ldots,N$), $Nh=H$, $N \geq 1$ and introduce values averaged with respect to the subinterval (h_i, h_{i+1}) in Reissner's sence, which correspond to beding state of plates : $u_\alpha^i(x_1, x_2)$, u_3^i, M_α^i, M_{12}^i, Q_α^i, R_α^i, $v^i = u^{i+1} - u^i$, $P^i = Q^{i+1} - Q^i$, $N^i = M^{i+1} - M^i$, $T^i = R^{i+1} - R^i$.
When $(x_1, x_2) \in D$ the functions $u_3^* = \sum_i u_3^i$, $Q = \sum_i Q^i$, v^i and P^i satisfies the following partial differential equations (suppose $g_\alpha^\pm = f_\alpha = 0$):

$$D\Delta^2 u_3^* = \left(1 - \frac{h^2(2-\nu)}{12(1-\nu)}\Delta\right)\left(g_3^+ - g_3^-\right) + \int_0^H f_3(x_1, x_2, t)\,dt -$$

$$- \frac{1}{1-\nu}\sum_{i=0}^{m-1}\int_{h_i}^{h_{i+1}}\left(\frac{1}{4}h^2 - \left(t - h_{i+\frac{1}{2}}\right)^2\right)\Delta f_3\,dt + \tau_1[u_3^*], \qquad (6)$$

$$Q_\alpha - \frac{h^2}{12}\Delta Q_\alpha = -D\Delta u^*_{3,\alpha} - \frac{h^2}{12(1-\nu)}\left(g^+_{3,\alpha} - g^-_{3,\alpha}\right) +$$

$$+\frac{1+\nu}{2(1-\nu)}\sum_{i=0}^{m-1}\int_{h_i}^{h_{i+1}}\left(\frac{1}{4}h^2-(t-h_{i+\frac{1}{2}})^2\right)f_{3,\alpha}dt + r_{1+\alpha}[Q_\alpha], \qquad (7)$$

$$D\Delta^2 v_3^i = \frac{-h}{2}\left(1-\frac{2-\nu}{1-\nu}\frac{h^2}{12}\Delta\right)P^i_{\alpha,\alpha} - \frac{h^3}{24(1-\nu)}\Delta\left[(2-\nu)(f_3^{i+1}-f_3^{i-1} -\right.$$

$$\left. - f_3^{i+1/2} + f_3^{i-1/2})\right] + r_4[v_3^i], \qquad (8)$$

$$P^i_\alpha - \frac{h^2}{12}\left(\Delta P^i_\alpha + \frac{h^2}{2(1-\nu)}P^i_{\beta,\alpha\beta}\right) = -D\Delta v^i_{3,\alpha} - \frac{h^3}{24(1-\nu)}\left(f^{i+1}_{3,\alpha} - f^{i-1}_{3,\alpha} -\right.$$

$$\left. - \frac{1+\nu}{2}\left(f^{i+1/2}_{3,\alpha} - f^{i-1/2}_{3,\alpha}\right)\right) + r_5[P^i_\alpha], \qquad (9)$$

where $D = Eh^3/12(1-\nu^2)$, r is an error members.

Equations (6)-(9) together with corresponding boundary conditions, defined by (2), constitute a closed system, from which it is possible to find functions u_3^* , Q_α , v_3^i , P_α^i . These values define u_α^i , M^i and $\sigma_{33}^i = \sigma_{33}(x_1, x_2, ih)$ by implicit formulae. Further if we introduce the functions $w_\alpha(x_1, x_2)$ -averaged with respect to the pseudolayers (h_{i-1} , h_{i+1}) characterazying tension of the plate, we shall have the following differential equations:

$$\mu\Delta w_+^i + (\lambda^* + \mu)\,\mathrm{grad\,div}\, w_+^i = \frac{\lambda}{2h(\lambda+2\mu)}\int_{h_{i-1}}^{h_{i+1}} \mathrm{grad}\,\sigma_{33}dt. \qquad (10)$$

The solution of (10) with corresponding boundary conditions defined by (2) gives the values $w_+ = (w_1, w_2)^T$, as solutions of the problem of generalized state plane stress.

Finally we consider the problem of justification of refined theories for elastic plates.

Starting from boundary value problem of nonlinear theory of elasticity (see f.e. [3]), the method is suggested free of hypotheses of physical and geometric character for constructing well-known (among them: von Karman equation, systems of differential equations of classical and refined theories, dynamic models of Rayleigh-Lamb...) models. The main equations in an isotropic case have the following form (for details see [5]) ;

$$D\Delta^2 u_3^* = h(g^+_{\alpha,\alpha} + g^-_{\alpha,\alpha}) + g_3^+ - g_3^- - \int_{-h}^{h} t f_{\alpha,\alpha} dt - \int_{-h}^{h} f_3 dt + \frac{1}{1-\nu}\int_{-h}^{h} (h^2 -$$

$$- t^2)\Delta f_3 dt + [u_3^*, F^*] - \frac{h^2(1+2\gamma)(2-\nu)}{3(1-\nu)}\Delta\left(g_3^+ - g_3^- + \frac{4h}{2-\nu}[u_3^*, F^*]\right) + r_4[u_3^*; \gamma], \qquad (11)$$

$$Q_\alpha - \frac{h^2(1+2\gamma)}{3}\Delta Q_\alpha = -D\Delta u^*_{3,\alpha} + \frac{h^2(1+2\gamma)}{3(1-\nu)}\partial_\alpha\left(g_3^+ - g_3^- + 2h(1+\nu)[u_3^*, F^*]\right) +$$

$$+ \frac{1+\nu}{2(1-\nu)}\int_{-h}^{h}(h^2 - t^2) f_{3,\alpha} dt + h(g_\alpha^+ + g_\alpha^-) - \int_{-h}^{h} t f_\alpha dt + r_{4+\alpha}[Q_\alpha; \gamma],$$

$$\Delta^2 F^* = -\frac{E}{2}[u_3^*, u_3^*] + \frac{\nu}{2}\Delta(g_3^+ - g_3^-) + r_7[F^*],$$

where

$$[u, v] = \partial_{11} u \partial_{22} v - 2\partial_{12} u \partial_{12} v + \partial_{11} v \partial_{22} u.$$

The rotations of the normals to the middle plane of the plate u_α^*, the bending and twising moments M_α, M_{12} satisfied the e.n.r., which depend implicity from deflection u_3^*, the intersecting forces Q_α, partial pressure $g_3^+ - g_3^-$ and also remainder terms. If e.n.r. are rewritten without r, then we obtain two-dimensional one-parameter system of nonlinear differential equations.

Let us make some conclusions, following from e.n.r.

1. Let the inicial boundary value problem is linear.

1.1. The remainder vector r depends only on the quatrature formulae of Simpson, trapezoids and Gauss (with one note). Then, as known, if functions σ_{i3} are sufficiently smooth, from e.n.r. an optimal scheme is the is the refined theory which corresponds when $\gamma = 0$, i.e. if we use Simpson rule.

I.2. When $\gamma = -0.5$ the e.n.r. give us the classical theory of plate bending and S.Germen-Lagrange biharmonic equation. Further, by an appropriate choice of γ from e.n.r. we obtain all-known refined theories(among them:the models of the Reissner, Hencky, Donell, Kromm and others that are important on the phisical sence).

2. Let us consider the nonlinear case.

2.I. Like in the linear case, the choice of the parameter γ leads with an accuracy of r to the equivalent models.

2.2. Let $\gamma = -0.5$, $g_\alpha = f_\alpha = 0$ then e.n.r. obtained von Karman equations.

2.3. When $\gamma = 0.1$ and there are some explicit simple fications, from e.n.r. follow the nonlinear equations von Karman-Reissner type without underlined summand in (11).

2.4. The statements of this part hold for the anisotropic nonhomogeneous case [5].

2.5. In the dynamic case using e.n.r. and assuming $f_3 = Z - \rho \partial_{tt} u\,(x,t)$, where $\partial_t = \partial / \partial t$, $x=(x_1, x_2, x_3)$ and ρ is the material density we lead only one with respect to deflection U^h from the main equations:

$$D(\Delta^2 + 2h\rho D^{-1}\partial_{tt} - \underline{2(1+\nu)E^{-1}\rho\partial_{tt}})U^h = g_3^+ - g_3^- - \frac{h^2(1+2\Upsilon)(2-\nu)}{3(1-\nu)}\Delta(g_3^+ -$$

$$- g_3^-) - \int_{-h}^{h}\left(1 - \frac{1}{1-\nu}\right)(h^2 - x_3^2)\,Z\,dx_3 .$$

From the reasoning section 1 [5] it follows that $2\rho(1-\nu)^{-1}$ is the exact coefficient of the underlined summand in the class of two-dimensional Rayleigh-Lamb models.

REFERENCES

1. Vorovich I.I.-Some mathematical problems of the theory of plates and shells. Proc. II USSR cong. in theor.appl.mech., III, 1965.
2. Vekua I.N.-Some general methods of constructing different versions of shell theory. M.Nauka, 1982.
3. Siarlet P., Rabier P. Von Karman equations, M. Mir, 1983.
4. Vashakmadze T.S.- To the construction of refined theories of anisotropic plates. Reports, Semin. I.Vekua Inst.Appl.Math., Tbilisi, 17, 1983, 18-23.
5. Vashakmadze T.S.-To the theory of plates. Reports, Semin. I.Vekua Inst. Appl.Math., Tbilisi, 18, 1984, 6-17.
6. Vashakmadze T.S. -On some numerical processes of solving linear problems of elasticity. Proc. IV All-Union Conf. Variational-Difference Meth. in Math. Physics, Novosibirsk, 1983, 5-13.

STRESS FUNCTIONS AND STRESS-FUNCTION SPACES FOR 3-DIMENSIONAL ELASTOSTATICS AND DYNAMICS

Sitiro MINAGAWA
Denkitsushin Daigaku - The University of Electro-Communications
Chofu, Tokyo 182, Japan

The investigation on the geometrical theory of stress functions was initiated by Schaefer [1]. The author re-examined his theory, attempting to establish the theory of stress-function spaces and its applications. The investigations were performed as a part of Kazuo Kondo's non-Riemannian plasticity theory [2]. The aim of this paper is to report some of the results of those investigations.

1. Riemannian stress-function space

Let χ_{ij} be the stress functions for a three-dimensional stress field, and put

$$\gamma_{ij}=\delta_{ij}+2\chi_{ij}, \tag{1}$$

where δ_{ij} is Kronecker's delta. Throughout this paper, i,j,k,... take 1, 2, 3 and Einstein's summation convention is used. The stress-function space is a Riemannian space having γ_{ij} as its fundamental metric tensor. Let Σ_{ij} and Σ be the Ricci tensor and the scalar curvature calculated from its Rieman-Christoffel curvature tensor. Then,

$$\sigma_{ij}=\Sigma_{ij}-\frac{1}{2}\gamma_{ij}\Sigma, \tag{2}$$

is compared with the stress tensor [1,3], because it satisfies automatically a non-divergent relation such as

$$\nabla_i\sigma_{ij}=0, \tag{3}$$

which is reduced to the equilibrium equation; viz.,

$$\partial_i\sigma_{ij}=0, \tag{4}$$

as far as small terms are neglected, where ∇_i is the covariant derivative with respect to χ_{ij}.

2. Non-Riemannian stress-function space

When concerning couple-stresses, we extend the theory into the geometry of non-Riemannian space [4,5]. Let $S_{ij}^{\cdot\cdot k}$ be the tortion tensor of a non-Riemannian space, then

$$M_{kj}^{\cdot\cdot i}=S_{kj}^{\cdot\cdot i}+2\delta^{i}_{[k}S_{j]}, \qquad Z^{ij}=\Sigma^{ij}-\frac{1}{2}\gamma^{ij}\Sigma, \tag{5}$$

are referred to the couple-stress and stress tensors, where $S_i = S_{ij}^{\cdot\cdot j}$.

This is confirmed as follows:- The Bianchi identity and the second identity are transformed into

$$\nabla_i Z_{\cdot m}^{\cdot i} = -S_{[mi}^{\cdot\cdot n} \Sigma_{k]nj}^{\cdot\cdot\cdot i} \gamma^{kj}, \quad \Sigma_{[ji]} = \nabla_m (S_{ji}^{\cdot\cdot m} + 2\delta_{[j}^{m} S_{i]}), \tag{6}$$

which are reduced to

$$\partial_j Z_{ij} = 0, \quad Z_{[ji]} = \partial_m M_{ij}^{\cdot\cdot m}, \tag{7}$$

as far as small terms are neglected.

The couple-stress field is also represented by Schouten's Q-tensor of a space of non-metric connection [6].

3. Duality principle

It is recognized that there exists a complete duality between the space of strain-incompatibility, called strain space, and the stress-function space such as follows:-

	strain space	stress-function space
metric tensor	strain	stress function
R.-C. curvature tensor	incompatibility disclination	stress
tortion tensor	dislocation	couple-stress
Bianchi identity The second identity	equations of continuity in the dislocation and disclination	equilibrium equation of stress and coupl-stress

It is expected to have a duality principle:- If we have an expression composed of the terms in the above table, and if we substitute each term by its corresponding term, we obtain the new expression, which has a certain physical meaning in most cases.

<u>Example</u>: The Peach-Koehler force is stated as $2s_{jni}Z_{ij}$, where s_{ijk} is the dislocation-density tensor. The provides the structure such as (dislocation)×(stress), which is converted into (couple-stress) ×(incompatibility). The author proved that the force exerted by the couple-stress field upon continuously distributed dislocations is stated as $M_{jni}\eta_{ij}$, η_{ij} being the incompatibility tensor [7]. This assumes the structure mentioned above.

4. Dual yielding

The duality principle leads the conception of the dual yielding. In Kondo's theory [8], the yielding of a 3-dimensional body is understood to be a phenomenon of instability such that, when the stress takes a certain critical value, the Riemannian strain space is curved out into an enveloping multi-dimensional Euclidean space. The dual conception is that, when the incompatibility takes a certain critical value, the Riemannian stress-function space is curved out into an enveloping multi-dimensional Euclidean space [9,10].

When the stress functions vary, the stresses do not alter the values, as far as the variations in the stress functions satisfy the gauge conditions. The gauge functions which escape the observation are compared with the genesis of the fracture, while the incompatibility with the atmosphare which is created through the process of fatigue. Therefore, the dual yielding can be compared with the process of fatigue fracture.

5. 4-dimensional stress-function space

The theory can be extended into the problem of elastodynamics [11]. In this case, the time appears as the fourth parameter such as $x_4=ict$, and the 4-stress tensor is given by

$$Z_{ij}=\sigma_{ij}+\rho v_i v_j, \quad Z_{i4}=Z_{4i}=-ic\rho v_i, \quad Z_{44}=c^2\rho, \tag{8}$$

where ρ is the mass density of the material, c the constant, and v_i is the velocity of displacement.

The 4-dimensional stress-function space is defined as a space having

$$\gamma_{\alpha\beta}=\delta_{\alpha\beta}+2\chi_{\alpha\beta}, \tag{9}$$

as the fundamental metric tensor. The 4-stress tensor is given in terms of its Einstein tensor such as

$$Z_{\alpha\beta}=\Sigma_{\alpha\beta}-\frac{1}{2}\gamma_{\alpha\beta}\Sigma, \tag{10}$$

where Greek indices take 1 to 4. Now, $\chi_{\alpha\beta}$'s play the role of the stress functions.

<u>Example</u>: Extended Kröner representation for a time dependent stress field is given by

$$Z_{ij}=2G[\Box\chi'_{ij}+\frac{1}{1-\nu}\{\partial_i\partial_j-\delta_{ij}(\Delta+\frac{1-\nu}{1-2\nu}\partial_4\partial_4)\}\chi'_{pp}], \tag{11}$$

where $\chi'_{\alpha\beta}$ is given through

$$\chi_{i\alpha}=2G[\chi'_{i\alpha}+\frac{\nu}{1-\nu}\delta_{\alpha i}\chi'_{pp}], \quad \chi_{44}=2G\chi'_{44},$$

satisfying the following subsidiary conditions,

$$\partial_\alpha\chi'_{i\alpha}-\frac{1}{2}\partial\chi'_{44}=0, \quad \frac{2\nu^2}{(1-2\nu)(1-\nu)}\partial_4\partial_4\chi'_{pp}+\partial_p\partial_4\chi'_{4p}-\frac{1}{2}\Delta\chi'_{44}=0. \tag{13}$$

From eqs. (11) to (13), one can obtain the expressions for the time-dependent stress field we have used in the analysis for the stress fields produced by moving dislocations [12].

References

[1] H. Schaefer, Die Spannungsfunktionen des dreidimensionalen Kontinuums und des elastischen Kropers, ZAMM, 33 (1953), 356
[2] K. Kondo, ed. RAAG memoirs of the unifying study of basic problems in engineering and physical sciences by means of geometry, I; II; III; IV, 1955; 1958; 1962; 1968, Gakujutsu Bunken Fukyukai, Tokyo
[3] S. Minagawa, Riemannian Three-Dimensional Stress-Function Space, RAAG Memoirs, 3(1962), 69
[4] R. Stojanovitch, Equilibrium conditions for internal stresses in non-Euclidean continua and stress space, Int. J. Engng. Sci. 1(1963), 323
[5] S. Minagawa, On Some Generalized Features of the Theory of Stress-Function Space with Reference to the Problem of Stresses in a Curved Membrane, RAAG Memoirs, 4(1968), 99
[6] S. Minagawa, Stress functions and stress-function spaces of non-metric connection for 3-dimensional micropolar elastostatics, Int. J. Engng. Sci. 12(1981), 1705
[7] S. Minagawa, On the Force Exerted upon Continuously Distributed Dislocations in a Cosserat Continuum, phys. stat. sol. 39(1970), 217
[8] K. Kondo, On the mathematical analysis of yield points, I; II; III, J. Jap. Soc. Appl. Mech. 3(1950), 188; 4(1951), 5; 35
[9] S. Minagawa and S.I.Amari, On the Dual Yielding and Related Problems, Mechanics of generalized continua,(Kroner, ed.), Springer, 1968, p.283
[10] S. Minagawa, Fatigue Limit Problem as Dual Yielding in Terms of Stress Functions, RAAG Memoirs, 4(1968), 162
[11] S. Minagawa, On the Stress Functions in Elastodynamics, Acta Mech. 24(1976), 209
[12] S. Minagawa and T. Nishida, A treatise on the stress-fields produced by moving dislocations, Int. J. Engng. Sci. 11(1973), 157

DISLOCATION DYNAMICS IN ANISOTROPIC THERMOELASTIC-PIEZOELECTRIC CRYSTALS

Sitiro MINAGAWA
Denkitsushin Daigaku - The University of Electro-Communications
Chofu, Tkyo 182, Japan

The aim of this paper is to report the results of the recent investigations on the piezoelectric and thermoelastic fields produced by a continuous distribution of moving dislocations in anisotropic crystals [1,2,3].

BASIC EQUATIONS

Using fixed rectangular Cartesian coordinate system x_i $(i=1,2,3)$, we state the basic equations of linear elastic thermo-piezoelectricity as follows:-

1. Divergence equations:

$$D_{i,i}=0, \quad \sigma_{ij,j}=\rho\ddot{u}_i, \quad \dot{q}-h_{i,i}=\rho C_v\dot{\theta}-T_o\lambda_{ij}\dot{u}_{j,i}, \tag{1}$$

where D_i is the electric displacement, σ_{ij} the stress tensor, ρ the mass density, u_i the displacement vector, q the heat generation per unit volume, h_i the heat flux density, C_v the specific heat at constant volume, θ the increment of temperature from the basic one (T_o), and λ_{ij} the tensor of thermal stress constant.

2. Gradient equations:

$$E_i=-\phi_{,i}, \quad u_{j,i}=\beta^E_{ij}+\beta^P_{ij}, \quad h_i=-\kappa_{ij}\theta_{,j}, \tag{2}$$

where E_i the electric intensity, β^E_{ij}, β^P_{ij} the elastic and plastic distortion tensors, and κ_{ij} the tensor of thermal conductivity.

3. Constitutive equations:

$$\sigma_{ij}=c_{ijpq}\beta^E_{pq}-c_{kij}E_k, \quad D_i=c_{ipq}\beta^E_{pq}+c_{ij}E_j \quad \text{(for PE-crystals)} \tag{3}$$

$$\sigma_{ij}=c_{ijpq}\beta^E_{pq}-\lambda_{ij}\theta, \quad \text{(for TE-crystals)} \tag{4}$$

where c_{ijpq} is the elastic constant tensor, c_{ijk} the piezoelectric tensor and c_{ij} the dielectric tensor.

4. Dislocation vs. plastic distortion relations:

$$\alpha_{ij}=-\varepsilon_{ipq}\beta^P_{qj,p}, \quad \dot{\beta}^P_{ij}=-\varepsilon_{imn}V_{mnj} \tag{5}$$

where α_{ij} is the dislocation density tensor, V_{ijk} the dislocation flux tensor, and ε_{ijk} Eddington's permutation symbol.

GENERAL SOLUTIONS

1. Piezoelectric fields

From eqs. (1) to (5),

$$c_{ijpq}u_{p,qj}-\rho\ddot{u}_i+c_{kij}\phi_{,kj}=c_{ijpq}\beta^P_{pq,j},$$
$$c_{jpq}u_{p,qj}-c_{jk}\phi_{,jk}=c_{jpq}\beta^P_{pq,j} \qquad (6)$$

which lead the solutions such as

$$u_m=-\iint(G^1_{mj}c_{jkpq}\beta^P_{pq,k'}+G^2_m c_{kpq}\beta^P_{pq,k'})dx'dt'$$
$$\phi=-\iint(F^1_j c_{jkpq}\beta^P_{pq,k'}+F^2 c_{kpq}\beta^P_{pq,k'})dx'dt' \qquad (7)$$

where G^1_{ij}, G^2_i $(=F^1_i)$, F^2 are Green's functions.

After a little rearrangement, we get the following equations,

$$\beta^E_{nm}=-\frac{\sqrt{-1}}{(2\pi)^4}\int\varepsilon_{skn}[\rho\omega\frac{D^1_{mq}}{D^1}V_{skq}+k_p(c_{jpkq}\frac{D^1_{jm}}{D^1}+c_{pkq}\frac{D^1_{4m}}{D^1})\alpha_{sq}]$$
$$\times\exp[\sqrt{-1}\{\vec{k}\cdot(\vec{x}-\vec{x}')-\omega(t-t')\}]d\vec{k}d\omega d\vec{x}'dt'$$

$$E_n=-\frac{\sqrt{-1}}{(2\pi)^4}\int\varepsilon_{skn}[\rho\omega\frac{D^1_{4q}}{D^1}V_{skq}+k_p(c_{jpkq}\frac{D^1_{j4}}{D^1}+c_{pkq}\frac{D^1_{44}}{D^1})\alpha_{sq}] \qquad (8)$$
$$\times\exp[\sqrt{-1}\{\vec{k}\cdot(\vec{x}-\vec{x}')-\omega(t-t')\}]d\vec{k}d\omega d\vec{x}'dt'$$

where k_i is the wave vector, ω the frequency, D^1 is the 4×4 determinant such as

$$\begin{vmatrix} c_{ipjq}k_pk_q-\rho\omega^2\delta_{ij} & c_{piq}k_pk_q \\ c_{pjq}k_pk_q & -c_{pq}k_pk_q \end{vmatrix}$$

and D^1_{IJ} $(I,J=1,\ldots,4)$ its IJ-cofactors.

2. Thermoelastic fields

Using

$$\frac{k_{11}}{c^2K_2},\quad \frac{k_{11}}{cK_2},\quad \frac{\lambda_{11}}{K_2},\quad \lambda_{11}$$

as the unit of time, displacement, temperature, and energy density, we get the following basic field equations,

$$\bar{c}_{ijpq}u_{q,pj}-\ddot{u}_i-\bar{\lambda}_{ij}\theta_{,j}=\bar{c}_{ijpq}\beta^P_{qp,j}$$
$$-\bar{\lambda}_{ij}\dot{u}_{i,j}+\bar{k}_{ij}\theta_{,ij}-\frac{\dot{\theta}}{\varepsilon_\theta}=-\dot{Q} \qquad (9)$$

where

$$k_{ij}=T_o^{-1}\kappa_{ij},\quad \alpha=T_o^{-1}\rho C_v,\quad Q=T_o^{-1}q,\quad K_2=\frac{\lambda^2_{11}}{\rho c^2},$$
$$\varepsilon_\theta=\frac{K_2}{\alpha},\quad \bar{c}_{ijpq}=\frac{c_{ijpq}}{\rho c^2},\quad \bar{\lambda}_{ij}=\frac{\lambda_{ij}}{\lambda_{11}},\quad \bar{k}_{ij}=\frac{k_{ij}}{k_{11}} \qquad (10)$$

c is the limiting velocity, and ε_θ the thermoelastic coupling coefficient. In the sequel, we use the dimentionless field quantities.

Eq. (9) leads the following solutions,

$$u_m = -\frac{\sqrt{-1}}{(2\pi)^4}\int\left(k_j \frac{D^2_{km}}{D^2}\bar{c}_{kjpq}\beta^P_{pq} + \varepsilon_\theta \frac{C_k}{D^2} Q\right) \times \exp[\sqrt{-1}\{\vec{k}\cdot(\vec{x}-\vec{x}') - \omega(t-t')\}]\, d\vec{k}\, d\omega\, d\vec{x}'\, dt'$$

$$\theta = -\frac{\varepsilon_\theta}{(2\pi)^4}\int\left(k_j \frac{C_k}{D^2}\bar{c}_{kjpq}\,{}^{P}_{pq} + \frac{D^2_{44}}{D^2} Q\right) \times \exp[\sqrt{-1}\{\vec{k}\cdot(\vec{x}-\vec{x}') - \omega(t-t')\}]\, d\vec{k}\, d\omega\, d\vec{x}'\, dt' \qquad (11)$$

where D^2 is the 4×4 determinant such as

$$\begin{vmatrix} \bar{c}_{ipjq}k_p k_q - \omega^2\delta_{ij} & \sqrt{\varepsilon_\theta}\,\bar{\lambda}_{ip}k_p \\ \sqrt{\varepsilon_\theta}\,\bar{\lambda}_{jp}k_p & -1-\varepsilon_\theta\frac{\sqrt{-1}}{\omega}\bar{k}_{pq}k_p k_q \end{vmatrix}$$

D^2_{IJ} its IJ-cofactors, and $C_k = D^2_{4k}/\sqrt{\varepsilon_\theta}$.

NUMERICAL EXAMPLE

We consider the case where an infinite straight dislocation is moved by a uniform subsonic velocity.

1. Piezoelectric field

The piezoelectric field in a garium arsenide is calculated. The dislocation is lying along the <112> orientation and moving in the <$\bar{1}$10> direction on the ($\bar{1}$11)-plane. Fig. 1 shows the variation of the longitudinal component of the electric field with the velocity of dislocation, where c_2 is the velocity of shear sound waves. Fig. 2 shows the contours of the constant tangential component for a didlocation moving with velocity $0.9c_2$. Contour values are $10^7 Vm^{-1}$ (outer) and $2\times10^7 vm^{-1}$(inner).

2. Reversible and irreversible thermal fields

The contour and zero lines of the reversible and irreversible thermal fields in copper are displayed in Figs. 3 and 4. In Fig. 3, the lines are plotted for θ=-1.0, 0.0, 1.0, and in Fig. 4 for $\theta/\sigma K_3$= 0.1, 1.0, 5.0, σ being the applied stress and $K_3 = K_1\lambda_{11}/T_o K_2$. The length of the Burgers vector (10^{-8}m) is taken as the unit of coordinates, and and the dislocation is placed at the origin of the coordinate system, moving into the direction of the abscissa. The irreversible generation of heat makes the temperature rise, while the reversible one c causes to rise the temperature in from the dislocation, and to lower its behaind, and the former is greater than the latter.

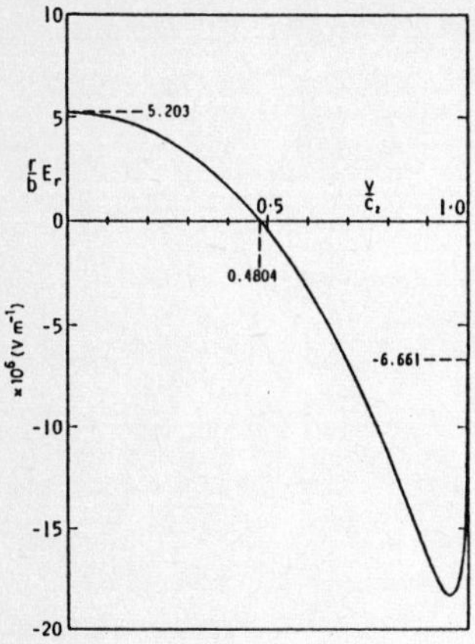

Fig. 1

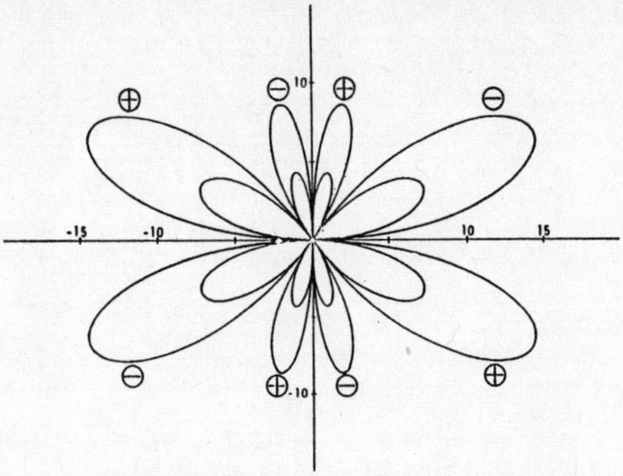

Fig. 2

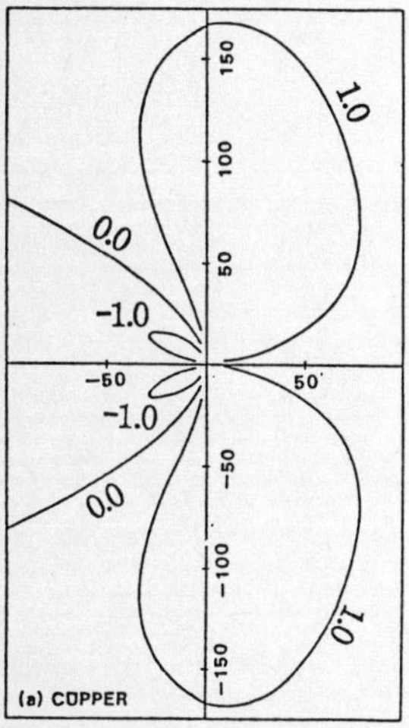

←Fig. 3

Fig. 4→

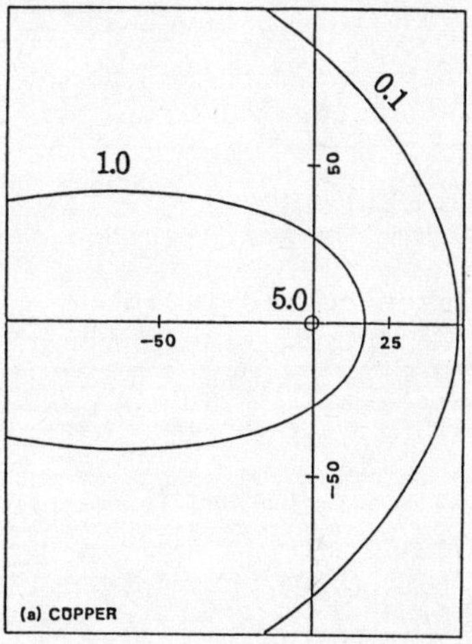

Acknowledgement: This work was supported by a Grant-in-Aid for fusion research from the Ministry of Education.

REFERENCES

[1] S. Minagawa, Mechanical, electrical and thermal fields produced by moving dislocations in thermo-piezoelectric continua, phys. stat. sol. (b), 124 (1984), 565

[2] S. Minagawa and K. Shintani, Dislocation dynamics in anisotropic piezoelectric crystals, Phil. Mag. A, 51 (1985), 277

[3] S. Minagwa, Reversible and irreversible generations of heat caused by moving dislocations in an elastic solid in a high temperature environment, to be printed

SOME RESULTS OF A BOUNDARY-LAYER THEORY FOR CURVED PHASE INTERFACES

Thorsten Alts

Hermann-Föttinger-Institut für Thermo- und Fluiddynamik
Technische Universität Berlin
Straße des 17. Juni 135
D - 1000 Berlin 12

ABSTRACT

Real phase interfaces are thin boundary-layers across which all bulk fields experience smooth though rapid changes when crossing from one phase to the other. This is modelled in a new boundary-layer theory for curved phase boundaries in non-equilibrium. Comparison with the theory of singular surfaces allows physical interpretation of the surface fields in terms of mean values of bulk fields, but it also requires satisfaction of dynamical consistency conditions for tangential momentum and surface stress. These yield new results for the curvature dependence of surface tension, for phase-change processes across the interface and for the dynamics of nucleation. A stability analysis proves the impossibility of certain nuclei.

An outline of the theory and some predictions on nuclei formation and on the phenomena of undercooling and superheating are presented.

1. Introduction

Real phase interfaces are thin boundary-layers across which all bulk fields experience smooth though rapid changes. Within the frame of statistical mechanics equilibrium properties of phase interfaces have been studied intensively[1]. Following the work of Gibbs[2] surface tension, adsorption, surface structure[3,4], electrostatic douple-layer[5,6] found theoretical and experimental attention[7,8]. It is now positive that the phase interface forms a smooth layer between adjacent bulk phases with a finite thickness of some molecular distances and own material properties.

Phase-boundary formation and transport of mass and heat across moving phase interfaces are non-equilibrium processes. Statistical theories for the description of these are not available. Progress in continuum physics, however, has originated a few papers on non-equilibrium properties of interfaces. One group of authors working in physical chemistry extended the ideas of Gibbs and modelled the phase boundary as a thin transition layer between the bulk phases[9,10,11]. Another group of authors coming from mathematics considered the interface as a mass, momentum and energy bearing singular surface[12,13,14,15,16]. Both groups of continuum theories allow the set up of balance laws for the motion of the interface. The transport properties of mass and heat across the interface, however, have not been considered at all or led to results which cannot be accepted for phase interfaces.

I compare the boundary-layer theory of phase interfaces with the singular-surface theory and find out that both theories are kinematically and dynamically consistend then and only then, if certain consistency conditions on the tangential motion and the normal shear stress are satisfied. Besides a physical interpretation of surface fields this comparison leads to new results for the curvature dependence of surface tension, for phase-change processes and the dynamics of nucleation.

A stability analysis of the boundary layer proves the impossibility of certain nuclei. Furthermore, the temperature dependences of surface tension and of boundary-layer thickness are determined explicitly and in accordance with observations.

Basic to the new boundary-layer theory is the concept of a curvature dependent "excentricity" between the mathematical surface of curvature and a physical surface of inertia within the layer. It emerges from the kinematic and dynamic consistency conditions and is quite general for all boundary layers of phase change of first order. It is a dynamical concept which cannot be obtained by equilibrium considerations, but it changes also equilibrium results. As a consequence, Gibbs' relation for surface entropy in equilibrium and some of its implications must be changed.

This short communication contains an outline of the basic ideas of this boundary-layer theory and a collection of some new results concerning surface tension, Clausius-Clapeyron's equation, nucleation, undercooling and superheating. For the details I refer to [17].

2. Phase Boundary as a Transition Layer

Physically the phase interface is a thin layer, across which all bulk fields change smoothly, though they may experience considerable changes over the thickness dimension of only a few molecular distances, see Fig. 1. The molecular structure and physical behaviour of matter inside this transition layer differ from those of the pure bulk phases.

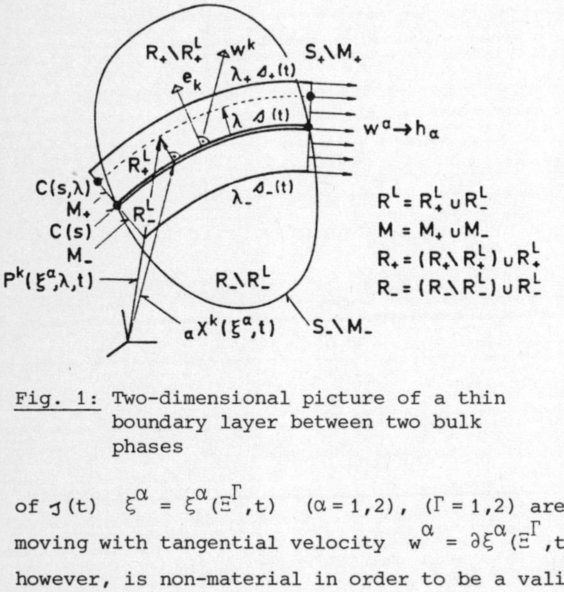

Fig. 1: Two-dimensional picture of a thin boundary layer between two bulk phases

The boundary layer may be devided by a non-material singular surface $\mathcal{S}(t)$ to which the physical properties of the layer may be attributed to. To this end the transition layer is split in a family of parallel surfaces

$$P^k(\xi^\alpha,\lambda,t) = {}_a\chi^k(\xi^\alpha,t) + \lambda e^k(\xi^\alpha,t)\,, \quad (2.1)$$

where ${}_a\chi^k(\xi^\alpha,t)$ is the motion of $\mathcal{S}(t)$ and $e^k(\xi^\alpha,t)$ its unit normal vector in Cartesian co-ordinates. The tangent vectors $\tau^k_\alpha = \partial\, {}_a\chi^k(\xi^\beta,t)/\partial\xi^\alpha$ and the normal vector form a right handed system. The actual surface co-ordinates of $\mathcal{S}(t)$ $\xi^\alpha = \xi^\alpha(\Xi^\Gamma,t)$ $(\alpha = 1,2)$, $(\Gamma = 1,2)$ are assumed to be material co-ordinates moving with tangential velocity $w^\alpha = \partial\xi^\alpha(\Xi^\Gamma,t)/\partial t$. The singular surface as a whole, however, is non-material in order to be a valid image of a phase boundary. It moves with a non-material normal velocity $w_n = e_k\, \partial\, {}_a\chi^k(\Xi^\alpha,t)/\partial t$. Metric, curvature and velocity of the singular surface are then given by

$$g_{\alpha\beta} = \tau^k_\alpha \tau_{\beta k} \ , \quad b_{\alpha\beta} = e_k \frac{\partial \tau^k_\alpha}{\partial \xi^\beta} \ , \quad w^k = \tau^k_\alpha w^\alpha + e^k w_n \ , \tag{2.2}$$

where summation convention is applied to diagonally repeated indices. Mean and Gaussian curvature are thus

$$K_M := \frac{1}{2} b^\alpha_\alpha = \frac{1}{2}(K_1 + K_2) \ , \quad K_G = \det\| b^\alpha_\beta \| = K_1 \cdot K_2 \ . \tag{2.3}$$

The general form of a balance law on the surface $\mathcal{J}(t)$ is

$$\frac{\partial \psi_{\mathcal{J}}}{\partial t} - 2K_M w_n \psi_{\mathcal{J}} + (\phi^\alpha_{\mathcal{J}} + \psi_{\mathcal{J}} w^\alpha)_{;\alpha} = - [\![\phi^k + \psi_v (v^k - w^k)]\!] e_k + \Pi_{\mathcal{J}} + \sigma_{\mathcal{J}} \ . \tag{2.4}$$

Here $\psi_{\mathcal{J}}(\xi^\beta,t)$ is a surface density per unit area on $\mathcal{J}(t)$. $\phi^\alpha_{\mathcal{J}}(\xi^\beta,t)$ is a non-convective influx within the tangent planes of $\mathcal{J}(t)$ per unit time and unit length of the material surface co-ordinates ξ^α and $\Pi_{\mathcal{J}}(\xi^\beta,t)$ and $\sigma_{\mathcal{J}}(\xi^\beta,t)$ are production density and supply densities within and to $\mathcal{J}(t)$ per unit area and unit time. Furtheron, ψ_v and ϕ^k denote the bulk density and non-convection bulk flux, where the flux is referred to a unit area in the bulk moving with the material bulk velocity v^k. Finally, the quantity $[\![H]\!] := H^+ - H^-$ defines the jump of adjacent bulk fields on the opposite banks of the singular surface. It accounts for the fact that mass, momentum and heat may be stuck to the phase interface.

These surface fields can be identified with mean values of bulk fields over the transition layer. This can be achieved by comparison of the balance law of the layer with the balance (2.4) of the singular surface. Extending $\psi^+_v := \psi_v(\xi^\alpha,\lambda^+,t)$ etc. into the part R^L_+ of the layer up the positive edge of $\mathcal{J}(t)$ and $\psi^-_v := \psi_v(\xi^\alpha,\lambda^-,t)$ etc. up the negative edge (c.p. Fig. 1), it can be shown that the surface fields are curvature weighted mean values over excess bulk fields within the layer, see [17]:

$$\begin{aligned}
\psi_{\mathcal{J}}(\xi^\alpha,t) &:= \int_{\lambda_-}^{\lambda_+} (\psi_v - \psi^\pm_v)(1 - 2\lambda K_M + \lambda^2 K_G)\,d\lambda \ , \\
\phi^\alpha_{\mathcal{J}}(\xi^\alpha,t) &:= \int_{\lambda_-}^{\lambda_+} \Big[(\phi^k - \phi^k_\pm) + (\psi_v v^k - \psi^\pm_v v^k_\pm) - w^k(\psi_v - \psi^\pm_v)\Big]\Big[\delta^\alpha_\beta - \lambda(2K_M\delta^\alpha_\beta - b^\alpha_\beta)\Big]\tau^\beta_k\,d\lambda \ , \\
\Pi_{\mathcal{J}}(\xi^\alpha,t) &:= \int_{\lambda_-}^{\lambda_+} (\Pi_v - \Pi^\pm_v)(1 - 2\lambda K_M + \lambda^2 K_G)\,d\lambda \ , \\
\sigma_{\mathcal{J}}(\xi^\alpha,t) &:= \int_{\lambda_-}^{\lambda_+} (\sigma_v - \sigma^\pm_v)(1 - 2\lambda K_M + \lambda^2 K_G)\,d\lambda \ .
\end{aligned} \tag{2.5}$$

Herein $\psi^\pm_v$ etc. are to be understood as follows

$$\psi^\pm_v = \begin{cases} \psi^+_v & \text{for} \quad \lambda > 0 \ , \\ \psi^-_v & \text{for} \quad \lambda < 0 \ , \end{cases} \qquad \text{etc.} \tag{2.6}$$

quantity	1 $\psi_{\mathcal{J}}$	2 $\phi^{\alpha}_{\mathcal{J}}$	3 $\pi_{\mathcal{J}}$	4 $\sigma_{\mathcal{J}}$	5 ψ_v	6 ϕ^m	7 Π_v	8 σ_v
mass	$\rho_{\mathcal{J}}$	0	0	0	ρ	0	0	0
momentum	$\rho_{\mathcal{J}} w^k$	$-t^{k\alpha}_{\mathcal{J}}$	0	$\rho_{\mathcal{J}} g^k$	ρv^k	$-t^{km}$	0	ρg^k
energy	$\rho_{\mathcal{J}}(u_{\mathcal{J}} + \frac{1}{2}\underline{w}^2)$	$q^{\alpha}_{\mathcal{J}} - w_k t^{k\alpha}_{\mathcal{J}}$	0	$\rho_{\mathcal{J}} w_k g^k + \rho_{\mathcal{J}} r_{\mathcal{J}}$	$\rho(u + \frac{1}{2}\underline{v}^2)$	$q^m - v_k t^{km}$	0	$\rho v_k g^k + \rho r$
entropy	$\rho_{\mathcal{J}} s_{\mathcal{J}}$	$\phi^{\alpha}_{\mathcal{J}}$	$\pi_{\mathcal{J}} \geqq 0$	$\sigma_{\mathcal{J}}$	ρs	ϕ^m	Π_v	σ_v

Table 1: Identification of fields for surface-balance equations and surface fields.

The special balance equations and surface fields for surface mass $\rho_{\mathcal{J}}$, momentum $\rho_{\mathcal{J}} w^k$, energy $\rho_{\mathcal{J}}(u_{\mathcal{J}} + \frac{1}{2}\underline{w}^2)$ and entropy $\rho_{\mathcal{J}} s_{\mathcal{J}}$ per unit area on $\mathcal{J}(t)$ can be obtained with the identifications of Table 1.

$q^{\alpha}_{\mathcal{J}}$ and $\phi^{\alpha}_{\mathcal{J}}$ are the heat flux and the entropy flux within $\mathcal{J}(t)$. The surface stress $t^{k\alpha}_{\mathcal{J}}$ can be decomposed according to

$$t^{k\alpha}_{\mathcal{J}} = \tau^k_{\beta} S^{\beta\alpha} + e^k S^{\alpha} , \tag{2.7}$$

where $S^{\beta\alpha}$ is the tensor of surface tension and S^{α} a normal surface shear. Scalar surface tension is then $\sigma = \frac{1}{2} S^{\alpha}_{\alpha}$. Phase interfaces should allow a membrane approximation, in which case

$$S^{\alpha\beta} = S^{\beta\alpha} , \quad S^{\alpha} = 0 \tag{2.8}$$

can be proven, see [17].

We decompose now the bulk velocity v^k, heat flux q^k etc. and Cauchy stress t^{kl} within the layer into tangential and normal components with respect to the singular surface:

$$\begin{aligned} v^k(\lambda) &= v^{\alpha}(\lambda)\tau^k_{\alpha} + v_n(\lambda)e^k , \quad \text{etc.} , \\ t^{kl}(\lambda) &= t^{\alpha\beta}(\lambda)\tau^k_{\alpha}\tau^l_{\beta} + t^{\alpha}_{\cdot n}(\lambda)\tau^k_{\alpha}e^l + t_n^{\cdot\alpha} e^k \tau^l_{\alpha} + t_n(\lambda)e^k e^l \end{aligned} \tag{2.9}$$

and insert the identifications of Table 1 into (2.5). The results are

i. <u>for surface mass density and surface mass flux:</u>

$$\rho_{\mathcal{J}} = \int_{\lambda_-}^{\lambda_+} (\rho - \rho^{\pm})(1 - 2\lambda K_M + \lambda^2 K_G)d\lambda , \tag{2.10}$$

$$0 = \int_{\lambda_-}^{\lambda_+} (\rho v^{\beta} - \rho^{\pm} v^{\beta}_{\pm}) \left[\delta^{\alpha}_{\beta} - \lambda(2K_M \delta^{\alpha}_{\beta} - b^{\alpha}_{\beta})\right] d\lambda - w^{\beta} \int_{\lambda_-}^{\lambda_+} (\rho - \rho^{\pm}) \left[\delta^{\alpha}_{\beta} - \lambda(2K_M \delta^{\alpha}_{\beta} - b^{\alpha}_{\beta})\right] d\lambda ; \tag{2.11}$$

ii. <u>for surface momentum density and surface stress:</u>

$$\rho_{\mathcal{J}} w^{\alpha} = \int_{\lambda_-}^{\lambda_+} (\rho v^{\alpha} - \overset{\pm}{\rho} v_{\pm}^{\alpha})(1 - 2\lambda K_M + \lambda^2 K_G)\, d\lambda \,,$$

$$\rho_{\mathcal{J}} w_n = \int_{\lambda_-}^{\lambda_+} (\rho v_n - \overset{\pm}{\rho} \overset{\pm}{v}_n)(1 - 2K_M + \lambda^2 K_G)\, d\lambda \,; \tag{2.12}$$

$$s^{\alpha\beta} = \int_{\lambda_-}^{\lambda_+} \Big\{ (t^{\alpha\gamma} - t_{\pm}^{\alpha\gamma}) - \Big[\rho (v^{\alpha} - w^{\alpha})(v^{\gamma} - w^{\gamma}) - \overset{\pm}{\rho} (v_{\pm}^{\alpha} - w^{\alpha})(v_{\pm}^{\gamma} - w^{\gamma}) \Big] \Big\} \Big[\delta_{\gamma}^{\beta} - \lambda (2K_M \delta_{\gamma}^{\beta} - b_{\gamma}^{\beta}) \Big] d\lambda \,,$$

$$s^{\alpha} = 0 = \int_{\lambda_-}^{\lambda_+} \Big\{ (t_n^{\cdot\gamma} - t_{n\pm}^{\cdot\gamma}) - \Big[\rho (v_n - w_n)(v^{\gamma} - w^{\gamma}) - \overset{\pm}{\rho} (\overset{\pm}{v}_n - w_n)(v_{\pm}^{\gamma} - w^{\gamma}) \Big] \Big\} \Big[\delta_{\gamma}^{\alpha} - \lambda (2K_M \delta_{\gamma}^{\alpha} - b_{\gamma}^{\alpha}) \Big] d\lambda \,; \tag{2.13}$$

iii. <u>for surface (internal) energy density and heat flux:</u>

$$\rho_{\mathcal{J}} u_{\mathcal{J}} = \int_{\lambda_-}^{\lambda_+} \Big\{ \rho \Big[u + \tfrac{1}{2} (\underset{\sim}{v} - \underset{\sim}{w})^2 \Big] - \overset{\pm}{\rho} \Big[\overset{\pm}{u} + \tfrac{1}{2} (\overset{\pm}{\underset{\sim}{v}} - \underset{\sim}{w})^2 \Big] \Big\} (1 - 2\lambda K_M + \lambda^2 K_G)\, d\lambda \,, \tag{2.14}$$

$$\begin{aligned} q_{\mathcal{J}}^{\alpha} = \int_{\lambda_-}^{\lambda_+} \Big\{ & (q^{\gamma} - q_{\pm}^{\gamma}) - \Big[(v_{\beta} - w_{\beta}) t^{\beta\gamma} - (\overset{\pm}{v}_{\beta} - w_{\beta}) t_{\pm}^{\beta\gamma} \Big] - \Big[(v_n - w_n) t_n^{\cdot\gamma} - (\overset{\pm}{v}_n - w_n) t_{n\pm}^{\cdot\gamma} \Big] \\ & + \Big[\rho \Big(u + \tfrac{1}{2} (\underset{\sim}{v} - \underset{\sim}{w})^2 \Big) (v^{\gamma} - w^{\gamma}) - \overset{\pm}{\rho} \Big(\overset{\pm}{u} + \tfrac{1}{2} (\overset{\pm}{\underset{\sim}{v}} - \underset{\sim}{w})^2 \Big) (v_{\pm}^{\gamma} - w^{\gamma}) \Big] \Big\} \times \\ & \times \Big[\delta_{\gamma}^{\alpha} - \lambda (2K_M \delta_{\gamma}^{\alpha} - b_{\gamma}^{\alpha}) \Big] d\lambda \,. \end{aligned} \tag{2.15}$$

Corresponding relations emerge for surface entropy, entropy flux, supply and production densities.

The membrane approximation (2.8) is physically motivated by the very small thickness of the transition layer and conservation of moment of momentum. Accordingly, the relations (2.13) contain three constraints on the possible profiles of Cauchy's stress within the phase-transition layer. These are called <u>dynamical consistency conditions</u>.

Another set of constraints is included in (2.11) and $(2.12)_1$. Both must give the same tangential velocity w^{α} of the singular surface. This is a consequence of the definition of moving materials lines (with velocity w^{α}) within the singular surface $\mathcal{J}(t)$, whose introduction is <u>necessary</u> in order that the concept of surface tension is physically significant. Elimination of w^{α} between (2.11) and $(2.12)_1$ yields thus the constraints

$$b^{\beta}_{\alpha}I^{\alpha} - K_G\delta^{\beta}_{\alpha}II^{\alpha} + K_G(2K_M\delta^{\beta}_{\alpha} - b^{\beta}_{\alpha})III^{\alpha} = 0 \qquad (\beta = 1,2) \ , \tag{2.16}$$

where I^{α}, II^{α} and III^{α} are abbreviations for

$$\begin{aligned} I^{\alpha} &:= \left[\int_{\lambda_-}^{\lambda_+} \varphi^{\alpha} d\lambda\right]\left[\int_{\lambda_-}^{\lambda_+} R\lambda d\lambda\right] - \left[\int_{\lambda_-}^{\lambda_+} \varphi^{\alpha}\lambda d\lambda\right]\left[\int_{\lambda_-}^{\lambda_+} R d\lambda\right] , \\ II^{\alpha} &= \left[\int_{\lambda_-}^{\lambda_+} \varphi^{\alpha} d\lambda\right]\left[\int_{\lambda_-}^{\lambda_+} R\lambda^2 d\lambda\right] - \left[\int_{\lambda_-}^{\lambda_+} \varphi^{\alpha}\lambda^2 d\lambda\right]\left[\int_{\lambda_-}^{\lambda_+} R d\lambda\right] , \\ III^{\alpha} &:= \left[\int_{\lambda_-}^{\lambda_+} \varphi^{\alpha}\lambda d\lambda\right]\left[\int_{\lambda_-}^{\lambda_+} R\lambda^2 d\lambda\right] - \left[\int_{\lambda_-}^{\lambda_+} \varphi^{\alpha}\lambda^2 d\lambda\right]\left[\int_{\lambda_-}^{\lambda_+} R\lambda d\lambda\right] , \end{aligned} \tag{2.17}$$

with

$$\varphi^{\alpha}(\lambda) := \rho(\lambda)v^{\alpha}(\lambda) - \rho^{\pm}v^{\alpha}_{\pm} \ , \quad R(\lambda) := \rho(\lambda) - \rho^{\pm} \ . \tag{2.18}$$

The relations (2.16) are constraints on the possible profiles $\rho(\lambda)$ and $v^{\alpha}(\lambda)$ of bulk density and bulk velocity within the layer; we call them the *kinematic consistency conditions*.

For flat phase interfaces relations (2.16) are trivially satisfied. For curved interfaces, however, they form two conditions on λ_+ and λ_- and fix the positions of $\mathcal{I}_+$ and $\mathcal{I}_-$ relative to the singular surface $\mathcal{I}(t)$ at $\lambda = 0$. Equivalent parameters are the thickness $d_{\mathcal{I}} := \lambda_+ - \lambda_-$ of the layer and $a_{\mathcal{I}} := \lambda_+ + \lambda_-$, a measure of "excentricity" between the mathematical surface of curvature and the physical surface of inertia within the layer. Given the position of the singular surface $\mathcal{I}$ (at $\lambda = 0$), its curvature, the distributions $\rho(\lambda)$ and $v^{\alpha}(\lambda)$ of density and tangential velocity within the layer and their values $\rho^{\pm}$ and $v^{\alpha}_{\pm}$ in the adjacent bulk phases, the thickness $d_{\mathcal{I}}$ and the excentricity $a_{\mathcal{I}}$ emerge as functions of the curvature and functionals of the distributions $\rho(\lambda)$ and $v^{\alpha}(\lambda)$. Because the actual distributions $\rho(\lambda)$ and $v^{\alpha}(\lambda)$ across the phase boundary layer are not known, they can be chosen such as to make $d_{\mathcal{I}}$ independent of curvature and $a_{\mathcal{I}}$ depending on the curvature only. This guarantees then that the tangential velocity w^{α} of the singular surface satisfies (2.11) *and* $(2.12)_1$. Under these conditions only does the singular surface represent a valid continuum physical image of the layer of phase transition.

3. Cubic Profiles

The phase-boundary layer between a solid and its fluid or a fluid and its vapour has constitutive properties of a non-simple material which are completely different from the properties of pure bulk materials. For instance the layer between a crystal and its melt must match between the anisotropic structure on the solid side and the isotropic one at the fluid side with a continuous loss of anisotropy over the thickness of the layer. Extremely large density gradients, velocity gradients and (possibly) temperature gradients in the normal direction may occur. Neither thermodynamic continuum theories exist nor are statistical approaches available that would embrace these peculiarities of phase-transition layers. To get a first insight we assume therefore cubic profiles for all bulk fields within the layer:

$$f(\xi^\alpha,\lambda,t) = f_0(\xi^\alpha,t) + f_1(\xi^\alpha,t)\cdot\lambda + f_2(\xi^\alpha,t)\cdot\lambda^3 \ . \tag{3.1}$$

$f_1 = (\partial f/\partial\lambda)\big|_{\lambda=0}$ is the normal slope at the location of $\mathcal{J}(t)$ and f_0 and f_2 are determined by the boundary conditions $f(\xi^\alpha,\lambda_\pm,t) = f^\pm(\xi^\alpha,t)$ at the sites of the adjacent bulk phases. These are the simplest profiles, with which the dynamic and kinematic consistency conditions can be fulfilled. Evaluation of (2.10) - (2.15) is tedious routine work, details of which may be found in [17]. Here only some results are cited:

The constraints (2.16) can be satisfied, if the slopes ρ_1 and v_1^α of the density and the tangential velocity profiles are proportional to the corresponding jumps

$$\rho_1 = \frac{X}{d_{\mathcal{J}}}[\![\rho]\!] \ , \qquad v_1^\alpha = \frac{X}{d_{\mathcal{J}}}[\![v^\alpha]\!] \tag{3.2}$$

with the same factor of proportionality. The identities (2.16) reduce then to two non-linear equations for the dimensionless factor X and the dimensionless excentricity $\alpha := a_{\mathcal{J}}/d_{\mathcal{J}}$ which require that

$$\alpha = \alpha(k_M,k_G) \ , \qquad X = X(k_M,k_G) \tag{3.3}$$

are functions of the dimensionless curvatures $k_M := K_M d_{\mathcal{J}}$ and $k_G := K_G d_{\mathcal{J}}^2$. Numerical solutions show, that within the physically permissible interval $-1 \leqq \alpha \leqq 1$ exactly two families, $\alpha_1(k_M,k_G) < 0$ and $\alpha_2(k_M,k_G) > 0$, exist; see Fig. 2.

Using these results, the surface density and tangential velocity follow from (2.10) and $(2.12)_1$ and are given by

$$\rho_{\mathcal{J}} = [\![\rho]\!]\ \frac{d_{\mathcal{J}}}{2}\ F_1(k_M,k_G) \ , \tag{3.4}$$

$$\rho_{\mathcal{J}} w^\alpha = [\![\rho v^\alpha]\!]\ \frac{d_{\mathcal{J}}}{2}\ F_1(k_M,k_G) + [\![\rho]\!]\cdot[\![v^\alpha]\!]\ \frac{d_{\mathcal{J}}}{6}\ F_2(k_M,k_G) \ . \tag{3.5}$$

The dimensionless functions F_1 and F_2 are explicitly known and are plotted in Fig. 2 for spherical geometry, $k_M^2 = k_G$.

Let us interpret (3.4) first. F_1 can be positive or negative. Its sign must be chosen such that $\rho_{\mathcal{J}} > 0$. Hence, two cases may be distinguished, which, for spherical geometries, are explained in Fig. 3. [Recall that the +side of the interfaces is always that side into which the surface normal vector points.] For an inclusion of the denser in the less dense phase (Case A: water in ice) the positive family $1 \geqq \alpha_2(k_M,k_G) > 0$ of excentricities must be chosen. Since for this family (2.16) possess solutions *for all* curvatures, inclusion of the denser phase in the

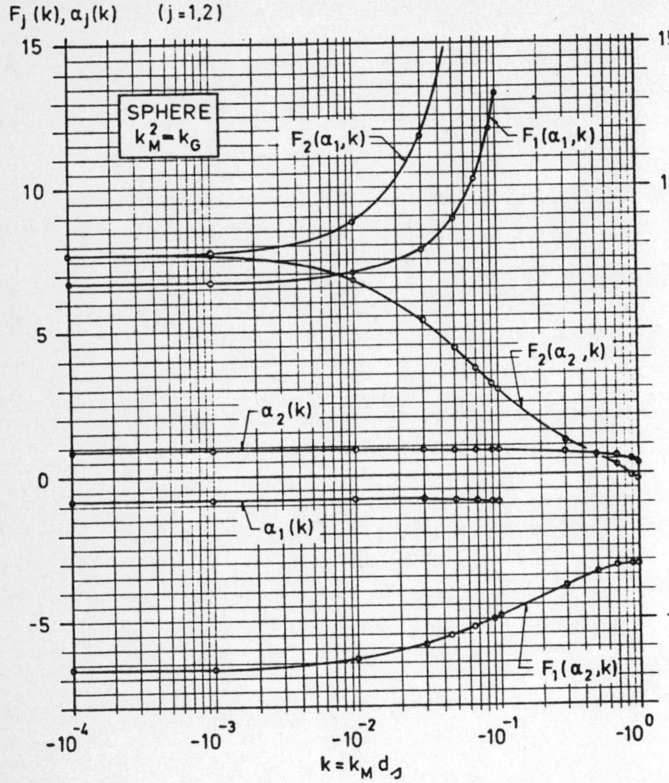

Fig. 2:
Solutions of equation (2.16) for spherical geometry, $K_M^2 = K_G$. Plotted on semi-logarithmic scale is $\alpha_1 < 0$ and $\alpha_2 > 0$ as a function of $k = K_M d_\vartheta$. For $|k| > 0.1$ one has $|\alpha_1| > 1$ which is unphysical and therefore not shown. Displayed are also the values of the functions F_1 and F_2 arising in (3.4) and (3.5). Dependence on curvature is significant only for $|k| > 10^{-3}$.

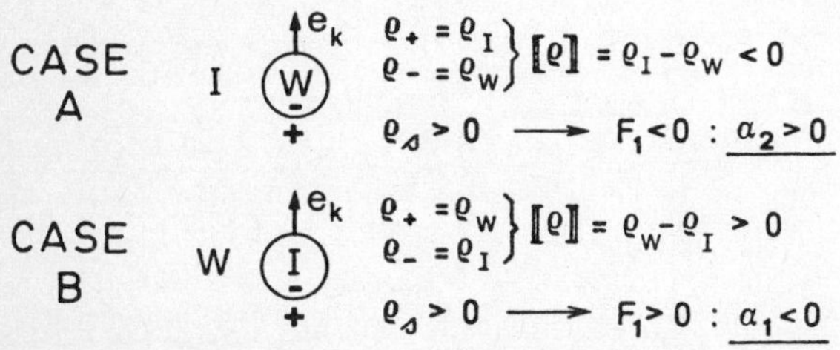

Fig. 3: Relative positions of the denser phase (water = W) and the less dense phase (ice = I) and choice of sign of the dimensionless excentricity α and function F_1.

less dense one (water in ice) do exist down to nucleation dimensions. The situation is different, when inclusions of the less dense phase in the dense one are considered (Case B: ice in water). In this case the negative family $-1 \leqq \alpha_1(k_M, k_G) < 0$ must be selected. Since for this family (2.16) possess no physically meaningful solution ($|\alpha_1| < 1$) when $|k_M| \geqq 0.1$, we conclude that very small inclusions of the less denser phase in the dense one (ice in water) cannot exist at nucleation dimensions. Ice nucleation in steam, however, is possible (case A is applicable!). From this we infer, that nucleation of ice from pure water must start from the water surface. This, in fact, is observed in nature. The non-existence of ice-nucleation in water may also explain the phenomenon of undercooling in clean water. Ice formation in water, however, is possible, when nucleation kernels are present with dimensions that are larger than ten times the boundary layer thickness, or $|k_M| \leqq 0.1$.

Measurements at the flat ice/water interface, Golecki & Jaccard[8], infer a thickness of $d_{\blacktriangleleft} = 100$ Å at the normal freezing point. Assuming the same magnitude for the water/vapour interface we obtain $\rho_{\blacktriangleleft} \cong 2.81 \cdot 10^{-7}\ \mathrm{g/cm^2}$ (Ice/Water) and $\rho_{\blacktriangleleft} \cong 3.35 \cdot 10^{-6}\ \mathrm{g/cm^2}$ (Water/Vapour). These are small numbers, which make the mass contribution at a single phase interface negligible. However, in two phase mixtures as in the mushy zone of temperate glaciers, or in dense fogs or clouds, the area of internal phase boundary per unit volume is considerable (up to $10^5\ \mathrm{cm^2/cm^3}$), and therefore the mass contribution of the phase interfaces to the total mass of the mixture cannot be ignored.

The boundary layer model is also valid for the fluid-vapour phase transitions. In this case the critical point can be reached, at which the phase boundary must disappear, since the phases are no longer distinguishable. This property is included in the foregoing results, for with $[\![\rho]\!] \to 0$ also $\rho_{\blacktriangleleft} \to 0$ for all curvatures: the surface field $\rho_{\blacktriangleleft}$ vanishes when the phase boundary layer disappears. All other surface fields must have similar behaviour at the critical point.

A brief discussion of formula (3.5) for surface momentum may be added. For adherence of the bulk material, viz $v_+^\alpha = v_-^\alpha$, we deduce $w^\alpha = v_+^\alpha = v_-^\alpha$: the tangential velocity of the interface equals the material velocities of the adjacent bulk materials. For $v_+^\alpha \neq v_-^\alpha$ the surface velocity w^α differs from both.

More complex are the results for surface tension, heat flux and internal energy. Neglecting all convective contributions in (2.13) - (2.15) and satisfying the membrane approximation we obtain for cubic profiles, c.p.[17]:

$$
\begin{aligned}
s^{\alpha\beta} - s^{\alpha\beta}\Big|_E &= - [\![t^{\gamma\delta} + p_E g^{\gamma\delta}]\!] \frac{d_{\blacktriangleleft}}{2} \left\{ G_1 \delta_\gamma^\alpha \delta_\delta^\beta + \frac{d_{\blacktriangleleft}}{4} G_2 \delta_\gamma^{(\alpha} b_\delta^{\beta)} + \frac{d_{\blacktriangleleft}^2}{16} G_3 b_\gamma^\alpha b_\delta^\beta \right\} , \\
q_{\blacktriangleleft}^\alpha &= - [\![q^\beta]\!] \frac{d_{\blacktriangleleft}}{2} \left\{ J_1 \delta_\beta^\alpha + \frac{d_{\blacktriangleleft}}{4} J_2 b_\beta^\alpha \right\} , \\
\rho_{\blacktriangleleft} u_{\blacktriangleleft} &= [\![\rho u]\!] \frac{d_{\blacktriangleleft}}{2} F_1 + [\![\rho]\!] \cdot [\![u]\!] \frac{d_{\blacktriangleleft}}{6} F_2 .
\end{aligned}
\tag{3.6}
$$

Corresponding relations hold for surface entropy and entropy flux. G_1, G_2, G_3, J_1, J_2 are dimensionless functions of the dimensionless curvatures k_M, k_G. Equilibrium surface tension $s^{\alpha\beta}\Big|_E = \sigma_E g^{\alpha\beta}$ satisfies the relations

$$
\begin{aligned}
\sigma_E &= \left[[\![p_E]\!] - (\rho_{\blacktriangleleft} e^k)\Big|_E g_k \right] \cdot \frac{d_{\blacktriangleleft}}{2} \frac{1}{k_M} , \\
\sigma_{E,\alpha} &= - (\rho_{\blacktriangleleft} \tau_\alpha^k)\Big|_E g_k .
\end{aligned}
\tag{3.7}
$$

$(3.7)_1$ is Laplace's formula in the gravitational field.

Common structure in all of these equations is, that the surface fields depend linearily on jumps of corresponding bulk fields, on the thickness of the boundary layer and on mean and Gaußian curvature.

4. Thermodynamic and Thermostatic Results

Evaluation of a generalization of Müller's entropy principle for phase interfaces leads to important restrictions. Some of them are [17]:

α. Non-equilibrium

Surface entropy $s_{\mathcal{I}} = s_{\mathcal{I}}(T_{\mathcal{I}}, \rho_{\mathcal{I}})$, internal energy $u_{\mathcal{I}} = u_{\mathcal{I}}(T_{\mathcal{I}}, \rho_{\mathcal{I}})$ and scalar surface tension $\sigma = \sigma(T_{\mathcal{I}}, \rho_{\mathcal{I}})$ are only functions of surface temperature $T_{\mathcal{I}}$ and density $\rho_{\mathcal{I}}$, and are related by a generalization of Gibbs' equation

$$ds_{\mathcal{I}} = \frac{1}{T_{\mathcal{I}}}\left(du_{\mathcal{I}} + \frac{\sigma}{\rho_{\mathcal{I}}^2}\, d\rho_{\mathcal{I}}\right) . \tag{4.1}$$

Surface entropy flux $\phi_{\mathcal{I}}^{\alpha}$ and heat flux $q_{\mathcal{I}}^{\alpha}$ satisfy the relation

$$\phi_{\mathcal{I}}^{\alpha} = \frac{1}{T_{\mathcal{I}}}\, q_{\mathcal{I}}^{\alpha} . \tag{4.2}$$

In classical theory of surface tension σ equals the density $\rho_{\mathcal{I}}(u_{\mathcal{I}} - T_{\mathcal{I}} s_{\mathcal{I}})$ of free energy. This does not follow from (4.1) for phase interfaces. (4.1) is thus a generalization of Gibbs' result with important new consequences for phase interfaces.

β. Equilibrium

Phase equilibrium requires equality of temperatures $T^+ = T^- = T_{\mathcal{I}}$ and of the specific free enthalpies $g_E^{\pm}$ and $g_{\mathcal{I}}^E$ of the adjacent bulk materials and the interface:

$$g^+(T_{\mathcal{I}}, p_E^+) = g^-(T_{\mathcal{I}}, p_E^-) = g_{\mathcal{I}}(T_{\mathcal{I}}, \sigma_E) , \tag{4.3}$$

where $p_E^{\pm}$ are the hydrostatic pressures of the adjacent bulk materials.

Mechanical equilibrium in the absence of gravitation requires in addition

$$p^+ - p^- = 2\sigma_E \cdot K_M , \qquad \sigma_{E,\alpha} = 0 . \tag{4.4}$$

(4.3) and $(4.4)_1$ are 4 equations for the determination of 5 fields $T_{\mathcal{I}}$, p_E^+, p_E^-, σ_E and K_M. Given $T_{\mathcal{I}}$ and K_M the fields p_E^+, p_E^- and σ_E emerge as functions of temperature and mean curvature

$$p_E^+ = p_E^+(T_{\mathcal{I}}, K_M) , \qquad p_E^- = p_E^-(T_{\mathcal{I}}, K_M) , \qquad \sigma_E = \sigma_E(T_{\mathcal{I}}, K_M) . \tag{4.5}$$

The same holds thus for the adjacent bulk densities $\rho_E^{\pm}$, internal energies $u_E^{\pm}$ and entropies $S_E^{\pm}$ and for surface energy $u_{\mathcal{I}}$ and entropy $s_{\mathcal{I}}$; all are functions of temperature and mean curvature.

The differential forms of the free enthalpies are

$$\begin{aligned} dg_E^{\pm} &= - s_E^{\pm} dT_{\mathcal{I}} + \frac{1}{\rho_E^{\pm}}\, dp_E^{\pm} , \\ dg_{\mathcal{I}}^E &= - s_{\mathcal{I}}^E dT_{\mathcal{I}} - \frac{1}{\rho_{\mathcal{I}}^E}\, d\sigma_E . \end{aligned} \tag{4.6}$$

γ. Change of curvature at fixed temperature

Neglecting gravity implies by $(4.4)_2$ that the mean curvature is constant along the equilibrium phase boundary. Isotropic phase boundaries that separate isotropic bulk phase are thus either spheres ($K_G = K_M^2$), circular cylinders ($K_G = 0$) or planes ($K_G = 0$, $K_M = 0$).

Differentiating the equilibrium conditions (4.3) and $(4.4)_1$ with respect to K_M, using (4.6) and the boundary layer result (3.4) yields after integration

$$\frac{\sigma_E}{\sigma_E^0} = \exp\left[-\int_0^{k_M} \frac{F_1(k)}{1 + kF_1(k)}\, dk\right] , \qquad (4.7)$$

where $\sigma_E^0 = \sigma_E(T_\Delta, k_M = 0)$ is the surface tension for the flat interface and where $F_1(k_M)$ equals either $F_1(k_M, k_G = k_M^2)$ for the sphere or $F_1(k_M, k_G = 0)$ for the circular cylinder.

Equation (4.7) exhibits a distinct and very interesting curvature dependence of the surface tension. For spherical geometry the result of numerical integration with the values for $F_1(k_M)$ from Fig. 2 is drawn in Fig. 4. Due to the different signs of $F_1(k_M)$ the surface tension for a water*) inclusion in ice*) increases with increasing curvature $|k_M|$ (case A); for an ice inclusion in water (case B), however, it decreases with increasing curvature and vanishes when $|k_M| = 0.09$. Vanishing surface tension, however, corresponds physically to disappearance of the phase boundary. We conclude, that ice inclusions in water of nucleation dimensions $|k_M| \gtrsim 0.09$ do not exist. This confirms our earlier deductions about the unlikelihood of ice nucleation in pure water.

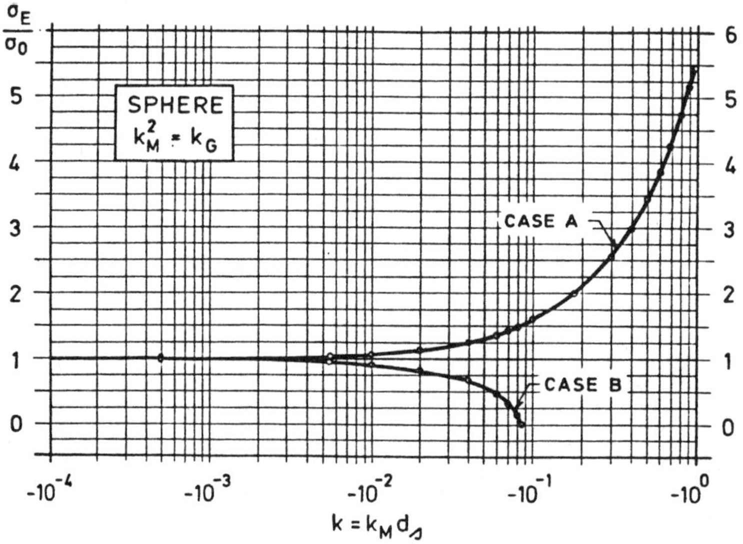

Fig. 4: Normalized surface tension σ_E/σ_E^0 as function of curvature according to (4.7) for spherical geometry. Case A: Inclusion of the denser in the less dense phase. Case B: Inclusion of the less dense in the denser phase.

*) The terms "water, ice or vapour" stand for the dense and the less dense phases. They are used to shorten the subsequent description in the following text and to combine it with a visualization. The results of this chapter, however, are independent of special choices of bulk materials.

δ. Change of temperature at fixed curvature

Differentiating (4.3) and $(4.4)_1$ with respect to T_Δ at fixed K_M yields using (4.6)

$$\left[\!\left[\frac{1}{\rho_E}\right]\!\right] \frac{\partial p_E^+}{\partial T}\Big|_{K_M} = [\![s_E]\!] - \frac{1}{\rho_E^-} \frac{\partial \sigma_E}{\partial T}\Big|_{K_M} \cdot 2K_M \,,$$
$$\left[\!\left[\frac{1}{\rho_E}\right]\!\right] \frac{\partial p_E^-}{\partial T}\Big|_{K_M} = [\![s_E]\!] - \frac{1}{\rho_E^+} \frac{\partial \sigma_E}{\partial T}\Big|_{K_M} \cdot 2K_M \,. \tag{4.8}$$

In the limit $K_M \to 0$, $s_E^\pm \to s_0^\pm$, $\rho_E^\pm \to \rho_0^\pm$, $p_E^\pm \to p_E^0$ relations (4.8) reduce to Clausius-Clapeyron equation

$$\left[\!\left[\frac{1}{\rho_E^0}\right]\!\right] \frac{dp_E^0}{dT} = [\![s_E^0]\!] \,. \tag{4.9}$$

For this reason (4.8) are generalized Clausius-Clapeyron equations. For the flat phase boundary $T\,[\![s_E^0]\!]$ is the latent heat of melting or evaporation. For the curved phase boundary, however, the quantity $T\,[\![s_E]\!]$ does not have this meaning. Details may be found in [17].

5. Thermostatic Stability

From the entropy inequality, the balance laws for mass, momentum and energy and the equilibrium conditions it can be proven that the *local thermostatic stability condition for phase boundaries* takes on the form

$$(g_\Delta - g_\Delta^E) + s_\Delta(T_\Delta - T_\Delta^E) + \frac{1}{\rho_\Delta}(\sigma - \sigma_E) \geq 0 \,, \tag{5.1}$$

where $g_\Delta = g_\Delta(T_\Delta, \sigma)$, $s_\Delta = s_\Delta(T_\Delta, \sigma)$, $\rho_\Delta = \rho_\Delta(T_\Delta, \sigma)$ are the free surface enthalpy, entropy and density in any virtual non-equilibrium state T_Δ, σ of a motionless phase boundary. The proof of (5.1) is akin to corresponding derivations for simple bulk materials, see [19], and may be found in [17].

Evaluation of the consequences of (5.1) is possible by transforming to the independent variables T_Δ and K_M. What emerges for the curved ice/water interphase are the following stability conditions:

(i) Conditions on geometry:

Case A: a spherical water inclusion in ice is *stable* for $0 \leq |k_M| \leq 1$.

Case B: a spherical ice inclusion in water is *stable* for $0 \leq |k_M| \leq 0.086$, unstable for $|k_M| \geq 0.086$. (5.2)

This proves the earlier statements on existence or non-existence of ice or water nuclei in the respective other phase.

(ii) Condition on boundary-layer thickness:

The phase interface is stable then and only then, if the temperature dependent thickness d_Δ satisfies

$$d_{\triangleleft}(T_{\triangleleft}) = d_{\triangleleft}(T_R) \cdot \frac{|[\![\rho_0(T_R)]\!]|}{|[\![\rho_0(T_{\triangleleft})]\!]|}\,, \tag{5.3}$$

where T_R (= 273 K) is some reference temperature. The boundary-layer thickness is inversely proportional to the absolute value of the density jump at the flat surface. Measurements of Golecki & Jaccard[8] indicate, that $d_{\triangleleft}(T_{\triangleleft})$ is decreasing with decreasing temperature below the normal freezing point. This means that $|[\![\rho_0(T_R)]\!]|/|[\![\rho_0(T_{\triangleleft})]\!]| \leqq 1$ whenever $T_{\triangleleft} \leqq T_R = 273$ K. This conclusion can be checked via density measurements in undercooled water, however, those measurements are not known to us.

(iii) Condition on surface tension:

Stability of the phase interface requires

$$\frac{d^2\sigma_0}{dT_{\triangleleft}^2} = 0\,. \tag{5.4}$$

Hence, surface tension of flat interfaces between ice and water must be a linear function of absolute temperature. This is confirmed by observations, see Hobbs[7].

The conclusion (5.2) about stable and unstable inclusions of ice in water provides a new understanding of nucleation phenomena.

Remark: Results (ii) and (iii) are restricted to the ice/water interphase, whereas (i) is also valid for the liquid/vapour transition layer.

REFERENCES

1. S. Ono & S. Kondo: "Molecular Theory of Surface Tension in Liquids", Handbuch der Physik, Bd. 10, p. 134 (Ed. S. Flügge), Berlin, Göttingen, Heidelberg 1960
2. J.W. Gibbs: "The Collected Work of J. Willard Gibbs", Vol. 1, Yale University Press, New Haven 1928
3. C.A. Croxton: "Statistical Mechanics of the Liquid Surface", John Wiley & Sons, Chichester, New York, Brisbane, Toronto 1980
4. N.H. Fletscher: Philos. Magazine, Eigth Ser., 18, 1287 (1966)
5. W.A. Weyl: J. Colloid Sci. 6, 389 (1951)
6. S.L. Carnie & G.M. Torrie: "The Statistical Mechanics of the Electrical Double Layer", Advances in Chemical Physics, Vol. LVI, 141 (1984)
7. P.V. Hobbs: "Ice Physics", Clarendon Press, Oxford 1974
8. I. Golecki & C. Jaccard: Physics Letters 63A, 374 (1977)
9. F.P. Buff: J. Chem. Phys. 25, 146 (1956)
10. F.P. Buff & H. Saltsburg: J. Chem. Phys. 26, 23, 1526 (1957)
11. A.D. Deemer & J.C. Slattery: Int. J. Multiphase Flow 4, 171 (1978)
12. L.E. Scriven: Chem. Engng. Sci. 12, 98 (1960)
13. G.P. Moeckel: Arch. Rat. Mech. Anal. 57, 255 (1974)
14. M.E. Gurtin & A.I. Murdoch: Arch. Rat. Mech. Anal. 57, 291 (1975)
15. K.A. Lindsay & B. Straughan: Arch. Rat. Mech. Anal. 71, 307 (1979)
16. A. Grauel. Physica 103 A, 468 (1980)
17. T. Alts & K. Hutter: "Towards a Theory of Temperate Glaciers, Part I: Dynamics and Thermodynamics of Phase Boundaries between Ice and Water". Mitteilungen der Versuchsanstalt für Wasserbau, Hydrologie und Glaziologie, ETH Zürich, 1986
18. I. Müller: "Thermodynamik - Die Grundlagen der Materialtheorie", Bertelsmann Universitätsverlag, Düsseldorf 1973
19. M.E. Gurtin: Arch. Rat. Mech. Anal. 59, 63 (1975)

SPECTRUM AND PERIODICITY FOR 0,1-FUNCTIONS

W. Möhring
Max-Planck-Institut für Strömungsforschung
Bunsenstr. 10
D 3400 Göttingen, Fed. Rep. of Germany

Introduction

For continuous (or L^2) functions there is a very close relation between their spectrum and their periodicity properties. The Fourier theorem states that a periodic function has a line spectrum with equidistant lines (an overtone series) and that a function which has equidistant spectral lines is periodic. The Fourier theorem shows also that a function which is a superposition of periodic functions with different periods has several series of equidistant spectral lines and conversely that a function which has several series of equidistant spectral lines is a superposition of several functions having different periods. The decomposition of such a function is however not unique if the ratios of the frequencies of the spectral lines are rational. Then some lines can be considered to belong to several series with a corresponding uncertainty in the periodic parts of the original function. We will assume that all frequency ratios are rational and that the functions considered are periodic with some period which is a multiple of the individual periods.

In some applications [1] one is interested in 0,1-functions (functions assuming only the values 0 and 1 in intervalls) which are, considered as L^2-functions, superpositions of periodic functions of different period lengths, have therefore several overtone series. To determine such functions one might assume that they are not only a superposition of periodic functions in L^2 but that they are also a superposition of periodic 0,1-functions. In [1] it was shown how such functions can be determined using only elementary concepts from number theory [2]. It was also shown that not all functions having several overtone series are actually a superposition of periodic 0,1-functions. A counterexample, based on Ramanujan sums [3], having three overtone series was mentioned. The purpose of this note is to show that the superposition principle used in [1] is valid for 0,1-functions having only two overtone series and that a slightly weaker principle is valid in the general case. This shows that all functions having only two overtone series can be determined by the elementary method of [1].

Spectra and generating polynomicals

Let us assume that the 0,1-function is of period N and that its discontinuities occur at integer values (Fig. 1). One can then describe f in terms of unit pulses φ_N of period

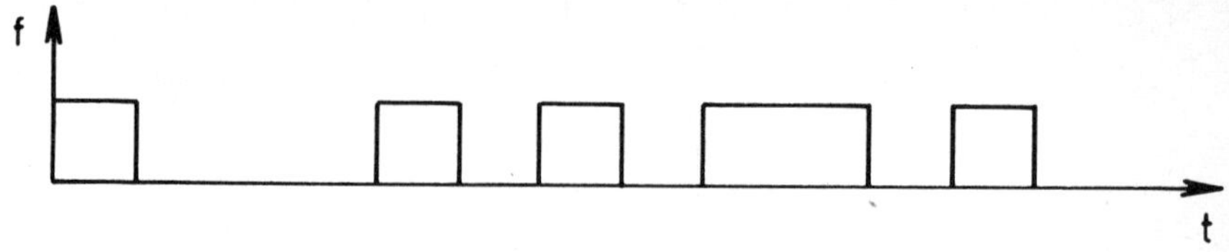

Fig. 1 A 0,1-function

N defined as

$$\varphi_N(t) = \begin{cases} 1 & 0 < t < 1 \\ 0 & 1 < t < N \end{cases}$$

$$f = \sum_{n \in M} \varphi_N(t - n) \tag{1}$$

where the set M contains those integers where a unit pulse begins. It is also possible to represent functions of different heights if some integers are included several times in M.

For the Fourier transform of f one obtains with (1)

$$\int_0^N f(t)\, e^{-2\pi i \frac{st}{N}}\, dt = A_s\, \alpha_s$$

with

$$A_s = \frac{N\, e^{-\frac{\pi i s}{N}}}{\pi s} \sin \pi \frac{s}{N}, \quad \alpha_s = \sum_{n \in M} e^{-2\pi i \frac{sn}{N}} \tag{2}$$

The A_s are different from zero for s = 0,1..., N - 1. The essential information of the spectrum is contained in the α_s. They are periodic with period length N, $\alpha_s = \alpha_{s+N}$. Some given f(t) is a function of frequency p (i.e. period $\frac{N}{p}$) if the α_s are nonzero at most for those s which are multiples of p. It is a superposition of three functions of frequencies p, q, r if $\alpha_s \neq 0$ at most for s which are multiples of p or q or r.

Now we consider the generating polynomial

$$F(x) = \prod_{n \in M} (x - e^{2\pi i \frac{n}{N}}) . \tag{3}$$

One observes that addition in eq. (1) corresponds to multiplication in eq. (3). If f is a superposition of the function f_1 and f_2 the polynomial F is a product of the associated polynomials F_1 and F_2. One observes also that all zeros of F are of first order if f is a 0,1-function. If M contains one integer twice, i.e. f assumes the value two, then F(x) has a second order zero at the corresponding position.

To relate the generating polynomial to the spectrum one forms its logarithmic derivative

$$\frac{F'(x)}{F(x)} = \sum_{n \in M} \frac{1}{x - e^{2\pi i \frac{n}{N}}} \tag{4}$$

Its power series expansion leads to

$$\frac{F'(x)}{F(x)} = - \sum_{s=0}^{\infty} \alpha_{s+1}\, x^s = - \frac{1}{1 - x^N} \sum_{s=1}^{N} \alpha_s\, x^{s-1} \tag{5}$$

where the periodicity of α_s has been taken into account.

Let us now consider a situation where f(t) contains only three overtone series with frequencies p, q, r which divide N. Then the α_s vanish for all s which are not multiples of p or q or r. This implies that the right hand side of eq. (5) admits a partial fraction decomposition

$$\frac{F'}{F} = \sum_{n=0}^{\frac{N}{p}-1} \frac{p\, b_n\, x^{p-1}}{x^p - e^{2\pi i \frac{np}{N}}} + \sum_{n=0}^{\frac{N}{q}-1} \frac{q\, c_n\, x^{q-1}}{x^q - e^{2\pi i \frac{nq}{N}}} + \sum_{n=0}^{\frac{N}{r}-1} \frac{r\, d_n\, x^{r-1}}{x^r - e^{2\pi i \frac{nr}{N}}} \tag{6}$$

where not all of the b_n, c_n, d_n vanish. The number of sums on the right hand side of eq. (6) agrees with the number of overtone series in f. There would be only two sums if f(t) contains only two overtone series. Eq. (6) is in a form which can be easily integrated

$$F = \prod_{n=0}^{\frac{N}{p}-1} (x^p - e^{2\pi i \frac{np}{N}})^{b_n} \prod_{n=0}^{\frac{N}{q}-1} (x^q - e^{2\pi i \frac{nq}{N}})^{c_n} \prod_{n=0}^{\frac{N}{r}-1} (x^r - e^{2\pi i \frac{nr}{N}})^{d_n} \tag{7}$$

Eq. (6) shows that F is a product of functions depending only on x^p, x^q and x^r. This would imply that f(t) is a superposition of three 0,1-functions of period lengths $\frac{N}{p}$, $\frac{N}{q}$, and $\frac{N}{r}$ if the functions were polynomials, which requires that the exponents b_n, c_n and d_n are nonnegative integers (and then necessarily either 0 or 1). Because of the counterexample of [1] this cannot generally be true. The exponents are related in a rather complicated way to the spectral components α_s. They can however also be determined directly from the set M, i.e. those integers where a unit pulse begins. If one writes eq. (4) as

$$\frac{F'(x)}{F(x)} = \sum_{n=0}^{N-1} \frac{a_n}{x - e^{2\pi i \frac{n}{N}}} \tag{8}$$

where the $a_n = 1$ if $n \in M$ and $a_n = 0$ otherwise, one obtains from a comparison between the eqs. (6) and (8) a set of N linear equations between the $\frac{N}{p} + \frac{N}{q} + \frac{N}{r}$ exponents b_n, c_n, d_n and the a_n. This set of equations can be written as

$$\begin{pmatrix} E_p & E_q & E_r \\ \vdots & \vdots & \vdots \\ E_p & E_q & E_r \end{pmatrix} \underset{\sim}{x} = \underset{\sim}{a} \tag{9}$$

where the E_p, E_q, E_r denote unit matrices of p, q, r rows and columns respectively. The vector $\underset{\sim}{x}$ contains the unknown exponents b_n, c_n, d_n and $\underset{\sim}{a}$ the a_n of eq. (8). The system of linear equations (9) is usually overdetermined and degenerate. It possesses solutions only for certain vectors $\underset{\sim}{a}$ and this solution is usually not uniquely determined. It is however obvious that the right hand side and the coefficients are either 0 or 1. It is then possible to transform this system into triagonal form where the right hand side and the coefficients are all integer. This form shows that the system possesses a solution in rational numbers if it possesses any solution at all. The exponents in eq. (7) can therefore be assumed rational. If B is the least common denominator of these rationals it follows further that F^B multiplied by some polynomials of x^p, x^q, x^r (those factors on the right hand side of eq. (7) having a negative exponent) is a polynomial of the same variables. This means that the original 0,1-function f(t) which has a spectrum with three overtone series of basic frequencies p, q, r admits a decomposition

$$B\, f(t) = \sum_{\nu} C_\nu\, f_\nu(t) \tag{10}$$

with integers B, C_ν and 0,1-functions $f_\nu(t)$ which are periodic of period $\frac{N}{p}$, $\frac{N}{q}$ or $\frac{N}{r}$. Equation (10) is the main result of this note. It shows that the relation between spectrum and periodicity which is valid for L^2-functions is in somewhat modified form valid also for 0,1-functions. It is obviously not restricted to three overtone series. It is valid for any finite number of overtone series. It can be considerably strengthened for 0,1-functions having only two overtone series, where it is possible to show that the B and C_ν can all be chosen as one. Let us assume that all $d_n = 0$ and let us select a primitive p-th root of unity e_p. Then we obtain from eq. (7)

$$\frac{F(e_p x)}{F(x)} = \frac{\sum_{n=0}^{\frac{N}{p}-1} (e_p{}^q x^q - e^{2\pi i \frac{nq}{N}})^{c_n}}{\sum_{n=0}^{\frac{N}{p}-1} (x^q - e^{2\pi i \frac{nq}{N}})^{c_n}} . \tag{11}$$

All factors on the right hand side have either identical or disjoint zeros or poles. If one combines all factors having identical poles or identical zeros all exponents become either zero or one because the left hand side of eq. (11) has only first order zeros and poles. All poles have to be zeros of F(x) because $F(e_p x)$ is a polynomial. This means that

$$\phi(x) = \frac{F(x)}{\prod (x^q - e^{2\pi i \frac{qn}{N}})^{c_n}}$$

is a rational function without poles, therefore a polynomial and because of $\phi(e_p x) = \phi(x)$ also a polynomial in x^p. This shows that the b_n and c_n in eq. (7) and therefore also the B and C_ν in eq. (10) are all zero or one if f(t) has only two overtone series.

Conclusion

It has been shown that some integer multiple of a 0,1-function having a finite number of overtone series can be decomposed into periodic 0,1-functions with integer coefficients. The problem of determining such functions is therefore only partially solved by the method of [1]. Only those functions having all coefficients equal to one can be determined. It solves completely the problem for functions with only two overtone series. The general problem has been reduced from the original linear system in terms of roots of unity to the linear system of eqs. (9) in terms of integers. Efficient methods for determining its solutions and especially those right hand sides for which nontrivial solutions exist are still lacking.

Acknowledgement: I am very grateful to Prof. S.J. Patterson and Prof. M.R. Schroeder for inspiring discussions.

References

1. W. Möhring, Rechteckfunktionen mit vorgegebenem Spektrum. Fortschritte der AKustik - DAGA '85. DPG-GmbH Bad Honnof, 451-454.
2. M.R. Schroeder, Number Theory in Science and Communication. Springer 1984.
3. G.H. Hardy, E.M. Wright, An introduction to the theory of numbers. Clarendon Press 1984.

Criticality in Nonlinear Elliptic Eigenvalue Problems

Dirk Meinköhn
DFVLR (German Aerospace Establishment)
7101 Hardthausen, Germany

1. Physical Motivation

For physical processes which are accompanied by heat liberation (e.g. energy dissipation due to internal friction, electrical resistance, combustion processes), the stationary state is characterized by a thermal balance between heat generated and heat transferred into the surroundings. As heat generation and heat transfer are two physically distinct processes, a situation may arise in which such a balance proves to be impossible whereupon the physical system is forced to embark upon an unstationary evolution. Its onset is traditionally termed "ignition", the whole process of the system losing its thermal balance "thermal explosion". It is the aim of this paper to provide some insight into the phenomenon of thermal explosion by investigating certain critical stationary states which serve as jumping-off points into such an essentially unstationary behaviour.

The stationary states of the system are uniquely given in terms of solutions (λ, y) of the following boundary value problem:

$$(1) \qquad \begin{aligned} L[y] + \lambda w(y) &= 0 \quad \text{in } D \\ B[y] &= 0 \quad \text{on } \partial D \end{aligned}$$

Here ,λ represents a so-called control parameter, which may be set independently, whereas the function y represents a dependent variable (termed "state variable"). L designates a differential operator uniformly elliptic in D, whereas the boundary condition on ∂D

$$(2) \qquad B[y] = \frac{\partial y}{\partial \nu} + \sigma y = 0 \qquad \sigma \geq 0$$

ensures that there is a transfer of the property y from within D into the surroundings, i.e. into the direction of the outward normal ν on ∂D.

The function w(y), w(0) > 0, with w possibly depending on additional control parameters β, ξ, ... designates the generation of property y.

Continuous changes in the independent control parameters may be regarded as quasi-stationary processes under the control of the experimenter as long as the system remains within the set of stationary states. Anyone of the stationary states may be classified according to its stability. Therefore, the set of stationary states decomposes into continuously connected branches of stable and unstable solutions. Branching points which join stable with unstable branches are of particular interest because they serve as jumping-off points into unstationary behaviour and are thus essentially connected with ignition (and possibly extinction). These branching points are therefore called critical points.

The solutions for Eq. (1) are sought in parametric representation

$$(3) \qquad \big(\lambda(\varepsilon)\,,\, y(\varepsilon)\big)$$

with a suitable parameter ε defined by a solvability condition. In consequence of this, Eq. (1) changes over into a nonlinear elliptic eigenvalue problem with the eigenvalue $\lambda(\varepsilon)$ representing the response curve associated with Eq. (1).

If (λ^*, y^*) designates a critical solution of Eq. (1), then it may be shown that neighbouring solutions only exist for λ-values of a left or right half-interval about λ^* on the λ-axis. For a critical point (λ^*, y^*), λ^* then necessarily constitutes an extremum of the response curve $\lambda(\varepsilon)$.

2. Shifted Linear Minorants and Majorants

On account of the maximum principle for elliptic differential equations, the state variable $y(x)$ from a solution (λ,y) of Eq. (1) possesses a unique maximum y_m in D.
If in-
vestigations are restricted to such problems which possess a center of symmetry $x_0 \in D$, then necessarily: $y_m = y(x_0)$. Consequently, it is possible to choose $\varepsilon \equiv y_m$ such that the parametric representation of Eq. (3) is obtained as $\big(\lambda(y_m), y(x;y_m)\big)$. The choice of y_m as an independent parameter leads to a restriction, for any given value of y_m, to the interval $0 \leq y \leq y_m$ of what needs to be known of the function $w(y)$ and, specifically, of the consequences which arise from approximations of $w(y)$. In particular, adaptable approximations are conceived which change upon a shift in the value of y_m. An example of this is provided by linear minorants and majorants of $w(y)$ for $0 \leq y \leq y_m$ which are adaptable to any change in y_m (cf. Meinköhn [1,2,3]). By continuously changing the value of y_m, continuous upper and lower bounds of the response curve $\lambda(y_m)$ may be derived. The question then arises whether it is possible to obtain from these bounding curves any specific information as to the extrema of $\lambda(y_m)$, i.e., the critical points. In order to pursue this investigation, a very simple type of linear majorant is chosen, namely a majorant which is constant for $0 \leq y \leq y_m$.

For a given y_m, a function $m(y,y_m)$ $\big[M(y,y_m)\big]$ is called a minorant (majorant) for $w(y)$ if on $0 \leq y \leq y_m$:

$$(4) \qquad m(y,y_m) \leq w(y) \leq M(y,y_m).$$

Two examples of the function $w(y)$ which figure prominently in the literature on critical behaviour (Kordylewski [4,5]) are given by

$$(5)\qquad \begin{aligned} w_1(y) &= \exp\left(y/(1+\beta y)\right) &&, \beta \geq 0 \\ w_2(y) &= (1-\xi y)\exp(y) &&, \xi \geq 0 \end{aligned}$$

$w_1(y)$ and $w_2(y)$ are convex-concave functions on the positive y-axis. It is easily shown that for criticality to appear, one must have:

$$(6)\qquad 0 \leq \beta < 1/4\,, \quad 0 \leq \xi < 1/4$$

Adaptable constant majorants may be defined by:

$$(7)\qquad \begin{aligned} M_1(y_m) &= \exp\left(y_m/(1+\beta y_m)\right), && 0 < y_m \\ M_2(y_m) &= (1-\xi y_m)\exp(y_m)\,, && 0 < y_m < (1-\xi)/\xi \end{aligned}$$

Choosing an appropriate constant majorant M of w(y) for a given value y_m, one may investigate the following eigenvalue problem:

$$(8)\qquad \begin{aligned} & L[u] + \lambda_M\, M(y_m) = 0 \quad \text{in } D \\ & u(x_0) = y_m \rightarrow \lambda_M = \lambda_M(y_m) \text{ such that } B[u] = 0 \text{ on } \partial D \end{aligned}$$

Eq. (1) may be subtracted from Eq. (8) and on account of

$$w(y) \leq M(y_m) \quad \text{for } 0 \leq y \leq y_m$$

the following inequality is verified in consequence of the maximum principle:

$$(9)\qquad \lambda_M(y_m) \leq \lambda(y_m)$$

By Eq. (9) a continuous curve of lower bounds results from the introduction of shifted constant majorants. Eq. (8) is easily solved with the help of the inverse operator L^{-1}:

$$(10)\qquad u(x) = -L^{-1}[\lambda_M M(y_m)] = -\lambda M(y_m) \cdot \int_D G(x,x')dx'$$

Here, G represents Green's function with:

$$(11)\qquad G(x,x') \leq 0 \quad \text{for } x,x' \in D$$

From $u(x_0) = y_m$ one obtains:

$$(12) \qquad \lambda_M = \frac{-y_m}{M(y_m)} \cdot \frac{1}{\bar{G}} > 0$$

Here, $\bar{G} = \int_D G(x_0, x')dx' < 0$ is independent of y_m.

The extrema of $\lambda_M(y_m)$ are obtained from:

$$(13) \qquad \frac{\partial}{\partial y_m} \lambda_M = 0 \rightarrow w(y_m) - y_m w'(y_m) = 0$$

Eq. (13) represents the condition for tangents of w(y) to intersect the y-axis at the origin. For any function w(y), w(0) > 0, there exists an interval on the positive y-axis, for which $w(y) - yw'(y) \geq 0$. This will be a finite interval $0 < y \leq \bar{y}$ with $\bar{y} < \infty$ if w(y) is of a convexity which is strong enough. For y with $0 < y \leq \bar{y}$, w(y) is called concave in the generalized sense and it has been shown in the literature (e.g., Aris [6]) that for such concave functions w(y) no critical solutions of Eq. (1) exist. Thus, critical solution may be expected only if $\bar{y}$ is finite and $y_m > \bar{y}$. $\bar{y}$ designating the upper limit of the concavity property in the generalized sense of the function w(y) entails: $w(\bar{y}) - \bar{y}\, w'(\bar{y}) = 0$. The requirement of convexity of w(y) at $\bar{y}$ implies: $w''(\bar{y}) > 0$. Therefore:

$$(14) \qquad \frac{\partial^2}{\partial y_m^2} \lambda_M = \bar{G}\,\frac{y_m w''}{w^2} < 0 \text{ for } y_m = \bar{y}$$

Thus λ_M assumes a maximum for $y_m = \bar{y}$. Therefore $\lambda_M(\bar{y})$ represents a lower bound of the first critical value λ^* of $\lambda(y_m)$ which is of necessity a maximum and represents an ignition phenomenon.

If the function w(y) becomes concave for large values of y (cf. Eq. (5)) then w will be concave for $\bar{\bar{y}}$, which designates the largest of the y_m-values which satisfy the condition of tangents intersecting the y-axis at the origin, i.e. $w(y_m) - y_m w'(y_m) = 0$. For $y_m = \bar{\bar{y}}$ then:

$$(15) \qquad \frac{\partial^2}{\partial y_m^2} \lambda_M = \bar{G}\,\frac{y_m w''}{w} > 0$$

i.e., $\lambda_M(y_m)$ will assume a minimum at $\bar{\bar{y}}$. It is then concluded that $\lambda_M(\bar{\bar{y}})$ represents a lower bound to the last critical value λ^{**} of $\lambda(y_m)$ which is of necessity a minimum and represents an extinction phenomenon. For the functions $w_1(y)$ and $w_2(y)$ (cf. Eq. (5)), $\bar{y}$ and $\bar{\bar{y}}$ have been indicated in Figs. (1a, 1b).

3. Disappearance of Criticality: an Example

Criticality for Eq. (1) will disappear if the convexity of w(y) is reduced by appropriate changes of the control parameters β, ξ.. governing the shape of w(y) [7,8,9] . For the sake of an example for which numerical results are available ([4,5]), functions $w_1(y)$ and $w_2(y)$ of Eq. (5) are investigated for a spherical domain D with $L = \Delta$(Laplacian) and $B[y] \equiv y = 0$ on ∂D. Then the center of the sphere represents the center x_0 of symmetry and thus the above analysis is applicable. Choosing the constant majorants of Eqs. (7), the extrema of $\lambda_M(y_m)$ (cf. Eqs. (8,13))are determined for $0 \leq \beta \leq 1/4$, $0 \leq \xi \leq 1/4$. It is found that for w_1 as well as for w_2, $\lambda_M(y_m)$ possesses two extrema at most which furnish the sought-for lower bounds of λ^* (ignition) and λ^{**} (extinction). In Figs. (2a, 2b) the bounds are marked "lower bounds-constants" and are compared with the exact numerical results by Kordylewski [4,5]. Also included are upper and improved lower bounds which are to be presented in a forthcoming publication [10]. In the case of a two-dimensional space of control parameters, disappearance of criticality is necessarily represented by a cusp. It is interesting to note that a cusp appears also in the curves of lower bounds.

Literature

1. Meinköhn, D., J. Chem. Phys. 70 (1979) 3209 - 3213
2. Meinköhn, D., Int. J. Heat Mass Transfer 23 (1980) 833 - 839
3. Meinköhn, D., J. Chem. Phys. 74 (1981) 3603 - 3608
4. Kordylewski, W., Comb. Flame 34 (1979) 109 - 117
5. Kordylewski, W., Comb. Flame 38 (198=) 103 - 105
6. Rutherford Aris, Chem. Eng. Sci. 24 (1969) 149 - 169
7. Gill, W./Donaldson, A.B./Shouman, A.R., Comb. Flame 36 (1979) 217 - 232
8. Boddington, T./Gray, P./Robinson, C., Proc. R.Soc. London A368 (1979) 441 - 461
9. Vega, J.M./Liñan, A., Comb. Flame 57 (1984) 247 - 253
10. Meinköhn, D., Comb. Flame to be published

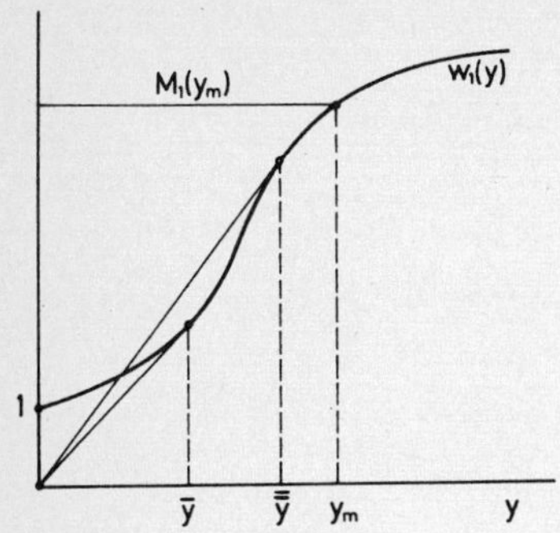

Fig. 1a

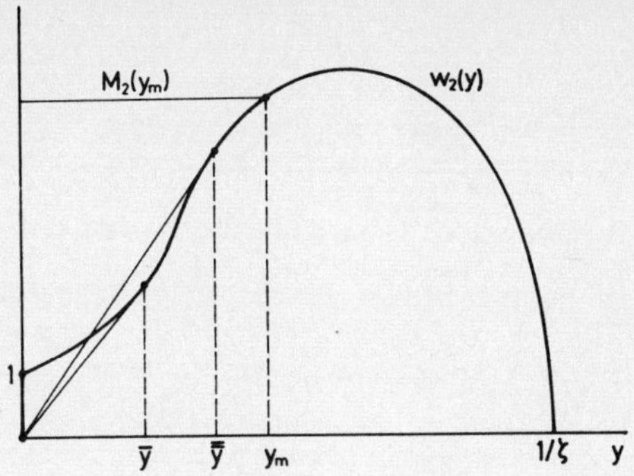

Fig. 1b

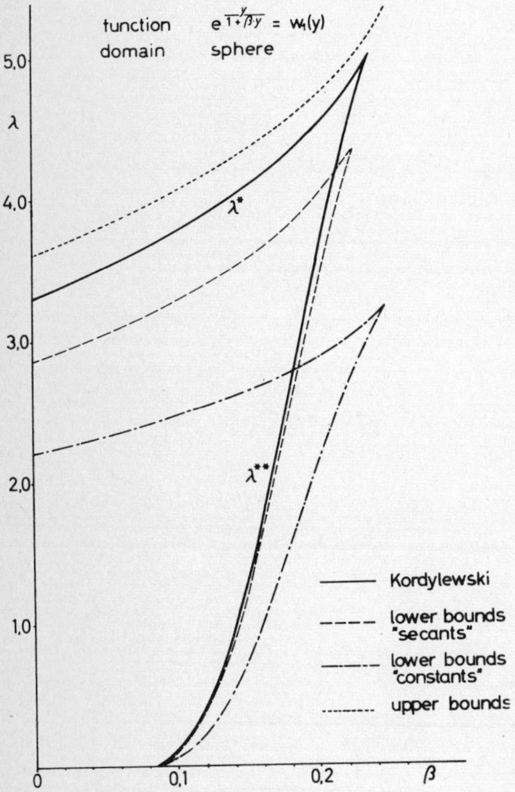

Fig. 2a

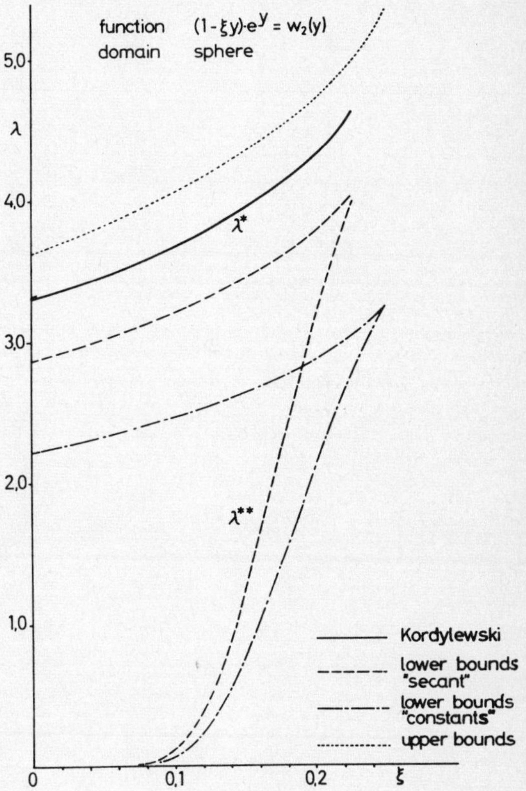

Fig. 2b

Lecture Notes in Physics